THE THEORY OF
MATRICES

F. R. GANTMACHER

VOLUME ONE

AMS CHELSEA PUBLISHING
American Mathematical Society • Providence, Rhode Island

The present work, published in two volumes, is an English translation by K. A. Hirsch of the Russian-language book TEORIYA MATRITS by F. R. Gantmacher (Гантмахер

2000 *Mathematics Subject Classification*. Primary 15-02.

Library of Congress Catalog Card Number 59-11779
International Standard Book Number 0-8218-1376-5 (Vol.I)
International Standard Book Number 0-8218-1393-5 (Set)

∞ The paper used in this book is acid-free and falls within the guidelines
established to ensure permanence and durability.
Visit the AMS home page at URL: `http://www.ams.org/`

10 9 8 7 6 5 4 3 2   04 03 02 01 00

PREFACE

THE MATRIX CALCULUS is widely applied nowadays in various branches of mathematics, mechanics, theoretical physics, theoretical electrical engineering, etc. However, neither in the Soviet nor the foreign literature is there a book that gives a sufficiently complete account of the problems of matrix theory and of its diverse applications. The present book is an attempt to fill this gap in the mathematical literature.

The book is based on lecture courses on the theory of matrices and its applications that the author has given several times in the course of the last seventeen years at the Universities of Moscow and Tiflis and at the Moscow Institute of Physical Technology.

The book is meant not only for mathematicians (undergraduates and research students) but also for specialists in allied fields (physics, engineering) who are interested in mathematics and its applications. Therefore the author has endeavoured to make his account of the material as accessible as possible, assuming only that the reader is acquainted with the theory of determinants and with the usual course of higher mathematics within the programme of higher technical education. Only a few isolated sections in the last chapters of the book require additional mathematical knowledge on the part of the reader. Moreover, the author has tried to keep the individual chapters as far as possible independent of each other. For example, Chapter V, *Functions of Matrices,* does not depend on the material contained in Chapters II and III. At those places of Chapter V where fundamental concepts introduced in Chapter IV are being used for the first time, the corresponding references are given. Thus, a reader who is acquainted with the rudiments of the theory of matrices can immediately begin with reading the chapters that interest him.

The book consists of two parts, containing fifteen chapters.

In Chapters I and III, information about matrices and linear operators is developed *ab initio* and the connection between operators and matrices is introduced.

Chapter II expounds the theoretical basis of Gauss's elimination method and certain associated effective methods of solving a system of n linear equations, for large n. In this chapter the reader also becomes acquainted with the technique of operating with matrices that are divided into rectangular 'blocks.'

In Chapter IV we introduce the extremely important 'characteristic' and 'minimal' polynomials of a square matrix, and the 'adjoint' and 'reduced adjoint' matrices.

In Chapter V, which is devoted to functions of matrices, we give the general definition of $f(A)$ as well as concrete methods of computing it—where $f(\lambda)$ is a function of a scalar argument λ and A is a square matrix. The concept of a function of a matrix is used in §§ 5 and 6 of this chapter for a complete investigation of the solutions of a system of linear differential equations of the first order with constant coefficients. Both the concept of a function of a matrix and this latter investigation of differential equations are based entirely on the concept of the minimal polynomial of a matrix and—in contrast to the usual exposition—do not use the so-called theory of elementary divisors, which is treated in Chapters VI and VII.

These five chapters constitute a first course on matrices and their applications. Very important problems in the theory of matrices arise in connection with the reduction of matrices to a normal form. This reduction is carried out on the basis of Weierstrass' theory of elementary divisors. In view of the importance of this theory we give two expositions in this book: an analytic one in Chapter VI and a geometric one in Chapter VII. We draw the reader's attention to §§ 7 and 8 of Chapter VI, where we study effective methods of finding a matrix that transforms a given matrix to normal form. In § 8 of Chapter VII we investigate in detail the method of A. N. Krylov for the practical computation of the coefficients of the characteristic polynomial.

In Chapter VIII certain types of matrix equations are solved. We also consider here the problem of determining all the matrices that are permutable with a given matrix and we study in detail the many-valued functions of matrices $\sqrt[m]{A}$ and $\ln A$.

Chapters IX and X deal with the theory of linear operators in a unitary space and the theory of quadratic and hermitian forms. These chapters do not depend on Weierstrass' theory of elementary divisors and use, of the preceding material, only the basic information on matrices and linear operators contained in the first three chapters of the book. In § 9 of Chapter X we apply the theory of forms to the study of the principal oscillations of a system with n degrees of freedom. In § 11 of this chapter we give an account of Frobenius' deep results on the theory of Hankel forms. These results are used later, in Chapter XV, to study special cases of the Routh-Hurwitz problem.

The last five chapters form the second part of the book [the second volume, in the present English translation]. In Chapter XI we determine normal forms for complex symmetric, skew-symmetric, and orthogonal matrices and establish interesting connections of these matrices with real matrices of the same classes and with unitary matrices.

In Chapter XII we expound the general theory of pencils of matrices of the form $A + \lambda B$, where A and B are arbitrary rectangular matrices of the same dimensions. Just as the study of regular pencils of matrices $A + \lambda B$ is based on Weierstrass' theory of elementary divisors, so the study of singular pencils is built upon Kronecker's theory of minimal indices, which is, as it were, a further development of Weierstrass's theory. By means of Kronecker's theory—the author believes that he has succeeded in simplifying the exposition of this theory—we establish in Chapter XII canonical forms of the pencil of matrices $A + \lambda B$ in the most general case. The results obtained there are applied to the study of systems of linear differential equations with constant coefficients.

In Chapter XIII we explain the remarkable spectral properties of matrices with non-negative elements and consider two important applications of matrices of this class: 1) homogeneous Markov chains in the theory of probability and 2) oscillatory properties of elastic vibrations in mechanics. The matrix method of studying homogeneous Markov chains was developed in the book [25] by V. I. Romanovskii and is based on the fact that the matrix of transition probabilities in a homogeneous Markov chain with a finite number of states is a matrix with non-negative elements of a special type (a 'stochastic' matrix).

The oscillatory properties of elastic vibrations are connected with another important class of non-negative matrices—the 'oscillation matrices.' These matrices and their applications were studied by M. G. Kreĭn jointly with the author of this book. In Chapter XIII, only certain basic results in this domain are presented. The reader can find a detailed account of the whole material in the monograph [7].

In Chapter XIV we compile the applications of the theory of matrices to systems of differential equations with variable coefficients. The central place (§§ 5-9) in this chapter belongs to the theory of the multiplicative integral (Produktintegral) and its connection with Volterra's infinitesimal calculus. These problems are almost entirely unknown in Soviet mathematical literature. In the first sections and in § 11, we study reducible systems (in the sense of Lyapunov) in connection with the problem of stability of motion; we also give certain results of N. P. Erugin. Sections 9-11 refer to the analytic theory of systems of differential equations. Here we clarify an inaccuracy in Birkhoff's fundamental theorem, which is usually applied to the investigation of the solution of a system of differential equations in the neighborhood of a singular point, and we establish a canonical form of the solution in the case of a regular singular point.

In § 12 of Chapter XIV we give a brief survey of some results of the fundamental investigations of I. A. Lappo-Danilevskiĭ on analytic functions of several matrices and their applications to differential systems.

The last chapter, Chapter XV, deals with the applications of the theory of quadratic forms (in particular, of Hankel forms) to the Routh-Hurwitz problem of determining the number of roots of a polynomial in the right half-plane ($\mathrm{Re}\, z > 0$). The first sections of the chapter contain the classical treatment of the problem. In § 5 we give the theorem of A. M. Lyapunov in which a stability criterion is set up which is equivalent to the Routh-Hurwitz criterion. Together with the stability criterion of Routh-Hurwitz we give, in § 11 of this chapter, the comparatively little known criterion of Liénard and Chipart in which the number of determinant inequalities is only about half of that in the Routh-Hurwitz criterion.

At the end of Chapter XV we exhibit the close connection between stability problems and two remarkable theorems of A. A. Markov and P. L. Chebyshev, which were obtained by these celebrated authors on the basis of the expansion of certain continued fractions of special types in series of decreasing powers of the argument. Here we give a matrix proof of these theorems.

This, then, is a brief summary of the contents of this book.

F. R. Gantmacher

PUBLISHERS' PREFACE

THE PUBLISHERS WISH TO thank Professor Gantmacher for his kindness in communicating to the translator new versions of several paragraphs of the original Russian-language book.

The Publishers also take pleasure in thanking the VEB Deutscher Verlag der Wissenschaften, whose many published translations of Russian scientific books into the German language include a counterpart of the present work, for their kind spirit of cooperation in agreeing to the use of their formulas in the preparation of the present work.

No material changes have been made in the text in translating the present work from the Russian except for the replacement of several paragraphs by the new versions supplied by Professor Gantmacher. Some changes in the references and in the Bibliography have been made for the benefit of the English-language reader.

CONTENTS

CHAPTER I

MATRICES AND OPERATIONS ON MATRICES

§ 1. Matrices. Basic Notation

1. Let F be a given number field.[1]

DEFINITION 1: *A rectangular array of numbers of the field* F

$$\left\| \begin{matrix} a_{11} & a_{12} & \dots & a_{1n} \\ a_{21} & a_{22} & \dots & a_{2n} \\ \dots & \dots & \dots & \dots \\ a_{m1} & a_{m2} & \dots & a_{mn} \end{matrix} \right\| \tag{1}$$

is called a matrix. When $m = n$, *the matrix is called square and the number* m, *equal to* n, *is called its order. In the general case the matrix is called rectangular* (*of dimension* $m \times n$). *The numbers that constitute the matrix are called its elements.*

NOTATION: In the double-subscript notation for the elements, the first subscript always denotes the row and the second subscript the column containing the given element.

As an alternative to the notation (1) for a matrix we shall also use the abbreviation

$$\| a_{ik} \| \quad (i = 1, 2, \dots, m; \ k = 1, 2, \dots, n). \tag{2}$$

Often the matrix (1) will also be denoted by a single letter, for example A. If A is a square matrix of order n, then we shall write $A = \| a_{ik} \|_1^n$. The determinant of a square matrix $A = \| a_{ik} \|_1^n$ will be denoted by $| a_{ik} |_1^n$ or by $| A |$.

[1] A *number field* is defined as an arbitrary collection of numbers within which the four operations of addition, subtraction, multiplication, and division by a non-zero number can always be carried out.

Examples of number fields are: the set of all rational numbers, the set of all real numbers, and the set of all complex numbers.

All the numbers that will occur in the sequel are assumed to belong to the number field given initially.

We introduce a concise notation for determinants formed from elements of the given matrix:

$$A\begin{pmatrix} i_1 & i_2 & \ldots & i_p \\ k_1 & k_2 & \ldots & k_p \end{pmatrix} = \begin{vmatrix} a_{i_1k_1} & a_{i_1k_2} & \ldots & a_{i_1k_p} \\ a_{i_2k_1} & a_{i_2k_2} & \ldots & a_{i_2k_p} \\ \cdots & \cdots & \cdots & \cdots \\ a_{i_pk_1} & a_{i_pk_2} & \ldots & a_{i_pk_p} \end{vmatrix}. \tag{3}$$

The determinant (3) is called a *minor* of A of order p, provided $1 \leqq i_1 < i_2 < \ldots < i_p \leqq m$ and $1 \leqq k_1 < k_2 < \ldots < k_p \leqq n$. A rectangular matrix $A = \| a_{ik} \|$ $(i = 1, 2, \ldots, m;\ k = 1, 2, \ldots, n)$ has $\binom{m}{p} \cdot \binom{n}{p}$ minors of order p

$$A\begin{pmatrix} i_1 & i_2 & \ldots & i_p \\ k_1 & k_2 & \ldots & k_p \end{pmatrix} \quad \left(\begin{matrix} 1 \leqq i_1 < i_2 < \cdots < i_p \leqq m \\ 1 \leqq k_1 < k_2 < \cdots < k_p \leqq n \end{matrix};\ p \leqq m, n \right). \tag{3'}$$

The minors (3′) in which $i_1 = k_1$, $i_2 = k_2$, $\ldots$, $i_p = k_p$, are called *principal* minors.

In the notation (3) the determinant of a square matrix $A = \| a_{ik} \|_1^n$ can be written as follows:

$$|A| = A\begin{pmatrix} 1 & 2 & \ldots & n \\ 1 & 2 & \ldots & n \end{pmatrix}.$$

The largest among the orders of the non-zero minors generated by a matrix is called the *rank* of the matrix. If r is the rank of a rectangular matrix A of dimension $m \times n$, then obviously $r \leqq \min(m, n)$.

A rectangular matrix consisting of a single column

$$\begin{Vmatrix} x_1 \\ x_2 \\ \vdots \\ x_n \end{Vmatrix}$$

is called a *column matrix* and will be denoted by $(x_1, x_2, \ldots, x_n)$.

A rectangular matrix consisting of a single row

$$\| z_1, z_2, \ldots, z_n \|$$

is called a *row matrix* and will be denoted by $[z_1, z_2, \ldots, z_n]$.

A square matrix in which all the elements outside the main diagonal are zero

$$\left\| \begin{matrix} d_1 & 0 & \dots & 0 \\ 0 & d_2 & \dots & 0 \\ \dots & \dots & \dots & \dots \\ 0 & 0 & \dots & d_n \end{matrix} \right\|$$

is called a *diagonal matrix* and is denoted by[2] $\| d_i \delta_{ik} \|_1^n$ or by

$$\{ d_1, d_2, \dots, d_n \}.$$

Suppose that m quantities $y_1, y_2, \dots, y_m$ have linear and homogeneous expressions in terms of n other quantities $x_1, x_2, \dots, x_n$:

$$\left. \begin{aligned} y_1 &= a_{11}x_1 + a_{12}x_2 + \cdots + a_{1n}x_n \\ y_2 &= a_{21}x_1 + a_{22}x_2 + \cdots + a_{2n}x_n \\ &\dots\dots\dots\dots\dots\dots\dots \\ y_m &= a_{m1}x_1 + a_{m2}x_2 + \cdots + a_{mn}x_n , \end{aligned} \right\} \tag{4}$$

or more concisely,

$$y_i = \sum_{k=1}^{n} a_{ik}x_k \qquad (i = 1, 2, \dots, m). \tag{4'}$$

The transformation of the quantities $x_1, x_2, \dots, x_n$ into the quantities $y_1, y_2, \dots, y_m$ by means of the formulas (4) is called a *linear transformation*.

The coefficients of this transformation form a rectangular matrix (1) of dimension $m \times n$.

The linear transformation (4) determines the matrix (1) uniquely, and vice versa.

In the next section we shall define the basic operations on rectangular matrices using the properties of the linear transformations (4) as our starting point.

§ 2. Addition and Multiplication of Rectangular Matrices

We shall define the basic operations on matrices: addition of matrices, multiplication of a matrix by a number, and multiplication of matrices.

1. Suppose that the quantities $y_1, y_2, \dots, y_m$ are expressed in terms of the quantities $x_1, x_2, \dots, x_n$ by means of the linear transformation

$$y_i = \sum_{k=1}^{n} a_{ik}x_k \qquad (i = 1, 2, \dots, m) \tag{5}$$

[2] Here δ_{ik} is the Kronecker symbol: $\delta_{ik} = \begin{cases} 1 & (i = k), \\ 0 & (i \neq k). \end{cases}$

and the quantities $z_1, z_2, \ldots, z_m$ in terms of the same quantities $x_1, x_2, \ldots, x_n$ by means of the transformation

$$z_i = \sum_{k=1}^{n} b_{ik} x_k \qquad (i = 1, 2, \ldots, m). \tag{6}$$

Then

$$y_i + z_i = \sum_{k=1}^{n} (a_{ik} + b_{ik})\, x_k \qquad (i = 1, 2, \ldots, m). \tag{7}$$

In accordance with this, we formulate the following definition.

DEFINITION 2: *The sum of two rectangular matrices $A = \| a_{ik} \|$ and $B = \| b_{ik} \|$, both of dimension $m \times n$, is the matrix $C = \| c_{ik} \|$, of the same dimension, whose elements are the sums of the corresponding elements of the given matrices:*

$$C = A + B,$$

where

$$c_{ik} = a_{ik} + b_{ik} \quad (i = 1, 2, \ldots, m;\ k = 1, 2, \ldots, n).$$

The operation of forming the sum of given matrices is called addition.

Example.

$$\left\| \begin{matrix} a_1 & a_2 & a_3 \\ b_1 & b_2 & b_3 \end{matrix} \right\| + \left\| \begin{matrix} c_1 & c_2 & c_3 \\ d_1 & d_2 & d_3 \end{matrix} \right\| = \left\| \begin{matrix} a_1 + c_1 & a_2 + c_2 & a_3 + c_3 \\ b_1 + d_1 & b_2 + d_2 & b_3 + d_3 \end{matrix} \right\|.$$

According to Definition 2, only rectangular matrices of equal dimension can be added.

By virtue of the same definition, the coefficient matrix of the transformation (7) is the sum of the coefficient matrices of the transformations (5) and (6).

From the definition of matrix addition it follows immediately that this operation has the properties of commutativity and associativity:

1. $A + B = B + A$;
2. $(A + B) + C = A + (B + C)$.

Here A, B, and C are arbitrary rectangular matrices all of equal dimension.

The operation of addition of matrices extends in a natural way to the case of an arbitrary finite number of summands.

2. Let us multiply the quantities $y_1, y_2, \ldots, y_m$ in the transformation (5) by some number α of F. Then

$$\alpha y_i = \sum_{k=1}^{n} (\alpha a_{ik})\, x_k \qquad (i = 1, 2, \ldots, m).$$

In accordance with this, we formulate the following definition.

DEFINITION 3. *The product of a matrix* $A = \| a_{ik} \|$ $(i = 1, 2, \ldots, m;$ $k = 1, 2, \ldots, n)$ *by a number* α *of* F *is the matrix* $C = \| c_{ik} \|$ $(i = 1, 2, \ldots, m;$ $k = 1, 2, \ldots, n)$ *whose elements are obtained from the corresponding elements of* A *by multiplication by* α:

$$C = \alpha A,$$

where

$$c_{ik} = \alpha a_{ik} \qquad (i = 1, 2, \ldots, m; k = 1, 2, \ldots, n).$$

The operation of forming the product of a matrix by a number is called multiplication of the matrix by the number.

Example.

$$\alpha \begin{Vmatrix} a_1 & a_2 & a_3 \\ b_1 & b_2 & b_3 \end{Vmatrix} = \begin{Vmatrix} \alpha a_1 & \alpha a_2 & \alpha a_3 \\ \alpha b_1 & \alpha b_2 & \alpha b_3 \end{Vmatrix}.$$

It is easy to see that

1. $\alpha(A + B) = \alpha A + \alpha B,$
2. $(\alpha + \beta) A = \alpha A + \beta A,$
3. $(\alpha\beta) A \quad = \alpha(\beta A).$

Here A and B are rectangular matrices of equal dimension and α and β are numbers of F.

The *difference* $A - B$ of two rectangular matrices of equal dimension is defined by

$$A - B = A + (-1) B.$$

If A is a square matrix of order n and α a number of F, then[3]

$$| \alpha A | = \alpha^n | A |.$$

3. Suppose that the quantities $z_1, z_2, \ldots, z_m$ are expressed in terms of the quantities $y_1, y_2, \ldots, y_n$ by the transformation

$$z_i = \sum_{k=1}^{n} a_{ik} y_k \qquad (i = 1, 2, \ldots, m) \tag{8}$$

and that the quantities $y_1, y_2, \ldots, y_n$ are expressed in terms of the quantities $x_1, x_2, \ldots, x_q$ by the formulas

$$y_k = \sum_{j=1}^{q} b_{kj} x_j \qquad (k = 1, 2, \ldots, n) \tag{9}$$

Then on substituting these expressions for the y_k $(k = 1, 2, \ldots, n)$ in (8) we can express $z_1, z_2, \ldots, z_m$ in terms of $x_1, x_2, \ldots, x_q$ by means of the composite transformation:

[3] Here the symbols $| A |$ and $| \alpha A |$ denote the determinants of the matrices A and αA (see p. 1).

$$z_i = \sum_{k=1}^{n} a_{ik} \sum_{j=1}^{q} b_{kj} x_j = \sum_{j=1}^{q} \left(\sum_{k=1}^{n} a_{ik} b_{kj} \right) x_j \qquad (i = 1, 2, \ldots, m). \tag{10}$$

In accordance with this we formulate the following definition.

DEFINITION 4. *The product of two rectangular matrices*

$$A = \begin{Vmatrix} a_{11} & a_{12} \ldots a_{1n} \\ a_{21} & a_{22} \ldots a_{2n} \\ \ldots\ldots & \ldots\ldots \\ a_{m1} & a_{m2} \ldots a_{mn} \end{Vmatrix}, \qquad B = \begin{Vmatrix} b_{11} & b_{12} \ldots b_{1q} \\ b_{21} & b_{22} \ldots b_{2q} \\ \ldots\ldots & \ldots\ldots \\ b_{n1} & b_{n2} \ldots b_{nq} \end{Vmatrix}$$

is the matrix

$$C = \begin{Vmatrix} c_{11} & c_{12} \ldots c_{1q} \\ c_{21} & c_{22} \ldots c_{2q} \\ \ldots\ldots & \ldots\ldots \\ c_{m1} & c_{m2} \ldots c_{mq} \end{Vmatrix}$$

in which the element c_{ij} *at the intersection of the i-th row and the j-th column is the 'product'*[4] *of the i-th row of the first matrix A into the j-th column of the second matrix B:*

$$c_{ij} = \sum_{k=1}^{n} a_{ik} b_{kj} \qquad (i = 1, 2, \ldots, m;\ j = 1, 2, \ldots, q). \tag{11}$$

The operation of forming the product of given matrices is called matrix multiplication.

Example.

$$\begin{Vmatrix} a_1 & a_2 & a_3 \\ b_1 & b_2 & b_3 \end{Vmatrix} \begin{Vmatrix} c_1 & d_1 & e_1 & f_1 \\ c_2 & d_2 & e_2 & f_2 \\ c_3 & d_3 & e_3 & f_3 \end{Vmatrix} =$$

$$= \begin{Vmatrix} a_1c_1 + a_2c_2 + a_3c_3 & a_1d_1 + a_2d_2 + a_3d_3 & a_1e_1 + a_2e_2 + a_3e_3 & a_1f_1 + a_2f_2 + a_3f_3 \\ b_1c_1 + b_2c_2 + b_3c_3 & b_1d_1 + b_2d_2 + b_3d_3 & b_1e_1 + b_2e_2 + b_3e_3 & b_1f_1 + b_2f_2 + b_3f_3 \end{Vmatrix}.$$

By Definition 4 the coefficient matrix of the transformation (10) is the product of the coefficient matrices of (8) and (9).

Note that *the operation of multiplication of two rectangular matrices can only be carried out when the number of columns of the first factor is equal to the number of rows of the second.* In particular, multiplication is always possible when both factors are square matrices of one and the same order.

---

[4] The product of two sequences of numbers $a_1, a_2, \ldots, a_n$ and $b_1, b_2, \ldots, b_n$ is defined as the sum of the products of the corresponding numbers: $\sum_{i=1}^{n} a_i b_i$.

The reader should observe that even in this special case the multiplication of matrices does not have the property of commutativity. For example,

$$\begin{Vmatrix} 1 & 2 \\ 3 & 4 \end{Vmatrix} \begin{Vmatrix} 2 & 0 \\ 3 & -1 \end{Vmatrix} = \begin{Vmatrix} 8 & -2 \\ 18 & -4 \end{Vmatrix}, \quad \text{but} \quad \begin{Vmatrix} 2 & 0 \\ 3 & -1 \end{Vmatrix} \begin{Vmatrix} 1 & 2 \\ 3 & 4 \end{Vmatrix} = \begin{Vmatrix} 2 & 4 \\ 0 & 2 \end{Vmatrix}.$$

If $AB = BA$, then the matrices A and B are called *permutable* or *commuting.*

Example. The matrices

$$A = \begin{Vmatrix} 1 & 2 \\ -2 & 0 \end{Vmatrix} \quad \text{and} \quad B = \begin{Vmatrix} -3 & 2 \\ -2 & -4 \end{Vmatrix}$$

are permutable, because

$$AB = \begin{Vmatrix} -7 & -6 \\ 6 & -4 \end{Vmatrix} \quad \text{and} \quad BA = \begin{Vmatrix} -7 & -6 \\ 6 & -4 \end{Vmatrix}.$$

It is very easy to verify the *associative* property of matrix multiplication and also the *distributive* property of multiplication with respect to addition:

1. $(AB)C = A(BC)$,
2. $(A + B)C = AC + BC$,
3. $A(B + C) = AB + AC$.

The definition of matrix multiplication extends in a natural way to the case of several factors.

When we make use of the multiplication of rectangular matrices, we can write the linear transformation

$$\begin{aligned} y_1 &= a_{11}x_1 + a_{12}x_2 + \cdots + a_{1n}x_n \\ y_2 &= a_{21}x_1 + a_{22}x_2 + \cdots + a_{2n}x_n \\ &\cdots\cdots\cdots\cdots\cdots\cdots\cdots\cdots \\ y_m &= a_{m1}x_1 + a_{m2}x_2 + \cdots + a_{mn}x_n \end{aligned}$$

as a single matrix equation

$$\begin{Vmatrix} y_1 \\ y_2 \\ \vdots \\ y_m \end{Vmatrix} = \begin{Vmatrix} a_{11} & a_{12} & \cdots & a_{1n} \\ a_{21} & a_{22} & \cdots & a_{2n} \\ \cdots & \cdots & \cdots & \cdots \\ a_{m1} & a_{m2} & \cdots & a_{mn} \end{Vmatrix} \begin{Vmatrix} x_1 \\ x_2 \\ \vdots \\ x_n \end{Vmatrix},$$

or in abbreviated form,

$$y = Ax.$$

Here $x = (x_1, x_2, \ldots, x_n)$ and $y = (y_1, y_2, \ldots, y_m)$ are column matrices and $A = \| a_{ik} \|$ is a rectangular matrix of dimension $m \times n$.

Let us treat the special case when in the product $C = AB$ the second factor is a square diagonal matrix $B = \{d_1, d_2, \ldots, d_n\}$. Then it follows from (11) that

$$c_{ij} = a_{ij} d_j \qquad (i = 1, 2, \ldots, m;\ j = 1, 2, \ldots, n),$$

i.e.,

$$\left\| \begin{matrix} a_{11} & a_{12} \ldots a_{1n} \\ a_{21} & a_{22} \ldots a_{2n} \\ \ldots & \ldots \\ a_{m1} & a_{m2} \ldots a_{mn} \end{matrix} \right\| \left\| \begin{matrix} d_1 & 0 \ldots 0 \\ 0 & d_2 \ldots 0 \\ \ldots & \ldots \\ 0 & 0 \ldots d_n \end{matrix} \right\| = \left\| \begin{matrix} a_{11}d_1 & a_{12}d_2 \ldots a_{1n}d_n \\ a_{21}d_1 & a_{22}d_2 \ldots a_{2n}d_n \\ \ldots & \ldots \\ a_{m1}d_1 & a_{m2}d_2 \ldots a_{mn}d_n \end{matrix} \right\|.$$

Similarly,

$$\left\| \begin{matrix} d_1 & 0 \ldots 0 \\ 0 & d_2 \ldots 0 \\ \ldots & \ldots \\ 0 & 0 \ldots d_m \end{matrix} \right\| \left\| \begin{matrix} a_{11} & a_{12} \ldots a_{1n} \\ a_{21} & a_{22} \ldots a_{2n} \\ \ldots & \ldots \\ a_{m1} & a_{m2} \ldots a_{mn} \end{matrix} \right\| = \left\| \begin{matrix} d_1a_{11} & d_1a_{12} \ldots d_1a_{1n} \\ d_2a_{21} & d_2a_{22} \ldots d_2a_{2n} \\ \ldots & \ldots \\ d_ma_{m1} & d_ma_{m2} \ldots d_ma_{mn} \end{matrix} \right\|.$$

Hence: *When a rectangular matrix A is multiplied on the right (left) by a diagonal matrix $\{d_1, d_2, \ldots\}$, then the columns (rows) of A are multiplied by $d_1, d_2, \ldots$, respectively.*

4. Suppose that a square matrix $C = \| c_{ij} \|_1^m$ is the product of two rectangular matrices $A = \| a_{ik} \|$ and $B = \| b_{kj} \|$ of dimension $m \times n$ and $n \times m$, respectively:

$$\left\| \begin{matrix} c_{11} \ldots c_{1m} \\ \ldots\ldots \\ c_{m1} \ldots c_{mm} \end{matrix} \right\| = \left\| \begin{matrix} a_{11}\ a_{12} \ldots a_{1n} \\ \ldots\ldots \\ a_{m1}\ a_{m2} \ldots a_{mn} \end{matrix} \right\| \left\| \begin{matrix} b_{11} \ldots b_{1m} \\ b_{21} \ldots b_{2m} \\ \ldots\ldots \\ b_{n1} \ldots b_{nm} \end{matrix} \right\|, \tag{12}$$

i.e.,

$$c_{ij} = \sum_{\alpha=1}^{n} a_{i\alpha} b_{\alpha j} \qquad (i, j = 1, 2, \ldots, m). \tag{13}$$

We shall establish the important *Binet-Cauchy formula,* which expresses the determinant $|C|$ in terms of the minors of A and B:

$$\begin{vmatrix} c_{11} & \dots & c_{1m} \\ \dots & \dots & \dots \\ c_{m1} & \dots & c_{mm} \end{vmatrix} = \sum_{1 \leqq k_1 < k_2 < \dots < k_m \leqq n} \begin{vmatrix} a_{1k_1} & \dots & a_{1k_m} \\ \dots & \dots & \dots \\ a_{mk_1} & \dots & a_{mk_m} \end{vmatrix} \begin{vmatrix} b_{k_1 1} & \dots & b_{k_1 m} \\ \dots & \dots & \dots \\ b_{k_m 1} & \dots & b_{k_m m} \end{vmatrix} \tag{14}$$

or, in the notation[5] of page 2,

$$C\begin{pmatrix} 1 & 2 & \dots & m \\ 1 & 2 & \dots & m \end{pmatrix} = \sum_{1 \leqq k_1 < k_2 < \dots < k_m \leqq n} A\begin{pmatrix} 1 & 2 & \dots & m \\ k_1 & k_2 & \dots & k_m \end{pmatrix} B\begin{pmatrix} k_1 & k_2 & \dots & k_m \\ 1 & 2 & \dots & m \end{pmatrix}. \tag{14'}$$

According to this formula *the determinant of C is the sum of the products of all possible minors of the maximal (m-th) order*[5] *of A into the corresponding minors of the same order of B.*

Derivation of the Binet-Cauchy formula. By (13) the determinant of C can be represented in the form

$$\begin{aligned} \begin{vmatrix} c_{11} & \dots & c_{1m} \\ \dots & \dots & \dots \\ c_{m1} & \dots & c_{mm} \end{vmatrix} &= \begin{vmatrix} \sum_{\alpha_1=1}^{n} a_{1\alpha_1} b_{\alpha_1 1} & \dots & \sum_{\alpha_m=1}^{n} a_{1\alpha_m} b_{\alpha_m m} \\ \dots & \dots & \dots \\ \sum_{\alpha_1=1}^{n} a_{m\alpha_1} b_{\alpha_1 1} & \dots & \sum_{\alpha_m=1}^{n} a_{m\alpha_m} b_{\alpha_m m} \end{vmatrix} \\ &= \sum_{\alpha_1, \dots, \alpha_m = 1}^{n} \begin{vmatrix} a_{1\alpha_1} b_{\alpha_1 1} & \dots & a_{1\alpha_m} b_{\alpha_m m} \\ \dots & \dots & \dots \\ a_{m\alpha_1} b_{\alpha_1 1} & \dots & a_{m\alpha_m} b_{\alpha_m m} \end{vmatrix} \\ &= \sum_{\alpha_1, \dots, \alpha_m = 1}^{n} A\begin{pmatrix} 1 & 2 & \dots & m \\ \alpha_1 & \alpha_2 & \dots & \alpha_m \end{pmatrix} b_{\alpha_1 1} b_{\alpha_2 2} \dots b_{\alpha_m m}. \end{aligned} \tag{15}$$

If $m > n$, then among the numbers $\alpha_1, \alpha_2, \dots, \alpha_m$ there are always at least two that are equal, so that every summand on the right-hand side of (15) is zero. Hence in this case $|C| = 0$.

Now let $m \leqq n$. Then in the sum on the right-hand side of (15) all those summands will be zero in which at least two of the subscripts $\alpha_1, \alpha_2, \dots, \alpha_m$ are equal. All the remaining summands of (15) can be split into groups of $m!$ terms each by combining into one group those summands that differ from each other only in the order of the subscripts $\alpha_1, \alpha_2, \dots, \alpha_m$ (so that

[5] When $m > n$, the matrices A and B do not have minors of order m. In that case the right-hand sides of (14) and (14′) are to be replaced by zero.

within each such group the subscripts $\alpha_1, \alpha_2, \ldots, \alpha_m$ have one and the same set of values). Now within one such group the sum of the corresponding terms is[6]

$$\sum \varepsilon(\alpha_1, \alpha_2, \ldots, \alpha_m) A\begin{pmatrix}1 & 2 & \ldots & m\\ k_1 & k_2 & \ldots & k_m\end{pmatrix} b_{\alpha_1 1}\, b_{\alpha_2 2} \cdots b_{\alpha_m m} =$$
$$= A\begin{pmatrix}1 & 2 & \ldots & m\\ k_1 & k_2 & \ldots & k_m\end{pmatrix} \sum \varepsilon(\alpha_1, \alpha_2, \ldots, \alpha_m)\, b_{\alpha_1 1} b_{\alpha_2 2} \cdots b_{\alpha_m m}$$
$$= A\begin{pmatrix}1 & 2 & \ldots & m\\ k_1 & k_2 & \ldots & k_m\end{pmatrix} B\begin{pmatrix}k_1 & k_2 & \ldots & k_m\\ 1 & 2 & \ldots & m\end{pmatrix}.$$

Hence from (15) we obtain (14′).

Example 1.

$$\left\|\begin{matrix} a_1c_1 + a_2c_2 + \cdots + a_nc_n & a_1d_1 + a_2d_2 + \cdots + a_nd_n \\ b_1c_1 + b_2c_2 + \cdots + b_nc_n & b_1d_1 + b_2d_2 + \cdots + b_nd_n \end{matrix}\right\| = \left\|\begin{matrix} a_1 & a_2 \ldots a_n \\ b_1 & b_2 \ldots b_n \end{matrix}\right\| \left\|\begin{matrix} c_1 & d_1 \\ c_2 & d_2 \\ \cdot & \cdot \\ \cdot & \cdot \\ c_n & d_n \end{matrix}\right\|.$$

Therefore formula (14) yields the so-called *Cauchy identity*

$$\begin{vmatrix} a_1c_1 + a_2c_2 + \cdots + a_nc_n & a_1d_1 + a_2d_2 + \cdots + a_nd_n \\ b_1c_1 + b_2c_2 + \cdots + b_nc_n & b_1d_1 + b_2d_2 + \cdots + b_nd_n \end{vmatrix} = \sum_{1 \le i < k \le n} \begin{vmatrix} a_i & a_k \\ b_i & b_k \end{vmatrix} \begin{vmatrix} c_i & d_i \\ c_k & d_k \end{vmatrix}. \quad (16)$$

Setting $a_i = c_i$, $b_i = d_i$ $(i = 1, 2, \ldots, n)$ in this identity, we obtain:

$$\begin{vmatrix} a_1^2 + a_2^2 + \cdots + a_n^2 & a_1b_1 + a_2b_2 + \cdots + a_nb_n \\ a_1b_1 + a_2b_2 + \cdots + a_nb_n & b_1^2 + b_2^2 + \cdots + b_n^2 \end{vmatrix} = \sum_{1 \le i < k \le n} \begin{vmatrix} a_i & a_k \\ b_i & b_k \end{vmatrix}^2.$$

If a_i and b_i $(i = 1, 2, \ldots, n)$ are real numbers, we deduce the well-known inequality

$$(a_1b_1 + a_2b_2 + \cdots + a_nb_n)^2 \le (a_1^2 + a_2^2 + \cdots + a_n^2)\,(b_1^2 + b_2^2 + \cdots + b_n^2). \quad (17)$$

Here the equality sign holds if and only if all the numbers a_i are proportional to the corresponding numbers b_i $(i = 1, 2, \ldots, n)$.

Example 2.

$$\left\|\begin{matrix} a_1c_1 + b_1d_1 & \ldots & a_1c_n + b_1d_n \\ \cdots & \cdots & \cdots \\ \cdots & \cdots & \cdots \\ a_nc_1 + b_nd_1 & \ldots & a_nc_n + b_nd_n \end{matrix}\right\| = \left\|\begin{matrix} a_1 & b_1 \\ \cdot & \cdot \\ \cdot & \cdot \\ a_n & b_n \end{matrix}\right\| \left\|\begin{matrix} c_1 \ldots c_n \\ d_1 \ldots d_n \end{matrix}\right\|.$$

[6] Here $k_1 < k_2 < \ldots < k_m$ is the normal order of the subscripts $\alpha_1, \alpha_2, \ldots, \alpha_m$ and $\varepsilon(\alpha_1, \alpha_2, \ldots, \alpha_m) = (-1)^N$, where N is the number of transpositions of the indices needed to put the permutation $\alpha_1, \alpha_2, \ldots, \alpha_m$ into normal order.

Therefore for $n > 2$

$$\begin{vmatrix} a_1c_1 + b_1d_1 & \dots & a_1c_n + b_1d_n \\ \dots & \dots & \dots \\ \dots & \dots & \dots \\ a_nc_1 + b_nd_1 & \dots & a_nc_n + b_nd_n \end{vmatrix} = 0.$$

Let us consider the special case where A and B are square matrices of one and the same order n. When we set $m = n$ in (14′), we arrive at the well-known multiplication theorem for determinants:

$$C\begin{pmatrix} 1 & 2 & \dots & n \\ 1 & 2 & \dots & n \end{pmatrix} = A\begin{pmatrix} 1 & 2 & \dots & n \\ 1 & 2 & \dots & n \end{pmatrix} B\begin{pmatrix} 1 & 2 & \dots & n \\ 1 & 2 & \dots & n \end{pmatrix}$$

or, in another notation,

$$|C| = |AB| = |A| \cdot |B|. \tag{18}$$

Thus, *the determinant of the product of two square matrices is equal to the product of the determinants of the factors.*

5. The Binet-Cauchy formula enables us, in the general case also, to express the minors of the product of two rectangular matrices in terms of the minors of the factors. Let

$$A = \|a_{ik}\|, \quad B = \|b_{kj}\|, \quad C = \|c_{ij}\|$$

$$(i = 1, 2, \dots, m;\ k = 1, 2, \dots, n;\ j = 1, 2, \dots, q)$$

and

$$C = AB.$$

We consider an arbitrary minor of C:

$$C\begin{pmatrix} i_1 & i_2 & \dots & i_p \\ j_1 & j_2 & \dots & j_p \end{pmatrix} \quad \left(1 \begin{matrix} \leqq i_1 < i_2 < \dots < i_p \leqq m \\ \leqq j_1 < j_2 < \dots < j_p \leqq q \end{matrix} ;\ p \leqq m \text{ and } p \leqq q \right).$$

The matrix formed from the elements of this minor is the product of two rectangular matrices

$$\begin{Vmatrix} a_{i_1 1} & a_{i_1 2} & \dots & a_{i_1 n} \\ \dots & \dots & \dots & \dots \\ a_{i_p 1} & a_{i_p 2} & \dots & a_{i_p n} \end{Vmatrix}, \quad \begin{Vmatrix} b_{1j_1} & \dots & b_{1j_p} \\ b_{2j_1} & \dots & b_{2j_p} \\ \dots & \dots & \dots \\ b_{nj_1} & \dots & b_{nj_p} \end{Vmatrix}.$$

Therefore, by applying the Binet-Cauchy formula, we obtain:[7]

$$C\begin{pmatrix} i_1 & i_2 & \dots & i_p \\ j_1 & j_2 & \dots & j_p \end{pmatrix} = \sum_{1 \leqq k_1 < k_2 < \dots < k_p \leqq n} A\begin{pmatrix} i_1 & i_2 & \dots & i_p \\ k_1 & k_2 & \dots & k_p \end{pmatrix} B\begin{pmatrix} k_1 & k_2 & \dots & k_p \\ j_1 & j_2 & \dots & j_p \end{pmatrix}. \tag{19}$$

For $p = 1$ formula (19) goes over into (11). For $p > 1$ formula (19) is a natural generalization of (11).

We mention another consequence of (19).

The rank of the product of two rectangular matrices does not exceed the rank of either factor.

If $C = AB$ and r_A, r_B, r_C are the ranks of A, B, C, then

$$r_C \leqq \min(r_A, r_B).$$

§ 3. Square Matrices

1. The square matrix of order n in which the main diagonal consists entirely of units and all the other elements are zero is called the *unit matrix* and is denoted by $E^{(n)}$ or simply by E. The name 'unit matrix' is connected with the following property of E: For every rectangular matrix

$$A = \| a_{ik} \| \qquad (i = 1, 2, \dots, m;\ k = 1, 2, \dots, n)$$

we have

$$E^{(m)}A = AE^{(n)} = A.$$

Clearly

$$E^{(n)} = \| \delta_{ik} \|_1^n.$$

Let $A = \| a_{ik} \|_1^n$ be a square matrix. Then the *power of the matrix* is defined in the usual way:

$$A^p = \underbrace{AA \cdots A}_{p \text{ times}} \qquad (p = 1, 2, \dots); \qquad A^0 = E.$$

From the associative property of matrix multiplication it follows that

$$A^p A^q = A^{p+q}.$$

Here p and q are arbitrary non-negative integers.

[7] It follows from the Binet-Cauchy formula that the minors of order p in C for $p > n$ (if minors of such orders exist) are all zero. In that case the right-hand side of (19) is to be replaced by zero. See footnote 5, p. 9.

We consider a polynomial (integral rational function) with coefficients in the field F:

$$f(t) = \alpha_0 t^m + \alpha_1 t^{m-1} + \cdots + \alpha_m.$$

Then by $f(A)$ we shall mean the matrix

$$f(A) = \alpha_0 A^m + \alpha_1 A^{m-1} + \cdots + \alpha_m E.$$

We define in this way a *polynomial in a matrix*.

Suppose that $f(t)$ is the product of two polynomials $g(t)$ and $h(t)$:

$$f(t) = g(t)h(t). \tag{21}$$

The polynomial $f(t)$ is obtained from $g(t)$ and $h(t)$ by multiplication term by term and collection of similar terms. In this we make use of the multiplication rule for powers: $t^p \cdot t^q = t^{p+q}$. Since all these operations remain valid when the scalar t is replaced by the matrix A, it follows from (21) that

$$f(A) = g(A)h(A).$$

Hence, in particular,[8]

$$g(A)h(A) = h(A)g(A); \tag{22}$$

i.e., *two polynomials in one and the same matrix are always permutable.*

Examples.

Let the sequence of elements a_{ik} for which $k - i = p$ $(i - k = p)$ in a rectangular matrix $A = \| a_{ik} \|$ be called the p-th *superdiagonal* (*subdiagonal*) of the matrix. We denote by $H^{(n)}$ the square matrix order n in which all the elements of the first superdiagonal are units and all the other elements are zero. The matrix $H^{(n)}$ will also be denoted simply by H. Then

$$H = H^{(n)} = \begin{Vmatrix} 0 & 1 & 0 & \dots & 0 \\ 0 & 0 & 1 & & \cdot \\ \cdot & \cdot & \cdot & \ddots & \cdot \\ \cdot & \cdot & \cdot & \ddots & \cdot \\ \cdot & \cdot & \cdot & & 1 \\ 0 & 0 & 0 & \dots & 0 \end{Vmatrix}, \qquad H^2 = \begin{Vmatrix} 0 & 0 & 1 & \dots & 0 \\ \cdot & & \ddots & & \cdot \\ \cdot & & & \ddots & \cdot \\ \cdot & & & & 1 \\ \cdot & & & & 0 \\ 0 & 0 & 0 & \dots & 0 \end{Vmatrix}, \dots,$$

$$H^p = 0 \qquad (p \geqq n).$$

---

[8] Since each of these products is equal to one and the same $f(A)$, by virtue of the fact that $h(t)g(t) = f(t)$. It is worth mentioning that the substitution of matrices in an algebraic identity in *several* variables is not valid. The substitution of matrices that commute with one another, however, is allowable in this case.

By these equations, if

$$f(t) = a_0 + a_1 t + a_2 t^2 + \cdots + a_{n-1} t^{n-1} + \cdots$$

is a polynomial in t, then

$$f(H) = a_0 E + a_1 H + a_2 H^2 + \cdots = \begin{Vmatrix} a_0 & a_1 & a_2 & \cdots & a_{n-1} \\ 0 & a_0 & a_1 & \ddots & \vdots \\ \vdots & & \ddots & \ddots & a_2 \\ \vdots & & & \ddots & a_1 \\ 0 & 0 & 0 & \cdots & a_0 \end{Vmatrix}.$$

Similarly, if F is the square matrix of order n in which all the elements of the first subdiagonal are units and all others are zero, then

$$f(F) = a_0 E + a_1 F + a_2 F^2 + \cdots = \begin{Vmatrix} a_0 & & 0 & \cdots & 0 \\ a_1 & a_0 & & \ddots & \vdots \\ \vdots & \ddots & \ddots & & \vdots \\ \vdots & & \ddots & \ddots & 0 \\ a_{n-1} & \cdots & & a_1 & a_0 \end{Vmatrix}.$$

We leave it to the reader to verify the following properties of the matrices H and F:

1. *When an arbitrary rectangular matrix A of dimension $m \times n$ is multiplied on the left by the matrix H (or F) of order m, then all the rows of A are shifted upward (or downward) by one place, the first (last) row of A disappears, and the last (first) row of the product is filled by zeros.* For example,

$$\begin{Vmatrix} 0 & 1 & 0 \\ 0 & 0 & 1 \\ 0 & 0 & 0 \end{Vmatrix} \begin{Vmatrix} a_1 & a_2 & a_3 & a_4 \\ b_1 & b_2 & b_3 & b_4 \\ c_1 & c_2 & c_3 & c_4 \end{Vmatrix} = \begin{Vmatrix} b_1 & b_2 & b_3 & b_4 \\ c_1 & c_2 & c_3 & c_4 \\ 0 & 0 & 0 & 0 \end{Vmatrix},$$

$$\begin{Vmatrix} 0 & 0 & 0 \\ 1 & 0 & 0 \\ 0 & 1 & 0 \end{Vmatrix} \begin{Vmatrix} a_1 & a_2 & a_3 & a_4 \\ b_1 & b_2 & b_3 & b_4 \\ c_1 & c_2 & c_3 & c_4 \end{Vmatrix} = \begin{Vmatrix} 0 & 0 & 0 & 0 \\ a_1 & a_2 & a_3 & a_4 \\ b_1 & b_2 & b_3 & b_4 \end{Vmatrix}.$$

2. *When an arbitrary rectangular matrix A of dimension $m \times n$ is multiplied on the right by the matrix H (or F) of order n, then all the columns of A are shifted to the right (left) by one place, the last (first) column of A disappears, and the first (last) column of the product is filled by zeros.* For example,

$$\begin{Vmatrix} a_1 & a_2 & a_3 & a_4 \\ b_1 & b_2 & b_3 & b_4 \\ c_1 & c_2 & c_3 & c_4 \end{Vmatrix} \begin{Vmatrix} 0 & 1 & 0 & 0 \\ 0 & 0 & 1 & 0 \\ 0 & 0 & 0 & 1 \\ 0 & 0 & 0 & 0 \end{Vmatrix} = \begin{Vmatrix} 0 & a_1 & a_2 & a_3 \\ 0 & b_1 & b_2 & b_3 \\ 0 & c_1 & c_2 & c_3 \end{Vmatrix},$$

$$\begin{Vmatrix} a_1 & a_2 & a_3 & a_4 \\ b_1 & b_2 & b_3 & b_4 \\ c_1 & c_2 & c_3 & c_4 \end{Vmatrix} \begin{Vmatrix} 0 & 0 & 0 & 0 \\ 1 & 0 & 0 & 0 \\ 0 & 1 & 0 & 0 \\ 0 & 0 & 1 & 0 \end{Vmatrix} = \begin{Vmatrix} a_2 & a_3 & a_4 & 0 \\ b_2 & b_3 & b_4 & 0 \\ c_2 & c_3 & c_4 & 0 \end{Vmatrix}.$$

2. A square matrix A is called *singular* if $|A| = 0$. Otherwise A is called *non-singular*.

Let $A = \| a_{ik} \|_1^n$ be a non-singular matrix ($|A| \neq 0$). Let us consider the linear transformation with coefficient matrix A

$$y_i = \sum_{k=1}^{n} a_{ik} x_k \quad (i = 1, 2, \ldots, n). \tag{23}$$

When we regard (23) as equations for $x_1, x_2, \ldots, x_n$ and observe that the determinant of the system of equations (23) is, by assumption, different from zero, then we can express $x_1, x_2, \ldots, x_n$ in terms of $y_1, y_2, \ldots, y_n$ by means of the well-known formulas:

$$x_i = \frac{1}{|A|} \begin{vmatrix} a_{11} & \cdots & a_{1,i-1} & y_1 & a_{1,i+1} & \cdots & a_{1n} \\ a_{21} & \cdots & a_{2,i-1} & y_2 & a_{2,i+1} & \cdots & a_{2n} \\ \cdot & \cdot & \cdot & \cdot & \cdot & \cdot & \cdot \\ a_{n1} & \cdots & a_{n,i-1} & y_n & a_{n,i+1} & \cdots & a_{nn} \end{vmatrix} \equiv \sum_{k=1}^{n}{}' a_{ik}^{(-1)} y_k \quad (i = 1, 2, \ldots, n). \tag{24}$$

We have thus obtained the 'inverse' transformation of the transformation (23). The coefficient matrix of this transformation

$$A^{-1} = \| a_{ik}^{(-1)} \|_1^n$$

will be called the *inverse matrix* of A. From (24) it is easy to see that

$$a_{ik}^{(-1)} = \frac{A_{ki}}{|A|} \quad (i, k = 1, 2, \ldots, n), \tag{25}$$

where A_{ki} is the algebraic complement (the cofactor) of the element a_{ki} in the determinant $|A|$ $(i, k = 1, 2, \ldots, n)$.

For example, if

$$A = \left\| \begin{matrix} a_1 & a_2 & a_3 \\ b_1 & b_2 & b_3 \\ c_1 & c_2 & c_3 \end{matrix} \right\| \text{ and } |A| \neq 0,$$

then

$$A^{-1} = \frac{1}{|A|} \left\| \begin{matrix} b_2c_3 - b_3c_2 & a_3c_2 - a_2c_3 & a_2b_3 - a_3b_2 \\ b_3c_1 - b_1c_3 & a_1c_3 - a_3c_1 & a_3b_1 - a_1b_3 \\ b_1c_2 - b_2c_1 & a_2c_1 - a_1c_2 & a_1b_2 - a_2b_1 \end{matrix} \right\|.$$

By forming the composite transformation of the given transformation (23) and the inverse (24), in either order, we obtain in both cases the identity transformation (with the unit matrix as coefficient matrix); therefore

$$AA^{-1} = A^{-1}A = E. \tag{26}$$

The validity of equation (26) can also be established by direct multiplication of the matrices A and A^{-1}. In fact, by (25) we have[9]

$$[AA^{-1}]_{ij} = \sum_{k=1}^{n} a_{ik}a_{kj}^{(-1)} = \frac{1}{|A|} \sum_{k=1}^{n} a_{ik}A_{jk} = \delta_{ij} \qquad (i, j = 1, 2, \ldots, n).$$

Similarly,

$$[A^{-1}A]_{ij} = \sum_{k=1}^{n} a_{ij}^{(-1)}a_{kj} = \frac{1}{|A|} \sum_{k=1}^{n} A_{ki}a_{kj} = \delta_{ij} \qquad (i, j = 1, 2, \ldots, n).$$

It is easy to see that the matrix equations

$$AX = E \text{ and } XA = E \qquad (|A| \neq 0) \tag{27}$$

have no solutions other than $X = A^{-1}$. For by multiplying both sides of the first (second) equation on the left (right) by A^{-1} and using the associative property of matrix multiplication we obtain from (26) in both cases:[10]

$$X = A^{-1}.$$

[9] Here we make use of the well-known property of determinants that the sum of the products of the elements of an arbitrary column into the cofactors of the elements of that column is equal to the value of the determinant and the sum of the products of the elements of a column into the cofactors of the corresponding element of another column is zero.

[10] If A is a singular matrix, then the equations (27) have no solution. For if one of these equations had a solution $X = \| x_{ik} \|_1^n$, then we would have by the multiplication theorem of determinants (see formula (18)) that $|A| \cdot |X| = |E| = 1$, and this is impossible when $|A| = 0$.

In the same way it can be shown that each of the matrix equations

$$AX = B,\ XA = B \qquad (|A| \neq 0), \tag{28}$$

where X and B are rectangular matrices of equal dimensions and A is a square matrix of appropriate order, have one and only one solution,

$$X = A^{-1}B \quad \text{and} \quad X = BA^{-1}, \tag{29}$$

respectively. The matrices (29) are the 'left' and the 'right' quotients on 'dividing' B by A. From (28) and (29) we deduce (see p. 12) that $r_B \leqq r_X$ and $r_X \leqq r_B$, so that $r_X = r_B$. On comparing this with (28), we have:

When a rectangular matrix is multiplied on the left or on the right by a non-singular matrix, the rank of the original matrix remains unchanged.

Note that (26) implies $|A| \cdot |A^{-1}| = 1$, i.e.

$$|A^{-1}| = \frac{1}{|A|}.$$

For any two non-singular matrices we have

$$(AB)^{-1} = B^{-1}A^{-1}. \tag{30}$$

3. All the matrices of order n form a ring[11] with unit element $E^{(n)}$.

Since in this ring the operation of multiplication by a number of F is defined, and since there exists a basis of n^2 linearly independent matrices in terms of which all the matrices of order n can be expressed linearly,[12] the ring of matrices of order n is an algebra.[13]

---

[11] A *ring* is a collection of elements in which two operations are defined and can always be carried out uniquely: the 'addition' of two elements (with the commutative and associative properties) and the 'multiplication' of two elements (with the associative and distributive properties with respect to addition); moreover, the addition is reversible. See, for example, van der Waerden, ***Modern Algebra***, § 14.

[12] For, an arbitrary matrix $A = \| a_{ik} \|_1^n$ with elements in F can be represented in the form $A = \sum_{i,k=1}^{n} a_{ik} E_{ik}$, where E_{ik} is the matrix of order n in which there is a 1 at the intersection of the i-th row and the k-th column and all the other elements are zeros.

[13] See, for example, van der Waerden, ***Modern Algebra***, § 17.

All the square matrices of order n form a commutative group with respect to the operation of addition.[14] All the non-singular matrices of order n form a (non-commutative) group with respect to the operation of multiplication.

A square matrix $A = \| a_{ik} \|_1^n$ is called *upper triangular* (*lower triangular*) if all the elements below (above) the main diagonal are zero:

$$A = \begin{Vmatrix} a_{11} & a_{12} & \dots & a_{1n} \\ 0 & a_{22} & \dots & a_{2n} \\ \cdot & & \cdot & \cdot \\ \cdot & & \cdot & \cdot \\ \cdot & & \cdot & \cdot \\ 0 & 0 & \dots & a_{nn} \end{Vmatrix}, \quad A = \begin{Vmatrix} a_{11} & 0 & \dots & 0 \\ a_{21} & a_{22} & \dots & 0 \\ \cdot & \cdot & & \\ \cdot & & \cdot & \\ \cdot & & & \cdot \\ a_{n1} & a_{n2} & \dots & a_{nn} \end{Vmatrix}.$$

A diagonal matrix is a special case both of an upper triangular matrix and a lower triangular matrix.

Since the determinant of a triangular matrix is equal to the product of its diagonal elements, a triangular (and, in particular, a diagonal) matrix is non-singular if and only if all its diagonal elements are different from zero.

It is easy to verify that the sum and the product of two diagonal (upper triangular, lower triangular) matrices is a diagonal (upper triangular, lower triangular) matrix and that the inverse of a non-singular diagonal (upper triangular, lower triangular) matrix is a matrix of the same type. Therefore:

1. All the diagonal matrices of order n form a commutative group under the operation of addition, as do all the upper triangular matrices or all the lower triangular matrices.

2. All the non-singular diagonal matrices form a commutative group under multiplication.

3. All the non-singular upper (lower) triangular matrices form a (non-commutative) group under multiplication.

4. We conclude this section with a further important operation on matrices —*transposition.*

---

[14] A *group* is a set of objects in which an operation is defined which associates with any two elements a and b of the set a well-defined third element $a * b$ of the same set provided that

1) the operation has the associative property $((a * b) * c = a * (b * c))$,

2) there exists a unit element e in the set $(a * e = e * a = a)$, and

3) for every element e of the set there exists an inverse element a^{-1} $(a * a^{-1} = a^{-1} * a = e)$.

A group is called *commutative*, or abelian, if the group operation has the commutative property. Concerning the group concept see, for example, [53], pp. 245ff.

If $A = \| a_{ik} \|$ $(i = 1, 2, \ldots, m;\ k = 1, 2, \ldots, n)$, then the *transpose* A^{T} is defined as $A^{\mathsf{T}} = \| a^{\mathsf{T}}_{ik} \|$, where $a^{\mathsf{T}}_{ki} = a_{ik}$ $(i = 1, 2, \ldots, m;\ k = 1, 2, \ldots, n)$. If A is of dimension $m \times n$, then A^{T} is of dimension $n \times m$.

It is easy to verify the following properties:[15]

1. $(A + B)^{\mathsf{T}} = A^{\mathsf{T}} + B^{\mathsf{T}}$,
2. $(\alpha A)^{\mathsf{T}} = \alpha A^{\mathsf{T}}$,
3. $(AB)^{\mathsf{T}} = B^{\mathsf{T}} A^{\mathsf{T}}$,
4. $(A^{-1})^{\mathsf{T}} = (A^{\mathsf{T}})^{-1}$.

If a square matrix $S = \| s_{ij} \|_1^n$ coincides with its transpose $(S^{\mathsf{T}} = S)$, then it is called *symmetric*. In a symmetric matrix elements that are symmetrically placed with respect to the main diagonal are equal. Note that the product of two symmetric matrices is not, in general, symmetric. By 3., this holds if and only if the two given symmetric matrices are permutable.

If a square matrix $K = \| k_{ij} \|_1^n$ differs from its transpose by a factor -1 $(K^{\mathsf{T}} = -K)$, then it is called *skew-symmetric*. In a skew-symmetric matrix any two elements that are symmetrical to the main diagonal differ from each other by a factor -1 and the diagonal elements are zero. From 3. it follows that the product of two permutable skew-symmetric matrices is a symmetric matrix.[16]

§ 4. Compound Matrices. Minors of the Inverse Matrix

1. Let $A = \| a_{ik} \|_1^n$ be a given matrix. We consider all possible minors of A of order p $(1 \leqq p \leqq n)$:

$$A\begin{pmatrix} i_1 & i_2 & \ldots & i_p \\ k_1 & k_2 & \ldots & k_p \end{pmatrix} \qquad \left(1 \leqq \begin{matrix} i_1 < i_2 < \cdots < i_p \\ k_1 < k_2 < \cdots < k_p \end{matrix} \leqq n\right). \tag{31}$$

The number of these minors is N^2, where $N = \binom{n}{p}$ is the number of combinations of n objects taken p at a time. In order to arrange the minors (31) in a square array, we enumerate in some definite order—lexicographic order, for example—all the N combinations of p indices selected from among the indices $1, 2, \ldots, n$.

[15] In formulas 1., 2., 3., A and B are arbitrary rectangular matrices for which the corresponding operations are feasible. In 4., A is an arbitrary square non-singular matrix.

[16] As regards the representation of a square matrix A in the form of a product of two symmetric matrices $(A = S_1 S_2)$ or two skew-symmetric matrices $(A = K_1 K_2)$, see [357].

If the combinations of indices $i_1 < i_2 < \dots < i_p$ and $k_1 < k_2 < \dots < k_p$ have the numbers α and β, then the minors (31) will also be denoted as follows:

$$\mathfrak{a}_{\alpha\beta} = A\begin{pmatrix} i_1 & i_2 & \dots & i_p \\ k_1 & k_2 & \dots & k_p \end{pmatrix}.$$

By giving to α and β independently all the values from 1 to N, we obtain all the minors of $A = \| a_{ik} \|_1^n$ of order p.

The square matrix of order N

$$\mathfrak{A}_p = \| \mathfrak{a}_{\alpha\beta} \|_1^N$$

is called the p-th *compound matrix* of $A = \| a_{ik} \|_1^n$; p can take the values $1, 2, \dots, n$. Here $\mathfrak{A}_1 = A$, and $\mathfrak{A}_n$ consists of the single element $|A|$.

Note. The order of enumeration of the combination of indices is fixed once and for all and does not depend on the choice of A.

Example. Let

$$A = \begin{Vmatrix} a_{11} & a_{12} & a_{13} & a_{14} \\ a_{21} & a_{22} & a_{23} & a_{24} \\ a_{31} & a_{32} & a_{33} & a_{34} \\ a_{41} & a_{42} & a_{43} & a_{44} \end{Vmatrix}.$$

We enumerate all combinations of the indices 1, 2, 3, 4 taken two at a time by arranging them in the following order:

$$(12) \quad (13) \quad (14) \quad (23) \quad (24) \quad (34).$$

Then

$$\mathfrak{A}_2 = \begin{Vmatrix} A\begin{pmatrix}1&2\\1&2\end{pmatrix} & A\begin{pmatrix}1&2\\1&3\end{pmatrix} & A\begin{pmatrix}1&2\\1&4\end{pmatrix} & A\begin{pmatrix}1&2\\2&3\end{pmatrix} & A\begin{pmatrix}1&2\\2&4\end{pmatrix} & A\begin{pmatrix}1&2\\3&4\end{pmatrix} \\ A\begin{pmatrix}1&3\\1&2\end{pmatrix} & A\begin{pmatrix}1&3\\1&3\end{pmatrix} & A\begin{pmatrix}1&3\\1&4\end{pmatrix} & A\begin{pmatrix}1&3\\2&3\end{pmatrix} & A\begin{pmatrix}1&3\\2&4\end{pmatrix} & A\begin{pmatrix}1&3\\3&4\end{pmatrix} \\ A\begin{pmatrix}1&4\\1&2\end{pmatrix} & A\begin{pmatrix}1&4\\1&3\end{pmatrix} & A\begin{pmatrix}1&4\\1&4\end{pmatrix} & A\begin{pmatrix}1&4\\2&3\end{pmatrix} & A\begin{pmatrix}1&4\\2&4\end{pmatrix} & A\begin{pmatrix}1&4\\3&4\end{pmatrix} \\ A\begin{pmatrix}2&3\\1&2\end{pmatrix} & A\begin{pmatrix}2&3\\1&3\end{pmatrix} & A\begin{pmatrix}2&3\\1&4\end{pmatrix} & A\begin{pmatrix}2&3\\2&3\end{pmatrix} & A\begin{pmatrix}2&3\\2&4\end{pmatrix} & A\begin{pmatrix}2&3\\3&4\end{pmatrix} \\ A\begin{pmatrix}2&4\\1&2\end{pmatrix} & A\begin{pmatrix}2&4\\1&3\end{pmatrix} & A\begin{pmatrix}2&4\\1&4\end{pmatrix} & A\begin{pmatrix}2&4\\2&3\end{pmatrix} & A\begin{pmatrix}2&4\\2&4\end{pmatrix} & A\begin{pmatrix}2&4\\3&4\end{pmatrix} \\ A\begin{pmatrix}3&4\\1&2\end{pmatrix} & A\begin{pmatrix}3&4\\1&3\end{pmatrix} & A\begin{pmatrix}3&4\\1&4\end{pmatrix} & A\begin{pmatrix}3&4\\2&3\end{pmatrix} & A\begin{pmatrix}3&4\\2&4\end{pmatrix} & A\begin{pmatrix}3&4\\3&4\end{pmatrix} \end{Vmatrix}$$

We mention some properties of compound matrices:

1. *From $C=AB$ it follows that $\mathfrak{C}_p=\mathfrak{A}_p\cdot\mathfrak{B}_p (p=1, 2, \dots, n)$.*

For when we express the minors of order p $(1\leq p\leq n)$ of the matrix product C, by formula (19), in terms of the minors of the same order of the factors, then we have:

$$C\begin{pmatrix} i_1 & i_2 & \dots & i_p \\ k_1 & k_2 & \dots & k_p \end{pmatrix} = \sum_{1\leq l_1<l_2<\dots<l_p\leq n} A\begin{pmatrix} i_1 & i_2 & \dots & i_p \\ l_1 & l_2 & \dots & l_p \end{pmatrix} B\begin{pmatrix} l_1 & l_2 & \dots & l_p \\ k_1 & k_2 & \dots & k_p \end{pmatrix}$$

$$\left(1\leq \begin{matrix} i_1<i_2<\dots<i_p \\ k_1<k_2<\dots<k_p \end{matrix} \leq n\right). \tag{32}$$

Obviously, in the notation of this section, equation (32) can be written as follows:

$$\mathfrak{c}_{\alpha\beta}=\sum_{\lambda=1}^{N}\mathfrak{a}_{\alpha\lambda}\mathfrak{b}_{\lambda\beta} \qquad (\alpha, \beta=1, 2, \dots, N)$$

(here α, β, and λ are the numbers of the combinations of indices $i_1<i_2<\dots<i_p$; $k_1<k_2<\dots<k_p$; $l_1<l_2<\dots<l_p$). Hence

$$\mathfrak{C}_p=\mathfrak{A}_p\mathfrak{B}_p \qquad (p=1, 2, \dots, n).$$

2. *From $B=A^{-1}$ it follows that $\mathfrak{B}_p=\mathfrak{A}_p^{-1}$ $(p=1, 2, \dots, n)$.*

This result follows immediately from the preceding one when we set $C=E$ and bear in mind that $\mathfrak{E}_p$ is the unit matrix of order $N=\binom{n}{p}$

From 2. there follows an important formula that expresses the minors of the inverse matrix in terms of the minors of the given matrix:

If $B=A^{-1}$, then for arbitrary $\left(1\leq \begin{matrix} i_1<i_2<\dots<i_p \\ k_1<k_2<\dots<k_p \end{matrix} \leq n\right)$

$$B\begin{pmatrix} i_1 & i_2 & \dots & i_p \\ k_1 & k_2 & \dots & k_p \end{pmatrix} = \frac{(-1)^{\sum_{\nu=1}^{p} i_\nu+\sum_{\nu=1}^{p} k_\nu} A\begin{pmatrix} k'_1 & k'_2 & \dots & k'_{n-p} \\ i'_1 & i'_2 & \dots & i'_{n-p} \end{pmatrix}}{A\begin{pmatrix} 1 & 2 & \dots & n \\ 1 & 2 & \dots & n \end{pmatrix}}, \tag{33}$$

where $i_1<i_2<\dots<i_p$ and $i'_1<i'_2<\dots<i'_{n-p}$ form a complete system of indices $1, 2, \dots, n$, as do $k_1<k_2<\dots<k_p$ and $k'_1<k'_2<\dots<k'_{n-p}$.

For it follows from $AB=E$ that

$$\mathfrak{A}_p\mathfrak{B}_p=\mathfrak{E}_p$$

or in more explicit form:

$$\sum_{\alpha=1}^{N} \mathfrak{a}_{\gamma\alpha} \mathfrak{b}_{\alpha\beta} = \delta_{\gamma\beta} = \begin{cases} 1 & (\gamma=\beta), \\ 0 & (\gamma\neq\beta). \end{cases} \tag{34}$$

Equations (34) can also be written as follows:

$$\sum_{1\leq i_1<i_2<\cdots<i_p\leq n} A\begin{pmatrix} j_1 j_2 \dots j_p \\ i_1 i_2 \dots i_p \end{pmatrix} B\begin{pmatrix} i_1 i_2 \dots i_p \\ k_1 k_2 \dots k_p \end{pmatrix} = \begin{cases} 1, & \text{if } \sum_{\nu=1}^{p} (j_\nu - k_\nu)^2 = 0, \\ 0, & \text{if } \sum_{\nu=1}^{p} (j_\nu - k_\nu)^2 > 0 \end{cases} \tag{34'}$$

$$\left(1 \leq \begin{matrix} j_1 < j_2 < \dots < j_p \\ k_1 < k_2 < \dots < k_p \end{matrix} \leq n\right).$$

On the other hand, when we apply the well-known Laplace expansion to the determinant $|A|$, we obtain

$$\sum_{1\leq i_1<i_2<\cdots<i_p\leq n} A\begin{pmatrix} j_1\, j_2 \dots j_p \\ i_1\, i_2 \dots i_p \end{pmatrix} \cdot (-1)^{\sum_{\nu=1}^{p} i_\nu + \sum_{\nu=1}^{p} k_\nu} A\begin{pmatrix} k'_1\, k'_2 \dots k'_{n-p} \\ i'_1\, i'_2 \dots i'_{n-p} \end{pmatrix} =$$

$$= \begin{cases} |A|, & \text{if } \sum_{\nu=1}^{p} (j_\nu - k_\nu)^2 = 0, \\ 0, & \text{if } \sum_{\nu=1}^{p} (j_\nu - k_\nu)^2 > 0, \end{cases} \tag{35}$$

where $i_1 < i_2 < \cdots < i_p$ and $i'_1 < i'_2 < \cdots < i'_{n-p}$ form a complete system of indices $1, 2, \dots, n$, as do $k_1 < k_2 < \cdots < k_p$ and $k'_1 < k'_2 < \cdots < k'_{n-p}$. Comparison of (35) with (34′) and (34) shows that the equations (34) are satisfied if we take together with $\mathfrak{b}_{\alpha\beta}$ not $B\begin{pmatrix} i_1\, i_2 \dots i_p \\ k_1\, k_2 \dots k_p \end{pmatrix}$ but rather

$$\frac{(-1)^{\sum_{\nu=1}^{p} i_\nu + \sum_{\nu=1}^{p} k_\nu} A\begin{pmatrix} k'_1\, k'_2 \dots k'_{n-p} \\ i'_1\, i'_2 \dots i'_{n-p} \end{pmatrix}}{A\begin{pmatrix} 1\ 2 \dots n \\ 1\ 2 \dots n \end{pmatrix}}.$$

Since the elements $\mathfrak{b}_{\alpha\beta}$ of the inverse matrix of $\mathfrak{A}_p$ are uniquely determined by (34), equation (33) must hold.

CHAPTER II

THE ALGORITHM OF GAUSS AND SOME OF ITS APPLICATIONS

§ 1. Gauss's Elimination Method

1. Let

$$\left.\begin{array}{l} a_{11}x_1 + a_{12}x_2 + \cdots + a_{1n}x_n = y_1 \\ a_{21}x_1 + a_{22}x_2 + \cdots + a_{2n}x_n = y_2 \\ \cdot \quad \cdot \quad \cdot \quad \cdot \quad \cdot \quad \cdot \quad \cdot \quad \cdot \quad \cdot \quad \cdot \quad \cdot \quad \cdot \quad \cdot \quad \cdot \\ a_{n1}x_1 + a_{n2}x_2 + \cdots + a_{nn}x_n = y_n \end{array}\right\} \tag{1}$$

be a system of n linear equations in n unknowns $x_1, x_2, \ldots, x_n$ with right-hand sides $y_1, y_2, \ldots, y_n$.

In matrix form this system may be written as

$$Ax = y. \tag{1'}$$

Here $x = (x_1, x_2, \ldots, x_n)$ and $y = (y_1, y_2, \ldots, y_n)$ are columns and $A = \| a_{ik} \|_1^n$ is the square coefficient matrix.

If A is non-singular, then we can rewrite this as

$$x = A^{-1}y, \tag{2}$$

or in explicit form:

$$x_i = \sum_{k=1}^{n} a_{ik}^{(-1)} y_k \qquad (i = 1, 2, \ldots, n). \tag{2'}$$

Thus, the task of computing the elements of the inverse matrix $A^{-1} = \| a_{ik}^{(-1)} \|_1^n$ is equivalent to the task of solving the system of equations (1) for *arbitrary* right-hand sides $y_1, y_2, \ldots, y_n$. The elements of the inverse matrix are determined by the formulas (25) of Chapter I. However, the actual computation of the elements of A^{-1} by these formulas is very tedious for large n. Therefore, effective methods of computing the elements of an inverse matrix—and hence of solving a system of linear equations—are of great practical value.[1]

---

[1] For a detailed account of these methods, we refer the reader to the book by Faddeev [15] and the group of papers that appeared in *Uspehi Mat. Nauk*, Vol. 5, 3 (1950).

In the present chapter we expound the theoretical basis of some of these methods; they are variants of Gauss's elimination method, whose acquaintance the reader first made in his algebra course at school.

2. Suppose that in the system of equations (1) we have $a_{11} \neq 0$. We eliminate x_1 from all the equations beginning with the second by adding to the second equation the first multipled by $-\frac{a_{21}}{a_{11}}$, to the third the first multiplied by $-\frac{a_{31}}{a_{11}}$, and so on. The system (1) has now been replaced by the equivalent system

$$\left.\begin{aligned} a_{11}x_1 + a_{12}x_2 + \cdots + a_{1n}x_n &= y_1 \\ a_{22}^{(1)}x_2 + \cdots + a_{2n}^{(1)}x_n &= y_2^{(1)} \\ \cdots\cdots\cdots\cdots\cdots & \\ a_{n2}^{(1)}x_2 + \cdots + a_{nn}^{(1)}x_n &= y_n^{(1)} \end{aligned}\right\} \tag{3}$$

The coefficients of the unknowns and the constant terms of the last $n-1$ equations are given by the formulas

$$a_{ij}^{(1)} = a_{ij} - \frac{a_{i1}}{a_{11}} a_{1j}, \qquad y_i^{(1)} = y_i - \frac{a_{i1}}{a_{11}} y_1 \qquad (i, j = 2, \ldots, n). \tag{3'}$$

Suppose that $a_{22}^{(1)} \neq 0$. Then we eliminate x_2 in the same way from the last $n-2$ equations of the system (3) and obtain the system

$$\left.\begin{aligned} a_{11}x_1 + a_{12}x_2 + a_{13}x_3 + \cdots + a_{1n}x_n &= y_1 \\ a_{22}^{(1)}x_2 + a_{23}^{(1)}x_3 + \cdots + a_{2n}^{(1)}x_n &= y_2^{(1)} \\ a_{33}^{(2)}x_3 + \cdots + a_{3n}^{(2)}x_n &= y_3^{(2)} \\ \cdots\cdots\cdots\cdots\cdots & \\ a_{n3}^{(2)}x_3 + \cdots + a_{nn}^{(2)}x_n &= y_n^{(2)}. \end{aligned}\right\} \tag{4}$$

The new coefficients and the new right-hand sides are connected with the preceding ones by the formulas:

$$a_{ij}^{(2)} = a_{ij}^{(1)} - \frac{a_{i2}^{(1)}}{a_{22}^{(1)}} a_{2j}^{(1)}, \quad y_i^{(2)} = y_i^{(1)} - \frac{a_{i2}^{(1)}}{a_{22}^{(1)}} y_2^{(1)} \qquad (i, j = 3, \ldots, n). \tag{5}$$

Continuing the algorithm, we go in $n-1$ steps from the original system (1) to the triangular recurrent system

$$\left.\begin{aligned} a_{11}x_1 + a_{12}x_2 + a_{13}x_3 + \cdots + a_{1n}x_n &= y_1 \\ a_{22}^{(1)}x_2 + a_{23}^{(1)}x_3 + \cdots + a_{2n}^{(1)}x_n &= y_2^{(1)} \\ a_{33}^{(2)}a_3 + \cdots + a_{3n}^{(2)}x_n &= y_3^{(2)} \\ \cdots\cdots\cdots\cdots\cdots & \\ a_{nn}^{(n-1)}x_n &= y_n^{(n-1)}. \end{aligned}\right\} \tag{6}$$

This reduction can be carried out if and only if in the process all the numbers $a_{11}, a_{22}^{(1)}, a_{33}^{(2)}, \ldots, a_{n-1,n-1}^{(n-2)}$ turn out to be different from zero.

This *algorithm of Gauss* consists of operations of a simple type such as can easily be carried out by present-day computing machines.

3. Let us express the coefficients and the right-hand sides of the reduced system in terms of the coefficients and the right-hand sides of the original system (1). We shall not assume here that in the reduction process all the numbers $a_{11}, a_{22}^{(1)}, a_{33}^{(2)}, \ldots, a_{n-1,n-1}^{(n-2)}$ turn out to be different from zero; we consider the general case, in which the first p of these numbers are different from zero:

$$a_{11} \neq 0, \quad a_{22}^{(1)} \neq 0, \quad \ldots, \quad a_{pp}^{(p-1)} \neq 0 \qquad (p \leq n-1). \tag{7}$$

This enables us (at the p-th step of the reduction) to put the original system of equations into the form

$$\left.\begin{array}{l} a_{11}x_1 + a_{12}x_2 + \cdots\cdots\cdots\cdots + a_{1n}x_n = y_1 \\ \quad a_{22}^{(1)}x_2 + \cdots\cdots\cdots\cdots + a_{2n}^{(1)}x_n = y_2^{(1)} \\ \cdots\cdots\cdots\cdots\cdots\cdots\cdots\cdots \\ \qquad a_{pp}^{(p-1)}x_p + \cdots\cdots\cdots + a_{pn}^{(p-1)}x_n = y_p^{(p-1)} \\ \qquad\quad a_{p+1,p+1}^{(p)}x_{p+1} + \cdots + a_{p+1,n}^{(p)}x_n = y_{p+1}^{(p)} \\ \qquad\quad \cdots\cdots\cdots\cdots\cdots\cdots \\ \qquad\quad a_{n,p+1}^{(p)}x_{p+1} + \cdots + a_{nn}^{(p)}x_n = y_n^{(p)}. \end{array}\right\} \tag{8}$$

We denote the coefficient matrix of this system of equations by G_p:

$$G_p = \left\| \begin{array}{cccccccc} a_{11} & a_{12} & \ldots & a_{1p} & a_{1,p+1} & \ldots & a_{1n} \\ 0 & a_{22}^{(1)} & \ldots & a_{2p}^{(1)} & a_{2,p+1}^{(1)} & \ldots & a_{2n}^{(1)} \\ \cdot & \cdot & \cdot & \cdot & \cdot & \cdot & \cdot \\ 0 & 0 & \ldots & a_{pp}^{(p-1)} & a_{p,p+1}^{(p-1)} & \ldots & a_{pn}^{(p-1)} \\ 0 & 0 & \ldots & 0 & a_{p+1,p+1}^{(p)} & \ldots & a_{p+1,n}^{(p)} \\ \cdot & \cdot & \cdot & \cdot & \cdot & \cdot & \cdot \\ 0 & 0 & \ldots & 0 & a_{n,p+1}^{(p)} & \ldots & a_{nn}^{(p)} \end{array} \right\|. \tag{9}$$

The transition from A to G_p is effected as follows: To every row of A in succession from the second to the n-th there are added some preceding rows (from the first p) multiplied by certain factors. Therefore all the minors of order h contained in the first h rows of A and G_p are equal:

$$A\begin{pmatrix} 1 & 2 & \ldots & h \\ k_1 & k_2 & \ldots & k_h \end{pmatrix} = G_p\begin{pmatrix} 1 & 2 & \ldots & h \\ k_1 & k_2 & \ldots & k_h \end{pmatrix} \quad \begin{pmatrix} 1 \leq k_1 < k_2 < \cdots < k_h \leq n \\ (h = 1, 2, \ldots, n) \end{pmatrix}. \tag{10}$$

From these formulas we find, by taking into account the structure (9) of G_p,

$$A\begin{pmatrix}1 & 2 & \dots & p\\ 1 & 2 & \dots & p\end{pmatrix}=a_{11}a_{22}^{(1)}\cdots a_{pp}^{(p-1)}, \tag{11}$$

$$A\begin{pmatrix}1 & 2 & \dots & p & i\\ 1 & 2 & \dots & p & k\end{pmatrix}=a_{11}a_{22}^{(1)}\cdots a_{pp}^{(p-1)}a_{ik}^{(p)} \quad (i,\, k=p+1,\dots,n), \tag{12}$$

When we divide the second of these equations by the first, we obtain the fundamental formulas[2]

$$a_{ik}^{(p)}=\frac{A\begin{pmatrix}1 & 2 & \dots & p & i\\ 1 & 2 & \dots & p & k\end{pmatrix}}{A\begin{pmatrix}1 & 2 & \dots & p\\ 1 & 2 & \dots & p\end{pmatrix}} \quad (i,\, k=p+1,\,\dots,\, n). \tag{13}$$

If the conditions (7) hold for a given value of p, then they also hold for every smaller value of p. Therefore the formulas (13) are valid not only for the given value of p but also for all smaller values of p. The same holds true of (11). Hence instead of this formula we can write the equations

$$A\begin{pmatrix}1\\ 1\end{pmatrix}=a_{11},\quad A\begin{pmatrix}1 & 2\\ 1 & 2\end{pmatrix}=a_{11}a_{22}^{(1)},\quad A\begin{pmatrix}1 & 2 & 3\\ 1 & 2 & 3\end{pmatrix}=a_{11}a_{22}^{(1)}a_{33}^{(2)},\ \dots. \tag{14}$$

Thus, the conditions (7), i.e., the necessary and sufficient conditions for the feasibility of the first p steps in Gauss's algorithm, can be written in the form of the following inequalities:

$$A\begin{pmatrix}1\\ 1\end{pmatrix}\neq 0,\quad A\begin{pmatrix}1 & 2\\ 1 & 2\end{pmatrix}\neq 0,\ \dots,\quad A\begin{pmatrix}1 & 2 & \dots & p\\ 1 & 2 & \dots & p\end{pmatrix}\neq 0 \tag{15}$$

From (14) we then find:

$$a_{11}=A\begin{pmatrix}1\\ 1\end{pmatrix},\quad a_{22}^{(1)}=\frac{A\begin{pmatrix}1 & 2\\ 1 & 2\end{pmatrix}}{A\begin{pmatrix}1\\ 1\end{pmatrix}},\quad a_{33}^{(2)}=\frac{A\begin{pmatrix}1 & 2 & 3\\ 1 & 2 & 3\end{pmatrix}}{A\begin{pmatrix}1 & 2\\ 1 & 2\end{pmatrix}},\ \dots,\quad a_{pp}^{(p-1)}=\frac{A\begin{pmatrix}1 & 2 & \dots & p\\ 1 & 2 & \dots & p\end{pmatrix}}{A\begin{pmatrix}1 & 2 & \dots & p-1\\ 1 & 2 & \dots & p-1\end{pmatrix}}. \tag{16}$$

In order to eliminàte $x_1, x_2, \dots, x_p$ *consecutively* by Gauss's algorithm it is necessary that all the values (16) should be different from zero, i.e., that the inequalities (15) should hold. However, the formulas for $a_{ik}^{(p)}$ make sense if only the last of the conditions (15). holds.

[2] See [181], p. 89.

4. Suppose the coefficient matrix of the system of equations (1) to be of rank r. Then, by a suitable permutation of the equations and a renumbering of the unknowns, we can arrange that the following inequalities hold:

$$A\begin{pmatrix}1 & 2 & \dots & j\\ 1 & 2 & \dots & j\end{pmatrix}\neq 0 \qquad (j=1, 2, \dots, r). \tag{17}$$

This enables us to eliminate $x_1, x_2, \dots, x_r$ consecutively and to obtain the system of equations

$$\left.\begin{array}{l}
a_{11}x_1 + a_{12}x_2 + \cdots\cdots + a_{1n}x_n = y_1\\
\qquad a_{22}^{(1)}x_2 + \cdots\cdots + a_{2n}^{(1)}x_n = y_2^{(1)}\\
\quad \cdots\cdots\cdots\cdots\cdots\cdots\\
\qquad\qquad a_{rr}^{(r-1)}x_r + \cdots + a_{rn}^{(r-1)}x_n = y_r^{(r-1)}\\
\qquad\qquad\quad a_{r+1,r+1}^{(r)}x_{r+1} + \cdots + a_{r+1,n}^{(r)}x_n = y_{r+1}^{(r)}\\
\quad \cdots\cdots\cdots\cdots\cdots\cdots\\
\qquad\qquad\quad a_{n,r+1}^{(r)}x_{r+1} + \cdots + a_{nn}^{(r)}x_n = y_n^{(r)}.
\end{array}\right\} \tag{18}$$

Here the coefficients are determined by the formulas (13). From these formulas it follows, because the rank of the matrix $A = \| a_{ik} \|_1^n$ is equal to r, that

$$a_{ik}^{(r)}=0 \qquad (i, k=r+1, \dots, n). \tag{19}$$

Therefore the last $n - r$ equations (18) reduce to the consistency conditions

$$y_i^{(r)}=0 \qquad (i=r+1, \dots, n). \tag{20}$$

Note that in the elimination algorithm the column of constant terms is subjected to the same transformations as the other columns, of coefficients. Therefore, by supplementing the matrix $A = \| a_{ik} \|_1^n$ with an $(n+1)$-th column of the constant terms we obtain:

$$y_i^{(p)}=\frac{A\begin{pmatrix}1 & \dots & p & i\\ 1 & \dots & p & n+1\end{pmatrix}}{A\begin{pmatrix}1 & \dots & p\\ 1 & \dots & p\end{pmatrix}} \qquad (i=1, 2, \dots, n;\ p=1, 2, \dots, r). \tag{21}$$

In particular, the consistency conditions (20) reduce to the well-known equations

$$A\begin{pmatrix}1 & \dots & r & r+j\\ 1 & \dots & r & n+1\end{pmatrix}=0 \qquad (j=1, 2, \dots, n-r). \tag{22}$$

If $n=r$, i.e. if the matrix $A=\|a_{ik}\|_1^n$ is non-singular, and

$$A\begin{pmatrix}1 & 2 & \dots & j\\ 1 & 2 & \dots & j\end{pmatrix}\neq 0 \qquad (j=1, 2, \dots, n),$$

then we can eliminate $x_1, x_2, \dots, x_{n-1}$ in succession by means of Gauss's algorithm and reduce the system of equations to the form (6).

§ 2. Mechanical Interpretation of Gauss's Algorithm

1. We consider an arbitrary elastic statical system S supported on edges (for example, a string, a rod, a multispan rod, a membrane, a lamina, or a discrete system) and choose n points (1), (2), . . . , (n) on it. We shall consider the displacements (sags) $y_1, y_2, \dots, y_n$ of the points (1), (2), . . . , (n) of S under the action of forces $F_1, F_2, \dots, F_n$ applied at these points.

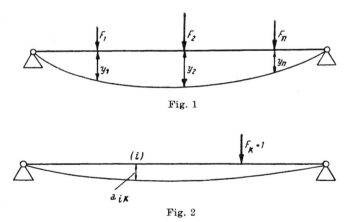

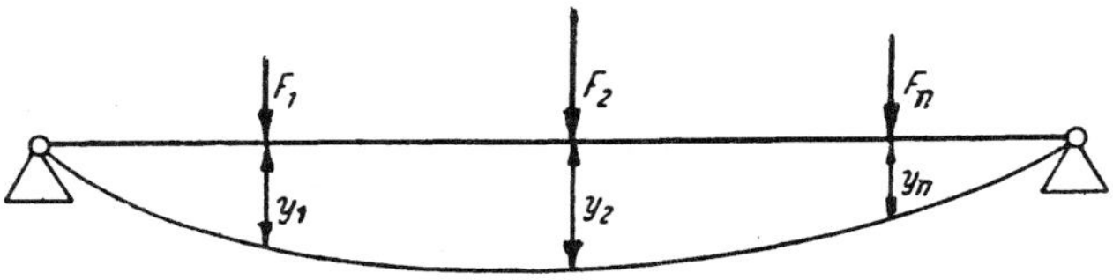

Fig. 1

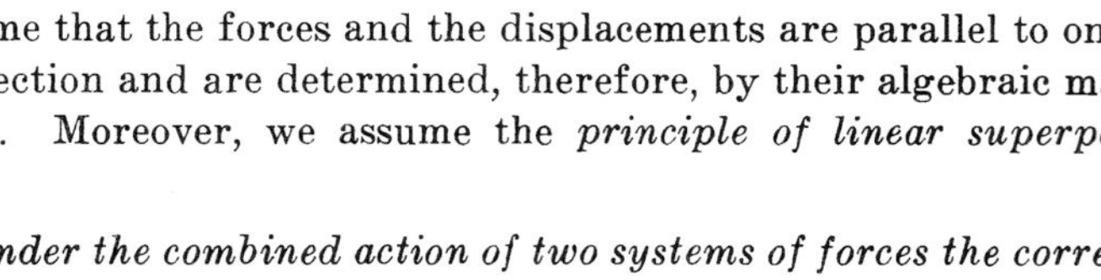

Fig. 2

We assume that the forces and the displacements are parallel to one and the same direction and are determined, therefore, by their algebraic magnitudes (Fig. 1). Moreover, we assume the *principle of linear superposition of forces*:

1. *Under the combined action of two systems of forces the corresponding displacements are added together.*

2. *When the magnitudes of all the forces are multiplied by one and the same real number, then all the displacements are multiplied by the same number.*

We denote by a_{ik} the coefficient of influence of the point (k) on the point (i), i.e., the displacement of (i) under the action of a unit force applied at (k) $(i, k = 1, 2, \ldots, n)$ (Fig. 2). Then under the combined action of the forces $F_1, F_2, \ldots, F_n$ the displacements $y_1, y_2, \ldots, y_n$ are determined by the formulas

$$\sum_{k=1}^{n} a_{ik} F_k = y_i \qquad (i = 1, 2, \ldots, n). \tag{23}$$

Comparing (23) with the original system (1), we can interpret the task of solving the system of equations (1) as follows:

The displacements $y_1, y_2, \ldots, y_n$ being given, we are required to find the corresponding forces $F_1, F_2, \ldots, F_n$.

We denote by S_p the statical system that is obtained from S by introducing p fixed hinged supports at the points (1), (2), ..., (p) $(p \leqq n)$. We denote the coefficients of influence for the remaining movable points $(p+1), \ldots, (n)$ of the system S_p by

$$a_{ik}^{(p)} \qquad (i, k = p+1, \ldots, n)$$

(see Fig. 3 for $p = 1$).

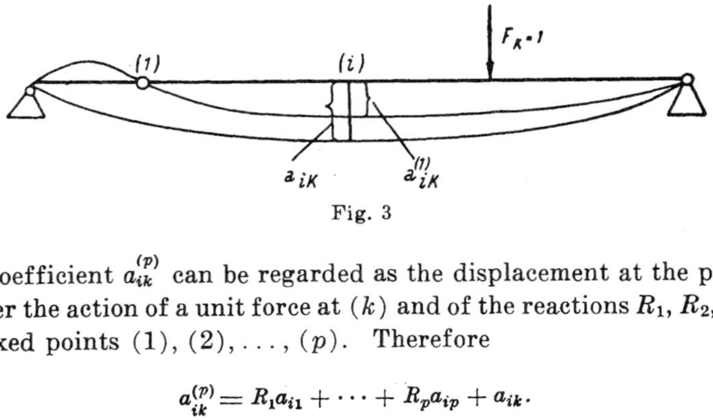

Fig. 3

The coefficient $a_{ik}^{(p)}$ can be regarded as the displacement at the point (i) of S under the action of a unit force at (k) and of the reactions $R_1, R_2, \ldots, R_p$ at the fixed points (1), (2), ..., (p). Therefore

$$a_{ik}^{(p)} = R_1 a_{i1} + \cdots + R_p a_{ip} + a_{ik}. \tag{24}$$

On the other hand, under the same forces the displacements of the system S at the points (1), (2), ..., (p) are zero:

$$\left.\begin{aligned} R_1 a_{11} + \cdots + R_p a_{1p} + a_{1k} &= 0 \\ \cdots\cdots\cdots\cdots\cdots\cdots & \\ R_1 a_{p1} + \cdots + R_p a_{pp} + a_{pk} &= 0. \end{aligned}\right\} \tag{25}$$

If

$$A\begin{pmatrix}1 & 2 & \dots & p\\ 1 & 2 & \dots & p\end{pmatrix} \neq 0,$$

then we can determine $R_1, R_2, \dots, R_p$ from (25) and substitute the expressions so obtained in (24). This elimination of $R_1, R_2, \dots, R_p$ can be carried out as follows. To the system of equations (25) we adjoin (24) written in the form

$$R_1 a_{i1} + \cdots + R_p a_{ip} + a_{ik} - a_{ik}^{(p)} = 0. \tag{24'}$$

Regarding (25) and (24′) as a system of $p+1$ homogeneous equations with non-zero solutions $R_1, R_2, \dots, R_p, R_{p+1} = 1$, we see that the determinant of the system must be zero:

$$\begin{vmatrix} a_{11} & \dots & a_{1p} & a_{1k} \\ \cdot & \cdot & \cdot & \cdot \\ a_{p1} & \dots & a_{pp} & a_{pk} \\ a_{i1} & \dots & a_{ip} & a_{ik} - a_{ik}^{(p)} \end{vmatrix} = 0.$$

Hence

$$a_{ik}^{(p)} = \frac{A\begin{pmatrix}1 & 2 \dots p & i\\ 1 & 2 \dots p & k\end{pmatrix}}{A\begin{pmatrix}1 & 2 \dots p\\ 1 & 2 \dots p\end{pmatrix}} \qquad (i, k = p+1, \dots, n). \tag{26}$$

These formulas express the coefficients of influence of the 'support' system S_p in terms of those of the original system S.

But formulas (26) coincide with formulas (13) of the preceding section. Therefore *for every p ($\leq n-1$) the coefficients $a_{ik}^{(p)}$ $(i, k = p+1, \dots, n)$ in the algorithm of Gauss are the coefficients of influence of the support system S_p.*

The truth of this fundamental proposition can also be ascertained by purely mechanical considerations without recourse to the algebraic derivation of formulas (13). For this purpose we consider, to begin with, the special case of a single support: $p = 1$ (Fig. 3). In this case, the coefficients of influence of the system S_1 are given by the formulas (we put $p = 1$ in (26)):

$$a_{ik}^{(1)} = \frac{A\begin{pmatrix}1 & i\\ 1 & k\end{pmatrix}}{A\begin{pmatrix}1\\ 1\end{pmatrix}} = a_{ik} - \frac{a_{i1}}{a_{11}} a_{1k} \qquad (i, k = 1, 2, \dots, n)$$

These formulas coincide with the formulas (3′).

Thus, if the coefficients a_{ik} $(i, k = 1, 2, \ldots, n)$ in the system of equations (1) are the coefficients of influence of the statical system S, then the coefficients $a_{ik}^{(1)}$ $(i, k = 2, \ldots, n)$ in Gauss's algorithm are the coefficients of influence of the system S_1. Applying the same reasoning to the system S_1 and introducing a second support at the point (2) in this system, we see that the coefficients $a_{ik}^{(2)}$ $(i, k = 3, \ldots, n)$ in the system of equations (4) are the coefficients of influence of the support system S_2 and, in general, for every p $(\leqq n-1)$ the coefficients $a_{ik}^{(p)}$ $(i, k = p+1, \ldots, n)$ in Gauss's algorithm are the coefficients of influence of the support system S_p.

From mechanical considerations it is clear that *the successive introduction of p supports is equivalent to the simultaneous introduction of these supports.*

Note. We wish to point out that in the mechanical interpretation of the elimination algorithm it was not necessary to assume that the points at which the displacements are investigated coincide with the points at which the forces $F_1, F_2, \ldots, F_n$ are applied. We can assume that $y_1, y_2, \ldots, y_n$ are the displacements of the points (1), (2), . . . , (n) and that the forces $F_1, F_2, \ldots, F_n$ are applied at the points (1′), (2′), . . . , (n'). Then a_{ik} is the coefficient of influence of the point (k') on the point (k). In that case we must consider instead of the support at the point (j) a generalized support at the points (j), (j') under which the displacement at the point (j) is maintained all the time equal to zero at the expense of a suitably chosen auxiliary force R_j at the point (j'). The conditions that allow us to introduce p generalized supports at the points (1), (1′) ; (2), (2′), . . . ; (p), (p'), i.e., that allow us to satisfy the conditions $y_1 = 0$, $y_2 = 0, \ldots, y_p = 0$ for arbitrary $F_{p+1}, \ldots, F_n$ at the expense of suitable $R_1 = F_1, \ldots, R_p = F_p$, can be expressed by the inequality

$$A\begin{pmatrix} 1 & 2 & \ldots & p \\ 1 & 2 & \ldots & p \end{pmatrix} \neq 0.$$

§ 3. Sylvester's Determinant Identity

1. In § 1, a comparison of the matrices A and G_p led to equations (10) and (11).

These equations enable us to give an easy proof of the important *determinant identity of Sylvester.* For from (10) and (11) we find:

$$|A| = A\begin{pmatrix} 1 & 2 & \ldots & n \\ 1 & 2 & \ldots & n \end{pmatrix} = A\begin{pmatrix} 1 & 2 & \ldots & p \\ 1 & 2 & \ldots & p \end{pmatrix} \begin{vmatrix} a_{p+1,p+1}^{(p)} & \ldots & a_{p+1,n}^{(p)} \\ \cdot & \cdot & \cdot \\ a_{n,p+1}^{(p)} & \ldots & a_{nn}^{(p)} \end{vmatrix}. \quad (27)$$

We introduce borderings of the minor $A\begin{pmatrix}1 & 2 & \dots & p\\ 1 & 2 & \dots & p\end{pmatrix}$ by the determinants

$$b_{ik} = A\begin{pmatrix}1 & 2 & \dots & p & i\\ 1 & 2 & \dots & p & k\end{pmatrix} \qquad (i, k = p+1, \dots, n).$$

The matrix formed from these determinants will be denoted by

$$B = \| b_{ik} \|_{p+1}^{n}.$$

Then by formulas (13)

$$\begin{vmatrix} a_{p+1,p+1}^{(p)} & \dots & a_{p+1,n}^{(p)} \\ \cdot & \cdot & \cdot \\ \cdot & \cdot & \cdot \\ a_{n,p+1}^{(p)} & \dots & a_{nn}^{(p)} \end{vmatrix} = \frac{\begin{vmatrix} b_{p+1,p+1} & \dots & b_{p+1,n} \\ \cdot & \cdot & \cdot \\ \cdot & \cdot & \cdot \\ b_{n,p+1} & \dots & b_{nn} \end{vmatrix}}{\left[A\begin{pmatrix}1 & 2 & \dots & p\\ 1 & 2 & \dots & p\end{pmatrix}\right]^{n-p}} = \frac{|B|}{\left[A\begin{pmatrix}1 & 2 & \dots & p\\ 1 & 2 & \dots & p\end{pmatrix}\right]^{n-p}}.$$

Therefore equation (27) can be rewritten as follows:

$$|B| = \left[A\begin{pmatrix}1 & 2 & \dots & p\\ 1 & 2 & \dots & p\end{pmatrix}\right]^{n-p-1} |A|. \tag{28}$$

This is Sylvester's determinant identity. It expresses the determinant $|B|$ formed from the bordered determinants in terms of the original determinant and the bordered minor.

We have established equation (28) for a matrix $A\begin{pmatrix}1 & 2 & \dots & p\\ 1 & 2 & \dots & p\end{pmatrix}$ whose elements satisfy the inequalities

$$A\begin{pmatrix}1 & 2 & \dots & j\\ 1 & 2 & \dots & j\end{pmatrix} \neq 0 \tag{29}$$

$$(j = 1, 2, \dots, p).$$

However, we can show by a 'continuity argument' that this restriction may be removed and that Sylvester's identity holds for an arbitrary matrix $A = \| a_{ik} \|_1^n$. For suppose that the inequalities (29) do not hold. We introduce the matrix

$$A_\varepsilon = A + \varepsilon E.$$

Obviously $\lim\limits_{\varepsilon \to 0} A_\varepsilon = A$. On the other hand, the minors

$$A_\varepsilon\begin{pmatrix}1 & 2 & \dots & j\\ 1 & 2 & \dots & j\end{pmatrix} = \varepsilon^j + \cdots$$

$$(j = 1, 2, \dots, p)$$

are p polynomials in ε that do not vanish identically. Therefore we can choose a sequence $\varepsilon_m \to 0$ such that

$$A_{\varepsilon_m}\begin{pmatrix}1 & 2 & \dots & j\\ 1 & 2 & \dots & j\end{pmatrix} \neq 0 \qquad (j=1, 2, \dots, p;\ m=1, 2, \dots).$$

We can write down the identity (28) for the matrices A_{ε_m}. Taking the limit $m \to \infty$ on both sides of this identity, we obtain Sylvester's identity for the limit matrix[3] $A = \lim\limits_{m\to\infty} A_{\varepsilon_m}$.

If we apply the identity (28) to the determinant

$$A\begin{pmatrix}1 & 2 & \dots & p & i_1 & i_2 & \dots & i_q\\ 1 & 2 & \dots & p & k_1 & k_2 & \dots & k_q\end{pmatrix} \qquad \left(p < \begin{matrix} i_1 < i_2 < \dots < i_q\\ k_1 < k_2 < \dots < k_q\end{matrix} \leqq n\right)$$

then we obtain a form of Sylvester's identity particularly convenient for applications

$$B\begin{pmatrix}i_1 & i_2 & \dots & i_q\\ k_1 & k_2 & \dots & k_q\end{pmatrix} = \left[A\begin{pmatrix}1 & 2 & \dots & p\\ 1 & 2 & \dots & p\end{pmatrix}\right]^{q-1} A\begin{pmatrix}1 & 2 & \dots & p & i_1 & i_2 & \dots & i_q\\ 1 & 2 & \dots & p & k_1 & k_2 & \dots & k_q\end{pmatrix}. \tag{30}$$

§ 4. The Decomposition of a Square Matrix into Triangular Factors

1. Let $A = \| a_{ik} \|_1^n$ be a given matrix of rank r. We introduce the following notation for the successive principal minors of the matrix

$$D_k = A\begin{pmatrix}1 & 2 & \dots & k\\ 1 & 2 & \dots & k\end{pmatrix} \qquad (k = 1, 2, \dots, n).$$

Let us assume that the conditions for the feasibility of Gauss's algorithm are satisfied:

$$D_k \neq 0 \qquad (k = 1, 2, \dots, r).$$

We denote by G the coefficient matrix of the system of equations (18) to which the system

$$\sum_{k=1}^{n} a_{ik} x_k = y_i \qquad (i = 1, 2, \dots, n)$$

[3] By the *limit* (for $p \to \infty$) *of a sequence of matrices* $X_p = \| x_{ik}^{(p)} \|_1^n$ we mean the matrix $X = \| x_{ik} \|_1^n$, where $x_{ik} = \lim\limits_{p\to\infty} x_{ik}^{(p)}$ $(i, k = 1, 2, \dots, n)$.

has been reduced by the elimination method of Gauss. The matrix G is of upper triangular form and the elements of its first r rows are determined by the formulas (13), while the elements of the last $n - r$ rows are all equal to zero:[4]

$$G = \left\| \begin{matrix} a_{11} & a_{12} \ldots a_{1r} & a_{1,r+1} \cdots a_{1n} \\ 0 & a_{22}^{(1)} \ldots a_{2r}^{(1)} & a_{2,r+1}^{(1)} \cdots a_{2n}^{(1)} \\ \cdot & \cdots\cdots\cdots & \cdots\cdots\cdots \\ 0 & 0 \quad \ldots a_{rr}^{(r-1)} & a_{r,r+1}^{(r-1)} \ldots a_{rn}^{(r-1)} \\ 0 & 0 \quad \ldots 0 & 0 \quad \ldots 0 \\ \cdot & \cdots\cdots\cdots & \cdots\cdots\cdots \\ 0 & 0 \quad \ldots 0 & 0 \quad \ldots 0 \end{matrix} \right\| .$$

The transition from A to G is effected by a certain number N of operations of the following type: to the i-th row of the matrix we add the j-th row ($j < i$), after a preliminary multiplication by some number α. Such an operation is equivalent to the multiplication on the left of the matrix to be transformed by the matrix

$$\begin{matrix} & & (j) & & (i) & & \\ \end{matrix}$$

$$\left\| \begin{matrix} 1 & \ldots & 0 & \ldots & 0 & \ldots & 0 \\ \cdot & \cdot & \cdot & & \cdot & & \cdot \\ \cdot & & 1 & & \cdot & & \cdot \\ \cdot & & \cdot & \cdot & \cdot & & \cdot \\ 0 & \ldots & \alpha & \ldots & 1 & \ldots & 0 \\ \cdot & & \cdot & & \cdot & \cdot & \cdot \\ 0 & \ldots & 0 & \ldots & 0 & \ldots & 1 \end{matrix} \right\| . \qquad (31)$$

In this matrix the main diagonal consists entirely of units, and all the remaining elements, except α, are zero.

Thus,

$$G = W_N \cdots W_2 W_1 A ,$$

where each matrix $W_1, W_2, \ldots, W_N$ is of the form (31) and is therefore a lower triangular matrix with diagonal elements equal to 1.

---

[4] See formulas (19). G coincides with the matrix G_p (p. 25) for $p = r$.

Let

$$W = W_N \cdots W_2 W_1 . \tag{32}$$

Then

$$G = WA . \tag{33}$$

We shall call W the *transforming* matrix for A in Gauss's elimination method. Both matrices G and W are uniquely determined by A. From (32) it follows that W is lower triangular with diagonal elements equal to 1.

Since W is non-singular, we obtain from (33):

$$A = W^{-1} G . \tag{33'}$$

We have thus represented A in the form of a product of a lower triangular matrix W^{-1} and an upper triangular matrix G. The problem of decomposing a matrix A into factors of this type is completely answered by the following theorem:

THEOREM 1: *Every matrix $A = \| a_{ik} \|_1^n$ of rank r in which the first r successive principal minors are different from zero*

$$D_k = A\begin{pmatrix} 1 & 2 & \dots & k \\ 1 & 2 & \dots & k \end{pmatrix} \neq 0 \quad \textit{for} \quad k = 1, 2 \dots, r \tag{34}$$

can be represented in the form of a product of a lower triangular matrix B and an upper triangular matrix C

$$A = BC = \begin{Vmatrix} b_{11} & 0 & \dots & 0 \\ b_{21} & b_{22} & \dots & 0 \\ \cdot & \cdot & \cdot & \cdot \\ b_{n1} & b_{n2} & \dots & b_{nn} \end{Vmatrix} \begin{Vmatrix} c_{11} & c_{12} & \dots & c_{1n} \\ 0 & c_{22} & \dots & c_{2n} \\ \cdot & \cdot & \cdot & \cdot \\ 0 & 0 & \dots & c_{nn} \end{Vmatrix} . \tag{35}$$

Here

$$b_{11} c_{11} = D_1, \quad b_{22} c_{22} = \frac{D_2}{D_1}, \quad \dots, \quad b_{rr} c_{rr} = \frac{D_r}{D_{r-1}} . \tag{36}$$

The values of the first r diagonal elements of B and C can be chosen arbitrarily subject to the conditions (36).

When the first r diagonal elements of B and C are given, then the elements of the first r rows of B and of the first r columns of C are uniquely determined, and are given by the following formulas:

$$b_{gk} = b_{kk} \frac{A\begin{pmatrix} 1 & 2 & \dots & k-1 & g \\ 1 & 2 & \dots & k-1 & k \end{pmatrix}}{A\begin{pmatrix} 1 & 2 & \dots & k \\ 1 & 2 & \dots & k \end{pmatrix}}, \quad c_{kg} = c_{kk} \frac{A\begin{pmatrix} 1 & 2 & \dots & k-1 & k \\ 1 & 2 & \dots & k-1 & g \end{pmatrix}}{A\begin{pmatrix} 1 & 2 & \dots & k \\ 1 & 2 & \dots & k \end{pmatrix}} \tag{37}$$

$$(g = k, k+1, \dots, n; \ k = 1, 2, \dots, r) .$$

If $r < n$ ($|A| = 0$), then all the elements in the last $n - r$ rows of C can be put equal to zero and all the elements of the last $n - r$ columns of B can be chosen arbitrarily; or, conversely, the last $n - r$ columns of B can be filled with zeros and the last $n - r$ rows of C can be chosen arbitrarily.

Proof. That a representation of a matrix satisfying conditions (34) can be given in the form of a product (35) has been proved above (see (33')).

Now let B and C be arbitrary lower and upper triangular matrices whose product is A. Making use of the formulas for the minors of the product of two matrices we find:

$$A\begin{pmatrix}1 & 2 & \dots & k-1 & g\\ 1 & 2 & \dots & k-1 & k\end{pmatrix} = \sum_{\alpha_1<\alpha_2<\dots<\alpha_k} B\begin{pmatrix}1 & 2 & \dots & k-1 & g\\ \alpha_1 & \alpha_2 & \dots & \alpha_{k-1} & \alpha_k\end{pmatrix} C\begin{pmatrix}\alpha_1 & \alpha_2 & \dots & \alpha_k\\ 1 & 2 & \dots & k\end{pmatrix} \tag{38}$$

$$(g = k, k+1, \dots, n;\ k = 1, 2, \dots, r).$$

Since C is an upper triangular matrix, the first k columns of C contain only one non-vanishing minor of order k, namely $C\begin{pmatrix}1 & 2 & \dots & k\\ 1 & 2 & \dots & k\end{pmatrix}$. Therefore, equation (38) can be written as follows:

$$A\begin{pmatrix}1 & 2 & \dots & k-1 & g\\ 1 & 2 & \dots & k-1 & k\end{pmatrix} = B\begin{pmatrix}1 & 2 & \dots & k-1 & g\\ 1 & 2 & \dots & k-1 & k\end{pmatrix} C\begin{pmatrix}1 & 2 & \dots & k\\ 1 & 2 & \dots & k\end{pmatrix}$$

$$= b_{11}b_{22}\cdots b_{k-1,k-1}b_{gk}c_{11}c_{22}\cdots c_{kk} \tag{39}$$

$$(g = k, k+1, \dots, n;\ k = 1, 2, \dots, r).$$

We put $g = k$ in this equation, obtaining

$$b_{11}b_{22}\cdots b_{kk}c_{11}c_{22}\cdots c_{kk} = D_k \qquad (k = 1, 2, \dots, r), \tag{40}$$

and relations (36) follow.

Without violating equation (35) we may multiply the matrix B in that equation on the right by an arbitrary non-singular diagonal matrix $M = \|\mu_i\delta_{ik}\|_1^n$, while multiplying C at the same time on the left by $M^{-1} = \|\mu_i^{-1}\delta_{ik}\|_1^n$. But this is equivalent to multiplying the columns of B by $\mu_1, \mu_2, \dots, \mu_n$, respectively, and the rows of C by $\mu_1^{-1}, \mu_2^{-1}, \dots, \mu_n^{-1}$. We may therefore give arbitrary values to the diagonal elements $b_{11}, b_{22}, \dots, b_{rr}$ and $c_{11}, c_{22}, \dots, c_{rr}$, provided they satisfy (36).

Further, from (39) and (40) we find:

$$b_{gk} = b_{kk}\frac{A\begin{pmatrix}1 & 2 & \dots & k-1 & g\\ 1 & 2 & \dots & k-1 & k\end{pmatrix}}{A\begin{pmatrix}1 & 2 & \dots & k\\ 1 & 2 & \dots & k\end{pmatrix}} \qquad (g = k, k+1, \dots, n;\ k = 1, 2, \dots, r),$$

i.e., the first formulas in (37). The second formulas in (37), for the elements of C, are established similarly.

We observe that in the multiplication of B and C the elements b_{kg} of the last $n-r$ columns of B and the elements c_{gk} of the last $n-r$ rows of C are multiplied only among each other. We have seen that all the elements of the last $n-r$ rows of C may be chosen to be zero.[5] But as a consequence, the elements of the last $n-r$ columns of B may be chosen arbitrarily. Clearly the product of B and C does not change if we choose the last $n-r$ columns of B to be zeros and choose the elements of the last $n-r$ rows of C arbitrarily.

This completes the proof of the theorem.

From this theorem there follow a number of interesting corollaries.

COROLLARY 1: *The elements of the first r columns of B and the first r rows of C are connected with the elements of A by the recurrence relations*

$$\left.\begin{aligned} b_{ik} &= \frac{a_{ik} - \sum_{j=1}^{k-1} b_{ij}c_{jk}}{c_{kk}} \qquad (i \geqq k;\ i = 1, 2, \ldots, n;\ k = 1, 2, \ldots, r), \\ c_{ik} &= \frac{a_{ik} - \sum_{j=1}^{i-1} b_{ij}c_{jk}}{b_{ii}} \qquad (i \leqq k;\ i = 1, 2, \ldots, r;\ k = 1, 2, \ldots, n). \end{aligned}\right\} \tag{41}$$

The relations (41) follow immediately from the matrix equation (35); they can be used to advantage in the actual computation of the elements of B and C.

COROLLARY 2: *If $A = \| a_{ik} \|_1^n$ is a non-singular matrix ($r = n$) satisfying* (34), *then the matrices B and C in the representation* (35) *are uniquely determined as soon as the diagonal elements of these matrices are chosen in accordance with* (36).

COROLLARY 3: *If $S = \| s_{ik} \|_1^n$ is a symmetric matrix of rank r and*

$$D_k = S\begin{pmatrix} 1 & 2 & \ldots & k \\ 1 & 2 & \ldots & k \end{pmatrix} \neq 0 \qquad (k = 1, 2, \ldots, r),$$

then

$$S = B\,B^{\mathsf{T}},$$

where $B = \| b_{ik} \|_1^n$ is a lower triangular matrix in which

[5] This follows from the representation (33′). Here, as we have shown already, arbitrary values may be given to the diagonal elements $b_{11}, \ldots, b_{rr}, c_{11}, \ldots, c_{rr}$ provided (36) is satisfied by the introduction of suitable factors $\mu_1, \mu_2, \ldots, \mu_r$.

$$b_{gk}=\begin{cases}\dfrac{1}{\sqrt{D_kD_{k-1}}}\,S\begin{pmatrix}1 & 2 & \dots & k-1 & g\\ 1 & 2 & \dots & k-1 & k\end{pmatrix} & (g=k,k+1,\dots,n;\ k=1,2,\dots,r),\\ 0 & (g=k,k+1,\dots,n;\ k=r+1,\dots,n).\end{cases} \tag{42}$$

2. In the representation (35) let the elements of the last $n-r$ columns of C be zero. Then we may set

$$B=F\begin{Vmatrix} b_{11} & & & & 0\\ & \ddots & & & \\ & & b_{rr} & & \\ & & & 0 & \\ & & & & \ddots \\ 0 & & & & 0\end{Vmatrix},\qquad C=\begin{Vmatrix} c_{11} & & & & 0\\ & \ddots & & & \\ & & c_{rr} & & \\ & & & 0 & \\ & & & & \ddots \\ 0 & & & & 0\end{Vmatrix}L, \tag{43}$$

where F and L are lower and upper triangular matrices respectively; the first r diagonal elements of F and L are 1 and the elements of the last $n-r$ columns of F and the last $n-r$ rows of L can be chosen completely arbitrarily. Substituting (43) for B and C in (35) and using (36), we obtain the following theorem:

THEOREM 2: *Every matrix $A=\|a_{ik}\|_1^n$ of rank r in which*

$$D_k=A\begin{pmatrix}1 & 2 & \dots & k\\ 1 & 2 & \dots & k\end{pmatrix}\neq 0 \qquad \textit{for } k=1,2,\dots,r$$

can be represented in the form of a product of a lower triangular matrix F, a diagonal matrix D, and an upper triangular matrix L:

$$A=FDL=\begin{Vmatrix} 1 & 0 & \dots & 0\\ f_{21} & 1 & \dots & 0\\ \cdot & \cdot & \cdot & \cdot\\ f_{n1} & f_{n2} & \dots & 1\end{Vmatrix}\begin{Vmatrix} D_1 & & & & & \\ & \dfrac{D_2}{D_1} & & & & \\ & & \ddots & & & \\ & & & \dfrac{D_r}{D_{r-1}} & & \\ & & & & 0 & \\ & & & & & \ddots \\ & & & & & & 0\end{Vmatrix}\begin{Vmatrix} 1 & l_{12} & \dots & l_{1n}\\ 0 & 1 & \dots & l_{2n}\\ \cdot & \cdot & \cdot & \cdot\\ 0 & 0 & \dots & 1\end{Vmatrix}, \tag{44}$$

where

$$f_{gk}=\frac{A\begin{pmatrix}1 & 2 & \dots & k-1 & g\\ 1 & 2 & \dots & k-1 & k\end{pmatrix}}{A\begin{pmatrix}1 & 2 & \dots & k\\ 1 & 2 & \dots & k\end{pmatrix}},\qquad l_{kg}=\frac{A\begin{pmatrix}1 & 2 & \dots & k-1 & k\\ 1 & 2 & \dots & k-1 & g\end{pmatrix}}{A\begin{pmatrix}1 & 2 & \dots & k\\ 1 & 2 & \dots & k\end{pmatrix}}, \tag{45}$$

$$(g=k+1,\dots,n;\ k=1,2,\dots,r),$$

and f_{gk} and l_{kg} are arbitrary for $g=k+1,\dots,n$; $k=r+1,\dots,n$.

3. The elimination method of Gauss, when applied to a matrix $A = \| a_{ik} \|_1^n$ of rank r for which $D_k \neq 0$ $(k = 1, 2, \dots, r)$, yields two matrices: a lower triangular matrix W with diagonal elements 1 and an upper triangular matrix G in which the first r diagonal elements are $D_1, \frac{D_2}{D_1}, \dots, \frac{D_r}{D_{r-1}}$ and the last $n - r$ rows consist entirely of zeros. G is the *Gaussian form* of the matrix A; W is the transforming matrix.

For actual computation of the elements of W we recommend the following device.

We obtain the matrix W when we apply to the unit matrix E all the transformations (given by $W_1, \dots, W_N$) that we have performed on A in the algorithm of Gauss (in this case we shall have instead of the product WA, equal to G, the product WE, equal to W). Let us, therefore, write the unit matrix E on the right of A:

$$\left\| \begin{matrix} a_{11} & \dots & a_{1n} & 1 & \dots & 0 \\ \cdot & \cdot & \cdot & \cdot & \cdot & \cdot \\ \cdot & \cdot & \cdot & \cdot & \cdot & \cdot \\ a_{n1} & \dots & a_{nn} & 0 & \dots & 1 \end{matrix} \right\|. \tag{46}$$

By applying all the transformations of the algorithm of Gauss to this rectangular matrix we obtain a rectangular matrix consisting of the two square matrices G and W:

$$(G, W).$$

Thus, *the application of Gauss's algorithm to the matrix* (46) *gives the matrices G and W simultaneously.*

If A is non-singular, so that $|A| \neq 0$, then $|G| \neq 0$ as well. In this case, (33) implies that $A^{-1} = G^{-1}W$. Since G and W are determined by means of the algorithm of Gauss, the task of finding the inverse matrix A^{-1} reduces to determining G^{-1} and multiplying G^{-1} by W.

Although there is no difficulty in finding the inverse matrix G^{-1} once the matrix G has been determined, because G is triangular, the operations involved can nevertheless be avoided. For this purpose we introduce, together with the matrices G and W, similar matrices G_1 and W_1 for the transposed matrix A^{T}. Then $A^{\mathsf{T}} = W_1^{-1}G_1$, i.e.,

$$A = G_1^{\mathsf{T}} W_1^{\mathsf{T}-1}. \tag{47}$$

Let us compare (33′) with (44):

$$A = W^{-1}G, \; A = FDL.$$

These equations may be regarded as two distinct decompositions of the form (35); here we take the product DL as the second factor C. Since the first r diagonal elements of the first factors are the same (they are equal to 1), their first r columns coincide. But then, since the last $n-r$ columns of F may be chosen arbitrarily, we chose them such that

$$F = W^{-1}. \tag{48}$$

On the other hand, a comparison of (47) with (44),

$$A = G_1^{\mathsf{T}} W_1^{\mathsf{T}-1}, \quad A = FDL,$$

shows that we may also select the arbitrary elements of L in such a way that

$$L = W_1^{\mathsf{T}-1}. \tag{49}$$

Replacing F and L in (44) by their expressions (48) and (49), we obtain

$$A = W^{-1} D W_1^{\mathsf{T}-1}. \tag{50}$$

Comparing this equation with (33′) and (47) we find:

$$G = D W_1^{\mathsf{T}-1}, \quad G_1^{\mathsf{T}} = W^{-1} D. \tag{51}$$

We now introduce the diagonal matrix

$$\hat{D} = \left\{ \frac{1}{D_1}, \frac{D_1}{D_2}, \ldots, \frac{D_{r-1}}{D_r}, 0, \ldots, 0 \right\}. \tag{52}$$

Since

$$D = D\hat{D}D,$$

it follows from (50) and (51) that

$$A = G_1^{\mathsf{T}} \hat{D} G. \tag{53}$$

Formula (53) shows that the decomposition of A into triangular factors can be obtained by applying the algorithm of Gauss to the matrices A and A^{T}.

Now let A be non-singular ($r = n$). Then $|D| \neq 0$, $\hat{D} = D^{-1}$. Therefore it follows from (50) that

$$A^{-1} = W_1^{\mathsf{T}} \hat{D} W. \tag{54}$$

This formula yields an effective computation of the inverse matrix A^{-1} by the application of Gauss's algorithm to the rectangular matrices

$$(A, E) \qquad (A^{\mathsf{T}}, E).$$

If, in particular, we take as our A a symmetrical matrix S, then G_1 coincides with G and W_1 with W, and therefore formulas (53) and (54) assume the form

$$S = G^{\mathsf{T}} \widehat{D} G, \tag{55}$$

$$S^{-1} = W^{\mathsf{T}} \widehat{D} W. \tag{56}$$

§ 5. The Partition of a Matrix into Blocks. The Technique of Operating with Partitioned Matrices. The Generalized Algorithm of Gauss

It often becomes necessary to use matrices that are partitioned into rectangular parts—'cells' or 'blocks.' In the present section we deal with such partitioned matrices.

1. Let a rectangular matrix

$$A = \| a_{ik} \| \qquad (i = 1, 2, \ldots, m; \; k = 1, 2, \ldots, n) \tag{57}$$

be given.

By means of horizontal and vertical lines we dissect A into rectangular blocks:

$$A = \begin{pmatrix} \overbrace{A_{11}}^{n_1} & \overbrace{A_{12}}^{n_2} & \ldots & \overbrace{A_{1t}}^{n_t} \\ A_{21} & A_{22} & \ldots & A_{2t} \\ \ldots & \ldots & \ldots & \ldots \\ A_{s1} & A_{s2} & \ldots & A_{st} \end{pmatrix} \begin{matrix} \} m_1 \\ \} m_2 \\ \vdots \\ \} m_s \end{matrix} . \tag{58}$$

We shall say of matrix (58) that it is *partitioned* into *st blocks,* or *cells* $A_{\alpha\beta}$ of dimensions $m_\alpha \times n_\beta$ $(\alpha = 1, 2, \ldots, s; \beta = 1, 2, \ldots, t)$, or that it is represented in the form of a *partitioned,* or blocked, matrix. Instead of (58) we shall simply write

$$A = (A_{\alpha\beta}) \qquad (\alpha = 1, 2, \ldots, s; \; \beta = 1, 2, \ldots, t). \tag{59}$$

In the case $s = t$ we shall use the following notation:

$$A = (A_{\alpha\beta})_1^s . \tag{60}$$

Operations on partitioned matrices are performed according to the same formal rules as in the case in which we have numerical elements instead of blocks. For example, let A and B be two rectangular matrices of equal dimensions partitioned into blocks in exactly the same way:

$$A = (A_{\alpha\beta}), \quad B = (B_{\alpha\beta}) \qquad (\alpha = 1, 2, \ldots, s;\ \beta = 1, 2, \ldots, t). \tag{61}$$

It is easy to verify that

$$A + B = (A_{\alpha\beta} + B_{\alpha\beta}) \qquad (\alpha = 1, 2, \ldots, s;\ \beta = 1, 2, \ldots, t). \tag{62}$$

We have to consider multiplication of partitioned matrices in more detail. We know (see Chapter I, p. 6) that for the multiplication of two rectangular matrices A and B the length of the rows of the first factor A must be the same as the height of the columns of the second factor B. For 'block' multiplication of these matrices we require, in addition, that the partitioning into blocks be such that the horizontal dimensions in the first factor are the same as the corresponding vertical dimensions in the second:

$$A = \begin{pmatrix} \overbrace{A_{11}}^{n_1} & \overbrace{A_{12}}^{n_2} & \ldots & \overbrace{A_{1t}}^{n_t} \\ A_{21} & A_{22} & \ldots & A_{2t} \\ \cdot & \cdot & \cdot & \cdot \\ A_{s1} & A_{s2} & \ldots & A_{st} \end{pmatrix} \begin{matrix} \}\, m_1 \\ \}\, m_2 \\ \vdots \\ \}\, m_s \end{matrix}, \qquad B = \begin{pmatrix} \overbrace{B_{11}}^{p_1} & \overbrace{B_{12}}^{p_2} & \ldots & \overbrace{B_{1u}}^{p_u} \\ B_{21} & B_{22} & \ldots & B_{2u} \\ \cdot & \cdot & \cdot & \cdot \\ B_{t1} & B_{t2} & \ldots & B_{tu} \end{pmatrix} \begin{matrix} \}\, n_1 \\ \}\, n_2 \\ \vdots \\ \}\, n_t \end{matrix}. \tag{63}$$

Then it is easy to verify that

$$AB = C = (C_{\alpha\beta}), \quad \text{where} \quad C_{\alpha\beta} = \sum_{\delta=1}^{t} A_{\alpha\delta} B_{\delta\beta} \qquad \begin{pmatrix} \alpha = 1, 2, \ldots, s \\ \beta = 1, 2, \ldots, u \end{pmatrix}. \tag{64}$$

We mention separately the special case in which one of the factors is a *quasi-diagonal* matrix. Let A be quasi-diagonal, i.e., let $s = t$ and $A_{\alpha\beta} = O$ for $\alpha \neq \beta$. In this case formula (64) gives

$$C_{\alpha\beta} = A_{\alpha\alpha} B_{\alpha\beta} \qquad (\alpha = 1, 2, \ldots, s;\ \beta = 1, 2, \ldots, u). \tag{65}$$

When a partitioned matrix is multiplied on the left by a quasi-diagonal matrix, then the rows of the matrix are multiplied on the left by the corresponding diagonal blocks of the quasi-diagonal matrix.

Now let B be a quasi-diagonal matrix, i.e., let $t = u$ and $B_{\alpha\beta} = O$ for $\alpha \neq \beta$. Then we obtain from (64):

$$C_{\alpha\beta} = A_{\alpha\beta} B_{\beta\beta} \qquad (\alpha = 1, 2, \ldots, s;\ \beta = 1, 2, \ldots, u). \tag{66}$$

When a partitioned matrix is multiplied on the right by a quasi-diagonal matrix, then all the columns of the partitioned matrix are multiplied on the right by the corresponding diagonal cells of the quasi-diagonal matrix.

Note that the multiplication of *square* partitioned matrices of one and the same order is always feasible if the factors are split into equal quadratic schemes of blocks and there are square matrices on the diagonal places in each factor.

The partitioned matrix (58) is called *upper* (*lower*) *quasi-triangular* if $s = t$ and all $A_{\alpha\beta} = O$ for $\alpha > \beta$ ($\alpha < \beta$). A quasi-diagonal matrix is a special case of a quasi-triangular matrix.

From the formulas (64) it is easy to see that:

The product of two upper (*lower*) *quasi-triangular matrices is itself an upper* (*lower*) *quasi-triangular matrix;*[6] *the diagonal cells of the product are obtained by multiplying the corresponding diagonal cells of the factors.*

For when we set $s = t$ in (64) and

$$A_{\alpha\beta} = O, \quad B_{\alpha\beta} = O \quad \text{for} \quad \alpha < \beta,$$

we find

and

$$\left.\begin{array}{ll} C_{\alpha\beta} = O & \text{for} \quad \alpha < \beta \\ C_{\alpha\alpha} = A_{\alpha\alpha} B_{\alpha\alpha} & \end{array}\right\} \quad (\alpha, \beta = 1, 2, \ldots, s).$$

The case of lower quasi-triangular matrices is treated similarly.

We mention a rule for the calculation of the determinant of a quasi-triangular matrix. This rule can be obtained from the Laplace expansion.

If A is a quasi-triangular matrix (*in particular, a quasi-diagonal matrix*), *then the determinant of the matrix is equal to the product of the determinant of the diagonal cells:*

$$|A| = |A_{11}|\,|A_{22}| \cdots |A_{ss}|. \tag{67}$$

2. Let a partitioned matrix

$$A = \begin{pmatrix} \overbrace{A_{11}}^{n_1} & \overbrace{A_{12}}^{n_2} & \ldots & \overbrace{A_{1t}}^{n_t} \\ A_{21} & A_{22} & \ldots & A_{2t} \\ \cdot & \cdot & \cdot & \cdot \\ A_{s1} & A_{s2} & \ldots & A_{st} \end{pmatrix} \begin{matrix} \}m_1 \\ \}m_2 \\ \vdots \\ \}m_s \end{matrix}. \tag{68}$$

[6] It is assumed here that the block multiplication is feasible.

be given. To the α-th row of submatrices we add the β-th row, multiplied on the left by a rectangular matrix X of dimension $m_\alpha \times m_\beta$. We obtain a partitioned matrix

$$B = \begin{pmatrix} A_{11} & \ldots & A_{1t} \\ \cdot\cdot\cdot\cdot\cdot & \cdot\cdot\cdot\cdot\cdot & \cdot\cdot\cdot\cdot\cdot \\ A_{\alpha 1} + XA_{\beta 1} & \ldots & A_{\alpha t} + XA_{\beta t} \\ \cdot\cdot\cdot\cdot\cdot & \cdot\cdot\cdot\cdot\cdot & \cdot\cdot\cdot\cdot\cdot \\ A_{\beta 1} & \ldots & A_{\beta t} \\ \cdot\cdot\cdot\cdot\cdot & \cdot\cdot\cdot\cdot\cdot & \cdot\cdot\cdot\cdot\cdot \\ A_{s1} & \ldots & A_{st} \end{pmatrix}. \tag{69}$$

We introduce an auxiliary square matrix V, which we give in the form of a square scheme of blocks:

$$V = \begin{array}{c} \begin{matrix} \overbrace{\quad}^{m_1} & \ldots & \overbrace{\quad}^{m_\alpha} & \ldots & \overbrace{\quad}^{m_\beta} & \ldots & \overbrace{\quad}^{m_s} \end{matrix} \\ \begin{pmatrix} E & \ldots & O & \ldots & O & \ldots & O \\ \cdot & \cdot & \cdot & \cdot & \cdot & \cdot & \cdot \\ O & \ldots & E & \ldots & X & \ldots & O \\ \cdot & \cdot & \cdot & \cdot & \cdot & \cdot & \cdot \\ O & \ldots & O & \ldots & E & \ldots & O \\ \cdot & \cdot & \cdot & \cdot & \cdot & \cdot & \cdot \\ O & \ldots & O & \ldots & O & \ldots & E \end{pmatrix} \end{array} \begin{matrix} \} m_1 \\ \vdots \\ \} m_\alpha \\ \vdots \\ \} m_\beta \\ \vdots \\ \} m_s \end{matrix} . \tag{70}$$

In the diagonal blocks of V there are unit matrices of order $m_1, m_2, \ldots, m_s$, respectively; all the non-diagonal blocks of V are equal to zero except the block X that lies at the intersection of the α-th row and β-th column.

It is easy to see that

$$VA = B. \tag{71}$$

As V is non-singular, we have[7] for the ranks of A and B:

$$r_A = r_B. \tag{72}$$

In the special case where A is a square matrix, we have from (71):

$$|V||A| = |B|. \tag{73}$$

But the determinant of the quasi-triangular matrix V is 1:

$$|V| = 1. \tag{74}$$

Hence

$$|A| = |B|. \tag{75}$$

---

[7] See p. 12.

The same conclusion holds when we add to an arbitrary column of (68) another column multiplied on the right by a rectangular matrix X of suitable dimensions.

The results obtained can be formulated as the following theorem.

THEOREM 3: *If to the α-th row (column) of the blocks of the partitioned matrix A we add the β-th row (column) multiplied on the left (right) by a rectangular matrix X of the corresponding dimensions, then the rank of A remains unchanged under this transformation and, if A is a square matrix, the determinant of A is also unchanged.*

3. We now consider the special case in which the diagonal block A_{11} in A is square and non-singular ($|A_{11}| \neq 0$).

To the α-th row of A we add the first row multiplied on the left by $-A_{\alpha 1}A_{11}^{-1}$ ($\alpha = 2, \ldots, s$). We thus obtain the matrix

$$B_1 = \begin{pmatrix} A_{11} & A_{12} & \ldots & A_{1t} \\ O & A_{22}^{(1)} & \ldots & A_{2t}^{(1)} \\ \cdot & \cdot & \cdot & \cdot \\ O & A_{s2}^{(1)} & \ldots & A_{st}^{(1)} \end{pmatrix}, \tag{76}$$

where

$$A_{\alpha\beta}^{(1)} = -A_{\alpha 1}A_{11}^{-1}A_{1\beta} + A_{\alpha\beta} \qquad (\alpha = 2, \ldots, s;\ \beta = 2, \ldots, t). \tag{77}$$

If the matrix $A_{22}^{(1)}$ is square and non-singular, then the process can be continued. In this way we arrive at the *generalized algorithm of Gauss*.

Let A be a square matrix. Then

$$|A| = |B_1| = |A_{11}| \begin{vmatrix} A_{22}^{(1)} & \ldots & A_{2t}^{(1)} \\ \cdot & \cdot & \cdot \\ A_{s2}^{(1)} & \ldots & A_{st}^{(1)} \end{vmatrix}. \tag{78}$$

Formula (78) reduces the computation of the determinant $|A|$, consisting of st blocks to the computation of a determinant of lower order consisting of $(s-1)\cdot(t-1)$ blocks.[8]

Let us consider a determinant Δ partitioned into four blocks:

$$\Delta = \begin{vmatrix} A & B \\ C & D \end{vmatrix}, \tag{79}$$

where A and D are square matrices.

Suppose $|A| \neq 0$. Then from the second row we subtract the first multiplied on the left by CA^{-1}. We obtain

[8] If $A_{22}^{(1)}$ is a square matrix and $\left|A_{22}^{(1)}\right| \neq 0$, then this determinant of $(s-1)\ (t-1)$ blocks can again be subjected to such a transformation, etc.

$$\Delta = \begin{vmatrix} A & B \\ O & D - CA^{-1}B \end{vmatrix} = |A|\ |D - CA^{-1}B|. \tag{I}$$

Similarly, if $|D| \neq 0$, we subtract from the first row in Δ the second multiplied on the left by BD^{-1}, obtaining

$$\Delta = \begin{vmatrix} A - BD^{-1}C & O \\ C & D \end{vmatrix} = |A - BD^{-1}C|\ |D|. \tag{II}$$

In the special case in which all four matrices A, B, C, D are square (of one and the same order n), we deduce from (I) and (II) the *formulas of Schur,* which reduce the computation of a determinant of order $2n$ to the computation of a determinant of order n:

$$\Delta = |AD - ACA^{-1}B| \qquad (|A| \neq 0), \tag{Ia}$$

$$\Delta = |AD - BD^{-1}CD| \qquad (|D| \neq 0). \tag{IIa}$$

If the matrices A and C are permutable, then it follows from (Ia) that

$$\Delta = |AD - CB| \qquad (\text{provided } AC = CA). \tag{Ib}$$

Similarly, if C and D are permutable, then

$$\Delta = |AD - BC| \qquad (\text{provided } CD = DC). \tag{IIb}$$

Formula (Ib) was obtained under the assumption $|A| \neq 0$, and (IIb) under the assumption $|D| \neq 0$. However, these restrictions can be removed by continuity arguments.

From formulas (I)-(IIb) we can obtain another six formulas by replacing A and D on the right-hand sides simultaneously by B and C.

Example.

$$\Delta = \begin{vmatrix} 1 & 0 & b_1 & b_2 \\ 0 & 1 & b_3 & b_4 \\ c_1 & c_2 & d_1 & d_2 \\ c_3 & c_4 & d_3 & d_4 \end{vmatrix}.$$

By formula (Ib),

$$\Delta = \begin{vmatrix} d_1 - c_1b_1 - c_2b_3 & d_2 - c_1b_2 - c_2b_4 \\ d_3 - c_3b_1 - c_4b_3 & d_4 - c_3b_2 - c_4b_4 \end{vmatrix}.$$

4. From Theorem 3 there follows also

THEOREM 4: *If a rectangular matrix R is represented in partitioned form*

$$R=\begin{pmatrix} A & B \\ C & D \end{pmatrix}, \tag{80}$$

where A is a square non-singular matrix of order n ($|A| \neq 0$), *then the rank of R is equal to n if and only if*

$$D=CA^{-1}B. \tag{81}$$

Proof. We subtract from the second row of blocks of R the first, multiplied on the left by CA^{-1}. Then we obtain the matrix

$$T=\begin{pmatrix} A & B \\ O & D-CA^{-1}B \end{pmatrix}. \tag{82}$$

By Theorem 3, the matrices R and T have the same rank. But the rank of T coincides with the rank of A (namely, n) if and only if $D-CA^{-1}B=O$, i.e., when (81) holds. This proves the theorem.

From Theorem 4 there follows an algorithm[9] for the construction of the inverse matrix A^{-1} and, more generally, the product $CA^{-1}B$, where B and C are rectangular matrices of dimensions $n \times p$ and $q \times n$.

By means of Gauss's algorithm,[10] we reduce the matrix

$$\begin{pmatrix} A & B \\ -C & O \end{pmatrix} \quad (|A| \neq 0) \tag{83}$$

to the form

$$\begin{pmatrix} G & B_1 \\ O & X \end{pmatrix}. \tag{84}$$

We will show that

$$X=CA^{-1}B. \tag{85}$$

For, the same transformation that was applied to the matrix (83) reduces the matrix

---

[9] See [181].

[10] We do not apply here the entire algorithm of Gauss to the matrix (83) but only the first n steps of the algorithm, where n is the order of the matrix. This can be done if the conditions (15) hold for $p=n$. But if these conditions do not hold, then, since $|A| \neq 0$, we may renumber the first n rows (or the first n columns) of the matrix (83) so that the n steps of Gauss's algorithm turn out to be feasible. Such a modified Gaussian algorithm is sometimes applied even when the conditions (15), with $p=n$, are satisfied.

$$\begin{pmatrix} A & B \\ -C & -CA^{-1}B \end{pmatrix} \tag{86}$$

to the form

$$\begin{pmatrix} G & B_1 \\ O & X-CA^{-1}B \end{pmatrix}. \tag{87}$$

By Theorem 4, the matrix (86) is of rank n (n is the order of A). But then (87) must also be of rank n. Hence $X - CA^{-1}B = O$, i.e., (85) holds.

In particular, if $B = y$, where y is a column matrix, and $C = E$, then

$$X = A^{-1}y.$$

Therefore, when we apply Gauss's algorithm to the matrix

$$\begin{pmatrix} A & y \\ -E & O \end{pmatrix},$$

we obtain the solution of the system of equations

$$Ax = y.$$

Further, if in (83) we set $B = C = E$, then by applying the algorithm of Gauss to the matrix

$$\begin{pmatrix} A & E \\ -E & O \end{pmatrix},$$

we obtain

$$\begin{pmatrix} G & W \\ O & X \end{pmatrix},$$

where

$$X = A^{-1}.$$

Let us illustrate this method by finding A^{-1} in the following example.

Example. Let

$$A = \begin{Vmatrix} 2 & 1 & 1 \\ 1 & 0 & 2 \\ 3 & 1 & 2 \end{Vmatrix}.$$

It is required to compute A^{-1}.

We apply a somewhat modified elimination method[11] to the matrix

[11] See the preceding footnote.

$$\begin{Vmatrix} 2 & 1 & 1 & 1 & 0 & 0 \\ 1 & 0 & 2 & 0 & 1 & 0 \\ 3 & 1 & 2 & 0 & 0 & 1 \\ -1 & 0 & 0 & 0 & 0 & 0 \\ 0 & -1 & 0 & 0 & 0 & 0 \\ 0 & 0 & -1 & 0 & 0 & 0 \end{Vmatrix}.$$

To all the rows we add certain multiples of the second row and we arrange that all the elements of the first column, except the second, become zero. Then we add to all the rows, except the second, the third row multiplied by certain factors and see to it that in the second column all the elements, except the second and third, become zero. Then we add to the last three rows the first row with suitable factors and obtain a matrix of the form

$$\begin{Vmatrix} * & * & * & * & * & * \\ * & * & * & * & * & * \\ * & * & * & * & * & * \\ 0 & 0 & 0 & -2 & -1 & 2 \\ 0 & 0 & 0 & 4 & 1 & -3 \\ 0 & 0 & 0 & 1 & 1 & -1 \end{Vmatrix}.$$

Therefore

$$A^{-1} = \begin{Vmatrix} -2 & -1 & 2 \\ 4 & 1 & -3 \\ 1 & 1 & -1 \end{Vmatrix}.$$

CHAPTER III

LINEAR OPERATORS IN AN n-DIMENSIONAL VECTOR SPACE

Matrices constitute the fundamental analytic apparatus for the study of linear operators in an n-dimensional space. The study of these operators, in turn, enables us to divide all matrices into classes and to exhibit the significant properties that all matrices of one and the same class have in common.

In the present chapter we shall expound the simpler properties of linear operators in an n-dimensional space. The investigation will be continued in Chapters VII and IX.

§ 1. Vector Spaces

1. Let $\boldsymbol{R}$ be a set of arbitrary elements $\boldsymbol{x}, \boldsymbol{y}, \boldsymbol{z}, \ldots$ in which two operations are defined:[1] the operation of 'addition' and the operation of 'multiplication by a number of the field F.' We postulate that these operations can always be performed uniquely in $\boldsymbol{R}$ and that the following rules hold for arbitrary elements $\boldsymbol{x}, \boldsymbol{y}, \boldsymbol{z}$ of $\boldsymbol{R}$ and numbers α, β of F:

1. $\boldsymbol{x} + \boldsymbol{y} = \boldsymbol{y} + \boldsymbol{x}$.
2. $(\boldsymbol{x} + \boldsymbol{y}) + \boldsymbol{z} = \boldsymbol{x} + (\boldsymbol{y} + \boldsymbol{z})$.
3. There exists an element $\boldsymbol{o}$ in $\boldsymbol{R}$ such that the product of the number 0 with any element $\boldsymbol{x}$ of $\boldsymbol{R}$ is equal to $\boldsymbol{o}$:

   $0 \cdot \boldsymbol{x} = \boldsymbol{o}$.
4. $1 \cdot \boldsymbol{x} = \boldsymbol{x}$.
5. $\alpha(\beta \boldsymbol{x}) = (\alpha\beta)\boldsymbol{x}$.
6. $(\alpha + \beta)\boldsymbol{x} = \alpha\boldsymbol{x} + \beta\boldsymbol{x}$.
7. $\alpha(\boldsymbol{x} + \boldsymbol{y}) = \alpha\boldsymbol{x} + \alpha\boldsymbol{y}$.

---

[1] These operations will be denoted by the usual signs '+' and '.'; the latter sign will sometimes be omitted.

DEFINITION 1: *A set* $\boldsymbol{R}$ *of elements in which two operations—'addition' of elements and 'multiplication of elements of* $\boldsymbol{R}$ *by a number of* F'*—can always be performed uniquely and for which postulates* 1.-7. *hold is called a vector space (over the field* F) *and the elements are called vectors.*[2]

DEFINITION 2. *The vectors* $\boldsymbol{x}, \boldsymbol{y}, \ldots, \boldsymbol{u}$ *of* $\boldsymbol{R}$, *are called linearly dependent if there exist numbers* $\alpha, \beta, \ldots, \delta$ *in* F, *not all zero, such that*

$$\alpha \boldsymbol{x} + \beta \boldsymbol{y} + \cdots + \delta \boldsymbol{u} = \boldsymbol{o}. \tag{1}$$

If such a linear dependence does not hold, then the vectors $\boldsymbol{x}, \boldsymbol{y}, \ldots, \boldsymbol{u}$ *are called linearly independent.*

If the vectors $\boldsymbol{x}, \boldsymbol{y}, \ldots, \boldsymbol{u}$ are linearly dependent, then one of the vectors can be repesented as a linear combination, with coefficients in R, of the remaining ones. For example, if $\alpha \neq 0$ in (1), then

$$\boldsymbol{x} = -\frac{\beta}{\alpha}\boldsymbol{y} - \cdots - \frac{\delta}{\alpha}\boldsymbol{u}.$$

DEFINITION 3. *The space* $\boldsymbol{R}$ *is called finite-dimensional and the number* n *is called the dimension of the space if there exist* n *linearly independent vectors in* $\boldsymbol{R}$, *while any* $n+1$ *vectors in* $\boldsymbol{R}$ *are linearly dependent. If the space contains linearly independent systems of an arbitrary number of vectors, then it is called infinite-dimensional.*

In this book we shall study mainly finite-dimensional spaces.

DEFINITION 4. *A system of* n *linearly independent vectors* $\boldsymbol{e}_1, \boldsymbol{e}_2, \ldots, \boldsymbol{e}_n$ *of an* n*-dimensional space, given in a definite order, is called a basis of the space.*

2. *Example* 1. The set of all ordinary vectors (directed geometrical segments) is a three-dimensional vector space. The part of this space that consists of the vectors parallel to some plane is a two-dimensional space, and all the vectors parallel to a given line form a one-dimensional vector space.

Example 2. Let us call a column $x = (x_1, x_2, \ldots, x_n)$ of n numbers of F a vector (where n is a fixed number). We define the basic operations as operations on column matrices:

---

[2] It is easy to see that all the usual properties of the operations of addition and of multiplication by a number follow from properties 1.-7. For example, for arbitrary $\boldsymbol{x}$ of $\boldsymbol{R}$ we have:

$$\boldsymbol{x} + \boldsymbol{o} = \boldsymbol{x}\ [\boldsymbol{x} + \boldsymbol{o} = 1 \cdot \boldsymbol{x} + 0 \cdot \boldsymbol{x} = (1+0) \cdot \boldsymbol{x} = 1 \cdot \boldsymbol{x} = \boldsymbol{x}];$$
$$\boldsymbol{x} + (-\boldsymbol{x}) = \boldsymbol{o}, \text{ where } -\boldsymbol{x} = (-1) \cdot \boldsymbol{x};$$

etc.

$$(x_1, x_2, \ldots, x_n) + (y_1, y_2, \ldots, y_n) = (x_1 + y_1, x_2 + y_2, \ldots, x_n + y_n),$$
$$\alpha (x_1, x_2, \ldots, x_n) = (\alpha x_1, \alpha x_2, \ldots, \alpha x_n).$$

The null vector is the column $(0, 0, \ldots, 0)$. It is easy to verify that all the postulates 1.-7. are satisfied. The vectors form an n-dimensional space. As a basis of the space we can take, for example, the column of unit matrices of order n:

$$(1, 0, \ldots, 0),\ (0, 1, \ldots, 0), \ldots, (0, 0, \ldots, 1).$$

The space thus defined is often called the *n-dimensional number space*.

Example 3. The set of all infinite sequences $(x_1, x_2, \ldots, x_n, \ldots)$ in which the operations are defined in a natural way, i.e.,

$$(x_1, x_2, \ldots, x_n, \ldots) + (y_1, y_2, \ldots, y_n, \ldots) = (x_1 + y_1, x_2 + y_2, \ldots, x_n + y_n, \ldots),$$
$$\alpha (x_1, x_2, \ldots, x_n, \ldots) = (\alpha x_1, \alpha x_2, \ldots, \alpha x_n, \ldots),$$

is an infinite-dimensional space.

Example 4. The set of polynomials $a_0 + a_1 t + \ldots + a_{n-1} t^{n-1}$ of degree $< n$ with coefficients in F is an n-dimensional vector space.[3] As a basis of this space we can take, say, the system of powers $t^0, t^1, \ldots, t^{n-1}$.

The set of all such polynomials (without a bound on the degree) form an infinite-dimensional space.

Example 5. The set of all functions defined on a closed interval $[a, b]$ form an infinite-dimensional space.

3. Let the vectors $\boldsymbol{e}_1, \boldsymbol{e}_2, \ldots, \boldsymbol{e}_n$ forms a basis of an n-dimensional vector space $\boldsymbol{R}$ and let $\boldsymbol{x}$ be an arbitrary vector of the space. Then the vectors $\boldsymbol{x}, \boldsymbol{e}_1, \boldsymbol{e}_2, \ldots, \boldsymbol{e}_n$ are linearly dependent (because there are $n + 1$ of them):

$$\alpha_0 \boldsymbol{x} + \alpha_1 \boldsymbol{e}_1 + \alpha_2 \boldsymbol{e}_2 + \cdots + \alpha_n \boldsymbol{e}_n = \boldsymbol{o},$$

where at least one of the numbers $\alpha_0, \alpha_1, \ldots, \alpha_n$ is different from zero. But in this case we must have $a_0 \neq 0$, since the vectors $\boldsymbol{e}_1, \boldsymbol{e}_2, \ldots, \boldsymbol{e}_n$ cannot be linearly dependent. Therefore

$$\boldsymbol{x} = x_1 \boldsymbol{e}_1 + x_2 \boldsymbol{e}_2 + \cdots + x_n \boldsymbol{e}_n \tag{2}$$

where $x_i = -a_i/a_0$ $(i = 1, 2, \ldots, n)$.

Note that the numbers $x_1, x_2, \ldots, x_n$ are uniquely determined when the vector $\boldsymbol{x}$ and the basis $\boldsymbol{e}_1, \boldsymbol{e}_2, \ldots, \boldsymbol{e}_n$ are given. For if there is another decomposition of $\boldsymbol{x}$ besides (2),

$$\boldsymbol{x} = x'_1 \boldsymbol{e}_1 + x'_2 \boldsymbol{e}_2 + \cdots + x'_n \boldsymbol{e}_n, \tag{3}$$

[3] The basic operations are taken to be ordinary addition of polynomials and multiplication of a polynomial by a number.

then, by subtracting (2) from (3), we obtain

$$(x_1' - x_1)\,\boldsymbol{e}_1 + (x_2' - x_2)\,\boldsymbol{e}_2 + \cdots + (x_n' - x_n)\,\boldsymbol{e}_n = \boldsymbol{o},$$

and since the vectors of a basis are linearly dependent, it follows that

$$x_1' - x_1 = x_2' - x_2 = \cdots = x_n' - x_n = 0,$$

i.e.,

$$x_1' = x_1,\; x_2' = x_2,\; \ldots,\; x_n' = x_n. \tag{4}$$

The numbers $x_1, x_2, \ldots, x_n$ are called the *coordinates* of $\boldsymbol{x}$ in the basis $\boldsymbol{e}_1, \boldsymbol{e}_2, \ldots, \boldsymbol{e}_n$.

If

$$\boldsymbol{x} = \sum_{i=1}^{n} x_i \boldsymbol{e}_i \quad \text{and} \quad \boldsymbol{y} = \sum_{i=1}^{n} y_i \boldsymbol{e}_i,$$

then

$$\boldsymbol{x} + \boldsymbol{y} = \sum_{i=1}^{n} (x_i + y_i)\,\boldsymbol{e}_i \quad \text{and} \quad \alpha\boldsymbol{x} = \sum_{i=1}^{n} \alpha x_i \boldsymbol{e}_i.$$

i.e., *the coordinates of a sum of vectors are obtained by addition of the corresponding coordinates of the summands and the product of a vector by a number α is obtained by multiplying all the coordinates of the vector by α.*

4. Let the vectors

$$\boldsymbol{x}_k = \sum_{i=1}^{n} x_{ik} \boldsymbol{e}_i$$

be linearly dependent, i.e.,

$$\sum_{k=1}^{m} c_k \boldsymbol{x}_k = \boldsymbol{o}, \tag{5}$$

where at least one of the numbers $c_1, c_2, \ldots, c_m$ is not equal to zero.

If a vector is the null vector, then all its components are zero. Hence the vector equation (5) is equivalent to the following system of scalar equations:

$$\left.\begin{array}{l} c_1 x_{11} + c_2 x_{12} + \cdots + c_m x_{1m} = 0 \\ c_1 x_{21} + c_2 x_{22} + \cdots + c_m x_{2m} = 0 \\ \cdots\cdots\cdots\cdots\cdots\cdots \\ c_1 x_{n1} + c_2 x_{n2} + \cdots + c_m x_{nm} = 0. \end{array}\right\} \tag{6}$$

As is well known, this system of homogeneous linear equations for $c_1, c_2, \ldots, c_m$ has a non-zero solution if and only if the rank of the coefficient matrix is less than the number of unknowns, i.e., less than m. A necessary and sufficient condition for the independence of the vectors $\boldsymbol{x}_1, \boldsymbol{x}_2, \ldots, \boldsymbol{x}_m$ is, therefore, that this rank should be m.

Thus, the following theorem holds:

THEOREM 1: *In order that the vectors* $\boldsymbol{x}_1, \boldsymbol{x}_2, \ldots, \boldsymbol{x}_m$ *be linearly independent it is necessary and sufficient that the rank* r *of the matrix formed from the coordinates of these vectors in an arbitrary basis*

$$\left\| \begin{matrix} x_{11} & x_{12} & \cdots & x_{1m} \\ x_{21} & x_{22} & \cdots & x_{2m} \\ \cdot & \cdot & \cdot & \cdot \\ \cdot & \cdot & \cdot & \cdot \\ x_{n1} & x_{n2} & \cdots & x_{nm} \end{matrix} \right\| \tag{7}$$

be equal to m, *i.e., to the number of vectors.*

Note. The linear independence of the vectors $\boldsymbol{x}_1, \boldsymbol{x}_2, \ldots, \boldsymbol{x}_m$ means that the columns of the matrix (7) are linearly independent, since the k-th column consists of the coordinates of x_k $(k = 1, 2, \ldots, m)$. By the theorem, therefore, if the columns of a matrix are linearly independent, then the rank of the matrix is equal to the number of columns. Hence it follows that in an arbitrary rectangular matrix the maximal number of linearly independent columns is equal to the rank of the matrix. Moreover, if we transpose the matrix, i.e., change the rows into columns and the columns into rows, then the rank obviously remains unchanged. Hence *in a rectangular matrix the number of linearly independent columns is always equal to the number of linearly independent rows and equal to the rank of the matrix.*[4]

5. If in an n-dimensional space a basis $\boldsymbol{e}_1, \boldsymbol{e}_2, \ldots, \boldsymbol{e}_n$ has been chosen, then to every vector $\boldsymbol{x}$ there corresponds uniquely the column $x = (x_1, x_2, \ldots, x_n)$, where $x_1, x_2, \ldots, x_n$ are the coordinates of $\boldsymbol{x}$ in the given basis. Thus, the choosing of a basis establishes a one-to-one correspondence between the vectors of an arbitrary n-dimensional vector space $\boldsymbol{R}$ and the vectors of the n-dimensional number space $\boldsymbol{R}'$ considered in Example 2. Here the sum of vectors in $\boldsymbol{R}$ corresponds to the sum of the corresponding vectors of $\boldsymbol{R}'$. The analogous correspondence holds for the product of a vector by a number a of F. In other words, an arbitrary n-dimensional vector space is *isomorphic* to the n-dimensional number space, and therefore *all vector spaces of the same number* n *of dimensions over the same number field* F *are isomorphic.* This means that to within isomorphism there exists only one n-dimensional vector space for a given number field.

[4] This proposition follows from Theorem 1, in the proof of which we have started from the well-known property of a system of linear homogeneous equations: a non-zero solution exists only when the rank of the coefficient matrix is less than the number of unknowns. For a proof of Theorem 1 independent of this property, see § 5.

The reader may ask why we have introduced an 'abstract' n-dimensional space if it coincides to within isomorphism with the n-dimensional number space. Indeed, we could have defined a vector as a system of n numbers given in a definite order and could have introduced the operations on these vectors in the very way it was done in Example 2. But we would then have mixed up properties of vectors that do not depend on the choice of a basis with properties of a particular basis. For example, the fact that all the coordinates of a vector are zero is a property of the vector itself; it does not depend on the choice of basis. But the equality of all its coordinates is not a property of the vector itself, because it disappears under a change of basis. *The axiomatic definition of a vector space immediately singles out the properties of vectors that do not depend on the choice of a basis.*

§ 2. A Linear Operator Mapping an n-Dimensional Space into an m-Dimensional Space

1. We consider a linear transformation

$$\left.\begin{aligned} y_1 &= a_{11}x_1 + a_{12}x_2 + \cdots + a_{1n}x_n \\ y_2 &= a_{21}x_1 + a_{22}x_2 + \cdots + a_{2n}x_n \\ &\cdots\cdots\cdots\cdots\cdots\cdots \\ y_m &= a_{m1}x_1 + a_{m2}x_2 + \cdots + a_{mn}x_n , \end{aligned}\right\} \qquad (8)$$

whose coefficients belong to the number field F as well as two vector spaces over F: an n-dimensional space $\boldsymbol{R}$ and an m-dimensional space $\boldsymbol{S}$. We choose a basis $\boldsymbol{e}_1, \boldsymbol{e}_2, \ldots, \boldsymbol{e}_n$ in $\boldsymbol{R}$ and a basis $\boldsymbol{g}_1, \boldsymbol{g}_2, \ldots, \boldsymbol{g}_m$ in $\boldsymbol{S}$. Then the transformation (8) associates with every vector $\boldsymbol{x} = \sum_{i=1}^{n} x_i\boldsymbol{e}_i$ of $\boldsymbol{R}$ a certain vector $\boldsymbol{y} = \sum_{k=1}^{m} y_k\boldsymbol{g}_k$ of $\boldsymbol{S}$, i.e., the transformation (8) determines a certain operator $\boldsymbol{A}$ that sets up a correspondence between the vector $\boldsymbol{x}$ and the vector $\boldsymbol{y} : \boldsymbol{y} = \boldsymbol{A}\boldsymbol{x}$. It is easy to see that this operator $\boldsymbol{A}$ has the property of linearity, which we formulate as follows:

DEFINITION 5: *An operator $\boldsymbol{A}$ mapping $\boldsymbol{R}$ into $\boldsymbol{S}$, i.e., associating with every vector $\boldsymbol{x}$ of $\boldsymbol{R}$ a certain vector $\boldsymbol{y} = \boldsymbol{A}\boldsymbol{x}$ of $\boldsymbol{S}$ is called linear if for arbitrary $\boldsymbol{x}_1, \boldsymbol{x}_2$ of $\boldsymbol{R}$ and α of* F

$$\boldsymbol{A}(\boldsymbol{x}_1 + \boldsymbol{x}_2) = \boldsymbol{A}\boldsymbol{x}_1 + \boldsymbol{A}\boldsymbol{x}_2, \quad \boldsymbol{A}(\alpha\boldsymbol{x}_1) = \alpha\boldsymbol{A}\boldsymbol{x}_1. \qquad (9)$$

Thus, the transformation (8), for a given basis in $\boldsymbol{R}$ and a given basis in $\boldsymbol{S}$, determines a linear operator mapping $\boldsymbol{R}$ into $\boldsymbol{S}$.

We shall now show the converse, i.e., that for an arbitrary linear operator $\boldsymbol{A}$ mapping $\boldsymbol{R}$ into $\boldsymbol{S}$ and arbitrary bases $\boldsymbol{e}_1, \boldsymbol{e}_2, \ldots, \boldsymbol{e}_n$ in $\boldsymbol{R}$ and $\boldsymbol{g}_1, \boldsymbol{g}_2, \ldots, \boldsymbol{g}_m$ in $\boldsymbol{S}$, there exists a rectangular matrix with elements in F

$$\left\| \begin{matrix} a_{11} & a_{12} & \ldots & a_{1n} \\ a_{21} & a_{22} & \ldots & a_{2n} \\ \cdot & \cdot & \cdot & \cdot \\ a_{m1} & a_{m2} & \ldots & a_{mn} \end{matrix} \right\| \tag{10}$$

such that the linear transformation (8) formed by means of this matrix expresses the coordinates of the transformed vector $\boldsymbol{y} = \boldsymbol{A}\boldsymbol{x}$ in terms of the coordinates of the original vector $\boldsymbol{x}$.

Let us, in fact, apply the operator $\boldsymbol{A}$ to the basis vector $\boldsymbol{e}_k$ and let the coordinates in the basis $\boldsymbol{g}_1, \boldsymbol{g}_2, \ldots, \boldsymbol{g}_m$ of the vector $\boldsymbol{A}\boldsymbol{e}_k$ thus obtained be denoted by $a_{1k}, a_{2k}, \ldots, a_{mk}$ $(k = 1, 2, \ldots, n)$:

$$\boldsymbol{A}\boldsymbol{e}_k = \sum_{i=1}^{m} a_{ik}\boldsymbol{g}_i \qquad (k = 1, 2, \ldots, n). \tag{11}$$

Multiplying both sides of (11) by x_k and summing from 1 to n, we obtain

$$\sum_{k=1}^{n} x_k \boldsymbol{A}\boldsymbol{e}_k = \sum_{i=1}^{m} \Big(\sum_{k=1}^{n} a_{ik}x_k\Big)\boldsymbol{g}_i;$$

hence

$$\boldsymbol{y} = \boldsymbol{A}\boldsymbol{x} = \boldsymbol{A}\Big(\sum_{k=1}^{n} x_k\boldsymbol{e}_k\Big) = \sum_{k=1}^{n} x_k\boldsymbol{A}\boldsymbol{e}_k = \sum_{i=1}^{m} y_i\boldsymbol{g}_i,$$

where

$$y_i = \sum_{k=1}^{n} a_{ik}x_k \qquad (i = 1, 2, \ldots, m),$$

and this is what we had to show.

Thus, for given bases of $\boldsymbol{R}$ and $\boldsymbol{S}$: *to every linear operator* $\boldsymbol{A}$ *mapping* $\boldsymbol{R}$ *into* $\boldsymbol{S}$ *there corresponds a rectangular matrix of dimension* $m \times n$ *and, conversely, to every such matrix there corresponds a linear operator mapping* $\boldsymbol{R}$ *into* $\boldsymbol{S}$.

Here, in the matrix A corresponding to the operator $\boldsymbol{A}$, *the k-th column consists of the coordinates of the vector* $\boldsymbol{A}\boldsymbol{e}_k$ $(k = 1, 2, \ldots, n)$.

We denote by $x = (x_1, x_2, \ldots, x_n)$ and $y = (y_1, y_2, \ldots, y_m)$ the coordinate columns of the vectors $\boldsymbol{x}$ and $\boldsymbol{y}$. Then the vector equation

$$\boldsymbol{y} = \boldsymbol{A}\boldsymbol{x}$$

corresponds to the matrix equation

$$y = Ax,$$

which is the matrix form of the transformation (8).

Example. We consider the set of all polynomials in t of degree $\leqq n-1$ with coefficients in F. This set forms an n-dimensional vector space $\boldsymbol{R}_n$ (see Example 4., p. 52). Similarly, the polynomials in t of degree $\leqq n-2$ with coefficients in F form a space $\boldsymbol{R}_{n-1}$. The differentiation operator $\frac{d}{dt}$ associates with every polynomial of $\boldsymbol{R}_n$ a certain polynomial in $\boldsymbol{R}_{n-1}$. Thus, this operator maps $\boldsymbol{R}_n$ into $\boldsymbol{R}_{n-1}$. The differentiation operator is linear, since

$$\frac{d}{dt}[\varphi(t)+\psi(t)]=\frac{d\varphi(t)}{dt}+\frac{d\psi(t)}{dt},\quad \frac{d}{dt}[\alpha\varphi(t)]=\alpha\,\frac{d\varphi(t)}{dt}.$$

In $\boldsymbol{R}_n$ and $\boldsymbol{R}_{n-1}$ we choose bases consisting of powers of t:

$$t^0=1, t, \dots, t^{n-1} \qquad \text{and} \qquad t^0=1, t, \dots, t^{n-2}.$$

Using formulas (11), we construct the rectangular matrix of dimension $(n-1\times n)$ corresponding to the differentiation operator $\frac{d}{dt}$ in these bases:

$$\left\|\begin{matrix} 0 & 1 & 0 \dots & 0 \\ 0 & 0 & 2 \dots & 0 \\ \dots & \dots & \dots & \dots \\ \dots & \dots & \dots & \dots \\ 0 & 0 & 0 \dots & n-1 \end{matrix}\right\|.$$

§ 3. Addition and Multiplication of Linear Operators

1. Let $\boldsymbol{A}$ and $\boldsymbol{B}$ be two linear operators mapping $\boldsymbol{R}$ into $\boldsymbol{S}$ and let the corresponding matrices be

$$A=\|a_{ik}\|,\quad B=\|b_{ik}\| \qquad (i=1,2,\dots,m;\ k=1,2,\dots,n).$$

DEFINITION 6: *The sum of the operators* $\boldsymbol{A}$ *and* $\boldsymbol{B}$ *is the operator* $\boldsymbol{C}$ *defined by the equation*[5]

$$\boldsymbol{C}\boldsymbol{x}=\boldsymbol{A}\boldsymbol{x}+\boldsymbol{B}\boldsymbol{x} \quad (\boldsymbol{x}\in\boldsymbol{R}). \tag{12}$$

On the basis of this definition it is easy to verify that the sum $\boldsymbol{C}=\boldsymbol{A}+\boldsymbol{B}$ of the linear operators $\boldsymbol{A}$ and $\boldsymbol{B}$ is itself a linear operator. Furthermore,

$$\boldsymbol{C}\boldsymbol{e}_k=\boldsymbol{A}\boldsymbol{e}_k+\boldsymbol{B}\boldsymbol{e}_k=\sum_{k=1}^{n}(a_{ik}+b_{ik})\,\boldsymbol{e}_k.$$

[5] $\boldsymbol{x}\in\boldsymbol{R}$ means that the element $\boldsymbol{x}$ belongs to the set $\boldsymbol{R}$. It is assumed that (12) holds for arbitrary $\boldsymbol{x}$ in $\boldsymbol{R}$.

Hence it follows that the operator $\boldsymbol{C}$ corresponds to the matrix $C = \| c_{ik} \|$, where $c_{ik} = a_{ik} + b_{ik}$ $(i = 1, 2, \ldots, m;\ k = 1, 2, \ldots, n)$, i.e., the operator $\boldsymbol{C}$ corresponds to the matrix

$$C = A + B. \tag{13}$$

We would come to the same conclusion starting from the matrix equation

$$Cx = Ax + Bx \tag{14}$$

(x is the coordinate column of the vector $\boldsymbol{x}$) corresponding to the vector equation (12). Since x is an arbitrary column, (13) follows from (14).

2. Let $\boldsymbol{R}$, $\boldsymbol{S}$, and $\boldsymbol{T}$ be three vector spaces of dimension q, n, and m, and let $\boldsymbol{A}$ and $\boldsymbol{B}$ be two linear operators, of which $\boldsymbol{B}$ maps $\boldsymbol{R}$ into $\boldsymbol{S}$ and $\boldsymbol{A}$ maps $\boldsymbol{S}$ into $\boldsymbol{T}$; in symbols:

$$\boldsymbol{R} \overset{\boldsymbol{B}}{\rightarrow} \boldsymbol{S} \overset{\boldsymbol{A}}{\rightarrow} \boldsymbol{T}.$$

DEFINITION 7. *The product of the operators* $\boldsymbol{A}$ *and* $\boldsymbol{B}$ *is the operator* $\boldsymbol{C}$ *for which*

$$\boldsymbol{Cx} = \boldsymbol{A}(\boldsymbol{Bx}) \quad (\boldsymbol{x} \in \boldsymbol{R}). \tag{15}$$

holds for every $\boldsymbol{x}$ *of* $\boldsymbol{R}$.

The operator $\boldsymbol{C}$ maps $\boldsymbol{R}$ into $\boldsymbol{T}$:

$$\boldsymbol{R} \xrightarrow{\boldsymbol{C} = \boldsymbol{AB}} \boldsymbol{T}.$$

From the linearity of the operators $\boldsymbol{A}$ and $\boldsymbol{B}$ follows the linearity of $\boldsymbol{C}$. We choose arbitrary bases in $\boldsymbol{R}$, $\boldsymbol{S}$, and $\boldsymbol{T}$ and denote by A, B, and C the matrices corresponding, in this choice of basis, to the operators $\boldsymbol{A}$, $\boldsymbol{B}$, and $\boldsymbol{C}$. Then the vector equations

$$\boldsymbol{z} = \boldsymbol{Ay},\ \boldsymbol{y} = \boldsymbol{Bx},\ \boldsymbol{z} = \boldsymbol{Cx} \tag{16}$$

correspond to the matrix equations:

$$z = Ay,\ y = Bx,\ z = Cx,$$

where x, y, z are the coordinate columns of the vectors $\boldsymbol{x}$, $\boldsymbol{y}$, $\boldsymbol{z}$. Hence

$$Cx = A(Bx) = (AB)x$$

and as the column x is arbitrary

$$C = AB. \tag{17}$$

Thus, the product $\boldsymbol{C} = \boldsymbol{AB}$ of the operators $\boldsymbol{A}$ and $\boldsymbol{B}$ corresponds to the matrix $C = \| c_{ij} \|$ $(i = 1, 2, \ldots, m;\ j = 1, 2, \ldots, q)$, which is the product of the matrices A and B.

We leave it to the reader to show that the operator[6]

$$\boldsymbol{C} = a\boldsymbol{A} \qquad (a \in F)$$

corresponds to the matrix

$$C = aA.$$

Thus we see that in Chapter I the operations on matrices were so defined that the sum $\boldsymbol{A} + \boldsymbol{B}$, the product $\boldsymbol{AB}$, and the product $a\boldsymbol{A}$ correspond to the matrices $A + B$, AB, and aA, respectively, where A and B are the matrices corresponding to the operators $\boldsymbol{A}$ and $\boldsymbol{B}$, and a is a number of F.

§ 4. Transformation of Coordinates

1. In an n-dimensional vector space we consider two bases: $\boldsymbol{e}_1, \boldsymbol{e}_2, \ldots, \boldsymbol{e}_n$ (the 'old' basis) and $\boldsymbol{e}_1^*, \boldsymbol{e}_2^*, \ldots, \boldsymbol{e}_n^*$ the 'new' basis).

The mutual disposition of the basis vectors is determined if the coordinates of the vectors of the basis are given relative to the other basis.

We set

$$\left.\begin{array}{l} \boldsymbol{e}_1^* = t_{11}\boldsymbol{e}_1 + t_{21}\boldsymbol{e}_2 + \cdots + t_{n1}\boldsymbol{e}_n \\ \boldsymbol{e}_2^* = t_{12}\boldsymbol{e}_1 + t_{22}\boldsymbol{e}_2 + \cdots + t_{n2}\boldsymbol{e}_n \\ \cdots\cdots\cdots\cdots\cdots\cdots\cdots\cdots \\ \boldsymbol{e}_n^* = t_{1n}\boldsymbol{e}_1 + t_{2n}\boldsymbol{e}_2 + \cdots + t_{nn}\boldsymbol{e}_n \end{array}\right\} \tag{18}$$

or in abbreviated form,

$$\boldsymbol{e}_k^* = \sum_{i=1}^{n} t_{ik}\boldsymbol{e}_i \quad (k = 1, 2, \ldots, n). \tag{18'}$$

We shall now establish the connection between the coordinates of one and the same vector in the two different bases.

Let $x_1, x_2, \ldots, x_n$ and $x_1^*, x_2^*, \ldots, x_n^*$ be the coordinates of the vector $\boldsymbol{x}$ relative to the 'old' and the 'new' bases, respectively:

$$\boldsymbol{x} = \sum_{i=1}^{n} x_i\boldsymbol{e}_i = \sum_{k=1}^{n} x_k^*\boldsymbol{e}_k^*. \tag{19}$$

In (19) we substitute for the vectors $\boldsymbol{e}_k^*$ the expressions given for them in (18). We obtain:

[6] I.e., the operator for which $\boldsymbol{Cx} = a\boldsymbol{Ax}$ $(\boldsymbol{x} \in \boldsymbol{R})$.

$$\boldsymbol{x} = \sum_{k=1}^{n} x_k^* \sum_{i=1}^{n} t_{ik} \boldsymbol{e}_i = \sum_{i=1}^{n} \left(\sum_{k=1}^{n} t_{ik} x_k^*\right) \boldsymbol{e}_i .$$

Comparing this with (19) and bearing in mind that the coordinates of a vector are uniquely determined when the vector and the basis are given, we find:

$$x_i = \sum_{k=1}^{n} t_{ik} x_k^* \quad (i = 1, 2, \ldots, n), \tag{20}$$

or in explicit form:

$$\left.\begin{aligned} x_1 &= t_{11} x_1^* + t_{12} x_2^* + \cdots + t_{1n} x_n^* \\ x_2 &= t_{21} x_1^* + t_{22} x_2^* + \cdots + t_{2n} x_n^* \\ &\cdots\cdots\cdots\cdots\cdots\cdots \\ x_n &= t_{n1} x_1^* + t_{n2} x_2^* + \cdots + t_{nn} x_n^* . \end{aligned}\right\} \tag{21}$$

Formulas (21) determine the transformation of the coordinates of a vector on transition from one basis to another. They express the 'old' coordinates in terms of the 'new' ones. The matrix

$$T = \| t_{ik} \|_1^n \tag{22}$$

is called the *matrix of the coordinate transformation* or the *transforming matrix*. Its k-th column consists of the 'old' coordinates of the k-th 'new' basis vector. This follows from formulas (18) or immediately from (21) if we set in the latter $x_k^* = 1$, $x_i^* = 0$ for $i \neq k$.

Note that the matrix T is non-singular, i.e.,

$$| T | \neq 0. \tag{23}$$

For when we set in (21) $x_1 = x_2 = \ldots = x_n = 0$, we obtain a system of n linear homogeneous equations in the n unknowns $x_1^*, x_2^*, \ldots, x_n^*$ with determinant $| T |$. This system can only have the zero solution $x_1^* = 0$, $x_2^* = 0$, $\ldots$, $x_n^* = 0$, since otherwise (19) would imply a linear dependence among the vectors $\boldsymbol{e}_1^*, \boldsymbol{e}_2^*, \ldots, \boldsymbol{e}_n^*$. Therefore $| T | \neq 0$.[7]

We now introduce the column matrices $x = (x_1, x_2, \ldots, x_n)$ and $x^* = (x_1^*, x_2^*, \ldots, x_n^*)$. Then the formulas (21) for the coordinate transformation can be written in the form of the following matrix equation:

$$x = T x^* . \tag{24}$$

Multiplying both sides of this equation by T^{-1}, we obtain the expression for the inverse transformation

$$x^* = T^{-1} x . \tag{25}$$

---

[7] The inequality (23) also follows from Theorem 1 (p. 54), because the elements of T are the 'old' coordinates of the linearly independent vectors $\boldsymbol{e}_1^*, \boldsymbol{e}_2^*, \ldots, \boldsymbol{e}_n^*$.

§ 5. Equivalent Matrices. The Rank of an Operator. Sylvester's Inequality

1. Let $\boldsymbol{R}$ and $\boldsymbol{S}$ be two vector spaces of dimension n and m, respectively, over the number field F and let $\boldsymbol{A}$ be a linear operator mapping $\boldsymbol{R}$ into $\boldsymbol{S}$. In the present section we shall make clear how the matrix A corresponding to the given linear operator $\boldsymbol{A}$ changes when the bases in $\boldsymbol{R}$ and $\boldsymbol{S}$ are changed.

We choose arbitrary bases $\boldsymbol{e}_1, \boldsymbol{e}_2, \ldots, \boldsymbol{e}_n$ in $\boldsymbol{R}$ and $\boldsymbol{g}_1, \boldsymbol{g}_2, \ldots, \boldsymbol{g}_m$ in $\boldsymbol{S}$. In these bases the operator $\boldsymbol{A}$ corresponds to a matrix $A = \| a_{ik} \|$ $(i = 1, 2, 3, \ldots, m;\ k = 1, 2, \ldots, n)$. To the vector equation

$$\boldsymbol{y} = \boldsymbol{A}\boldsymbol{x} \tag{26}$$

there corresponds the matrix equation

$$y = Ax, \tag{27}$$

where x and y are the coordinate columns for the vectors $\boldsymbol{x}$ and $\boldsymbol{y}$ in the bases $\boldsymbol{e}_1, \boldsymbol{e}_2, \ldots, \boldsymbol{e}_n$ and $\boldsymbol{g}_1, \boldsymbol{g}_2, \ldots, \boldsymbol{g}_m$.

We now choose other bases $\boldsymbol{e}_1^*, \boldsymbol{e}_2^*, \ldots, \boldsymbol{e}_n^*$ and $\boldsymbol{g}_1^*, \boldsymbol{g}_2^*, \ldots, \boldsymbol{g}_m^*$ in $\boldsymbol{R}$ and $\boldsymbol{S}$. In the new bases we shall have x^*, y^*, A^* instead of x, y, A. Here

$$y^* = A^* x^*. \tag{28}$$

Let us denote by Q and N the non-singular square matrices of order n and m, respectively, that realize the coordinate tranformations in the spaces $\boldsymbol{R}$ and $\boldsymbol{S}$ on transition from the old bases to the new ones (see § 4):

$$x = Qx^*, \quad y = Ny^*. \tag{29}$$

Then we obtain from (27) and (29):

$$y^* = N^{-1}y = N^{-1}Ax = N^{-1}AQx^*. \tag{30}$$

Setting $P = N^{-1}$, we find from (28) and (30):

$$A^* = PAQ. \tag{31}$$

DEFINITION 8: *Two rectangular matrices A and B of the same dimension are called equivalent if there exist two non-singular matrices P and Q such that*[8]

$$B = PAQ. \tag{32}$$

[8] If the matrices A and B are of dimension $m \times n$, then in (32) the square matrix P is of order m, and Q of order n. If the elements of the equivalent matrices A and B belong to some number field, then P and Q may be chosen such that their elements belong to the same number field.

From (31) it follows that two matrices corresponding to one and the same linear operator $\boldsymbol{A}$ for different choices of bases in $\boldsymbol{R}$ and $\boldsymbol{S}$ are always equivalent. It is easy to see that, conversely, if a matrix A corresponds to the operator $\boldsymbol{A}$ for certain bases in $\boldsymbol{R}$ and S, and if a matrix B is equivalent to A, then it corresponds to the same linear operator for certain other bases in $\boldsymbol{R}$ and S.

Thus, to every linear operator mapping $\boldsymbol{R}$ into $\boldsymbol{S}$ there corresponds a class of equivalent matrices with elements in F.

2. The following theorem establishes a criterion for the equivalence of two matrices:

THEOREM 2: *Two rectangular matrices of the same dimension are equivalent if and only if they have the same rank.*

Proof. The condition is *necessary*. When a rectangular matrix is multiplied by an arbitrary non-singular square matrix (on the right or left), then its rank does not change (see Chapter I, p. 17). Therefore it follows from (32) that

$$r_A = r_B .$$

The condition is *sufficient*. Let A be a rectangular matrix of dimension $m \times n$. It determines a linear operator $\boldsymbol{A}$ mapping the space $\boldsymbol{R}$ with the basis $\boldsymbol{e}_1, \boldsymbol{e}_2, \ldots, \boldsymbol{e}_n$ into the space $\boldsymbol{S}$ with the basis $\boldsymbol{g}_1, \boldsymbol{g}_2, \ldots, \boldsymbol{g}_m$. Let r denote the number of linearly independent vectors among the vectors $\boldsymbol{Ae}_1, \boldsymbol{Ae}_2, \ldots, \boldsymbol{Ae}_n$. Without loss of generality we may assume that the vectors $\boldsymbol{Ae}_1, \boldsymbol{Ae}_2, \ldots, \boldsymbol{Ae}_r$ are linearly independent[9] and that the remaining $\boldsymbol{Ae}_{r+1}, \boldsymbol{Ae}_{r+2}, \ldots, \boldsymbol{Ae}_n$ are expressed linearly in terms of them:

$$\boldsymbol{Ae}_k = \sum_{j=1}^{r} c_{kj} \boldsymbol{Ae}_j \qquad (k = r+1, \ldots, n). \tag{33}$$

We define a new basis in $\boldsymbol{R}$ as follows:

$$\boldsymbol{e}_i^* = \begin{cases} \boldsymbol{e}_i & (i = 1, 2, \ldots, r), \\ \boldsymbol{e}_i - \sum_{j=1}^{r} c_{ij} \boldsymbol{e}_j & (i = r+1, \ldots, n). \end{cases} \tag{34}$$

Then by (33),

$$\boldsymbol{Ae}_k^* = \boldsymbol{o} \qquad (k = r+1, \ldots, n). \tag{35}$$

Next, we set

$$\boldsymbol{Ae}_j^* = \boldsymbol{g}_j^* \qquad (j = 1, 2, \ldots, r). \tag{36}$$

[9] This can be achieved by a suitable numbering of the basis vectors $\boldsymbol{e}_1, \boldsymbol{e}_2, \ldots, \boldsymbol{e}_n$.

The vectors $\boldsymbol{g}_1^*, \boldsymbol{g}_2^*, \ldots, \boldsymbol{g}_r^*$ are linearly independent. We supplement them with suitable vectors $\boldsymbol{g}_{r+1}^*, \boldsymbol{g}_{r+2}^*, \ldots, \boldsymbol{g}_m^*$ to obtain a basis $\boldsymbol{g}_1^*, \boldsymbol{g}_2^*, \ldots, \boldsymbol{g}_m^*$ of $\boldsymbol{S}$.

The matrix corresponding to the same operator $\boldsymbol{A}$ in the new bases $\boldsymbol{e}_1^*, \boldsymbol{e}_2^*, \ldots, \boldsymbol{e}_n^*$, $\boldsymbol{g}_1^*, \boldsymbol{g}_2^*, \ldots, \boldsymbol{g}_m^*$ has now, by (35) and (36), the form

$$I_r = \left\| \begin{array}{cccccc} \overbrace{1 & 0 \ldots & 0}^{r} & 0 \ldots & 0 \\ 0 & 1 \ldots & 0 & 0 \ldots & 0 \\ \cdot & \cdot & \cdot & \cdot & \cdot \\ 0 & 0 \ldots & 1 & 0 \ldots & 0 \\ 0 & 0 \ldots & 0 & 0 \ldots & 0 \\ \cdot & \cdot & \cdot & \cdot & \cdot \\ 0 & 0 \ldots & 0 & 0 \ldots & 0 \end{array} \right\| . \tag{37}$$

Along the main diagonal of I_r, starting at the top, there are r units; all the remaining elements of I_r are zeros. Since the matrices A and I_r correspond to one and the same operator $\boldsymbol{A}$, they are equivalent. As we have proved, equivalent matrices have the same rank. Hence the rank of the original matrix A is r.

We have shown that an arbitrary rectangular matrix of rank r is equivalent to the 'canonical' matrix I_r. But I_r is completely determined by specifying its dimensions $m \times n$ and the number r. Therefore all rectangular matrices of given dimension $m \times n$ and of given rank r are equivalent to one and the same matrix I_r and consequently to each other. This completes the proof of the theorem.

3. Let $\boldsymbol{A}$ be a linear operator mapping an n-dimensional space $\boldsymbol{R}$ into an n-dimensional space $\boldsymbol{S}$. The set of all vectors of the form $\boldsymbol{A}\boldsymbol{x}$, where $\boldsymbol{x} \in \boldsymbol{R}$, forms a vector space.[10] This space will be denoted by $\boldsymbol{AR}$; it is part of the space $\boldsymbol{S}$ or, as we shall say, is a *subspace* of $\boldsymbol{S}$.

Together with the subspace $\boldsymbol{AR}$ of $\boldsymbol{S}$ we consider the set of all vectors $\boldsymbol{x} \in \boldsymbol{R}$ that satisfy the equation

$$\boldsymbol{A}\boldsymbol{x} = \boldsymbol{o} \tag{38}$$

These vectors also form a subspace of $\boldsymbol{R}$, which we shall denote by $\boldsymbol{N}_{\boldsymbol{A}}$.

---

[10] The set of vectors of the form $\boldsymbol{A}\boldsymbol{x}$ $(\boldsymbol{x} \in \boldsymbol{R})$ satisfies the postulates 1.-7. of § 1, because the sum of two such vectors and the product of such a vector by a number are also vectors of this form.

DEFINITION 9: *If a linear operator $\boldsymbol{A}$ maps $\boldsymbol{R}$ into $\boldsymbol{S}$, then the dimension r of the space $\boldsymbol{AR}$ is called the rank of $\boldsymbol{A}$,*[11] *and the dimension d of the space $\boldsymbol{N_A}$ consisting of all vectors $\boldsymbol{x} \in \boldsymbol{R}$ that satisfy the condition* (38) *is called the defect, or nullity, of $\boldsymbol{A}$.*

Among all the equivalent rectangular matrices that describe a given operator $\boldsymbol{A}$ in distinct bases there occurs the canonical matrix I_r (see (37)). We denote the corresponding bases of $\boldsymbol{R}$ and $\boldsymbol{S}$ by $\boldsymbol{e}_1^*, \boldsymbol{e}_2^*, \ldots, \boldsymbol{e}_n^*$ and $\boldsymbol{g}_1^*, \boldsymbol{g}_2^*, \ldots, \boldsymbol{g}_m^*$. Then

$$\boldsymbol{Ae}_1^* = \boldsymbol{g}_1^*, \ \ldots, \ \boldsymbol{Ae}_r^* = \boldsymbol{g}_r^*, \quad \boldsymbol{Ae}_{r+1}^* = \ldots = \boldsymbol{Ae}_n^* = \boldsymbol{o}.$$

From the definition of $\boldsymbol{AR}$ and $\boldsymbol{N_A}$ it follows that the vectors $\boldsymbol{g}_1^*, \boldsymbol{g}_2^*, \ldots, \boldsymbol{g}_r^*$ form a basis of $\boldsymbol{AR}$ and that the vectors $\boldsymbol{e}_{r+1}^*, \boldsymbol{e}_{r+2}^*, \ldots, \boldsymbol{e}_n^*$ form a basis of $\boldsymbol{N_A}$. Hence it follows that r is the rank of the operator $\boldsymbol{A}$ and that

$$d = n - r. \tag{39}$$

If A is an arbitrary matrix corresponding to $\boldsymbol{A}$, then it is equivalent to I_r and therefore has the same rank r. Thus, *the rank of an operator $\boldsymbol{A}$ coincides with the rank of the rectangular matrix A*

$$A = \left\| \begin{matrix} a_{11} & a_{12} & \ldots & a_{1n} \\ a_{21} & a_{22} & \ldots & a_{2n} \\ \cdot & \cdot & \cdot & \cdot \\ a_{m1} & a_{m2} & \ldots & a_{mn} \end{matrix} \right\|$$

determined by $\boldsymbol{A}$ in arbitrary bases $\boldsymbol{e}_1, \boldsymbol{e}_2, \ldots, \boldsymbol{e}_n \in \boldsymbol{R}$ and $\boldsymbol{g}_1, \boldsymbol{g}_2, \ldots, \boldsymbol{g}_m \in \boldsymbol{S}$.

The columns of A are formed by the coordinate vectors $\boldsymbol{A}_1\boldsymbol{e}_1, \ldots, \boldsymbol{A}_n\boldsymbol{e}_n$. Since it follows from $\boldsymbol{x} = \sum_{i=1}^{n} x_i\boldsymbol{e}_i$ that $\boldsymbol{Ax} = \sum_{i=1}^{n} x_i\boldsymbol{Ae}_i$, the rank of $\boldsymbol{A}$, i.e., the dimension of $\boldsymbol{RA}$, is equal to the maximal number of linearly independent vectors among $\boldsymbol{Ae}_1, \boldsymbol{Ae}_2, \ldots, \boldsymbol{Ae}_n$. Thus:

The rank of a matrix coincides with the number of linearly independent columns of the matrix.

Since under transposition the rows of a matrix become its columns and the rank remains unchanged:

---

[11] The dimension of the space $\boldsymbol{AR}$ never exceeds the dimension of $\boldsymbol{R}$, so that $r \leqq n$. This follows from the fact that the equation $\boldsymbol{x} = \sum_{i=1}^{n} x_i\boldsymbol{e}_i$ (where $\boldsymbol{e}_1, \boldsymbol{e}_2, \ldots, \boldsymbol{e}_n$ is a basis of $\boldsymbol{R}$) implies the equation $\boldsymbol{Ax} = \sum_{i=1}^{n} x_i\boldsymbol{Ae}_i$.

The number of linearly independent rows of a matrix is also equal to the rank of the matrix.[12]

4. Let $\boldsymbol{A}$ and $\boldsymbol{B}$ be two linear operators and let $\boldsymbol{C}=\boldsymbol{AB}$ be their product. Suppose that the operator $\boldsymbol{B}$ maps $\boldsymbol{R}$ into $\boldsymbol{S}$ and that the operator $\boldsymbol{A}$ maps $\boldsymbol{S}$ into $\boldsymbol{T}$. Then the operator $\boldsymbol{C}$ maps $\boldsymbol{R}$ into $\boldsymbol{T}$:

$$\boldsymbol{R}\xrightarrow{\boldsymbol{B}}\boldsymbol{S}\xrightarrow{\boldsymbol{A}}\boldsymbol{T},\quad \boldsymbol{R}\xrightarrow{\boldsymbol{C}}\boldsymbol{T}.$$

We introduce the matrices A, B, C corresponding to $\boldsymbol{A}$, $\boldsymbol{B}$, $\boldsymbol{C}$ in some choice of bases in $\boldsymbol{R}$, $\boldsymbol{S}$, and $\boldsymbol{T}$. Then the matrix equation $C=AB$ will correspond to the operator equation $\boldsymbol{C}=\boldsymbol{AB}$.

We denote by r_A, r_B, r_C the ranks of the operators $\boldsymbol{A}$, $\boldsymbol{B}$, $\boldsymbol{C}$ or, what is the same, of the matrices A, B, C. These numbers determine the dimensions of the subspaces $\boldsymbol{AS}$, $\boldsymbol{BR}$, $\boldsymbol{A}(\boldsymbol{BR})$. Since $\boldsymbol{BR}\subset\boldsymbol{S}$, we have $\boldsymbol{A}(\boldsymbol{BR})\subset\boldsymbol{AS}$.[13] Moreover, the dimension of $\boldsymbol{A}(\boldsymbol{BR})$ cannot exceed the dimension of $\boldsymbol{BR}$.[14] Therefore

$$r_C\leqq r_A,\quad r_C\leqq r_B.$$

These inequalities were obtained in Chapter I, § 2 from the formula for the minors of a product of two matrices.

Let us regard $\boldsymbol{A}$ as an operator mapping $\boldsymbol{BR}$ into $\boldsymbol{T}$. Then the rank of this operator is equal to the dimension of the space $\boldsymbol{A}(\boldsymbol{BR})$, i.e., to r_C. Therefore, by applying (39) we obtain

$$r_C=r_B-d_1, \tag{40}$$

where d_1 is the maximal number of linearly independent vectors of $\boldsymbol{BR}$ that satisfy the equation

$$\boldsymbol{Ax}=\boldsymbol{o}. \tag{41}$$

But all the solutions of this equation that belong to $\boldsymbol{S}$ form a subspace of dimension d, where

$$d=n-r_A \tag{42}$$

is the defect of the operator $\boldsymbol{A}$ mapping $\boldsymbol{S}$ into $\boldsymbol{T}$. Since $\boldsymbol{BR}\subset\boldsymbol{S}$,

$$d_1\leqq d. \tag{43}$$

From (40), (42), and (43) we find:

$$r_A+r_B-n\leqq r_C.$$

---

[12] In § 1 we reached these conclusions on the basis of other arguments (see p. 54).

[13] $\boldsymbol{R}\subset\boldsymbol{S}$ means that the set $\boldsymbol{R}$ forms part of the set $\boldsymbol{S}$.

[14] See Footnote 11.

Thus we have obtained *Sylvester's inequality* for the rank of the product of two rectangular matrices A and B of dimensions $m \times n$ and $n \times q$:

$$r_A + r_B - n \leqq r_{AB} \leqq \min(r_A, r_B). \tag{44}$$

§ 6. Linear Operators Mapping an n-Dimensional Space into Itself

1. A linear operator mapping the n-dimensional vector space $\boldsymbol{R}$ into itself (here $\boldsymbol{R} \equiv \boldsymbol{S}$ and $n = m$) will be referred to simply as a *linear operator in* $\boldsymbol{R}$.

The sum of two linear operators in $\boldsymbol{R}$ and the product of such an operator by a number are also linear operators in $\boldsymbol{R}$. Multiplication of two such operators is always feasible, and this product is also a linear operator in $\boldsymbol{R}$. Hence the linear operators in $\boldsymbol{R}$ form a ring.[15] This ring has an identity operator, namely the operator $\boldsymbol{E}$ for which

$$\boldsymbol{E}\boldsymbol{x} = \boldsymbol{x} \qquad (\boldsymbol{x} \in \boldsymbol{R}). \tag{45}$$

For every operator $\boldsymbol{A}$ in $\boldsymbol{R}$ we have

$$\boldsymbol{E}\boldsymbol{A} = \boldsymbol{A}\boldsymbol{E} = \boldsymbol{A}.$$

If $\boldsymbol{A}$ is a linear operator in $\boldsymbol{R}$, then the powers $\boldsymbol{A}^2 = \boldsymbol{A}\boldsymbol{A}$, $\boldsymbol{A}^3 = \boldsymbol{A}\boldsymbol{A}\boldsymbol{A}$, and in general $\boldsymbol{A}^m = \underbrace{\boldsymbol{A}\boldsymbol{A}\cdots\boldsymbol{A}}_{m \text{ times}}$ have a meaning. We set $\boldsymbol{A}^0 = \boldsymbol{E}$. Then it is easy to see that for all non-negative integers p and q we have

$$\boldsymbol{A}^p\boldsymbol{A}^q = \boldsymbol{A}^{p+q}.$$

Let $f(t) = \alpha_0 t^m + \alpha_1 t^{m-1} + \cdots + \alpha_{m-1}t + \alpha_m$ be a polynomial in a scalar argument t with coefficients in the field F. Then we set:

$$f(\boldsymbol{A}) = \alpha_0\boldsymbol{A}^m + \alpha_1\boldsymbol{A}^{m-1} + \cdots + \alpha_{m-1}\boldsymbol{A} + \alpha_m\boldsymbol{E}. \tag{46}$$

Here $f(\boldsymbol{A})g(\boldsymbol{A}) = g(\boldsymbol{A})f(\boldsymbol{A})$ for any two polynomials $f(t)$ and $g(t)$.

Let

$$\boldsymbol{y} = \boldsymbol{A}\boldsymbol{x} \qquad (\boldsymbol{x}, \boldsymbol{y} \in \boldsymbol{R}).$$

We denote by $x_1, x_2, \ldots, x_n$ the coordinates of the vector $\boldsymbol{x}$ in an arbitrary basis $\boldsymbol{e}_1, \boldsymbol{e}_2, \ldots, \boldsymbol{e}_n$ and by $y_1, y_2, \ldots, y_n$ the coordinates of $\boldsymbol{y}$ in the same basis. Then

$$y_i = \sum_{k=1}^{n} a_{ik}x_k \qquad (i = 1, 2, \ldots, n). \tag{47}$$

[15] This ring is in fact an algebra. See Chapter I, p. 17.

In the basis $\boldsymbol{e}_1, \boldsymbol{e}_2, \ldots, \boldsymbol{e}_n$ the linear operator $\boldsymbol{A}$ corresponds to a square matrix $A = \| a_{ik} \|_1^n$.[16] We remind the reader (see § 2.) that in the k-th column of this matrix are to be found the coordinates of the vector $\boldsymbol{A}\boldsymbol{e}_k$ $(k = 1, 2, \ldots, n)$. Introducing the coordinate columns $x = (x_1, x_2, \ldots, x_n)$ and $y = (y_1, y_2, \ldots, y_n)$, we can write the transformation (47) in matrix form

$$y = Ax. \tag{48}$$

The sum and product of two operators $\boldsymbol{A}$ and $\boldsymbol{B}$ correspond to the sum and product of the corresponding square matrices $A = \| a_{ik} \|_1^n$ and $B = \| b_{ik} \|_1^n$. The product $\alpha\boldsymbol{A}$ corresponds to the matrix αA. The identity operator $\boldsymbol{E}$ corresponds to the square unit matrix $E = \| \delta_{ik} \|_1^n$. Thus, the choice of a basis establishes *an isomorphism between the ring of linear operators in* $\boldsymbol{R}$ *and the ring of square matrices of order* n *with elements in* F. In this isomorphism the polynomial $f(\boldsymbol{A})$ corresponds to the matrix $f(A)$.

Let us consider, apart from the basis $\boldsymbol{e}_1, \boldsymbol{e}_2, \ldots, \boldsymbol{e}_n$, another basis $\boldsymbol{e}_1^*, \boldsymbol{e}_2^*, \ldots, \boldsymbol{e}_n^*$ of $\boldsymbol{R}$. Then, in analogy with (48), we have

$$y^* = A^* x^*, \tag{49}$$

where x^*, y^* are the column matrices formed from the coordinates of the vectors $\boldsymbol{x}, \boldsymbol{y}$ in the basis $\boldsymbol{e}_1^*, \boldsymbol{e}_2^*, \ldots, \boldsymbol{e}_n^*$ and $A^* = \| a_{ik}^* \|_1^n$ is the square matrix corresponding to the operator $\boldsymbol{A}$ in this basis. We rewrite in matrix form the formulas for the transformation of coordinates

$$x = Tx^*, \quad y = Ty^*. \tag{50}$$

Then from (48) and (50) we find:

$$y^* = T^{-1}ATx^*;$$

and a comparison with (49) gives:

$$A^* = T^{-1}AT. \tag{51}$$

Formula (51) is a special case of (31) on p. 61 (namely, $P = T^{-1}$ and $Q = T$).

DEFINITION 10: *Two matrices A and B connected by the relation*

$$B = T^{-1}AT. \tag{51'}$$

where T is a non-singular matrix, are called similar.[17]

---

[16] See § 2 of this chapter. In this case the spaces $\boldsymbol{R}$ and $\boldsymbol{S}$ coincide; in the same way, the bases $\boldsymbol{e}_1, \boldsymbol{e}_2, \ldots, \boldsymbol{e}_n$ and $\boldsymbol{g}_1, \boldsymbol{g}_2, \ldots, \boldsymbol{g}_m$ of these spaces are identified.

Thus, we have shown that *two matrices corresponding to one and the same linear operator in* $\boldsymbol{R}$ *for distinct bases are similar* and the matrix T linking these matrices coincides with the matrix of the coordinate transformation in the transition from the first basis to the second (see (50)).

In other words, to a linear operator in $\boldsymbol{R}$ there corresponds a whole class of similar matrices; they represent the given operator in various bases.

In studying properties of a linear operator in $\boldsymbol{R}$, we are at the same time studying the matrix properties that are common to the whole class of similar matrices, that is, that remain unchanged, or invariant, under transition from a given matrix to a similar one.

We note at once that two similar matrices always have the same determinant. For it follows from (51′) that

$$B = |T|^{-1} |A| |T| = |A|. \tag{52}$$

The equation $|B| = |A|$ is a necessary, but not a sufficient condition for the similarity of the matrices A and B.

In Chapter VI we shall establish a criterion for the similarity of two matrices, i.e., we shall give necessary and sufficient conditions for two square matrices of order n to be similar.

In accordance with (52) we may define the determinant $|A|$ of a linear operator $\boldsymbol{A}$ in $\boldsymbol{R}$ as the determinant of an arbitrary matrix corresponding to the given operator.

If $|\boldsymbol{A}| = 0$ ($\neq 0$), then the operator $\boldsymbol{A}$ is called *singular* (*non-singular*). In accordance with this definition a singular (non-singular) operator corresponds to a singular (non-singular) matrix in any basis. For a singular operator:

1) There always exists a vector $\boldsymbol{x} \neq \boldsymbol{o}$ such that $\boldsymbol{A}\boldsymbol{x} = \boldsymbol{o}$;
2) $\boldsymbol{A}\boldsymbol{R}$ is a proper part of $\boldsymbol{R}$.

For a non-singular operator:

1) $\boldsymbol{A}\boldsymbol{x} = \boldsymbol{o}$ implies that $\boldsymbol{x} = \boldsymbol{o}$;
2) $\boldsymbol{A}\boldsymbol{R} \equiv \boldsymbol{R}$, i.e., the vectors of the form $\boldsymbol{A}\boldsymbol{x}$ ($\boldsymbol{x} \in \boldsymbol{R}$) fill out the whole space $\boldsymbol{R}$.

In other words, a linear operator in $\boldsymbol{R}$ is singular or non-singular depending on whether its defect is positive or zero.

---

[17] The matrix T can always be chosen such that its elements belong to the same basic number field F as those of A and B. It is easy to verify the three properties of similar matrices:

Reflexivity (a matrix A is always similar to itself);

Symmetry (if A is similar to B, then B is similar to A); and

Transitivity (if A is similar to B, and B to C, then A is similar to C).

§ 7. Characteristic Values and Characteristic Vectors of a Linear Operator

1. An important role in the study of the structure of a linear operator $\boldsymbol{A}$ in $\boldsymbol{R}$ is played by the vectors $\boldsymbol{x}$ for which

$$\boldsymbol{A}\boldsymbol{x} = \lambda \boldsymbol{x} \qquad (\lambda \in F, \quad \boldsymbol{x} \neq \boldsymbol{o}) \tag{53}$$

Such vectors are called *characteristic vectors* and the numbers λ corresponding to them are called *characteristic values* or *characteristic roots* of the operator $\boldsymbol{A}$ (or of the matrix A).†

In order to find the characteristic values and characteristic vectors of an operator $\boldsymbol{A}$ we choose an arbitrary basis $\boldsymbol{e}_1, \boldsymbol{e}_2, \ldots, \boldsymbol{e}_n$ in $\boldsymbol{R}$. Let $\boldsymbol{x} = \sum_{i=1}^{n} x_i \boldsymbol{e}_i$ and let $A = \| a_{ik} \|_1^n$ be the matrix corresponding to $\boldsymbol{A}$ in the basis $\boldsymbol{e}_1, \boldsymbol{e}_2, \ldots, \boldsymbol{e}_n$. Then if we equate the corresponding coordinates of the vectors on the left-hand and right-hand sides of (53), we obtain a system of scalar equations

$$\left.\begin{array}{l} a_{11}x_1 + a_{12}x_2 + \cdots + a_{1n}x_n = \lambda x_1 \\ a_{21}x_1 + a_{22}x_2 + \cdots + a_{2n}x_n = \lambda x_2 \\ \dots\dots\dots\dots\dots\dots\dots \\ a_{n1}x_1 + a_{n2}x_2 + \cdots + a_{nn}x_n = \lambda x_n , \end{array}\right\} \tag{54}$$

which can also be written as

$$\left.\begin{array}{l} (a_{11} - \lambda)\, x_1 + a_{12}x_2 + \cdots + a_{1n}x_n = 0 \\ a_{21}x_1 + (a_{22} - \lambda)\, x_2 + \cdots + a_{2n}x_n = 0 \\ \dots\dots\dots\dots\dots\dots\dots\dots \\ a_{n1}x_1 + a_{n2}x_2 + \cdots + (a_{nn} - \lambda)\, x_n = 0 \end{array}\right\} \tag{55}$$

Since the required vector must not be the null vector, at least one of its coordinates $x_1, x_2, \ldots, x_n$ must be different from zero.

In order that the system of linear homogeneous equations (55) should have a non-zero solution it is necessary and sufficient that the determinant of the system be zero:

$$\begin{vmatrix} a_{11} - \lambda & a_{12} & \ldots & a_{1n} \\ a_{21} & a_{22} - \lambda & \ldots & a_{2n} \\ \dots & \dots & \dots & \dots \\ a_{n1} & a_{n2} & \ldots & a_{nn} - \lambda \end{vmatrix} = 0. \tag{56}$$

† Other terms in use for the former are: *proper vector, latent vector, eigenvector.* Other terms for the latter are: *proper value, latent value, latent root, latent number, characteristic number, eigenvalue,* etc.

The equation (56) is an algebraic equation of degree n in λ. Its coefficients belong to the same number field F as the elements of the matrix $A = \| a_{ik} \|_1^n$.

Equation (56) occurs in various problems of geometry, mechanics, astronomy, and physics and is known as the *characteristic equation* or the *secular equation*[18] of the matrix $A = \| a_{ik} \|_1^n$ (the left-hand side is called the *characteristic polynomial*).

Thus, every characteristic value λ of a linear operator $\boldsymbol{A}$ is a root of the characteristic equation (56). And conversely, if a number λ is a root of (56), then for this value λ the system (55) and hence (54) has a non-zero solution $x_1, x_2, \ldots, x_n$, i.e., to this number λ there corresponds a characteristic vector $\boldsymbol{x} = \sum x_i \boldsymbol{e}_i$ of the operator $\boldsymbol{A}$.

From what we have shown, it follows that every linear operator $\boldsymbol{A}$ in $\boldsymbol{R}$ has not more than n distinct characteristic values.

If F is the field of complex numbers, then every linear operator in $\boldsymbol{R}$ always has at least one characteristic vector in $\boldsymbol{R}$ corresponding to a characteristic value λ.[19] This follows from the fundamental theorem of algebra, according to which an algebraic equation (56) in the field of complex numbers always has at least one root.

Let us write (56) in explicit form

$$| A - \lambda E | \equiv (-\lambda)^n + S_1 (-\lambda)^{n-1} + S_2 (-\lambda)^{n-2} + \cdots + S_{n-1} (-\lambda) + S_n = 0. \quad (57)$$

It is easy to see that here

$$S_1 = \sum_{i=1}^{n} a_{ii}, \quad S_2 = \sum_{1 \leq i < k \leq n} A \begin{pmatrix} i & k \\ i & k \end{pmatrix}, \quad \cdots \quad (58)$$

and, in general, S_p is the sum of the principal minors of order p of the matrix $A = \| a_{ik} \|_1^n$ $(p = 1, 2, \ldots, n)$.[20] In particular, $S_n = | A |$.

We denote by $\tilde{A}$ the matrix corresponding to the same operator $\boldsymbol{A}$ in another basis. $\tilde{A}$ is similar to A:

---

[18] The name is due to the fact that this equation occurs in the study of secular perturbations of the planets.

[19] This proposition is valid even in the more general case in which F is an arbitrary algebraically closed field, i.e., a field that contains the roots of all algebraic equations with coefficients in the field.

[20] The power $(-\lambda)^{n-p}$ occurs only in those terms of the characteristic determinant (56) that contain precisely $n - p$ of the diagonal elements, say,

$$a_{j_1 j_1} - \lambda, \; a_{j_2 j_2} - \lambda, \; \ldots, \; a_{j_{n-p} j_{n-p}} - \lambda.$$

The product of these diagonal elements occurs in the expansion of the determinant (56)

$$\tilde{A} = T^{-1} A T.$$

Hence

$$\tilde{A} - \lambda E = T^{-1} (A - \lambda E) T$$

and therefore

$$|\tilde{A} - \lambda E| = |A - \lambda E|. \tag{59}$$

Thus, similar matrices A and $\tilde{A}$ have the same characteristic polynomial. This polynomial is sometimes called the characteristic polynomial of the operator $\boldsymbol{A}$ and is denoted by $|\boldsymbol{A} - \lambda \boldsymbol{E}|$

If $\boldsymbol{x}, \boldsymbol{y}, \boldsymbol{z}, \ldots$ are linearly independent characteristic vectors of an operator $\boldsymbol{A}$ corresponding to one and the same characteristic λ, and $\alpha, \beta, \gamma, \ldots$ are arbitrary numbers of F, *then the vector $\alpha \boldsymbol{x} + \beta \boldsymbol{y} + \gamma \boldsymbol{z} + \cdots$ is either equal to zero or is also a characteristic vector of $\boldsymbol{A}$ corresponding to the same λ.*

For from

$$\boldsymbol{A}\boldsymbol{x} = \lambda \boldsymbol{x}, \quad \boldsymbol{A}\boldsymbol{y} = \lambda \boldsymbol{y}, \quad \boldsymbol{A}\boldsymbol{z} = \lambda \boldsymbol{z}, \ \ldots$$

it follows that

$$\boldsymbol{A}(\alpha \boldsymbol{x} + \beta \boldsymbol{y} + \gamma \boldsymbol{z} + \cdots) = \lambda (\alpha \boldsymbol{x} + \beta \boldsymbol{y} + \gamma \boldsymbol{z} + \cdots).$$

In other words, linearly independent characteristic vectors corresponding to one and the same characteristic value λ form a basis of a 'characteristic' subspace each vector of which is a characteristic vector for the same λ. In particular, each characteristic vector generates a one-dimensional subspace, a 'characteristic' direction.

However, if characteristic vectors of a linear operator $\boldsymbol{A}$ correspond to distinct characteristic values, then a linear combination of these characteristic vectors is not, in general, a characteristic vector of $\boldsymbol{A}$.

The significance of the characteristic vectors and characteristic numbers for the study of linear operators will be illustrated in the next section by the example of operators of simple structure.

---

with a factor in which the term free of λ is the principal minor

$$A \begin{pmatrix} i_1 & i_2 & \ldots & i_p \\ i_1 & i_2 & \ldots & i_p \end{pmatrix},$$

where $i_1, i_2, \ldots, i_p$ together with $j_1, j_2, \ldots, j_{n-p}$ forms a complete set of indices $1, 2, \ldots, n$; hence in the development of (56) we have

$$|A - \lambda E| = (a_{j_1 j_1} - \lambda)(a_{j_2 j_2} - \lambda) \cdots (a_{j_{n-p} j_{n-p}} - \lambda) \, A \begin{pmatrix} i_1 & i_2 & \ldots & i_p \\ i_1 & i_2 & \ldots & i_p \end{pmatrix} + (^*)$$

When we take all possible combinations $j_1, j_2, \ldots, j_{n-p}$ of $n - p$ of the indices $1, 2, \ldots, n$, we obtain for the coefficient S_p of $(-\lambda)^{n-p}$ the sum of all principal minors of order p in A.

§ 8. Linear Operators of Simple Structure

1. We begin with the following lemma.

LEMMA: *Characteristic vectors belonging to pairwise distinct characteristic values are always linearly independent.*

Proof. Let

$$\boldsymbol{A}\boldsymbol{x}_i = \lambda_i \boldsymbol{x}_i \quad (\boldsymbol{x}_i \neq \boldsymbol{o};\ \lambda_i \neq \lambda_k \text{ for } i \neq k;\ i, k = 1, 2, \ldots, m) \tag{60}$$

and

$$\sum_{i=1}^{m} c_i \boldsymbol{x}_i = \boldsymbol{o}. \tag{61}$$

Applying the operator $\boldsymbol{A}$ to both sides we obtain:

$$\sum_{i=1}^{m} c_i \lambda_i \boldsymbol{x}_i = \boldsymbol{o}. \tag{62}$$

We multiply both sides of (61) by λ_1 and subtract (61) from (62) term by term. Then we obtain

$$\sum_{i=2}^{m} c_i (\lambda_i - \lambda_1) \boldsymbol{x}_i = \boldsymbol{o}. \tag{63}$$

We can say that (63) is obtained from (61) by termwise application of the operator $\boldsymbol{A} - \lambda_1 \boldsymbol{E}$. If we apply the operators $\boldsymbol{A} - \lambda_2 \boldsymbol{E}, \ldots, \boldsymbol{A} - \lambda_{m-1} \boldsymbol{E}$ to (63) term by term, we are led to the following equation:

$$c_m (\lambda_m - \lambda_{m-1})(\lambda_m - \lambda_{m-2}) \cdots (\lambda_m - \lambda_1) \boldsymbol{x}_m = \boldsymbol{o},$$

so that $c_m = 0$. Since any of the summands in (61) can be put last, we have in (61)

$$c_1 = c_2 = \ldots = c_m = 0,$$

i.e., there is no linear dependence among the vectors $\boldsymbol{x}_1, \boldsymbol{x}_2, \ldots, \boldsymbol{x}_m$. This proves the lemma.

If the characteristic equation of an operator has n distinct roots and these roots belong to F, then by the lemma the characteristic vectors belonging to these roots are linearly independent.

DEFINITION 11: *A linear operator $\boldsymbol{A}$ in $\boldsymbol{R}$ is said to be an operator of simple structure if $\boldsymbol{A}$ has n linearly independent characteristic vectors in $\boldsymbol{R}$, where n is the dimension of $\boldsymbol{R}$.*

Thus, a linear operator in $\boldsymbol{R}$ has simple structure if all the roots of the characteristic equation are distinct and belong to F. However, these conditions are not necessary. There exist linear operators of simple structure whose characteristic polynomial has multiple roots.

Let us consider an arbitrary linear operator $\boldsymbol{A}$ of simple structure. We denote by $\boldsymbol{g}_1, \boldsymbol{g}_2, \dots, \boldsymbol{g}_n$ a basis of $\boldsymbol{R}$ consisting of characteristic vectors of the operator, i.e.,

$$\boldsymbol{A}\boldsymbol{g}_k = \lambda_k \boldsymbol{g}_k \quad (k = 1, 2, \dots, n).$$

If

$$\boldsymbol{x} = \sum_{k=1}^{n} x_k \boldsymbol{g}_k ,$$

then

$$\boldsymbol{A}\boldsymbol{x} = \sum_{k=1}^{n} x_k \boldsymbol{A}\boldsymbol{g}_k = \sum_{k=1}^{n} \lambda_k x_k \boldsymbol{g}_k .$$

The effect of the operator $\boldsymbol{A}$ of simple structure on the vector $\boldsymbol{x} = \sum_{k=1}^{n} x_k \boldsymbol{g}_k$ may be put into words as follows:

In the n-dimensional space $\boldsymbol{R}$ *there exist n linearly independent 'directions' along which the operator* $\boldsymbol{A}$ *of simple structure realizes a 'dilatation' with coefficients* $\lambda_1, \lambda_2, \dots, \lambda_n$. *An arbitrary vector* $\boldsymbol{x}$ *may be decomposed into components along these characteristic directions. These components are subject to the corresponding 'dilatations' and their sum then gives the vector* $\boldsymbol{A}\boldsymbol{x}$.

It is easy to see that to the operator $\boldsymbol{A}$ in a 'characteristic' basis $\boldsymbol{g}_1, \boldsymbol{g}_2, \dots, \boldsymbol{g}_n$ there corresponds the diagonal matrix

$$\tilde{A} = \| \lambda_i \delta_{ik} \|_1^n .$$

If we denote by A the matrix corresponding to $\boldsymbol{A}$ in an arbitrary basis $\boldsymbol{e}_1, \boldsymbol{e}_2, \dots, \boldsymbol{e}_n$, then

$$A = T \| \lambda_i \delta_{ik} \|_1^n T^{-1} . \tag{64}$$

A matrix that is similar (p. 68) to a diagonal matrix is called a *matrix of simple structure*. Thus, to an operator of simple structure there corresponds in any basis a matrix of simple structure, and vice versa.

2. The matrix T in (64) realizes the transition from the basis $\boldsymbol{e}_1, \boldsymbol{e}_2, \dots, \boldsymbol{e}_n$ to the basis $\boldsymbol{g}_1, \boldsymbol{g}_2, \dots, \boldsymbol{g}_n$. The k-th column of T contains the coordinates of a characteristic vector $\boldsymbol{g}_k$ (with respect to $\boldsymbol{e}_1, \boldsymbol{e}_2, \dots, \boldsymbol{e}_n$) that corresponds to the characteristic value λ_k of A $(k = 1, 2, \dots, n)$. The matrix T is called the *fundamental* matrix for A.

We rewrite (64) as follows:

$$A = TLT^{-1} \quad (L = \{\lambda_1, \lambda_2, \ldots, \lambda_n\}). \tag{64'}$$

On going over to the p-th compound matrices ($1 \leqq p \leqq n$), we obtain (see Chapter I, § 4):

$$\mathfrak{A}_p = \mathfrak{T}_p \mathfrak{L}_p \mathfrak{T}_p^{-1}. \tag{65}$$

$\mathfrak{L}_p$ is a diagonal matrix of order N ($N = \binom{n}{p}$) along whose main diagonal are all the possible products of $\lambda_1, \lambda_2, \ldots, \lambda_n$ taken p at a time. A comparison of (65) with (64') yields the following theorem:

THEOREM 3: *If a matrix $A = \| a_{ik} \|_1^n$ has simple structure, then for every $p \leqq n$ the compound matrix $\mathfrak{A}_p$ also has simple structure; moreover, the characteristic values of $\mathfrak{A}_p$ are all the possible products $\lambda_{i_1}\lambda_{i_2}\cdots\lambda_{i_p}$ $(1 \leqq i_1 < i_2 < \ldots < i_p \leqq n)$ of p of the characteristic values $\lambda_1, \lambda_2, \ldots, \lambda_n$ of A, and the fundamental matrix of $\mathfrak{A}_p$ is the compound $\mathfrak{T}_p$ of the fundamental matrix T of A.*

COROLLARY: *If a characteristic value λ_k of a matrix of simple structure $A = \| a_{ik} \|_1^n$ corresponds to a characteristic vector with the coordinates $t_{1k}, t_{2k}, \ldots, t_{nk}$ $(k = 1, 2, \ldots, n)$ and if $T = \| t_{ik} \|_1^n$, then the characteristic value $\lambda_{k_1}\lambda_{k_2}\cdots\lambda_{k_p}$ $(1 \leqq k_1 < k_2 < \ldots < k_p \leqq n)$ of $\mathfrak{A}_p$ corresponds to the characteristic vector with coordinates*

$$T\begin{pmatrix} i_1 & i_2 & \ldots & i_p \\ k_1 & k_2 & \ldots & k_p \end{pmatrix} \quad (1 \leqq i_1 < i_2 < \cdots < i_p \leqq n). \tag{66}$$

An arbitrary matrix $A = \| a_{ik} \|_1^n$ may be represented in the form of a sequence of matrices A_m $(m \to \infty)$ each of which does not have multiple characteristic values and, therefore, has simple structure. The characteristic values $\lambda_1^{(m)}, \lambda_2^{(m)}, \ldots, \lambda_n^{(m)}$ of the matrix A_m converge for $m \to \infty$ to the characteristic values $\lambda_1, \lambda_2, \ldots, \lambda_n$ of A,

$$\lim_{m \to \infty} \lambda_k^{(m)} = \lambda_k \qquad (k = 1, 2, \ldots, n).$$

Hence

$$\lim_{m \to \infty} \lambda_{k_1}^{(m)} \lambda_{k_2}^{(m)} \cdots \lambda_{k_p}^{(m)} = \lambda_{k_1}\lambda_{k_2} \cdots \lambda_{k_p} \quad (1 \leqq k_1 < k_2 < \cdots < k_p \leqq n).$$

Moreover, since $\lim\limits_{m \to \infty} \mathfrak{A}_{(m)p} = \mathfrak{A}_p$, we deduce from Theorem 3:

THEOREM 4 (Kronecker): *If* $\lambda_1, \lambda_2, \ldots, \lambda_n$ *is a complete system of characteristic values of an arbitrary matrix* A, *then a complete system of characteristic values of the compound matrix* $\mathfrak{A}_p$ *consists of all possible products of the numbers* $\lambda_1, \lambda_2, \ldots, \lambda_n$ *taken* p *at a time* $(p = 1, 2, \ldots, n)$.

In the present section we have investigated operators and matrices of simple structure. The study of the structure of operators and matrices of general type will be resumed in Chapters VI and VII.

CHAPTER IV

THE CHARACTERISTIC POLYNOMIAL AND THE MINIMAL POLYNOMIAL OF A MATRIX

Two polynomials are associated with every square matrix: the characteristic polynomial and the minimal polynomial. These polynomials play an important role in various problems of the theory of matrices. For example, the concept of a function of a matrix, which we shall introduce in the next chapter, will be based entirely on the concept of the minimal polynomial. In the present chapter, the properties of the characteristic polynomial and the minimal polynomial are studied. A prerequisite to this investigation is some basic information about polynomials with matrix coefficients and operations on them.

§ 1. Addition and Multiplication of Matrix Polynomials

1. We consider a square *polynomial* matrix $A(\lambda)$, i.e., a square matrix whose elements are polynomials in λ (with coefficients in the given number field F):

$$A(\lambda) = \| a_{ik}(\lambda) \|_1^n = \| a_{ik}^{(0)}\lambda^m + a_{ik}^{(1)}\lambda^{m-1} + \cdots + a_{ik}^{(m)} \|_1^n . \qquad (1)$$

The matrix $A(\lambda)$ can be represented in the form of a polynomial with matrix coefficients arranged with respect to the powers of λ:

$$A(\lambda) = A_0\lambda^m + A_1\lambda^{m-1} + \cdots + A_m , \qquad (2)$$

where

$$A_j = \| a_{ik}^{(j)} \|_1^n \qquad (j = 0,\ 1,\ \ldots,\ m). \qquad (3)$$

The number m is called the *degree* of the polynomial, provided $A_0 \neq O$. The number n is called the *order* of the polynomial. The polynomial (1) is called *regular* if $| A_0 | \neq 0$.

A polynomial with matrix coefficients will sometimes be called a *matrix polynomial.* In contrast to a matrix polynomial an ordinary polynomial with scalar coefficients will be called a *scalar polynomial.*

We shall now consider the fundamental operations on matrix polynomials. Let two matrix polynomials $A(\lambda)$ and $B(\lambda)$ of the same order be given. We denote by m the larger of their degrees. These polynomials can be written in the form

$$A(\lambda) = A_0\lambda^m + A_1\lambda^{m-1} + \cdots + A_m,$$
$$B(\lambda) = B_0\lambda^m + B_1\lambda^{m-1} + \cdots + B_m.$$

Then

$$A(\lambda) \pm B(\lambda) = (A_0 \pm B_0)\lambda^m + (A_1 \pm B_1)\lambda^{m-1} + \cdots + (A_m \pm B_m),$$

i.e.: *The sum (difference) of two matrix polynomials of the same order can be represented in the form of a polynomial whose degree does not exceed the larger of the degrees of the given polynomials.*

Let $A(\lambda)$ and $B(\lambda)$ be two matrix polynomials of the same order n and of respective degrees m and p:

$$A(\lambda) = A_0\lambda^m + A_1\lambda^{m-1} + \cdots + A_m \qquad (A_0 \neq O),$$
$$B(\lambda) = B_0\lambda^p + B_1\lambda^{p-1} + \cdots + B_p \qquad (B_0 \neq O).$$

Then

$$A(\lambda)B(\lambda) = A_0B_0\lambda^{m+p} + (A_0B_1 + A_1B_0)\lambda^{m+p-1} + \cdots + A_mB_p. \tag{4}$$

If we multiply $B(\lambda)$ by $A(\lambda)$ (i.e., interchange the order of the factors), then we obtain, in general, a different polynomial.

2. The multiplication of matrix polynomials has a specific property. In contrast to the product of scalar polynomials, the product (4) of matrix polynomials may have a degree less than $m + p$, i.e., less than the sum of the degrees of the factors. For, in (4) the product A_0B_0 may be the null matrix even though $A_0 \neq O$, $B_0 \neq O$. However, if at least one of the matrices A_0 and B_0 is non-singular, then it follows from $A_0 \neq O$ and $B_0 \neq O$ that $A_0B_0 \neq O$. Thus: *The product of two matrix polynomials is a polynomial whose degree is less than or equal to the sum of the degrees of the factors. If at least one of the two factors is regular, then the degree of the product is always equal to the sum of the degrees of the factors.*

§ 2. Right and Left Division of Matrix Polynomials

1. Let $A(\lambda)$ and $B(\lambda)$ be two matrix polynomials of the same order n, and let $B(\lambda)$ be regular:

$$A(\lambda) = A_0\lambda^m + A_1\lambda^{m-1} + \cdots + A_m \qquad (A_0 \neq O),$$
$$B(\lambda) = B_0\lambda^p + B_1\lambda^{p-1} + \cdots + B_p \qquad (|B_0| \neq 0).$$

We shall say that the matrix polynomials $Q(\lambda)$ and $R(\lambda)$ are the *right quotient* and the *right remainder,* respectively, of $A(\lambda)$ on division by $B(\lambda)$ if

$$A(\lambda) = Q(\lambda)\,B(\lambda) + R(\lambda) \tag{5}$$

and if the degree of $R(\lambda)$ is less than that of $B(\lambda)$.

Similarly, we shall call the polynomials $\hat{Q}(\lambda)$ and $\hat{R}(\lambda)$ the *left quotient* and the *left remainder* of $A(\lambda)$ on division by $B(\lambda)$ if

$$A(\lambda) = B(\lambda)\,\hat{Q}(\lambda) + \hat{R}(\lambda) \tag{6}$$

and if the degree of $\hat{R}(\lambda)$ is less than that of $B(\lambda)$.

The reader should note that in the '*right*' division (i.e., when the right quotient and the right remainder are to be found) in (5) the quotient $Q(\lambda)$ is multiplied by the 'divisor' $B(\lambda)$ on the *right*, and in the '*left*' division in (6) the quotient $\hat{Q}(\lambda)$ is multiplied by the divisor $B(\lambda)$ on the *left*. The polynomials $\hat{Q}(\lambda)$ and $\hat{R}(\lambda)$ do not, in general, coincide with $Q(\lambda)$ and $R(\lambda)$.

2. We shall now show that *both right and left division of matrix polynomials of the same order are always possible and unique, provided the divisor is a regular polynomial.*

Let us consider the right division of $A(\lambda)$ by $B(\lambda)$. If $m < p$, we can set $Q(\lambda) = O$, $R(\lambda) = A(\lambda)$. If $m \geqq p$, we apply the usual scheme for the division of a polynomial by a polynomial in order to find the quotient $Q(\lambda)$ and the remainder $R(\lambda)$. We 'divide' the highest term of the dividend $A_0\lambda^m$ by the highest term of the divisor $B_0\lambda^p$. We obtain the highest term $A_0B_0^{-1}\lambda^{m-p}$ of the required quotient. We multiply this term on the right by the divisor $B(\lambda)$ and subtract the product so obtained from $A(\lambda)$. Thus we find the 'first remainder' $A^{(1)}(\lambda)$:

$$A(\lambda) = A_0B_0^{-1}\lambda^{m-p}\,B(\lambda) + A^{(1)}(\lambda). \tag{7}$$

The degree $m^{(1)}$ of $A^{(1)}(\lambda)$ is less than m:

$$A^{(1)}(\lambda) = A_0^{(1)}\,\lambda^{m^{(1)}} + \cdots \qquad (A_0^{(1)} \neq O,\ m^{(1)} < m). \tag{8}$$

If $m^{(1)} \geqq p$, then we repeat the process and obtain:

$$\left.\begin{aligned} A^{(1)}(\lambda) &= A_0^{(1)}\,B_0^{-1}\,\lambda^{m^{(1)}-p}\,B(\lambda) + A^{(2)}(\lambda),\\ A^{(2)}(\lambda) &= A_0^{(2)}\,\lambda^{m^{(2)}} + \cdots \qquad (m^{(2)} < m^{(1)}),\end{aligned}\right\} \tag{9}$$

etc.

If A and B commute $Q = \hat{Q}$, but $Q = \hat{Q} \not\Rightarrow$ A commutes w/ B

Since the degrees of $A(\lambda)$, $A^{(1)}(\lambda)$, $A^{(2)}(\lambda)$, ... decrease, at some stage we arrive at a remainder $R(\lambda)$ whose degree is less than p. Then it follows from (7) and (9) that

$$A(\lambda) = Q(\lambda)B(\lambda) + R(\lambda),$$

where

$$Q(\lambda) = A_0 B_0^{-1} \lambda^{m-p} + A_0^{(1)} B_0^{-1} \lambda^{m^{(1)}-p} + \cdots. \tag{10}$$

We shall now prove the *uniqueness* of the right division. Suppose we have simultaneously

$$A(\lambda) = Q(\lambda)\, B(\lambda) + R(\lambda) \tag{11}$$

and

$$A(\lambda) = Q^*(\lambda)\, B(\lambda) + R^*(\lambda), \tag{12}$$

where the degrees of $R(\lambda)$ and $R^*(\lambda)$ are less than that of $B(\lambda)$, i.e., less than p. Subtracting (11) from (12) term by term we obtain

$$[Q(\lambda) - Q^*(\lambda)]\, B(\lambda) = R^*(\lambda) - R(\lambda). \tag{13}$$

If we had $Q(\lambda) - Q^*(\lambda) \not\equiv O$, then the degree on the left-hand side of (13) would be the sum of the degrees of $B(\lambda)$ and $Q(\lambda) - Q^*(\lambda)$, because $|B_0| \neq 0$, and would therefore be at least equal to p. This is impossible, since the degree of the polynomial on the right-hand side of (13) is less than p. Thus, $Q(\lambda) - Q^*(\lambda) \equiv O$, and then it follows from (13) that $R(\lambda) - R^*(\lambda) \equiv O$, i.e.,

$$Q(\lambda) = Q^*(\lambda), \qquad R(\lambda) = R^*(\lambda).$$

The existence and uniqueness of the left quotient and left remainder is established similarly.[1]

---

[1] Note that the possibility and uniqueness of the left division of $A(\lambda)$ by $B(\lambda)$ follows from that of the right division of the transposed matrices $A^{\mathsf{T}}(\lambda)$ and $B^{\mathsf{T}}(\lambda)$. (The regularity of $B(\lambda)$ implies that of $B^{\mathsf{T}}(\lambda)$.) For from

$$A^{\mathsf{T}}(\lambda) = Q_1(\lambda)\, B^{\mathsf{T}}(\lambda) + R_1(\lambda)$$

it follows (see Chapter I, p. 19) that

$$A(\lambda) = B(\lambda)\, Q_1^{\mathsf{T}}(\lambda) + R_1^{\mathsf{T}}(\lambda). \tag{6'}$$

By the same reasoning, the left division of $A(\lambda)$ by $B(\lambda)$ is unique; for if it were not, then the right division of $A^{\mathsf{T}}(\lambda)$ by $B^{\mathsf{T}}(\lambda)$ would not be unique.

Comparison of (6) and (6′) gives

$$\hat{Q}(\lambda) = Q_1^{\mathsf{T}}(\lambda), \quad \hat{R}(\lambda) = R_1^{\mathsf{T}}(\lambda).$$

Example.

$$A(\lambda) = \begin{Vmatrix} \lambda^3 + \lambda & 2\lambda^3 + \lambda^2 \\ -\lambda^3 - 2\lambda^2 + 1 & 3\lambda^3 + \lambda \end{Vmatrix}$$

$$= \overbrace{\begin{Vmatrix} 1 & 2 \\ -1 & 3 \end{Vmatrix}}^{A_0} \lambda^3 + \begin{Vmatrix} 0 & 1 \\ -2 & 0 \end{Vmatrix} \lambda^2 + \begin{Vmatrix} 1 & 0 \\ 0 & 1 \end{Vmatrix} \lambda + \begin{Vmatrix} 0 & 0 \\ 1 & 0 \end{Vmatrix},$$

$$B(\lambda) = \begin{Vmatrix} 2\lambda^2 + 3 & -\lambda^2 + 1 \\ -\lambda^2 - 1 & \lambda^2 + 2 \end{Vmatrix} = \overbrace{\begin{Vmatrix} 2 & -1 \\ -1 & 1 \end{Vmatrix}}^{B_0} \lambda^2 + \begin{Vmatrix} 3 & 1 \\ -1 & 2 \end{Vmatrix},$$

$$|B_0| = 1,\ B_0^{-1} = \begin{Vmatrix} 1 & 1 \\ 1 & 2 \end{Vmatrix},\ A_0 B_0^{-1} = \begin{Vmatrix} 3 & 5 \\ 2 & 5 \end{Vmatrix},\ A_0 B_0^{-1} B(\lambda) = \begin{Vmatrix} \lambda^2 + 4 & 2\lambda^2 + 13 \\ -\lambda^2 + 1 & 3\lambda^2 + 12 \end{Vmatrix},$$

$$A^{(1)}(\lambda) = \begin{Vmatrix} \lambda^3 + \lambda & 2\lambda^3 + \lambda^2 \\ -\lambda^3 - 2\lambda^2 + 1 & 3\lambda^3 + \lambda \end{Vmatrix} - \begin{Vmatrix} \lambda^3 + 4\lambda & 2\lambda^3 + 13\lambda \\ -\lambda^3 + \lambda & 3\lambda^3 + 12\lambda \end{Vmatrix}$$

$$= \begin{Vmatrix} -3\lambda & \lambda^2 - 13\lambda \\ -2\lambda^2 - \lambda + 1 & -11\lambda \end{Vmatrix},$$

$$A^{(1)}(\lambda) = \begin{Vmatrix} 0 & 1 \\ -2 & 0 \end{Vmatrix} \lambda^2 + \begin{Vmatrix} -3 & -13 \\ -1 & -11 \end{Vmatrix} \lambda + \begin{Vmatrix} 0 & 0 \\ 1 & 0 \end{Vmatrix},$$

$$A_0^{(1)} B_0^{-1} = \begin{Vmatrix} 0 & 1 \\ -2 & 0 \end{Vmatrix} \cdot \begin{Vmatrix} 1 & 1 \\ 1 & 2 \end{Vmatrix} = \begin{Vmatrix} 1 & 2 \\ -2 & -2 \end{Vmatrix},$$

$$A_0^{(1)} B_0^{-1} B(\lambda) = \begin{Vmatrix} 1 & 2 \\ -2 & -2 \end{Vmatrix} \cdot \begin{Vmatrix} 2\lambda^2 + 3 & -\lambda^2 + 1 \\ -\lambda^2 - 1 & \lambda^2 + 2 \end{Vmatrix} = \begin{Vmatrix} 1 & \lambda^2 + 5 \\ -2\lambda^2 - 4 & -6 \end{Vmatrix},$$

$$R(\lambda) = A^{(1)}(\lambda) - A_0^{(1)} B_0^{-1} B(\lambda)$$

$$= \begin{Vmatrix} -3\lambda & \lambda^2 - 13\lambda \\ -2\lambda^2 - \lambda + 1 & -11\lambda \end{Vmatrix} - \begin{Vmatrix} 1 & \lambda^2 + 5 \\ -2\lambda^2 - 4 & -6 \end{Vmatrix} = \begin{Vmatrix} -3\lambda - 1 & -13\lambda - 5 \\ -\lambda + 5 & -11\lambda + 6 \end{Vmatrix},$$

$$Q(\lambda) = A_0 B_0^{-1} \lambda + A_0^{(1)} B_0^{-1} = \begin{Vmatrix} 3 & 5 \\ 2 & 5 \end{Vmatrix} \lambda + \begin{Vmatrix} 1 & 2 \\ -2 & -2 \end{Vmatrix} = \begin{Vmatrix} 3\lambda + 1 & 5\lambda + 2 \\ 2\lambda - 2 & 5\lambda - 2 \end{Vmatrix}.$$

As an exercise, the reader should verify that

$$A(\lambda) = Q(\lambda) B(\lambda) + R(\lambda).$$

§ 3. The Generalized Bézout Theorem

1. We consider an arbitrary matrix polynomial of order n

$$F(\lambda) = F_0 \lambda^m + F_1 \lambda^{m-1} + \cdots + F_m \qquad (F_0 \neq O). \tag{14}$$

This polynomial can also be written as follows:

$$F(\lambda) = \lambda^m F_0 + \lambda^{m-1} F_1 + \cdots + F_m. \tag{15}$$

For a scalar λ, both ways of writing give the same result. However, if we substitute for the scalar argument λ a square matrix A of order n, then the results of the substitution in (14) and (15) will, in general, be distinct, since the powers of A need not be permutable with the matrix coefficients $F_0, F_1, \ldots, F_m$.

We set

$$F(A) = F_0 A^m + F_1 A^{m-1} + \cdots + F_m \tag{16}$$

and

$$\widehat{F}(A) = A^m F_0 + A^{m-1} F_1 + \cdots + F_m, \tag{17}$$

and call $F(A)$ the *right value* and $\widehat{F}(A)$ the *left value* of $F(\lambda)$ on substitution of A for λ.[2]

We divide $F(\lambda)$ by the binomial $\lambda E - A$. In this case the right remainder $R(\lambda)$ and left remainder $\widehat{R}(\lambda)$ will not depend on λ. To determine the right remainder we use the usual division scheme:

$$\begin{aligned}
F(\lambda) &= F_0\lambda^m + F_1\lambda^{m-1} + \cdots + F_m \\
&= F_0\lambda^{m-1}(\lambda E - A) + (F_0 A + F_1)\lambda^{m-1} + F_2\lambda^{m-2} + \cdots \\
&= [F_0\lambda^{m-1} + (F_0 A + F_1)\lambda^{m-2}](\lambda E - A) + (F_0 A^2 + F_1 A + F_2)\lambda^{m-2} + F_3\lambda^{m-3} + \cdots \\
&= [F_0\lambda^{m-1} + (F_0 A + F_1)\lambda^{m-2} + \cdots \\
&\qquad + F_0 A^{m-1} + F_1 A^{m-2} + \cdots + F_{m-1}](\lambda E - A) \\
&\qquad + F_0 A^m + F_1 A^{m-1} + \cdots + F_m.
\end{aligned}$$

Thus we have found that

$$R = F_0 A^m + F_1 A^{m-1} + \cdots + F_m = F(A). \tag{18}$$

Similarly

$$\widehat{R} = \widehat{F}(A). \tag{19}$$

This proves

THEOREM 1 (The Generalized Bézout Theorem): *When the matrix polynomial $F(\lambda)$ is divided on the right by the binomial $\lambda E - A$, the remainder is $F(A)$; when it is divided on the left, the remainder is $\widehat{F}(A)$.*

---

[2] In the 'right' value $F(A)$ the powers of A are at the right of the coefficients; in the 'left' value $\widehat{F}(A)$, at the left.

2. From this theorem it follows that:

A polynomial $F(\lambda)$ is divisible by the binomial $\lambda E - A$ on the right (left) without remainder if and only if $F(A) = O$ ($\hat{F}(A) = O$).

Example. Let $A = \| a_{ik} \|_1^n$ and let $f(\lambda)$ be a polynomial in λ. Then

$$F(\lambda) = f(\lambda)E - f(A)$$

is divisible by $\lambda E - A$ (both on the right and on the left) without remainder. This follows immediately from the generalized Bézout Theorem, because in this case $F(A) = \hat{F}(A) = O$.

§ 4. The Characteristic Polynomial of a Matrix. The Adjoint Matrix

1. We consider a matrix $A = \| a_{ik} \|_1^n$. The *characteristic matrix* of A is $\lambda E - A$. The determinant of the characteristic matrix

$$\Delta(\lambda) = | \lambda E - A | = | \lambda \delta_{ik} - a_{ik} |_1^n ,$$

is a scalar polynomial in λ and is called the *characteristic polynomial* of A (see Chapter III, § 7).[3]

The matrix $B(\lambda) = \| b_{ik}(\lambda) \|_1^n$, where $b_{ik}(\lambda)$ is the algebraic complement of the element $\lambda \delta_{ik} - a_{ik}$ in the determinant $\Delta(\lambda)$ is called the *adjoint matrix* of A.

By way of example, for the matrix

$$A = \begin{Vmatrix} a_{11} & a_{12} & a_{13} \\ a_{21} & a_{22} & a_{23} \\ a_{31} & a_{32} & a_{33} \end{Vmatrix}$$

we have:

$$\lambda E - A = \begin{Vmatrix} \lambda - a_{11} & -a_{12} & -a_{13} \\ -a_{21} & \lambda - a_{22} & -a_{23} \\ -a_{31} & -a_{32} & \lambda - a_{33} \end{Vmatrix},$$

$$\Delta(\lambda) = | \lambda E - A | = \lambda^3 - (a_{11} + a_{22} + a_{33})\lambda^2 + \dots,$$

$$B(\lambda) = \begin{Vmatrix} \lambda^2 - (a_{22} + a_{33})\lambda + a_{22}a_{33} - a_{23}a_{32} & * & * \\ a_{21}\lambda + a_{23}a_{31} - a_{21}a_{33} & * & * \\ a_{31}\lambda + a_{21}a_{32} - a_{22}a_{31} & * & * \end{Vmatrix}.$$

[3] This polynomial differs by the factor $(-1)^n$ from the polynomial $\Delta(\lambda)$ introduced in Chapter III, § 7.

These definitions imply the following identities in λ:

$$(\lambda E - A)\,B(\lambda) = \Delta(\lambda)\,E\,, \tag{20}$$

$$B(\lambda)\,(\lambda E - A) = \Delta(\lambda)\,E\,. \tag{20'}$$

The right-hand sides of these equations can be regarded as polynomials with matrix coefficients (each of these coefficients is the product of a scalar and the unit matrix E). The polynomial matrix $B(\lambda)$ can also be represented in the form of a polynomial arranged with respect to the powers of λ. Equations (20) and (20′) show that $\Delta(\lambda)E$ is divisible on the right and on the left by $\lambda E - A$ without remainder. By the Generalized Bézout Theorem, this is only possible when the remainder $\Delta(A)E = \Delta(A)$ is the null matrix. Thus we have proved:

THEOREM 2 (Hamilton-Cayley): *Every square matrix A satisfies its characteristic equation,* i.e.

$$\Delta(A) = O\,. \tag{21}$$

Example.

$$A = \begin{Vmatrix} 2 & 1 \\ -1 & 3 \end{Vmatrix},$$

$$\Delta(\lambda) = \begin{vmatrix} \lambda - 2 & -1 \\ 1 & \lambda - 3 \end{vmatrix} = \lambda^2 - 5\lambda + 7\,,$$

$$\Delta(A) = A^2 - 5A + 7E = \begin{Vmatrix} 3 & 5 \\ -5 & 8 \end{Vmatrix} - 5\begin{Vmatrix} 2 & 1 \\ -1 & 3 \end{Vmatrix} + 7\begin{Vmatrix} 1 & 0 \\ 0 & 1 \end{Vmatrix} = \begin{Vmatrix} 0 & 0 \\ 0 & 0 \end{Vmatrix} = O\,.$$

2. We denote by $\lambda_1, \lambda_2, \ldots, \lambda_n$ all the characteristic values of A, i.e., all the roots of the characteristic polynomial $\Delta(\lambda)$ (each λ_i is repeated as often as its multiplicity as a root of $\Delta(\lambda)$ requires). Then

$$\Delta(\lambda) = |\lambda E - A| = (\lambda - \lambda_1)(\lambda - \lambda_2)\cdots(\lambda - \lambda_n)\,. \tag{22}$$

Let $g(\mu)$ be an arbitrary scalar polynomial. We wish to find the characteristic values of $g(A)$. For this purpose we split $g(\mu)$ into linear factors

$$g(\mu) = a_0(\mu - \mu_1)(\mu - \mu_2)\cdots(\mu - \mu_l)\,. \tag{23}$$

On both sides of this identity we substitute the matrix A for μ:

$$g(A) = a_0(A - \mu_1 E)(A - \mu_2 E)\cdots(A - \mu_l E)\,. \tag{24}$$

Passing to determinants on both sides of (24) and using (22) and (23) we find

$$\begin{aligned}|g(A)| &= a_0^n |A-\mu_1 E|\,|A-\mu_2 E| \cdots |A-\mu_l E| \\ &= (-1)^{nl} a_0^n \Delta(\mu_1)\Delta(\mu_2)\cdots\Delta(\mu_l) \\ &= (-1)^{nl} a_0^n \prod_{i=1}^{l}\prod_{k=1}^{n}(\mu_i-\lambda_k) = g(\lambda_1)g(\lambda_2)\cdots g(\lambda_n).\end{aligned}$$

If in the equation

$$|g(A)| = g(\lambda_1)g(\lambda_2)\cdots g(\lambda_n) \tag{25}$$

we replace the polynomial $g(\mu)$ by $\lambda - g(\mu)$, where λ is some parameter, we find:

$$|\lambda E - g(A)| = [\lambda - g(\lambda_1)]\,[\lambda - g(\lambda_2)] \cdots [\lambda - g(\lambda_n)]. \tag{26}$$

This leads to the following theorem.

Theorem 3: *If $\lambda_1, \lambda_2, \ldots, \lambda_n$ are all the characteristic values (with the proper multiplicities) of a matrix A and if $g(\mu)$ is a scalar polynomial, then $g(\lambda_1), g(\lambda_2), \ldots, g(\lambda_n)$ are the characteristic values of $g(A)$.*

In particular, if A has the characteristic values $\lambda_1, \lambda_2, \ldots, \lambda_n$, then A^k has the characteristic values $\lambda_1^k, \lambda_2^k, \ldots, \lambda_n^k$ $(k = 0, 1, 2, \ldots)$.

3. We shall now derive an effective formula expressing the adjoint matrix $B(\lambda)$ in terms of the characteristic polynomial $\Delta(\lambda)$.

Let

$$\Delta(\lambda) = \lambda^n - p_1\lambda^{n-1} - p_2\lambda^{n-2} - \cdots - p_n. \tag{27}$$

The difference $\Delta(\lambda) - \Delta(\mu)$ is divisible by $\lambda - \mu$ without remainder Therefore

$$\delta(\lambda, \mu) = \frac{\Delta(\lambda) - \Delta(\mu)}{\lambda - \mu} = \lambda^{n-1} + (\mu - p_1)\lambda^{n-2} + (\mu^2 - p_1\mu - p_2)\lambda^{n-3} + \cdots \tag{28}$$

is a polynomial in λ and μ.

The identity

$$\Delta(\lambda) - \Delta(\mu) = \delta(\lambda, \mu)(\lambda - \mu) \tag{29}$$

will still hold if we replace λ and μ by the permutable matrices λE and A. Since by the Hamilton-Cayley Theorem $\Delta(A) = O$,

$$\Delta(\lambda)E = \delta(\lambda E, A)(\lambda E - A). \tag{30}$$

Comparing (20′) with (30), we obtain by virtue of the uniqueness of the quotient the required formula

$$B(\lambda) = \delta(\lambda E, A). \tag{31}$$

Hence by (28)

$$B(\lambda)=E\lambda^{n-1}+B_1\lambda^{n-2}+B_2\lambda^{n-3}+\cdots+B_{n-1}, \tag{32}$$

where

$$B_1=A-p_1E, \quad B_2=A^2-p_1A-p_2E, \quad \ldots$$

and, in general,

$$B_k=A^k-p_1A^{k-1}-p_2A^{k-2}-\cdots-p_kE \qquad (k=1,2,\ldots,n-1). \tag{33}$$

The matrices $B_1, B_2, \ldots, B_{n-1}$ can be computed in succession, starting from the recurrence relation

$$B_k=AB_{k-1}-p_kE \qquad (k=1,2,\ldots,n-1;\ B_0=E). \tag{34}$$

Moreover,[4]

$$AB_{n-1}-p_nE=O. \tag{35}$$

The relations (34) and (35) follow immediately from (20) if we equate the coefficients of equal powers of λ on both sides.

If A is non-singular, then

$$p_n=(-1)^{n-1}|A|\neq 0,$$

and it follows from (35) that

$$A^{-1}=\frac{1}{p_n}B_{n-1}. \tag{36}$$

Let λ_0 be a characteristic value of A, so that $\Delta(\lambda_0)=0$. Substituting the value λ_0 in (20), we find:

$$(\lambda_0E-A)\,B(\lambda_0)=O. \tag{37}$$

Let us assume that $B(\lambda_0)\neq O$ and denote by b an arbitrary non-zero column of this matrix. Then from (37) we have $(\lambda_0E-A)b=O$ or

$$Ab=\lambda_0b. \tag{38}$$

Therefore every non-zero column of $B(\lambda_0)$ determines a characteristic vector corresponding to the characteristic value λ_0.[5]

Thus:

---

[4] From (34) follows (33). If we substitute in (35) the expression for B_{n-1} given in (33), we obtain $\Delta(A)=0$. This approach to the Hamilton-Cayley Theorem does not require the Generalized Bézout Theorem explicitly, but contains this theorem implicitly.

[5] See Chapter III, § 7. If to the characteristic value λ_0 there correspond d_0 linearly independent characteristic vectors ($n-d_0$ is the rank of λ_0E-A), then the rank of $B(\lambda_0)$ does not exceed d_0. In particular, if only one characteristic direction corresponds to λ_0, then in $B(\lambda_0)$ the elements of any two columns are proportional.

If the coefficients of the characteristic polynomial are known, then the adjoint matrix can be found by formula (31). *If the given matrix A is non-singular, then the inverse matrix A^{-1} can be found by formula* (36). *If λ_0 is a characteristic value of A, then the non-zero columns of $B(\lambda_0)$ are characteristic vectors of A for $\lambda = \lambda_0$.*

Example.

$$A = \left\| \begin{matrix} 2 & -1 & 1 \\ 0 & 1 & 1 \\ -1 & 1 & 1 \end{matrix} \right\|,$$

$$\Delta(\lambda) = |\lambda E - A| = \begin{vmatrix} \lambda - 2 & 1 & -1 \\ 0 & \lambda - 1 & -1 \\ 1 & -1 & \lambda - 1 \end{vmatrix} = \lambda^3 - 4\lambda^2 + 5\lambda - 2,$$

$$\delta(\lambda, \mu) = \frac{\Delta(\lambda) - \Delta(\mu)}{\lambda - \mu} = \lambda^2 + \lambda(\mu - 4) + \mu^2 - 4\mu + 5,$$

$$B(\lambda) = \delta(\lambda E, A) = \lambda^2 E + \lambda \underbrace{(A - 4E)}_{B_1} + \underbrace{A^2 - 4A + 5E}_{B_2}.$$

But

$$B_1 = A - 4E = \left\| \begin{matrix} -2 & -1 & 1 \\ 0 & -3 & 1 \\ -1 & 1 & -3 \end{matrix} \right\|, \quad B_2 = AB_1 + 5E = \left\| \begin{matrix} 0 & 2 & -2 \\ -1 & 3 & -2 \\ 1 & -1 & 2 \end{matrix} \right\|,$$

$$B(\lambda) = \left\| \begin{matrix} \lambda^2 - 2\lambda & -\lambda + 2 & \lambda - 2 \\ -1 & \lambda^2 - 3\lambda + 3 & \lambda - 2 \\ -\lambda + 1 & \lambda - 1 & \lambda^2 - 3\lambda + 2 \end{matrix} \right\|,$$

$$|A| = 2, \quad A^{-1} = \frac{1}{2} B_2 = \left\| \begin{matrix} 0 & 1 & -1 \\ -\frac{1}{2} & \frac{3}{2} & -1 \\ \frac{1}{2} & -\frac{1}{2} & 1 \end{matrix} \right\|.$$

Furthermore,

$$\Delta(\lambda) = (\lambda - 1)^2 (\lambda - 2).$$

The first column of the matrix $B(+1)$ gives the characteristic vector $(+1, +1, 0)$ for the characteristic value $\lambda = 1$.

The first column of the matrix $B(+2)$ gives the characteristic vector $(0, +1, +1)$ corresponding to the characteristic value $\lambda = 2$.

§ 5. The Method of Faddeev for the Simultaneous Computation of the Coefficients of the Characteristic Polynomial and of the Adjoint Matrix

1. D. K. Faddeev[6] has suggested a method for the simultaneous determination of the scalar coefficients $p_1, p_2, \ldots, p_n$ of the characteristic polynomial

$$\Delta(\lambda) = \lambda^n - p_1\lambda^{n-1} - p_2\lambda^{n-2} - \cdots - p_n \tag{39}$$

and of the matrix coefficients $B_1, B_2, \ldots, B_{n-1}$ of the adjoint matrix $B(\lambda)$.

In order to explain the method of Faddeev[7] we introduce the concept of the trace (or spur) of a matrix.

By the *trace* $\operatorname{tr} A$ of a matrix $A = \| a_{ik} \|_1^n$ we mean the sum of the diagonal elements of the matrix:

$$\operatorname{tr} A = \sum_{i=1}^{n} a_{ii}. \tag{40}$$

It is easy to see that

$$\operatorname{tr} A = p_1 = \sum_{i=1}^{n} \lambda_i, \tag{41}$$

if $\lambda_1, \lambda_2, \ldots, \lambda_n$ are the characteristic values of A, i.e., if

$$\Delta(\lambda) = (\lambda - \lambda_1)(\lambda - \lambda_2) \cdots (\lambda - \lambda_n). \tag{42}$$

Since by Theorem 3 A^k has the characteristic values $\lambda_1^k, \lambda_2^k, \ldots, \lambda_n^k$ $(k = 0, 1, 2, \ldots,)$, we have

$$\operatorname{tr} A^k = s_k = \sum_{i=1}^{n} \lambda_i^k \qquad (k = 0, 1, 2, \ldots). \tag{43}$$

The sums s_k $(k = 1, 2, \ldots, n)$ of powers of the roots of the polynomial (39) are connected with the coefficients by Newton's formulas[8]

$$kp_k = s_k - p_1 s_{k-1} - \cdots - p_{k-1}s_1 \qquad (k = 1, 2, \ldots, n). \tag{44}$$

If the traces $s_1, s_2, \ldots, s_n$ of the matrices $A, A^2, \ldots, A^n$ are computed, then the coefficients $p_1, p_2, \ldots, p_n$ can be determined from (44). This is the method of Leverrier for the determination of the coefficients of the characteristic polynomial from the traces of the powers of the matrix.

2. Faddeev has proposed to compute successively, instead of the traces of the powers $A, A^2, \ldots, A^n$, the traces of certain other matrices $A_1, A_2, \ldots, A_n$

---

[6] See [14], p. 160.

[7] In Chapter VII, § 8, we shall discuss another effective method, due to A. N. Krylov, of computing the coefficients of the characteristic polynomial.

[8] See, for example, G. Chrystal, *Textbook of Algebra*, Vol. I, pp. 436ff.

and so to determine $p_1, p_2, \ldots, p_n$ and $B_1, B_2, \ldots, B_n$ by the following formulas:

$$\left.\begin{array}{lll} A_1 = A, & p_1 = \operatorname{tr} A_1, & B_1 = A_1 - p_1 E, \\ A_2 = AB_1, & p_2 = \frac{1}{2} \operatorname{tr} A_2, & B_2 = A_2 - p_2 E, \\ \cdots & \cdots & \cdots \\ A_{n-1} = AB_{n-2}, & p_{n-1} = \frac{1}{n-1} \operatorname{tr} A_{n-1}, & B_{n-1} = A_{n-1} - p_{n-1} E, \\ A_n = AB_{n-1}, & p_n = \frac{1}{n} \operatorname{tr} A_n, & B_n = A_n - p_n E = O. \end{array}\right\} \quad (45)$$

The last equation $B_n = A_n - p_n E = O$ may be used to check the computation.

In order to convince ourselves that the numbers $p_1, p_2, \ldots, p_n$ and the matrices $B_1, B_2, \ldots, B_{n-1}$ that are determined successively by (45) are, in fact, the coefficients of $\Delta(\lambda)$ and $B(\lambda)$, we note that the following formulas for A_k and B_k $(k = 1, 2, \ldots, n)$ follow from (45):

$$A_k = A^k - p_1 A^{k-1} - \cdots - p_{k-1} A, \quad B_k = A^k - p_1 A^{k-1} - \cdots - p_{k-1} A - p_k E. \quad (46)$$

Equating the traces on the left-hand and right-hand sides of the first of these formulas, we obtain

$$kp_k = s_k - p_1 s_{k-1} - \cdots - p_{k-1} s_1 .$$

But these formulas coincide with Newton's formulas (44) by which the coefficients of the characteristic polynomial $\Delta(\lambda)$ are determined successively. Therefore the numbers $p_1, p_2, \ldots, p_n$ determined by (45) are also the coefficients of $\Delta(\lambda)$. But then the second of formulas (46) coincide with formulas (33) by which the matrix coefficients $B_1, B_2, \ldots, B_{n-1}$ of the adjoint matrix $B(\lambda)$ are determined. Therefore, formulas (45) also determine the coefficients $B_1, B_2, \ldots, B_{n-1}$ of the matrix polynomial $B(\lambda)$.

Example.[9]

$$A = \left\| \begin{array}{rrrr} 2 & -1 & 1 & 2 \\ 0 & 1 & 1 & 0 \\ -1 & 1 & 1 & 1 \\ 1 & 1 & 1 & 0 \end{array} \right\| , \quad p_1 = \operatorname{tr} A = 4, \quad B_1 = A - 4E = \left\| \begin{array}{rrrr} -2 & -1 & 1 & 2 \\ 0 & -3 & 1 & 0 \\ -1 & 1 & -3 & 1 \\ 1 & 1 & 1 & -4 \end{array} \right\| ;$$
$$\begin{array}{rrrr} \mathit{2} & \mathit{2} & \mathit{4} & \mathit{3} \end{array}$$

[9] As a check on the computation, we write under each matrix A_1, A_2, A_3 a row whose elements are the sums of the elements above it. The product of this row of 'column-sums' of the first factor into the columns of the second factor must give the elements of the column-sum of the product.

$$A_2 = AB_1 = \left\|\begin{array}{rrrr} -3 & 4 & 0 & -3 \\ -1 & -2 & -2 & 1 \\ 2 & 0 & -2 & -5 \\ -3 & -3 & -1 & 3 \end{array}\right\|, \; p_2 = \frac{1}{2}\operatorname{tr} A_2 = -2, \; B_2 = A_2 + 2E = \left\|\begin{array}{rrrr} -1 & 4 & 0 & -3 \\ -1 & 0 & -2 & 1 \\ 2 & 0 & 0 & -5 \\ -3 & -3 & -1 & 3 \end{array}\right\|;$$

$$\begin{array}{rrrr} \mathit{-5} & \mathit{-1} & \mathit{-5} & \mathit{-4} \end{array}$$

$$A_3 = AB_2 = \left\|\begin{array}{rrrr} -5 & 2 & 0 & -2 \\ 1 & 0 & -2 & -4 \\ -1 & -7 & -3 & 4 \\ 0 & 4 & -2 & -7 \end{array}\right\|, \; p_3 = \frac{1}{3}\operatorname{tr} A_3 = -5, \; B_3 = A_3 + 5E = \left\|\begin{array}{rrrr} 0 & 2 & 0 & -2 \\ 1 & 5 & -2 & 4 \\ -1 & -7 & 2 & 4 \\ 0 & 4 & -2 & -2 \end{array}\right\|;$$

$$\begin{array}{rrrr} \mathit{-5} & \mathit{-1} & \mathit{-7} & \mathit{-9} \end{array}$$

$$A_4 = AB_3 = \left\|\begin{array}{rrrr} -2 & 0 & 0 & 0 \\ 0 & -2 & 0 & 0 \\ 0 & 0 & -2 & 0 \\ 0 & 0 & 0 & -2 \end{array}\right\|, \quad p_4 = -2.$$

$$\Delta(\lambda) = \lambda^4 - 4\lambda^3 + 2\lambda^2 + 5\lambda + 2,$$

$$|A| = 2, \quad A^{-1} = \frac{1}{p_4} B_3 = \left\|\begin{array}{cccc} 0 & -1 & 0 & 1 \\ -\frac{1}{2} & -\frac{5}{2} & 1 & -2 \\ \frac{1}{2} & \frac{7}{2} & -1 & -2 \\ 0 & -2 & 1 & 1 \end{array}\right\|.$$

Note. If we wish to determine p_1, p_2, p_3, p_4 and only the first columns of B_1, B_2, B_3, it is sufficient to compute in A_2 the elements of the first column and only the diagonal elements of the remaining columns, in A_3 only the elements of the first column, and in A_4 only the first two elements of the first column.

§ 6. The Minimal Polynomial of a Matrix

1. DEFINITION 1: *A scalar polynomial $f(\lambda)$ is called an annihilating polynomial of the square matrix A if*

$$f(A) = O.$$

An annihilating polynomial $\psi(\lambda)$ of least degree with highest coefficient 1 is called a *minimal polynomial* of A.

By the Hamilton-Cayley Theorem the characteristic polynomial $\Delta(\lambda)$ is an annihilating polynomial of A. However, as we shall show below, it is not, in general, a minimal polynomial.

Let us divide an arbitrary annihilating polynomial $f(\lambda)$ by a minimal polynomial

$$f(\lambda) = \psi(\lambda)\, q(\lambda) + r(\lambda),$$

where the degree of $r(\lambda)$ is less than that of $\psi(\lambda)$. Hence we have:

$$f(A) = \psi(A)\,q(A) + r(A).$$

Since $f(A) = O$ and $\psi(A) = O$, it follows that $r(A) = O$. But the degree of $r(\lambda)$ is less than that of the minimal polynomial $\psi(\lambda)$. Therefore $r(\lambda) \equiv 0$.[10] Hence: *Every annihilating polynomial of a matrix is divisible without remainder by the minimal polynomial.*

Let $\psi_1(\lambda)$ and $\psi_2(\lambda)$ be two minimal polynomials of one and the same matrix. Then each is divisible without remainder by the other, i.e., the polynomials differ by a constant factor. This constant factor must be **1**, because the highest coefficients in $\psi_1(\lambda)$ and $\psi_2(\lambda)$ are 1. Thus we have proved the *uniqueness of the minimal polynomial* of a given matrix A.

2. We shall now derive a formula connecting the minimal polynomial with the characteristic polynomial.

We denote by $D_{n-1}(\lambda)$ the greatest common divisor of all the minors of order $n-1$ of the characteristic matrix $\lambda E - A$, i.e., of all the elements of the matrix $B(\lambda) = \| b_{ik}(\lambda) \|_1^n$ (see the preceding section). Then

$$B(\lambda) = D_{n-1}(\lambda)\,C(\lambda), \tag{47}$$

where $C(\lambda)$ is a certain polynomial matrix, the *'reduced' adjoint matrix* of $\lambda E - A$. From (20) and (47) we have:

$$\Delta(\lambda)\,E = (\lambda E - A)\,C(\lambda)\,D_{n-1}(\lambda). \tag{48}$$

Hence it follows that $\Delta(\lambda)$ is divisible without remainder by $D_{n-1}(\lambda)$:[11]

$$\frac{\Delta(\lambda)}{D_{n-1}(\lambda)} = \psi(\lambda), \tag{49}$$

where $\psi(\lambda)$ is some polynomial. The factor $D_{n-1}(\lambda)$ in (48) may be cancelled on both sides:[12]

$$\psi(\lambda)\,E = (\lambda E - A)\,C(\lambda). \tag{50}$$

---

[10] Otherwise there would exist an annihilating polynomial of degree less than that of the minimal polynomial.

[11] We could also verify this immediately by expanding the characteristic determinant $\Delta(\lambda)$ with respect to the elements of an arbitrary row.

[12] In this case we have, apart from (50), also the identity (see (20′))

$$\psi(\lambda)\,E = C(\lambda)\,(\lambda E - A),$$

i.e., $C(\lambda)$ is at one and the same time the left quotient and right quotient of $\psi(\lambda)E$ on division by $\lambda E - A$.

Since $\psi(\lambda)E$ is divisible on the left without remainder by $\lambda E - A$, it follows by the Generalized Bézout Theorem that

$$\psi(A) = O.$$

Thus, the polynomial $\psi(\lambda)$ defined by (49) is an annihilating polynomial of A. Let us show that it is the minimal polynomial.

We denote the minimal polynomial by $\psi^*(\lambda)$. Then $\psi(\lambda)$ is divisible by $\psi^*(\lambda)$ without remainder:

$$\psi(\lambda) = \psi^*(\lambda)\chi(\lambda). \tag{51}$$

Since $\psi^*(A) = O$, by the Generalized Bézout Theorem the matrix polynomial $\psi^*(\lambda)E$ is divisible on the left by $\lambda E - A$ without remainder:

$$\psi^*(\lambda)E = (\lambda E - A)C^*(\lambda). \tag{52}$$

From (51) and (52) it follows that

$$\psi(\lambda)E = (\lambda E - A)C^*(\lambda)\chi(\lambda). \tag{53}$$

The identities (50) and (53) show that $C(\lambda)$ as well as $C^*(\lambda)\chi(\lambda)$ are left quotients of $\psi(\lambda)E$ on division by $\lambda E - A$. By the uniqueness of division

$$C(\lambda) = C^*(\lambda)\chi(\lambda).$$

Hence it follows that $\chi(\lambda)$ is a common divisor of all the elements of the polynomial matrix $C(\lambda)$. But, on the other hand, the greatest common divisor of all the elements of the reduced adjoint matrix $C(\lambda)$ is equal to 1, because the matrix was obtained from $B(\lambda)$ by division by $D_{n-1}(\lambda)$. Therefore $\chi(\lambda) = \text{const}$. Since the highest coefficients of $\psi(\lambda)$ and $\psi^*(\lambda)$ are equal, we have in (51) $\chi(\lambda) = 1$, i.e., $\psi(\lambda) = \psi^*(\lambda)$, and this is what we had to prove.

We have established the following formula for the minimal polynomial:

$$\psi(\lambda) = \frac{\Delta(\lambda)}{D_{n-1}(\lambda)}. \tag{54}$$

3. For the reduced adjoint matrix $C(\lambda)$ we have a formula analogous to (31) (p. 84):

$$C(\lambda) = \Psi(\lambda E, A); \tag{55}$$

where the polynomial $\Psi(\lambda, \mu)$ is defined by the equation[13]

[13] Formula (55) can be deduced in the same way as (31). On both sides of the identity $\psi(\lambda) - \psi(\mu) = (\lambda - \mu)\Psi(\lambda, \mu)$ we substitute for λ and μ the matrices λE and A and compare the matrix equation so obtained with (50).

$$\Psi(\lambda, \mu) = \frac{\psi(\lambda) - \psi(\mu)}{\lambda - \mu}. \tag{56}$$

Moreover,

$$(\lambda E - A)\, C(\lambda) = \psi(\lambda)\, E. \tag{57}$$

Going over to determinants on both sides of (57), we obtain

$$\Delta(\lambda)\, |C(\lambda)| = [\psi(\lambda)]^n. \tag{58}$$

Thus, $\Delta(\lambda)$ is divisible without remainder by $\psi(\lambda)$ and some power of $\psi(\lambda)$ is divisible without remainder by $\Delta(\lambda)$, i.e., the sets of all the distinct roots of the polynomials $\Delta(\lambda)$ and $\psi(\lambda)$ are equal. In other words: *All the distinct characteristic values of A are roots of $\psi(\lambda)$.*

If

$$\begin{gathered} \Delta(\lambda) = (\lambda - \lambda_1)^{n_1} (\lambda - \lambda_2)^{n_2} \cdots (\lambda - \lambda_s)^{n_s} \\ (\lambda_i \neq \lambda_j \text{ for } i \neq j;\ n_i > 0,\ i, j = 1, 2, \ldots, s), \end{gathered} \tag{59}$$

then

$$\psi(\lambda) = (\lambda - \lambda_1)^{m_1} (\lambda - \lambda_2)^{m_2} \cdots (\lambda - \lambda_s)^{m_s}, \tag{60}$$

where

$$0 < m_k \leq n_k \qquad (k = 1, 2, \ldots, s). \tag{61}$$

4. We mention one further property of the matrix $C(\lambda)$. Let λ_0 be an arbitrary characteristic value of $A = \| a_{ik} \|_1^n$. Then $\psi(\lambda_0) = 0$ and therefore, by (57),

$$(\lambda_0 E - A)\, C(\lambda_0) = O. \tag{62}$$

Note that $C(\lambda_0) \neq O$ always holds, for otherwise all the elements of the reduced adjoint matrix $C(\lambda)$ would be divisible without remainder by $\lambda - \lambda_0$, and this is impossible.

We denote by c an arbitrary non-zero column of $C(\lambda_0)$. Then from (62)

$$(\lambda_0 E - A)\, c = o,$$

i.e.,

$$Ac = \lambda_0 c. \tag{63}$$

In other words, every non-zero column of $C(\lambda_0)$ (and such a column always exists) determines a characteristic vector for $\lambda = \lambda_0$.

Example.

$$A = \left\| \begin{matrix} 3 & -3 & 2 \\ -1 & 5 & -2 \\ -1 & 3 & 0 \end{matrix} \right\|,$$

$$\Delta(\lambda) = \begin{vmatrix} \lambda - 3 & 3 & -2 \\ 1 & \lambda - 5 & 2 \\ 1 & -3 & \lambda \end{vmatrix} = \lambda^3 - 8\lambda^2 + 20\lambda - 16 = (\lambda - 2)^2 (\lambda - 4),$$

$$\delta(\lambda,\mu) = \frac{\Delta(\mu)-\Delta(\lambda)}{\mu-\lambda} = \mu^2 + \mu(\lambda-8) + \lambda^2 - 8\lambda + 20,$$

$$B(\lambda) = A^2 + (\lambda-8)A + (\lambda^2-8\lambda+20)E$$

$$= \begin{Vmatrix} 10 & -18 & 12 \\ -6 & 22 & -12 \\ -6 & 18 & -8 \end{Vmatrix} + (\lambda-8)\begin{Vmatrix} 3 & -3 & 2 \\ -1 & 5 & -2 \\ -1 & 3 & 0 \end{Vmatrix} + (\lambda^2-8\lambda+20)\begin{Vmatrix} 1 & 0 & 0 \\ 0 & 1 & 0 \\ 0 & 0 & 1 \end{Vmatrix}$$

$$= \begin{Vmatrix} \lambda^2-5\lambda+6 & -3\lambda+6 & 2\lambda-4 \\ -\lambda+2 & \lambda^2-3\lambda+2 & -2\lambda+4 \\ -\lambda+2 & 3\lambda-6 & \lambda^2-8\lambda+2 \end{Vmatrix}.$$

All the elements of the matrix $B(\lambda)$ are divisible by $D_2(\lambda) = \lambda - 2$. Cancelling this factor, we have:

$$C(\lambda) = \begin{Vmatrix} \lambda-3 & -3 & 2 \\ -1 & \lambda-1 & -2 \\ -1 & 3 & \lambda-6 \end{Vmatrix}$$

and

$$\psi(\lambda) = \frac{\Delta(\lambda)}{\lambda-2} = (\lambda-2)(\lambda-4).$$

In $C(\lambda)$ we substitute for λ the value $\lambda_0 = 2$:

$$C(2) = \begin{Vmatrix} -1 & -3 & 2 \\ -1 & 1 & -2 \\ -1 & 3 & -4 \end{Vmatrix}.$$

The first column gives us the characteristic vector $(1, 1, 1,)$ for $\lambda_0 = 2$. The second column gives us the characteristic vector $(-3, 1, 3)$ for the same characteristic value $\lambda_0 = 2$. The third column is a linear combination of the first two.

Similarly, setting $\lambda_0 = 4$, we find from the first column of the matrix $C(4)$ the characteristic vector $(1, -1, -1)$ corresponding to the characteristic value $\lambda_0 = 4$.

The reader should note that $\psi(\lambda)$ and $C(\lambda)$ could have been determined by a different method.

To begin with, let us find $D_2(\lambda)$. $D_2(\lambda)$ can only have 2 and 4 as its roots. For $\lambda = 4$ the second order minor

$$\begin{vmatrix} 1 & \lambda-5 \\ 1 & -3 \end{vmatrix} = -\lambda + 2$$

of $\Delta(\lambda)$ does not vanish. Therefore $D_2(4) \neq 0$. For $\lambda = 2$ the columns of $\Delta(\lambda)$ become proportional. Therefore all the minors of order two in $\Delta(\lambda)$

vanish for $\lambda = 2 : D_2(2) = 0$. Since the minor to be computed is of the first degree, $D_2(\lambda)$ cannot be divisible by $(\lambda - 2)^2$. Therefore

$$D_2(\lambda) = \lambda - 2.$$

Hence

$$\psi(\lambda) = \frac{\Delta(\lambda)}{\lambda - 2} = (\lambda - 2)(\lambda - 4) = \lambda^2 - 6\lambda + 8,$$

$$\Psi(\lambda, \mu) = \frac{\psi(\mu) - \psi(\lambda)}{\mu - \lambda} = \mu + \lambda - 6,$$

$$C(\lambda) = \Psi(\lambda E, A) = A + (\lambda - 6)E = \left\| \begin{matrix} \lambda - 3 & -3 & 2 \\ -1 & \lambda - 1 & -2 \\ -1 & 3 & \lambda - 6 \end{matrix} \right\|.$$

CHAPTER V

FUNCTIONS OF MATRICES

§ 1. Definition of a Function of a Matrix

1. Let $A = \| a_{ik} \|_1^n$ be a square matrix and $f(\lambda)$ a function of a scalar argument λ. We wish to define what is to be meant by $f(A)$, i.e., we wish to extend the function $f(\lambda)$ to a matrix value of the argument.

We already know the solution of this problem in the simplest special case where $f(\lambda) = \gamma_0 \lambda^l + \gamma_1 \lambda^{l-1} + \cdots + \gamma_l$ is a polynomial in λ. In this case, $f(A) = \gamma_0 A^l + \gamma_1 A^{l-1} + \cdots + \gamma_l E$. Starting from this special case, we shall obtain a definition of $f(A)$ in the general case.

We denote by

$$\psi(\lambda) = (\lambda - \lambda_1)^{m_1} (\lambda - \lambda_2)^{m_2} \cdots (\lambda - \lambda_s)^{m_s} \tag{1}$$

the minimal polynomial[1] of A (where $\lambda_1, \lambda_2, \ldots, \lambda_s$ are all the distinct characteristic values of A). The degree of this polynomial is $m = \sum_{k=1}^{s} m_k$.

Let $g(\lambda)$ and $h(\lambda)$ be two polynomials such that

$$g(A) = h(A). \tag{2}$$

Then the difference $d(\lambda) = g(\lambda) - h(\lambda)$, as an annihilating polynomial for A, is divisible by $\psi(\lambda)$ without remainder; we shall write this as follows:

$$g(\lambda) \equiv h(\lambda) \;(\operatorname{mod} \psi(\lambda)). \tag{3}$$

Hence by (1)

$$d(\lambda_k) = 0, \quad d'(\lambda_k) = 0, \quad \ldots, \quad d^{(m_k - 1)}(\lambda_k) = 0 \qquad (k = 1, 2, \ldots, s),$$

i.e.,

$$g(\lambda_k) = h(\lambda_k), \quad g'(\lambda_k) = h'(\lambda_k), \ldots, g^{(m_k - 1)}(\lambda_k) = h^{(m_k - 1)}(\lambda_k) \qquad (k = 1, 2, \ldots, s). \tag{4}$$

[1] See Chapter IV, § 6.

The m numbers

$$f(\lambda_k), \quad f'(\lambda_k), \quad \ldots, \quad f^{(m_k-1)}(\lambda_k) \qquad (k=1, 2, \ldots, s) \tag{5}$$

will be called *the values of the function* $f(\lambda)$ *on the spectrum of the matrix* A and the set of all these values will be denoted symbolically by $f(\Lambda_A)$. If for a function $f(\lambda)$ the values (5) exist (i.e., have meaning), then we shall say that *the function* $f(\lambda)$ *is defined on the spectrum of the matrix* A.

Equation (4) shows that the polynomials $g(\lambda)$ and $h(\lambda)$ have the same values on the spectrum of A. In symbols:

$$g(\Lambda_A) = h(\Lambda_A).$$

Our argument is reversible: from (4) follows (3) and therefore (2).

Thus, given a matrix A, the values of the polynomial $g(\lambda)$ on the spectrum of A determine the matrix $g(A)$ completely, i.e., all polynomials $g(\lambda)$ that assume the same values on the spectrum of A have one and the same matrix value $g(A)$.

We postulate that the definition of $f(A)$ in the general case be subject to the same principle: *The values of the function* $f(\lambda)$ *on the spectrum of the matrix* A *must determine* $f(A)$ *completely, i.e., all functions* $f(\lambda)$ *having the same values on the spectrum of* A *must have the same matrix value* $f(A)$.

But then it is obvious that for the general definition of $f(A)$ it is sufficient to look for a polynomial[2] $g(\lambda)$ that assumes the same values on the spectrum of A as $f(\lambda)$ does and to set:

$$f(A) = g(A).$$

We are thus led to the following definition:

DEFINITION 1: *If the function* $f(\lambda)$ *is defined on the spectrum of the matrix* A, *then*

$$f(A) = g(A),$$

where $g(\lambda)$ *is an arbitrary polynomial that assumes on the spectrum of* A *the same values as does* $f(\lambda)$:

$$f(\Lambda_A) = g(\Lambda_A).$$

Among all the polynomials with complex coefficients that assume on the spectrum of A the same values as $f(\lambda)$ there is one and only one polynomial

[2] It will be proved in § 2 that such an interpolation polynomial always exists and an algorithm for the computation of the coefficients of the interpolation polynomial of least degree will be given.

$r(\lambda)$ that is of degree less than m.[3] This polynomial $r(\lambda)$ is uniquely determined by the interpolation conditions:

$$r(\lambda_k)=f(\lambda_k),\quad r'(\lambda_k)=f'(\lambda_k),\quad \dots,\quad r^{(m_k-1)}(\lambda_k)=f^{(m_k-1)}(\lambda_k) \qquad (k=1,2,\dots,s). \tag{6}$$

The polynomial $r(\lambda)$ is called the *Lagrange-Sylvester interpolation polynomial* for $f(\lambda)$ on the spectrum of A. Definition 1 can also be formulated as follows:

DEFINITION 1′: *Let $f(\lambda)$ be a function defined on the spectrum of a matrix A and $r(\lambda)$ the corresponding Lagrange-Sylvester interpolation polynomial. Then*

$$f(A)=r(A).$$

Note. If the minimal polynomial $\psi(\lambda)$ of a matrix A has no multiple roots[4] (in (1) $m_1=m_2=\dots=m_s=1$; $s=m$), then for $f(A)$ to have a meaning it is sufficient that $f(\lambda)$ be defined at the characteristic values $\lambda_1, \lambda_2, \dots, \lambda_m$. But if $\psi(\lambda)$ has multiple roots, then for some characteristic values the derivatives of $f(\lambda)$ up to a certain order (see (6)) must be defined as well.

Example 1: Let us consider the matrix[5]

$$H=\left\|\overbrace{\begin{matrix}0 & 1 & 0\dots 0\\ 0 & 0 & 1\dots 0\\ \dots & \dots & \dots\\ 0 & 0 & 0\dots 1\\ 0 & 0 & 0\dots 0\end{matrix}}^{n}\right\|.$$

Its minimal polynomial is λ^n. Therefore the values of $f(\lambda)$ on the spectrum of H are the numbers $f(0), f'(0), \dots, f^{(n-1)}(0)$, and the polynomial $r(\lambda)$ is of the form

$$r(\lambda)=f(0)+\frac{f'(0)}{1!}\lambda+\dots+\frac{f^{(n-1)}(0)}{(n-1)!}\lambda^{n-1}.$$

Therefore

---

[3] This polynomial is obtained from any other polynomial having the same spectral values by taking the remainder on division by $\psi(\lambda)$ of that polynomial.

[4] In Chapter VI it will be shown that A is a matrix of simple structure (see Chapter III, § 8) in this case, and this case only.

[5] The properties of the matrix H were worked out in the example on pp. 13-14.

$$f(H) = f(0)E + \frac{f'(0)}{1!}H + \cdots + \frac{f^{(n-1)}(0)}{(n-1)!}H^{n-1} = \left\| \begin{matrix} f(0) & \frac{f'(0)}{1!} & \cdots & \frac{f^{(n-1)}(0)}{(n-1)!} \\ 0 & f(0) & \ddots & \vdots \\ \vdots & & \ddots & \frac{f'(0)}{1!} \\ 0 & 0 & \cdots & f(0) \end{matrix} \right\|$$

Example 2: Let us consider the matrix

$$J = \left\| \begin{matrix} \lambda_0 & 1 & 0 & \cdots & 0 \\ 0 & \lambda_0 & 1 & \cdots & 0 \\ \vdots & & & & \vdots \\ 0 & 0 & 0 & \cdots & 1 \\ 0 & 0 & 0 & \cdots & \lambda_0 \end{matrix} \right\| \quad (n \text{ columns}).$$

Note that $J = \lambda_0 E + H$, so that $J - \lambda_0 E = H$. The minimal polynomial of J is clearly $(\lambda - \lambda_0)^n$. The interpolation polynomial $r(\lambda)$ of $f(\lambda)$ is given by the equation

$$r(\lambda) = f(\lambda_0) + \frac{f'(\lambda_0)}{1!}(\lambda - \lambda_0) + \cdots + \frac{f^{(n-1)}(\lambda_0)}{(n-1)!}(\lambda - \lambda_0)^{n-1}$$

Therefore

$$f(J) = r(J) = f(\lambda_0)E + \frac{f'(\lambda_0)}{1!}H + \cdots + \frac{f^{(n-1)}(\lambda_0)}{(n-1)!}H^{n-1}$$

$$= \left\| \begin{matrix} f(\lambda_0) & \frac{f'(\lambda_0)}{1!} & \cdots & \frac{f^{(n-1)}(\lambda_0)}{(n-1)!} \\ 0 & f(\lambda_0) & \ddots & \vdots \\ \vdots & & \ddots & \frac{f'(\lambda_0)}{1!} \\ 0 & 0 & \cdots & f(\lambda_0) \end{matrix} \right\|.$$

2. We mention two properties of functions of matrices.

1. *If two matrices A and B are similar and T transforms A into B,*

$$B = T^{-1}AT,$$

then the matrices $f(A)$ and $f(B)$ are also similar and T transforms $f(A)$ into $f(B)$,

$$f(B) = T^{-1}f(A)T.$$

For two similar matrices have equal minimal polynomials,[6] so that $f(\lambda)$ assumes the same values on the spectrum of A and of B. Therefore there exists an interpolation polynomial $r(\lambda)$ such that $f(A) = r(A)$ and $f(B) = r(B)$. But then it follows[6] from the equation $r(B) = T^{-1}r(A)T$ that

$$f(B) = T^{-1}f(A)T.$$

2. *If A is a quasi-diagonal matrix*

$$A = \{A_1, A_2, \ldots, A_u\},$$

then

$$f(A) = \{f(A_1), f(A_2), \ldots, f(A_u)\}.$$

Let us denote by $r(\lambda)$ the Lagrange-Sylvester interpolation polynomial of $f(\lambda)$ on the spectrum of A. Then it is easy to see that

$$f(A) = r(A) = \{r(A_1), r(A_2), \ldots, r(A_u)\}. \tag{7}$$

On the other hand, the minimal polynomial $\psi(\lambda)$ of A is an annihilating polynomial for each of the matrices $A_1, A_2, \ldots, A_u$. Therefore it follows from the equation

$$f(A_A) = r(A_A)$$

that

$$f(A_{A_1}) = r(A_{A_1}), \ldots, f(A_{A_u}) = r(A_{A_u}).$$

Therefore

$$f(A_1) = r(A_1), \ldots, f(A_u) = r(A_u),$$

and equation (7) can be written as follows:

$$f(A) = \{f(A_1), f(A_2), \ldots, f(A_u)\}. \tag{8}$$

Example 1: If the matrix A is of simple structure

$$A = T\{\lambda_1, \lambda_2, \ldots, \lambda_n\}T^{-1},$$

then

$$f(A) = T\{f(\lambda_1), f(\lambda_2), \ldots, f(\lambda_n)\}T^{-1}.$$

$f(A)$ has meaning if the function $f(\lambda)$ is defined at $\lambda_1, \lambda_2, \ldots, \lambda_n$.

---

[6] From $B = T^{-1}AT$ it follows that $B^k = T^{-1}A^kT$ $(k = 0, 1, 2, \ldots)$. Hence for every polynomial $g(\lambda)$ we have $g(B) = T^{-1}g(A)T$. Therefore it follows from $g(A) = O$ that $g(B) = O$, and vice versa.

Example 2: Let J be a matrix of the following quasi-diagonal form

$$J = \left\| \begin{array}{ccc} \overbrace{\begin{matrix} \lambda_1 & 1 & 0 & \ldots & 0 \\ 0 & \lambda_1 & 1 & \ldots & 0 \\ \cdot & \cdot & \cdot & \cdots & \cdot \\ 0 & 0 & 0 & \ldots & 1 \\ 0 & 0 & 0 & \ldots & \lambda_1 \end{matrix}}^{\nu_1} & & \\ & \ddots & \\ & & \overbrace{\begin{matrix} \lambda_u & 1 & 0 & \ldots & 0 \\ 0 & \lambda_u & 1 & \ldots & 0 \\ \cdot & \cdot & \cdot & \cdots & \cdot \\ 0 & 0 & 0 \ldots \lambda_u & & 1 \\ 0 & 0 & 0 \ldots 0 & & \lambda_u \end{matrix}}^{\nu_u} \end{array} \right\| .$$

All the elements in the non-diagonal blocks are zero. By (8) (see also the example on pp. 12-13),

$$f(J) = \left\| \begin{array}{ccc} \begin{matrix} f(\lambda_1) & \frac{f'(\lambda_1)}{1!} & \cdots & \frac{f^{(\nu_1-1)}(\lambda_1)}{(\nu_1-1)!} \\ 0 & f(\lambda_1) & \ddots & \vdots \\ \vdots & & \ddots & \frac{f'(\lambda_1)}{1!} \\ 0 & 0 & \cdots & f(\lambda_1) \end{matrix} & & \\ & \ddots & \\ & & \begin{matrix} f(\lambda_u) & \frac{f'(\lambda_u)}{1!} & \cdots & \frac{f^{(\nu_u-1)}(\lambda_u)}{(\nu_u-1)!} \\ 0 & (f\,\lambda_u) & \ddots & \vdots \\ \vdots & & \ddots & \frac{f'(\lambda_u)}{1!} \\ 0 & 0 & \cdots & f(\lambda_u) \end{matrix} \end{array} \right\| .$$

Here, as in the matrix J, all the elements in the non-diagonal blocks are also zero.[7]

[7] It will be established later (Chapter VI, § 6 or Chapter VII, § 7) that an arbitrary matrix $A = \| a_{ik} \|_1^n$ is always similar to some matrix of the form $J : A = TJT^{-1}$. Therefore (see 1, on p. 98) we always have $f(A) = Tf(J)T^{-1}$.

§ 2. The Lagrange-Sylvester Interpolation Polynomial

1. To begin with, we consider the case in which the characteristic equation $|\lambda E - A| = 0$ has no multiple roots. The roots of this equation—the characteristic values of the matrix A—will be denoted by $\lambda_1, \lambda_2, \ldots, \lambda_n$. Then

$$\psi(\lambda) = |\lambda E - A| = (\lambda - \lambda_1)(\lambda - \lambda_2)\cdots(\lambda - \lambda_n),$$

and condition (6) can be written as follows:

$$r(\lambda_k) = f(\lambda_k) \qquad (k = 1, 2, \ldots, n).$$

In this case, $r(\lambda)$ is the ordinary Lagrange interpolation polynomial for the function $f(\lambda)$ at the points $\lambda_1, \lambda_2, \ldots, \lambda_n$:

$$r(\lambda) = \sum_{k=1}^{n} \frac{(\lambda - \lambda_1)\cdots(\lambda - \lambda_{k-1})(\lambda - \lambda_{k+1})\cdots(\lambda - \lambda_n)}{(\lambda_k - \lambda_1)\cdots(\lambda_k - \lambda_{k-1})(\lambda_k - \lambda_{k+1})\cdots(\lambda_k - \lambda_n)} f(\lambda_k).$$

By Definition 1′

$$f(A) = r(A) = \sum_{k=1}^{n} \frac{(A - \lambda_1 E)\cdots(A - \lambda_{k-1}E)(A - \lambda_{k+1}E)\cdots(A - \lambda_n E)}{(\lambda_k - \lambda_1)\cdots(\lambda_k - \lambda_{k-1})(\lambda_k - \lambda_{k+1})\cdots(\lambda_k - \lambda_n)} f(\lambda_k).$$

2. Let us assume now that the characteristic polynomial has multiple roots, but that the minimal polynomial, which is a divisor of the characteristic polynomial, has only simple roots:[8]

$$\psi(\lambda) = (\lambda - \lambda_1)(\lambda - \lambda_2)\cdots(\lambda - \lambda_m).$$

In this case (as in the preceding one) all the exponents m_k in (1) are equal to 1, and the equation (6) takes the form

$$r(\lambda_k) = f(\lambda_k) \qquad (k = 1, 2, \ldots, m).$$

$r(\lambda)$ is again the ordinary Lagrange interpolation polynomial and

$$f(A) = \sum_{k=1}^{m} \frac{(A - \lambda_1 E)\cdots(A - \lambda_{k-1}E)(A - \lambda_{k+1}E)\cdots(A - \lambda_m E)}{(\lambda_k - \lambda_1)\cdots(\lambda_k - \lambda_{k-1})(\lambda_k - \lambda_{k+1})\cdots(\lambda_k - \lambda_m)} f(\lambda_k).$$

3. We now consider the general case:

$$\psi(\lambda) = (\lambda - \lambda_1)^{m_1}(\lambda - \lambda_2)^{m_2}\cdots(\lambda - \lambda_s)^{m_s} \qquad (m_1 + m_2 + \cdots + m_s = m).$$

We represent the rational function $\frac{r(\lambda)}{\psi(\lambda)}$, where the degree of $r(\lambda)$ is less than the degree of $\psi(\lambda)$, as a sum of partial fractions:

---

[8] See footnote 4.

$$\frac{r(\lambda)}{\psi(\lambda)}=\sum_{k=1}^{s}\left[\frac{\alpha_{k1}}{(\lambda-\lambda_k)^{m_k}}+\frac{\alpha_{k2}}{(\lambda-\lambda_k)^{m_k-1}}+\cdots+\frac{\alpha_{k\,m_k}}{\lambda-\lambda_k}\right], \tag{9}$$

where α_{kj} ($j=1, 2, \ldots, m_k$; $k=1, 2, \ldots, s$) are certain constants.

In order to determine the numerators α_{kj} of the partial fractions we multiply both sides of (9) by $(\lambda-\lambda_k)^{m_k}$ and denote by $\psi_k(\lambda)$ the polynomial $\frac{\psi(\lambda)}{(\lambda-\lambda_k)^{m_k}}$. Then we obtain:

$$\frac{r(\lambda)}{\psi_k(\lambda)}=\alpha_{k1}+\alpha_{k2}(\lambda-\lambda_k)+\cdots+\alpha_{k\,m_k}(\lambda-\lambda_k)^{m_k-1}+$$
$$+(\lambda-\lambda_k)^{m_k}\varrho_k(\lambda) \qquad (k=1, 2, \ldots, s), \tag{10}$$

where $\varrho_k(\lambda)$ is a rational function, regular for $\lambda=\lambda_k$.[9]

Hence

$$\left.\begin{aligned}\alpha_{k1}&=\left[\frac{r(\lambda)}{\psi_k(\lambda)}\right]_{\lambda=\lambda_k},\\ \alpha_{k2}&=\left[\frac{r(\lambda)}{\psi_k(\lambda)}\right]'_{\lambda=\lambda_k}=r(\lambda_k)\left[\frac{1}{\psi_k(\lambda)}\right]'_{\lambda=\lambda_k}+r'(\lambda_k)\frac{1}{\psi_k(\lambda_k)},\quad \ldots\ (k=1, 2, \ldots s).\end{aligned}\right\} \tag{11}$$

Formulas (11) show that the numerators α_{kj} on the right-hand side of (9) are expressible in terms of the values of the polynomial $r(\lambda)$ on the spectrum of A, and these values are known: they are equal to the corresponding values of the function $f(\lambda)$ and its derivatives. Therefore

$$\alpha_{k1}=\frac{f(\lambda_k)}{\psi_k(\lambda_k)},\ \alpha_{k2}=f(\lambda_k)\left[\frac{1}{\psi_k(\lambda)}\right]'_{\lambda=\lambda_k}+f'(\lambda_k)\frac{1}{\psi_k(\lambda_k)},\ \ldots \qquad (k=1, 2, \ldots, s). \tag{12}$$

Formulas (12) may be abbreviated as follows:

$$\alpha_{kj}=\frac{1}{(j-1)!}\left[\frac{f(\lambda)}{\psi_k(\lambda)}\right]^{(j-1)}_{\lambda=\lambda_k} \qquad (j=1, 2, \ldots, m_k;\quad k=1, 2, \ldots, s). \tag{13}$$

When all the α_{kj} have been found, we can determine $r(\lambda)$ from the following formula, which is obtained from (9) by multiplying both sides by $\psi(\lambda)$:

$$r(\lambda)=\sum_{k=1}^{s}[\alpha_{k1}+\alpha_{k2}(\lambda-\lambda_k)+\cdots+\alpha_{k\,m_k}(\lambda-\lambda_k)^{m_k-1}]\,\psi_k(\lambda). \tag{14}$$

In this formula the expression in brackets that multiplies $\psi_k(\lambda)$ is, by (13), equal to the sum of the first m_k terms of the Taylor expansion of $f(\lambda)$ in powers of $(\lambda-\lambda_k)$.

[9] I.e., that does not become infinite for $\lambda=\lambda_k$.

Note. The Lagrange-Sylvester interpolation polynomial can be obtained by a limiting process from the Lagrange interpolation polynomial.

Let

$$\psi(\lambda) = (\lambda - \lambda_1)^{m_1} (\lambda - \lambda_2)^{m_2} \cdots (\lambda - \lambda_s)^{m_s} \quad \left(m = \sum_{k=1}^{s} m_k\right).$$

We denote the Lagrange interpolation polynomial constructed for the m points

$$\lambda_1^{(1)}, \lambda_1^{(2)}, \ldots, \lambda_1^{(m_1)}; \lambda_2^{(1)}, \lambda_2^{(2)}, \ldots, \lambda_2^{(m_2)}; \ldots; \lambda_s^{(1)}, \lambda_s^{(2)}, \ldots, \lambda_s^{(m_s)}$$

by

$$L(\lambda) = L\begin{pmatrix} \lambda_1^{(1)}, \ldots, \lambda_1^{(m_1)}; \ldots; \lambda_s^{(1)}, \ldots, \lambda_s^{(m_s)}; & \\ f(\lambda_1^{(1)}), \ldots, f(\lambda_1^{(m_1)}); \ldots; f(\lambda_s^{(1)}), \ldots, f(\lambda_s^{(m_s)}); & \lambda \end{pmatrix}.$$

Then it is not difficult to show that the required Lagrange-Sylvester polynomial is determined by the formula

$$r(\lambda) = \lim_{\substack{\lambda_1^{(1)}, \ldots, \lambda_1^{(m_1)} \to \lambda_1 \\ \cdots\cdots\cdots\cdots \\ \lambda_s^{(1)}, \ldots, \lambda_s^{(m_s)} \to \lambda_s}} L(\lambda).$$

Example:

$$\psi(\lambda) = (\lambda - \lambda_1)^2 (\lambda - \lambda_2)^3 \qquad (m = 5).$$

Then

$$\frac{r(\lambda)}{\psi(\lambda)} = \frac{\alpha}{(\lambda - \lambda_1)^2} + \frac{\beta}{\lambda - \lambda_1} + \frac{\gamma}{(\lambda - \lambda_2)^3} + \frac{\delta}{(\lambda - \lambda_2)^2} + \frac{\varepsilon}{\lambda - \lambda_2}.$$

Hence

$$r(\lambda) = [\alpha + \beta(\lambda - \lambda_1)](\lambda - \lambda_2)^3 + [\gamma + \delta(\lambda - \lambda_2) + \varepsilon(\lambda - \lambda_2)^2](\lambda - \lambda_1)^2$$

and therefore

$$r(A) = [\alpha E + \beta(A - \lambda_1 E)](A - \lambda_2 E)^3 + [\gamma E + \delta(A - \lambda_2 E) + \varepsilon(A - \lambda_2 E)^2](A - \lambda_1 E)^2.$$

α, β, γ, δ, and ε can be found from the following formulas:

$$\alpha = \frac{f(\lambda_1)}{(\lambda_1 - \lambda_2)^3}, \quad \beta = -\frac{3}{(\lambda_1 - \lambda_2)^4} f(\lambda_1) + \frac{1}{(\lambda_1 - \lambda_2)^3} f'(\lambda_1),$$

$$\gamma = \frac{f(\lambda_2)}{(\lambda_2 - \lambda_1)^2}, \quad \delta = -\frac{2}{(\lambda_2 - \lambda_1)^3} f(\lambda_2) + \frac{1}{(\lambda_2 - \lambda_1)^2} f'(\lambda_2),$$

$$\varepsilon = \frac{3}{(\lambda_2 - \lambda_1)^4} f(\lambda_2) - \frac{2}{(\lambda_2 - \lambda_1)^3} f'(\lambda_2) + \frac{1}{2} \frac{1}{(\lambda_2 - \lambda_1)^2} f''(\lambda_2).$$

§ 3. Other Forms of the Definition of $f(A)$. The Components of the Matrix A

1. Let us return to the formula (14) for $r(\lambda)$. When we substitute in (14) the expressions (12) for the coefficients α and combine the terms that contain one and the same value of the function $f(\lambda)$ or of one of its derivatives, we represent $r(\lambda)$ in the form

$$r(\lambda)=\sum_{k=1}^{s}\left[f(\lambda_k)\,\varphi_{k1}(\lambda)+f'(\lambda_k)\,\varphi_{k2}(\lambda)+\cdots+f^{(m_k-1)}(\lambda_k)\,\varphi_{k\,m_k}(\lambda)\right]. \qquad (15)$$

Here $\varphi_{kj}(\lambda)$ $(j=1,2,\ldots,m_k;\ k=1,2,\ldots,s)$ are easily computable polynomials in λ of degree less than m. These polynomials are completely determined when $\psi(\lambda)$ is given and do not depend on the choice of the function $f(\lambda)$. The number of these polynomials is equal to the number of values of the function $f(\lambda)$ on the spectrum of A, i.e., equal to m (m is the degree of the minimal polynomial $\psi(\lambda)$). The functions $\varphi_{kj}(\lambda)$ represent the Lagrange-Sylvester interpolation polynomial for the function whose values on the spectrum of A are all equal to zero with the exception of $f^{(j-1)}(\lambda_k)$, which is equal to 1.

All the polynomials $\varphi_{kj}(\lambda)$ $(j=1,2,\ldots,m_k;\ k=1,2,\ldots,s)$ are linearly independent. For suppose that

$$\sum_{k=1}^{s}\sum_{j=1}^{m_k} c_{kj}\varphi_{kj}(\lambda)=0.$$

Let us determine the interpolation polynomial $r(\lambda)$ from the m conditions:

$$r^{(j-1)}(\lambda_k)=c_{kj} \qquad (j=1,2,\ldots,m_k;\ k=1,2,\ldots,s). \qquad (16)$$

Then by (15) and (16)

$$r(\lambda)=\sum_{k=1}^{s}\sum_{j=1}^{m_k} c_{kj}\varphi_{kj}(\lambda)=0$$

and, therefore, by (16)

$$c_{kj}=0 \qquad (j=1,2,\ldots,m_k;\ k=1,2,\ldots,s).$$

From (15) we deduce the *fundamental formula for* $f(A)$:

$$f(A)=\sum_{k=1}^{s}\left[f(\lambda_k)\,Z_{k1}+f'(\lambda_k)\,Z_{k2}+\cdots+f^{(m_k-1)}(\lambda_k)\,Z_{km_k}\right], \qquad (17)$$

where

$$Z_{kj}=\varphi_{kj}(A) \quad (j=1,2,\ldots,m_k;\ k=1,2,\ldots,s). \qquad (18)$$

The matrices Z_{kj} are completely determined when A is given and do not depend on the choice of the function $f(\lambda)$. On the right-hand side of (17) the function $f(\lambda)$ is represented only by its values on the spectrum of A.

The matrices Z_{kj} $(j=1, 2, \ldots, m_k;\ k=1, 2, \ldots, s)$ will be called the *constituent matrices* or *components* of the given matrix A.

The components Z_{kj} are linearly independent.

For suppose that

$$\sum_{k=1}^{s}\sum_{j=1}^{m_k} c_{kj}Z_{kj}=O.$$

Then by (18)

$$\chi(A)=O, \tag{19}$$

where

$$\chi(\lambda)=\sum_{k=1}^{s}\sum_{j=1}^{m_k} c_{kj}\varphi_{kj}(\lambda). \tag{20}$$

Since by (20) the degree of $\chi(\lambda)$ is less than m, the degree of the minimal polynomial $\psi(\lambda)$, it follows from (19) that

$$\chi(\lambda)=0.$$

But then, since the m functions $\varphi_{kj}(\lambda)$ are linearly independent, (20) implies that

$$c_{kj}=0 \qquad (j=1, 2, \ldots, m_k;\ k=1, 2, \ldots, s),$$

and this is what we had to prove.

2. From the linear independence of the constituent matrices Z_{kj} it follows, among other things, that none of these matrices can be zero. Let us also note that any two components Z_{kj} are permutable among each other and with A, because they are all scalar polynomials in A.

The formula (17) for $f(A)$ is particularly convenient to use when it is necessary to deal with several functions of one and the same matrix A, or when the function $f(\lambda)$ depends not only on λ, but also on some parameter t. In the latter case, the components Z_{kj} on the right-hand side of (17) do not depend on t, and the parameter t enters only into the scalar coefficients of the matrices.

In the example at the end of § 2, where $\psi(\lambda)=(\lambda-\lambda_1)^2(\lambda-\lambda_2)^3$, we may represent $r(\lambda)$ in the form

$$r(\lambda)=f(\lambda_1)\varphi_{11}(\lambda)+f'(\lambda_1)\varphi_{12}(\lambda)+f(\lambda_2)\varphi_{21}(\lambda)+f'(\lambda_2)\varphi_{22}(\lambda)+f''(\lambda_2)\varphi_{23}(\lambda),$$

where

$$\varphi_{11}(\lambda)=\left(\frac{\lambda-\lambda_2}{\lambda_1-\lambda_2}\right)^3\left[1-\frac{3(\lambda-\lambda_1)}{\lambda_1-\lambda_2}\right],\qquad \varphi_{12}(\lambda)=\frac{(\lambda-\lambda_1)(\lambda-\lambda_2)^3}{(\lambda_1-\lambda_2)^3},$$

$$\varphi_{21}(\lambda)=\left(\frac{\lambda-\lambda_1}{\lambda_2-\lambda_1}\right)^2\left[1-\frac{2(\lambda-\lambda_2)}{\lambda_2-\lambda_1}+\frac{3(\lambda-\lambda_2)^2}{(\lambda_2-\lambda_1)^2}\right],$$

$$\varphi_{22}(\lambda)=\frac{(\lambda-\lambda_1)^2(\lambda-\lambda_2)}{(\lambda_2-\lambda_1)^2}\left[1-\frac{2(\lambda-\lambda_2)}{\lambda_2-\lambda_1}\right],$$

$$\varphi_{23}(\lambda)=\frac{(\lambda-\lambda_1)^2(\lambda-\lambda_2)^2}{2(\lambda_2-\lambda_1)^2}.$$

Therefore

$$f(A)=f(\lambda_1)Z_{11}+f'(\lambda_1)Z_{12}+f(\lambda_2)Z_{21}+f'(\lambda_2)Z_{22}+f''(\lambda_2)Z_{23}$$

where

$$Z_{11}=\varphi_{11}(A)=\frac{1}{(\lambda_1-\lambda_2)^3}(A-\lambda_2E)^3\left[E-\frac{3}{\lambda_1-\lambda_2}(A-\lambda_1E)\right],$$

$$Z_{12}=\varphi_{12}(A)=\frac{1}{(\lambda_1-\lambda_2)^3}(A-\lambda_1E)(A-\lambda_2E)^3,\ \ldots .$$

3. When the matrix A is given and its components have actually to be found, we can set in the fundamental formula (17) $f(\mu)=\frac{1}{\lambda-\mu}$, where λ is a parameter. Then we obtain

$$(\lambda E-A)^{-1}=\frac{C(\lambda)}{\psi(\lambda)}=\sum_{k=1}^{s}\left[\frac{Z_{k1}}{\lambda-\lambda_k}+\frac{1!\,Z_{k2}}{(\lambda-\lambda_k)^2}+\cdots+\frac{(m_k-1)!\,Z_{km_k}}{(\lambda-\lambda_k)^{m_k}}\right], \tag{21}$$

where $C(\lambda)$ is the reduced adjoint matrix of $\lambda E-A$ (Chapter IV, § 6).[10]

The matrices $(j-1)!\,Z_{kj}$ are the numerators of the partial fractions in the decomposition (21), and by analogy with (9) they may be expressed by the values of $C(\lambda)$ on the spectrum of A by formulas similar to (11):

$$(m_k-1)!\,Z_{km_k}=\frac{C(\lambda_k)}{\psi_k(\lambda)},\qquad (m_k-2)!\,Z_{k\,m_k-1}=\left[\frac{C(\lambda)}{\psi_k(\lambda)}\right]'_{\lambda=\lambda_k},\quad \ldots .$$

Hence

$$Z_{kj}=\frac{1}{(j-1)!\,(m_k-j)!}\left[\frac{C(\lambda)}{\psi_k(\lambda)}\right]^{(m_k-j)}_{\lambda=\lambda_k}\quad (j=1,2,\ldots,m_k;\ k=1,2,\ldots,s). \tag{22}$$

When we replace the constituent matrices in (17) by their expressions (22), we can represent the fundamental formula (17) in the form

---

[10] For $f(\mu)=\frac{1}{\lambda-\mu}$ we have $f(A)=(\lambda E-A)^{-1}$. For $f(A)=r(A)$, where $r(\mu)$ is the Lagrange-Sylvester interpolation polynomial. From the fact that $f(\mu)$ and $r(\mu)$ coincide on the spectrum of A it follows that $(\lambda-\mu)\,r(\mu)$ and $(\lambda-\mu)\,f(\mu)=1$ coincide on this spectrum. Hence $(\lambda E-A)\,r(A)=(\lambda E-A)\,f(A)=E$.

$$f(A)=\sum_{k=1}^{s}\frac{1}{(m_k-1)!}\left[\frac{C(\lambda)}{\psi_k(\lambda)}f(\lambda)\right]^{(m_k-1)}_{\lambda=\lambda_k}. \tag{23}$$

Example 1:[11]

$$A=\left\|\begin{array}{rrr|r} 2 & -1 & 1 & \mathit{2}\\ 0 & 1 & 1 & \mathit{2}\\ -1 & 1 & 1 & \mathit{1}\end{array}\right\|, \quad \lambda E-A=\left\|\begin{array}{ccc} \lambda-2 & 1 & -1\\ 0 & \lambda-1 & -1\\ 1 & -1 & \lambda-1\end{array}\right\|.$$

In this case $\Delta(\lambda)=|\lambda E-A|=(\lambda-1)^2(\lambda-2)$. Since the minor of the element in the first row and second column of $\lambda E-A$ is equal to 1, we have $D_2(\lambda)=1$ and, therefore,

$$\psi(\lambda)=\Delta(\lambda)=(\lambda-1)^2(\lambda-2)=\lambda^3-4\lambda^2+5\lambda-2,$$

$$\Psi(\lambda,\mu)=\frac{\psi(\mu)-\psi(\lambda)}{\mu-\lambda}=\mu^2+(\lambda-4)\mu+\lambda^2-4\lambda+5$$

and

$$C(\lambda)=\Psi(\lambda E, A)=A^2+(\lambda-4)A+(\lambda^2-4\lambda+5)E$$

$$=\left\|\begin{array}{rrr|r} 3 & -2 & 2 & \mathit{3}\\ -1 & 2 & 2 & \mathit{3}\\ -3 & 3 & 1 & \mathit{1}\end{array}\right\|+(\lambda-4)\left\|\begin{array}{rrr} 2 & -1 & 1\\ 0 & 1 & 1\\ -1 & 1 & 1\end{array}\right\|+(\lambda^2-4\lambda+5)\left\|\begin{array}{ccc} 1 & 0 & 0\\ 0 & 1 & 0\\ 0 & 0 & 1\end{array}\right\|.$$

The fundamental formula has in this case the form

$$f(A)=f(1)Z_{11}+f'(1)Z_{12}+f(2)Z_{21}. \tag{24}$$

Setting $f(\mu)=\frac{1}{\lambda-\mu}$, we find:

$$(\lambda E-A)^{-1}=\frac{C(\lambda)}{\psi(\lambda)}=\frac{Z_{11}}{\lambda-1}+\frac{Z_{12}}{(\lambda-1)^2}+\frac{Z_{21}}{\lambda-2};$$

hence

$$Z_{11}=-C(1)-C'(1), \qquad Z_{12}=-C(1), \qquad Z_{21}=C(2).$$

We now use the above expression for $C(\lambda)$, compute Z_{11}, Z_{12}, Z_{21}, and substitute the results obtained in (24):

[11] The elements of the sum column are printed in italics and are used for checking the computation. When we multiply the rows of A into the sum column of B we obtain the sum column of AB.

$$f(A) = f(1)\begin{Vmatrix} 1 & 0 & 0 \\ 1 & 0 & 0 \\ 1 & -1 & 1 \end{Vmatrix} + f'(1)\begin{Vmatrix} 1 & -1 & 1 \\ 1 & -1 & 1 \\ 0 & 0 & 0 \end{Vmatrix} + f(2)\begin{Vmatrix} 0 & 0 & 0 \\ -1 & 1 & 0 \\ -1 & 1 & 0 \end{Vmatrix}$$

$$= \begin{Vmatrix} f(1) + f'(1) & -f'(1) & f'(1) \\ f(1) + f'(1) - f(2) & -f'(1) + f(2) & f'(1) \\ f(1) - f(2) & -f(1) + f(2) & f(1) \end{Vmatrix}. \tag{25}$$

Example 2: Let us show that we can determine $f(A)$ starting only from the fundamental formula. Again let

$$A = \begin{Vmatrix} 2 & -1 & 1 \\ 0 & 1 & 1 \\ -1 & 1 & 1 \end{Vmatrix}, \quad \psi(\lambda) = (\lambda - 1)^2 (\lambda - 2).$$

Then

$$f(A) = f(1) Z_1 + f'(1) Z_2 + f(2) Z_3. \tag{24'}$$

In (24′) we substitute for $f(\lambda)$ in succession $1,\ \lambda - 1,\ (\lambda - 1)^2$:

$$Z_1 + Z_3 = E = \begin{Vmatrix} 1 & 0 & 0 \\ 0 & 1 & 0 \\ 0 & 0 & 1 \end{Vmatrix},$$

$$Z_2 + Z_3 = A - E = \begin{Vmatrix} 1 & -1 & 1 \\ 0 & 0 & 1 \\ -1 & 1 & 0 \end{Vmatrix} \begin{matrix} \mathit{1} \\ \mathit{1}, \\ \mathit{0} \end{matrix}$$

$$Z_3 = (A - E)^2 = \begin{Vmatrix} 0 & 0 & 0 \\ -1 & 1 & 0 \\ -1 & 1 & 0 \end{Vmatrix} \begin{matrix} \mathit{0} \\ \mathit{0}. \\ \mathit{0} \end{matrix}$$

Computing the third equation from the first two term by term, we can determine all the Z. Substituting in (24′), we obtain the expression for $f(A)$.

4. The examples we have analyzed illustrate three methods of practical computation of $f(A)$. In the first method, we found the interpolation polynomial $r(\lambda)$ and put $f(A) = r(A)$. In the second method, we made use of the decomposition (21) and expressed the components Z_{kj} in (17) by the values of the reduced adjoint matrix $C(\lambda)$ on the spectrum of A. In the third method, we started from the fundamental formula (17) and substituted in succession certain simple polynomials for $f(\lambda)$; from the linear equations so obtained we determined the constituent matrices Z_{kj}.

The third method is perhaps the most convenient for practical purposes. In the general case it can be stated as follows:

In (17) we substitute for $f(\lambda)$ successively certain polynomials $g_1(\lambda)$, $g_2(\lambda), \ldots, g_m(\lambda)$:

$$g_i(A)=\sum_{k=1}^{s}\left[g_i(\lambda_k)\,Z_{k1}+g_i'(\lambda_k)\,Z_{k2}+\cdots+g_i^{(m_k-1)}(\lambda_k)\,Z_{k\,m_k}\right]$$
$$(i=1,2,\ldots,m). \tag{26}$$

From the m equations (26) we determine the matrices Z_{kj} and substitute the expressions so obtained in (17).

The result of eliminating Z_{kj} from the $(m+1)$ equations (26) and (17) can be written in the form

$$\begin{vmatrix} f(A) & f(\lambda_1)\ldots f^{(m_1-1)}(\lambda_1)\ldots f(\lambda_s)\ldots f^{(m_s-1)}(\lambda_s) \\ g_1(A) & g_1(\lambda_1)\ldots g_1^{(m_1-1)}(\lambda_1)\ldots g_1(\lambda_s)\ldots g_1^{(m_s-1)}(\lambda_s) \\ \cdot & \cdot \\ \cdot & \cdot \\ \cdot & \cdot \\ g_m(A) & g_m(\lambda_1)\ldots g_m^{(m_1-1)}(\lambda_1)\ldots g_m(\lambda_s)\ldots g_m^{(m_s-1)}(\lambda_s) \end{vmatrix}=0.$$

Expanding this determinant with respect to the elements of the first column, we obtain the required expression for $f(A)$. As the factor of $f(A)$ we have here the determinant $\Delta=|\,g_i^{(j)}(\lambda_k)\,|$ (in the i-th row of Δ there are found the values of the polynomial $g_i(\lambda)$ on the spectrum of A; $i=1,2,\ldots,m$). In order to determine $f(A)$ we must have $\Delta\neq 0$. This will be so if no linear combination[12] of the polynomials vanishes completely on the spectrum of A, i.e., is divisible by $\psi(\lambda)$.

The condition $\Delta\neq 0$ is always satisfied when the degrees of the polynomial $g_1(\lambda), g_2(\lambda), \ldots, g_m(\lambda)$ are $0, 1, \ldots, m-1$, respectively.[13]

5. In conclusion, we mention that high powers of a matrix A^n can be conveniently computed by formula (17) by setting $f(\lambda)$ equal to λ^n.[14]

Example: Given the matrix $A=\left\|\begin{matrix}5 & -4\\ 4 & -3\end{matrix}\right\|$ it is required to compute the elements of A^{100}. The minimal polynomial of the matrix is $\psi(\lambda)=(\lambda-1)^2$.

---

[12] With coefficients not all equal to zero.

[13] In the last example, $m=3$, $g_1(\lambda)=1$, $g_2(\lambda)=\lambda-1$, $g_3(\lambda)=(\lambda-1)^2$.

[14] Formula (17) may also be used to compute the inverse matrix A^{-1}, by setting $f(\lambda)=\frac{1}{\lambda}$ or, what is the same, by setting $\lambda=0$ in (21).

The fundamental formula is

$$f(A) = f(1) Z_1 + f'(1) Z_2 .$$

Replacing $f(\lambda)$ successively by 1 and $\lambda - 1$, we obtain:

$$Z_1 = E, \quad Z_2 = A - E .$$

Therefore

$$f(A) = f(1) E + f'(1)(A - E) .$$

Setting $f(\lambda) = \lambda^{100}$, we find

$$A^{100} = E + 100(A - E) = \begin{Vmatrix} 1 & 0 \\ 0 & 1 \end{Vmatrix} + 100 \begin{Vmatrix} 4 & -4 \\ 4 & -4 \end{Vmatrix} = \begin{Vmatrix} 401 & -400 \\ 400 & -399 \end{Vmatrix} .$$

§ 4. Representation of Functions of Matrices by means of Series

1. Let $A = \| a_{ik} \|_1^n$ be a matrix with the minimal polynomial (1):

$$\psi(\lambda) = (\lambda - \lambda_1)^{m_1} (\lambda - \lambda_2)^{m_2} \cdots (\lambda - \lambda_s)^{m_s} \quad (m = \sum_{k=1}^{s} m_k).$$

Furthermore, let $f(\lambda)$ be a function and let $f_1(\lambda), f_2(\lambda), \ldots, f_p(\lambda), \ldots$ be a sequence of functions defined on the spectrum of A.

We shall say that *the sequence of functions $f_p(\lambda)$ converges for $p \to \infty$ to some limit on the spectrum of A* if the limits

$$\lim_{p\to\infty} f_p(\lambda_k), \quad \lim_{p\to\infty} f_p'(\lambda_k), \quad \ldots, \quad \lim_{p\to\infty} f_p^{(m_k-1)}(\lambda_k) \qquad (k = 1, 2, \ldots, s)$$

exist.

We shall say that *the sequence of functions $f_p(\lambda)$ converges for $p \to \infty$ to the function $f(\lambda)$ on the spectrum of A,* and we shall write

$$\lim_{p\to\infty} f_p(\Lambda_A) = f(\Lambda_A)$$

if

$$\lim_{p\to\infty} f_p(\lambda_k) = f(\lambda_k), \quad \lim_{p\to\infty} f_p'(\lambda_k) = f'(\lambda_k), \quad \ldots, \quad \lim_{p\to\infty} f_p^{(m_k-1)}(\lambda_k) = f^{(m_k-1)}(\lambda_k)$$
$$(k = 1, 2, \ldots, s).$$

The fundamental formula

$$f(A) = \sum_{k=1}^{s} \left[f(\lambda_k) Z_{k1} + f'(\lambda_k) Z_{k2} + \cdots + f^{(m_k-1)}(\lambda_k) Z_{km_k} \right]$$

expresses $f(A)$ in terms of the values of $f(\lambda)$ on the spectrum of A. If we regard the matrix as a vector in a space $\boldsymbol{R}_{n^2}$ of dimension n^2, then it follows from the fundamental formula, by the linear independence of the matrices Z_{kj}, that all the $f(A)$ (for given A) form an m-dimensional subspace of $\boldsymbol{R}_{n^2}$

with basis Z_{kj} ($j=1, 2, \dots, m_k$; $k=1, 2, \dots, s$). In this basis the 'vector' $f(A)$ has as its coordinates the m values of the function $f(\lambda)$ on the spectrum of A.

These considerations make the following theorem perfectly obvious:

THEOREM 1: *A sequence of matrices $f_p(A)$ converges for $p \to \infty$ to some limit if and only if the sequence $f_p(\lambda)$ converges for $p \to \infty$ on the spectrum of A to a limit, i.e., the limits*

$$\lim_{p\to\infty} f_p(A) \quad \textit{and} \quad \lim_{p\to\infty} f_p(\Lambda_A)$$

always exist simultaneously. Moreover, the equation

$$\lim_{p\to\infty} f_p(\Lambda_A) = f(\Lambda_A) \tag{27}$$

implies that

$$\lim_{p\to\infty} f_p(A) = f(A) \tag{28}$$

and conversely.

Proof. 1) If the values of $f_p(\lambda)$ converge on the spectrum of A for $p \to \infty$ to limit values, then from the formulas

$$f_p(A) = \sum_{k=1}^{s} \left[f_p(\lambda_k) Z_{k1} + f_p'(\lambda_k) Z_{k2} + \cdots + f_p^{(m_k-1)}(\lambda_k) Z_{km_k} \right] \tag{29}$$

there follows the existence of the limit $\lim_{p\to\infty} f_p(A)$. On the basis of this formula and of (17) we deduce (28) from (27).

2) Suppose, conversely, that $\lim_{p\to\infty} f_p(A)$ exists. Since the m constituent matrices Z are linearly independent, we can express, by (29), the m values of $f_p(\lambda)$ on the spectrum of A (as a linear form) by the m elements of the matrix $f_p(A)$. Hence the existence of the limit $\lim_{p\to\infty} f_p(\Lambda_A)$ follows, and (27) holds in the presence of (28).

According to this theorem, if a sequence of polynomials $g_p(\lambda)$ ($p=1, 2, 3, \dots$) converges to the function $f(\lambda)$ on the spectrum of A, then

$$\lim_{p\to\infty} g_p(A) = f(A).$$

2. This formula underlines the naturalness and generality of our definition of $f(A)$. $f(A)$ is always obtained from the $g_p(A)$ by passing to the limit $p \to \infty$, provided only that the sequence of polynomials $g_p(\lambda)$ converges to $f(\lambda)$ on the spectrum of A. The latter condition is necessary for the existence of the limit $\lim_{p\to\infty} g_p(A)$.

We shall say that *the series* $\sum_{p=0}^{\infty} u_p(\lambda)$ *converges on the spectrum of* A *to the function* $f(\lambda)$ and we shall write

$$f(\Lambda_A) = \sum_{p=0}^{\infty} u_p(\Lambda_A), \tag{30}$$

if all the functions occurring here are defined on the spectrum of A and the following equations hold:

$$f(\lambda_k) = \sum_{p=0}^{\infty} u_p(\lambda_k), \quad f'(\lambda_k) = \sum_{p=0}^{\infty} u'_p(\lambda_k), \quad \dots, \quad f^{(m_k-1)}(\lambda_k) = \sum_{p=0}^{\infty} u_p^{(m_k-1)}(\lambda_k)$$
$$(k = 1, 2, \dots, s),$$

where the series on the right-hand sides of these equations converge. In other words, if we set

$$s_p(\lambda) = \sum_{q=0}^{p} u_q(\lambda) \qquad (p = 0, 1, 2, \dots),$$

then (30) is equivalent to

$$f(\Lambda_A) = \lim_{p \to \infty} s_p(\Lambda_A). \tag{31}$$

It is obvious that the theorem just proved can be stated in the following equivalent form:

THEOREM 1′: *The series* $\sum_{p=0}^{\infty} u_p(A)$ *converges to a matrix if and only if the series* $\sum_{p=0}^{\infty} u_p(\lambda)$ *converges on the spectrum of* A. *Moreover, the equation*

$$f(\Lambda_A) = \sum_{p=0}^{\infty} u_p(\Lambda_A)$$

implies that

$$f(A) = \sum_{p=0}^{\infty} u_p(A),$$

and conversely.

3. Suppose a power series is given with the circle of convergence $|\lambda - \lambda_0| < R$ and the sum $f(\lambda)$:

$$f(\lambda) = \sum_{p=0}^{\infty} \alpha_p (\lambda - \lambda_0)^p \qquad (|\lambda - \lambda_0| < R). \tag{32}$$

Since a power series may be differentiated term by term any number of times within the circle of convergence, (32) converges on the spectrum of any matrix whose characteristic values lie within the circle of convergence.

Thus we have:

THEOREM 2: *If the function* $f(\lambda)$ *can be expanded in a power series in the circle* $|\lambda - \lambda_0| < r$,

$$f(\lambda) = \sum_{p=0}^{\infty} \alpha_p (\lambda - \lambda_0)^p, \tag{33}$$

then this expansion remains valid when the scalar argument λ *is replaced by a matrix* A *whose characteristic values lie within the circle of convergence.*

Note. In this theorem we may allow a characteristic value λ_k of A to fall on the circumference of the circle of convergence; but we must then postulate in addition that the series (33), differentiated $m_k - 1$ times term by term, should converge at the point $\lambda = \lambda_k$. It is well known that this already implies the convergence of the j times differentiated series (33) at the point λ_k to $f^{(j)}(\lambda_k)$ for $j = 0, 1, \ldots, m_k - 1$.

The theorem just proved leads, for example, to the following expansions:[15]

$$e^A = \sum_{p=0}^{\infty} \frac{A^p}{p!}, \quad \cos A = \sum_{p=0}^{\infty} \frac{(-1)^p}{(2p)!} A^{2p}, \quad \sin A = \sum_{p=0}^{\infty} (-1)^p \frac{A^{2p+1}}{(2p+1)!},$$

$$\cosh A = \sum_{p=0}^{\infty} \frac{A^{2p}}{(2p)!}, \quad \sinh A = \sum_{p=0}^{\infty} \frac{A^{2p+1}}{(2p+1)!},$$

$$(E - A)^{-1} = \sum_{p=0}^{\infty} A^p \qquad (|\lambda_k| < 1;\ k = 1, 2, \ldots, s),$$

$$\ln A = \sum_{p=1}^{\infty} \frac{(-1)^{p-1}}{p} (A - E)^p \qquad (|\lambda_k - 1| < 1;\ k = 1, 2, \ldots, s)$$

(by $\ln \lambda$ we mean here the so-called principal value of the many-valued function $\text{Ln}\,\lambda$, i.e., that branch for which $\text{Ln}\,1 = 0$).

Let $G(u_1, u_2, \ldots, u_l)$ be a polynomial in $u_1, u_2, \ldots, u_l$; let $f_1(\lambda), f_2(\lambda), \ldots, f_l(\lambda)$ be functions of λ defined on the spectrum of the matrix A, and let

$$g(\lambda) \equiv G[f_1(\lambda), f_2(\lambda), \ldots, f_l(\lambda)].$$

Then from

$$g(\Lambda_A) = 0 \tag{34}$$

there follows:

$$G[f_1(A), f_2(A), \ldots, f_l(A)] = O. \tag{35}$$

[15] The expansions in the first two rows hold for an arbitrary matrix A.

For let us denote by $r_1(\lambda), r_2(\lambda), \ldots, r_l(\lambda)$ the Lagrange-Sylvester interpolation polynomials for $f_1(\lambda), f_2(\lambda), \ldots, f_l(\lambda)$, and let us set:

$$h(\lambda) = G[r_1(\lambda), r_2(\lambda), \ldots, r_l(\lambda)].$$

Then (34) implies

$$h(\Lambda_A) = 0. \tag{36}$$

Hence it follows that

$$G[f_1(A), f_2(A), \ldots, f_l(A)] = G[r_1(A), r_2(A), \ldots, r_l(A)] = h(A) = O,$$

and this is what we had to show.

This result allows us to extend identities between functions of a scalar variable to matrix values of the argument.

For example, from

$$\cos^2 \lambda + \sin^2 \lambda = 1$$

we obtain for an arbitrary matrix A

$$\cos^2 A + \sin^2 A = E$$

(in this case $G(u_1, u_2) = u_1^2 + u_2^2 - 1$, $f_1(\lambda) = \cos \lambda$, and $f_2(\lambda) = \sin \lambda$).

Similarly, for every matrix A

$$e^A e^{-A} = E,$$

i.e.,

$$e^{-A} = (e^A)^{-1}$$

Further, for every matrix A

$$e^{iA} = \cos A + i \sin A$$

Let A be a non-singular matrix ($|A| \neq 0$). We denote by $\sqrt{\lambda}$ the single-valued branch of the many-valued function $\sqrt{\lambda}$ that is defined in a domain not containing the origin and containing all the characteristic values of A. Then $\sqrt{A}$ has a meaning. From $(\sqrt{\lambda})^2 - \lambda = 0$ it now follows that

$$(\sqrt{A})^2 = A.$$

Let $f(\lambda) = \frac{1}{\lambda}$ and let $A = \| a_{ik} \|_1^n$ be a non-singular matrix. Then $f(\lambda)$ is defined as the spectrum of A, and in the equation

$$\lambda f(\lambda) = 1$$

we can therefore replace λ by A:

$$A \cdot f(A) = E,$$

i.e.,[16]

$$f(A) = A^{-1}.$$

Denoting by $r(\lambda)$ the interpolation polynomial for the function $1/\lambda$ we may represent the inverse matrix A^{-1} in the form of a polynomial in A:

---

[16] We have already made use of this on p. 109. See footnote 10.

$$A^{-1}=r(A).$$

Let us consider a rational function $\varrho(\lambda)=\frac{g(\lambda)}{h(\lambda)}$, where $g(\lambda)$ and $h(\lambda)$ are co-prime polynomials in λ. This function is defined on the spectrum of A if and only if the characteristic values of A are not roots of $h(\lambda)$, i.e.,[17] if $|h(A)|\neq 0$. Under this assumption we may replace λ by A in the identity

$$\varrho(\lambda)h(\lambda)=g(\lambda),$$

obtaining:

$$\varrho(A)h(A)=g(A).$$

Hence

$$\varrho(A)=g(A)[h(A)]^{-1}=[h(A)]^{-1}g(A). \tag{37}$$

Notes. 1) If $\boldsymbol{A}$ is a linear operator in an n-dimensional space $\boldsymbol{R}$, then $f(\boldsymbol{A})$ is defined exactly like $f(A)$:

$$f(\boldsymbol{A})=r(\boldsymbol{A}),$$

where $r(\lambda)$ is the Lagrange-Sylvester interpolation polynomial for $f(\lambda)$ on the spectrum of the operator $\boldsymbol{A}$ (the spectrum of $\boldsymbol{A}$ is determined by the minimal annihilating polynomial $\psi(\lambda)$ of $\boldsymbol{A}$).

According to this definition, if the matrix $A=\|a_{ik}\|_1^n$ corresponds to the operator $\boldsymbol{A}$ in some basis of the space, then in the same basis the matrix $f(A)$ corresponds to the operator $f(\boldsymbol{A})$. All the statements of this chapter in which there occurs a matrix A remain valid after replacement of the matrix A by the operator $\boldsymbol{A}$.

2) We can also define[18] a function of a matrix $f(A)$ starting from the characteristic polynomial

$$\Delta(\lambda)=\prod_{k=1}^{s}(\lambda-\lambda_k)^{n_k}$$

instead of the minimal polynomial

$$\psi(\lambda)=\prod_{k=1}^{s}(\lambda-\lambda_k)^{m_k}.$$

---

[17] See (25) on p. 84.

[18] See, for example, MacMillan, W. D., *Dynamics of Rigid Bodies* (New York, 1936).

We have then to set $f(A)=g(A)$, where $g(\lambda)$ is an interpolation polynomial of degree less than n modulo $\Delta(\lambda)$ of the function $f(\lambda)$.[19] The formulas (17), (21), and (23) are to be replaced by the following[20]

$$f(A)=\sum_{k=1}^{s}\left[f(\lambda_k)\hat{Z}_{k1}+f'(\lambda_k)\hat{Z}_{k2}+\cdots+f^{(n_k-1)}(\lambda_k)\hat{Z}_{kn_k}\right], \tag{17'}$$

$$(\lambda E-A)^{-1}=\frac{B(\lambda)}{\Delta(\lambda)}=\sum_{k=1}^{s}\left[\frac{\hat{Z}_{k1}}{\lambda-\lambda_k}+\frac{1!\hat{Z}_{k2}}{(\lambda-\lambda_k)^2}+\cdots+\frac{(n_k-1)!\hat{Z}_{kn_k}}{(\lambda-\lambda_k)^{n_k-1}}\right], \tag{21'}$$

$$f(A)=\sum_{k=1}^{s}\frac{1}{(n_k-1)!}\left[\frac{B(\lambda)}{\Delta_k(\lambda)}f(\lambda)\right]_{\lambda=\lambda_k}^{(n_k-1)}, \tag{23'}$$

where

$$\Delta_k(\lambda)=\frac{\Delta(\lambda)}{(\lambda-\lambda_k)^{n_k}} \qquad (k=1,2,\ldots,s).$$

However, in (17′) the values $f^{(m_k)}(\lambda_k), f^{(m_k+1)}(\lambda_k),\ldots,f^{(n_k-1)}(\lambda_k)$ occur only fictitiously, because a comparison of (21) with (21′) yields:

$$\hat{Z}_{k1}=Z_{k1},\ \ldots,\ \hat{Z}_{km_k},\ \hat{Z}_{km_k+1}=\cdots=\hat{Z}_{kn_k}=O.$$

§ 5. Application of a Function of a Matrix to the Integration of a System of Linear Differential Equations with Constant Coefficients

1. We begin by considering a system of homogeneous linear differential equations of the first order with constant coefficients:

$$\left.\begin{aligned}\frac{dx_1}{dt}&=a_{11}x_1+a_{12}x_2+\cdots+a_{1n}x_n\\ \frac{dx_2}{dt}&=a_{21}x_1+a_{22}x_2+\cdots+a_{2n}x_n\\ &\cdots\cdots\cdots\cdots\cdots\cdots\\ \frac{dx_n}{dt}&=a_{n1}x_1+a_{n2}x_2+\cdots+a_{nn}x_n,\end{aligned}\right\} \tag{38}$$

where t is the independent variable, $x_1, x_2, \ldots, x_n$ are unknown functions of t, and a_{ik} $(i, k=1,2,\ldots,n)$ are complex numbers.

We introduce the square matrix $A=\|a_{ik}\|_1^n$ of the coefficients and the column matrix $x=(x_1, x_2, \ldots, x_n)$. Then the system (38) can be written in the form of a single matrix differential equation

[19] The polynomial $g(\lambda)$ is not uniquely determined by the equation $f(A)=g(A)$ and the condition 'degree less than n.'

[20] The special case of (23′) in which $f(\lambda)\equiv\lambda^n$ is sometimes called Perron's formula (see [40], pp. 25-27).

$$\frac{dx}{dt} = Ax. \tag{39}$$

Here, and in what follows, we mean by the *derivative of a matrix* that matrix which is obtained from the given one by replacing all its elements by their derivatives. Therefore $\frac{dx}{dt}$ is the column matrix with the elements $\frac{dx_1}{dt}, \frac{dx_2}{dt}, \dots, \frac{dx_n}{dt}$.

We shall seek a solution of the system of differential equations satisfying the following initial conditions:

$$x_1 |_{t=0} = x_{10}, \quad x_2 |_{t=0} = x_{20}, \quad \dots, \quad x_n |_{t=0} = x_{n0}$$

or, briefly,

$$x |_{t=0} = x_0. \tag{40}$$

Let us expand the unknown column x into a MacLaurin series in powers of t:

$$x = x_0 + \dot{x}_0 t + \ddot{x}_0 \frac{t^2}{2!} + \cdots \quad \left(\dot{x}_0 = \frac{dx}{dt}\Big|_{t=0}, \ \ddot{x}_0 = \frac{d^2x}{dt^2}\Big|_{t=0}, \ \dots\right). \tag{41}$$

Then by successive differentiations we find from (39):

$$\frac{d^2x}{dt^2} = A\frac{dx}{dt} = A^2 x, \quad \frac{d^3x}{dt^3} = A\frac{d^2x}{dt^2} = A^3 x, \quad \dots . \tag{42}$$

Substituting the value $t = 0$ in (39) and (42), we obtain:

$$\dot{x}_0 = Ax_0, \quad \ddot{x}_0 = A^2 x_0, \ \dots .$$

Now the series (41) can be written as follows:

$$x = x_0 + tAx_0 + \frac{t^2}{2!} A^2 x_0 + \cdots = e^{At} x_0. \tag{43}$$

By direct substitution in (39) we see[21] that (43) is a solution of the differential equation (39). Setting $t = 0$ in (43), we find:

$$x |_{t=0} = x_0.$$

Thus, the formula (43) gives us the solution of the given system of differential equations satisfying the initial conditions (40).

Let us set $f(\lambda) = e^{\lambda t}$ in (17). Then

$$e^{At} = \| q_{ik}(t) \|_1^n = \sum_{k=1}^{s} (Z_{k1} + Z_{k2} t + \cdots + Z_{km_k} t^{m_k - 1}) e^{\lambda_k t}. \tag{44}$$

---

[21] $\frac{d}{dt}(e^{At}) = \frac{d}{dt}\left(E + At + \frac{A^2 t^2}{2!} + \cdots\right) = A + A^2 t + \frac{A^3 t^2}{2!} + \cdots = Ae^{At}.$

The solution (43) may then be written in the following form:

$$\left.\begin{aligned} x_1 &= q_{11}(t)\,x_{10} + q_{12}(t)\,x_{20} + \cdots + q_{1n}(t)\,x_{n0} \\ x_2 &= q_{21}(t)\,x_{10} + q_{22}(t)\,x_{20} + \cdots + q_{2n}(t)\,x_{n0} \\ &\cdots\cdots\cdots\cdots\cdots\cdots\cdots\cdots \\ x_n &= q_{n1}(t)\,x_{10} + q_{n2}(t)\,x_{20} + \cdots + q_{nn}(t)\,x_{n0}, \end{aligned}\right\} \qquad (45)$$

where $x_{10}, x_{20}, \ldots, x_{n0}$ are constants equal to the initial values of the unknown functions $x_1, x_2, \ldots, x_n$.

Thus, the integration of the given system of differential equations reduces to the computation of the elements of the matrix e^{At}.

If $t = t_0$ is taken as the initial value of the argument, then (43) is to be replaced by the formula

$$x = e^{A(t-t_0)}\,x_0. \qquad (46)$$

Example.

$$\frac{dx_1}{dt} = 3x_1 - x_2 + x_3,$$

$$\frac{dx_2}{dt} = 2x_1 \qquad + x_3,$$

$$\frac{dx_3}{dt} = x_1 \quad - x_2 + 2x_3.$$

The coefficient matrix is

$$A = \left\| \begin{matrix} 3 & -1 & 1 \\ 2 & 0 & 1 \\ 1 & -1 & 2 \end{matrix} \right\|.$$

We form the characteristic determinant

$$\Delta(\lambda) = -\begin{vmatrix} 3-\lambda & 1 & 1 \\ 2 & -\lambda & 1 \\ 1 & -1 & 2-\lambda \end{vmatrix} = (\lambda - 1)(\lambda - 2)^2.$$

The greatest common divisor of the minors of order 2 is $D_2(\lambda) = 1$. Therefore

$$\psi(\lambda) = \Delta(\lambda) = (\lambda - 1)(\lambda - 2)^2.$$

The fundamental formula is

$$f(A) = f(1)\,Z_1 + f(2)\,Z_2 + f'(2)\,Z_3.$$

For $f(\lambda)$ we choose in succession 1, $\lambda - 2$, $(\lambda - 2)^2$. We obtain:

$$Z_1 + Z_2 = E = \left\| \begin{matrix} 1 & 0 & 0 \\ 0 & 1 & 0 \\ 0 & 0 & 1 \end{matrix} \right\|,$$

$$-Z_1 + Z_3 = A - 2E = \left\| \begin{matrix} 1 & -1 & 1 \\ 2 & -2 & 1 \\ 1 & -1 & 0 \end{matrix} \right\| \begin{matrix} 1 \\ 1, \\ 0 \end{matrix}$$

$$Z_1 = (A - 2E)^2 = \left\| \begin{matrix} 0 & 0 & 0 \\ -1 & 1 & 0 \\ -1 & 1 & 0 \end{matrix} \right\| \begin{matrix} 0 \\ 0. \\ 0 \end{matrix}$$

Hence we determine Z_1, Z_2, Z_3 and substitute in the fundamental formula

$$f(A) = f(1) \left\| \begin{matrix} 0 & 0 & 0 \\ -1 & 1 & 0 \\ -1 & 1 & 0 \end{matrix} \right\| + f(2) \left\| \begin{matrix} 1 & 0 & 0 \\ 1 & 0 & 0 \\ 1 & -1 & 1 \end{matrix} \right\| + f'(2) \left\| \begin{matrix} 1 & -1 & 1 \\ 1 & -1 & 1 \\ 0 & 0 & 0 \end{matrix} \right\|.$$

If we now replace $f(\lambda)$ by $e^{\lambda t}$, we obtain:

$$e^{At} = e^t \left\| \begin{matrix} 0 & 0 & 0 \\ -1 & 1 & 0 \\ -1 & 1 & 0 \end{matrix} \right\| + e^{2t} \left\| \begin{matrix} 1 & 0 & 0 \\ 1 & 0 & 0 \\ 1 & -1 & 1 \end{matrix} \right\| + te^{2t} \left\| \begin{matrix} 1 & -1 & 1 \\ 1 & -1 & 1 \\ 0 & 0 & 0 \end{matrix} \right\| = \left\| \begin{matrix} (1+t)e^{2t} & -te^{2t} & te^{2t} \\ -e^t + (1+t)e^{2t} & e^t - te^{2t} & te^{2t} \\ -e^t + e^{2t} & e^t - e^{2t} & e^{2t} \end{matrix} \right\|.$$

Thus

$$\begin{aligned} x_1 &= C_1 (1+t) e^{2t} - C_2 te^{2t} + C_3 te^{2t}, \\ x_2 &= C_1 [-e^t + (1+t) e^{2t}] + C_2 (e^t - te^{2t}) + C_3 te^{2t}, \\ x_3 &= C_1 (-e^t + e^{2t}) + C_2 (e^t - e^{2t}) + C_3 e^{2t}, \end{aligned}$$

where

$$C_1 = x_{10}, \quad C_2 = x_{20}, \quad C_3 = x_{30}.$$

2. We now consider a system of inhomogeneous linear differential equations with constant coefficients:

$$\left. \begin{aligned} \frac{dx_1}{dt} &= a_{11}x_1 + a_{12}x_2 + \cdots + a_{1n}x_n + f_1(t) \\ \frac{dx_2}{dt} &= a_{21}x_1 + a_{22}x_2 + \cdots + a_{2n}x_n + f_2(t) \\ &\cdots\cdots\cdots\cdots\cdots\cdots\cdots\cdots \\ \frac{dx_n}{dt} &= a_{n1}x_1 + a_{n2}x_2 + \cdots + a_{nn}x_n + f_n(t), \end{aligned} \right\} \tag{47}$$

where $f_i(t)$ $(i=1, 2, \ldots, n)$ are continuous functions in the interval $t_0 \leqq t \leqq t_1$. Denoting by $f(t)$ the column matrix with the elements $f_1(t), f_2(t), \ldots, f_n(t)$ and again setting $A = \| a_{ik} \|_1^n$, we write the system (47) as follows:

$$\frac{dx}{dt} = Ax + f(t). \tag{48}$$

We replace x by a new column z of unknown functions, connected with x by the relation

$$x = e^{At}z \tag{49}$$

Differentiating (49) term by term and substituting the expression for $\frac{dx}{dt}$ in (48) we find:[22]

$$e^{At}\frac{dz}{dt} = f(t). \tag{50}$$

Hence[23]

$$z(t) = c + \int_{t_0}^{t} e^{-A\tau} f(\tau)\, d\tau \tag{51}$$

and so by (49)

$$x = e^{At}\left[c + \int_{t_0}^{t} e^{-A\tau} f(\tau)\, d\tau\right] = e^{At}c + \int_{t_0}^{t} e^{A(t-\tau)} f(\tau)\, d\tau\,; \tag{52}$$

where c is a column with arbitrary constant elements.

When we give to the argument t in (52) the value t_0 we find $c = e^{-At_0}x_0$; so that (52) can be written in the following form:

$$x = e^{A(t-t_0)}x_0 + \int_{t_0}^{t} e^{A(t-\tau)} f(\tau)\, d\tau. \tag{53}$$

---

[22] See footnote 21.

[23] If a matrix function of a scalar argument is given, $B(\tau) = \| b_{ik}(\tau) \|$ $(i = 1, 2, \ldots, m;\ k = 1, 2, \ldots, n;\ t_1 \leqq \tau \leqq t_2)$, then the *integral* $\int_{t_1}^{t_2} B(\tau)\, d\tau$ is defined in the natural way:

$$\int_{t_1}^{t_2} B(\tau)\, d\tau = \left\| \int_{t_1}^{t_2} b_{ik}(\tau)\, d\tau \right\| \qquad (i = 1, 2, \ldots, m;\ k = 1, 2, \ldots, n).$$

Setting $e^{At} = \| q_{ik}(t) \|_1^n$, we can write the solution (53) in expanded form:

$$\left.\begin{aligned} x_1 &= q_{11}(t-t_0)\,x_{10} + \cdots + q_{1n}(t-t_0)\,x_{n0} + \\ &\quad + \int_{t_0}^{t} [q_{11}(t-\tau) f_1(\tau) + \cdots + q_{1n}(t-\tau) f_n(\tau)]\, d\tau \\ &\cdots\cdots\cdots\cdots\cdots\cdots \\ x_n &= q_{n1}(t-t_0)\,x_{10} + \cdots + q_{nn}(t-t_0)\,x_{n0} + \\ &\quad + \int_{t_0}^{t} [q_{n1}(t-\tau) f_1(\tau) + \cdots + q_{nn}(t-\tau) f_n(\tau)]\, d\tau . \end{aligned}\right\} \tag{54}$$

3. As an example we consider *the motion of a heavy material point in a vacuum near the surface of the earth, taking the motion of the earth into account.* It is known[24] that in this case the acceleration of the point relative to the earth is determined by the constant force of gravity $m\boldsymbol{g}$ and the inertial Coriolis force $-2m\boldsymbol{\omega} \times \boldsymbol{v}$ ($\boldsymbol{v}$ is the velocity of the point relative to the earth, $\boldsymbol{\omega}$ the constant angular velocity of the earth). Therefore the differential equation of motion of the point has the form[25]

$$\frac{d\boldsymbol{v}}{dt} = \boldsymbol{g} - 2\boldsymbol{\omega} \times \boldsymbol{v}. \tag{55}$$

We define a linear operator $\boldsymbol{A}$ in three-dimensional euclidean space by the equation

$$\boldsymbol{A}\boldsymbol{x} = -2\boldsymbol{\omega} \times \boldsymbol{x} \tag{56}$$

and write instead of (55)

$$\frac{d\boldsymbol{v}}{dt} = \boldsymbol{A}\boldsymbol{v} + \boldsymbol{g}. \tag{57}$$

Comparing (57) with (48), we easily find from (53):

$$\boldsymbol{v} = e^{\boldsymbol{A}t}\boldsymbol{v}_0 + \int_0^t e^{\boldsymbol{A}\tau}\, d\tau \cdot \boldsymbol{g} \qquad (\boldsymbol{v}_0 = \boldsymbol{v}|_{t=0}).$$

Integrating term by term, we determine the radius vector of the motion of the point:

$$\boldsymbol{r} = \boldsymbol{r}_0 + \int_0^t e^{\boldsymbol{A}\tau}\, d\tau\, \boldsymbol{v}_0 + \int_0^t\!\!\int_0^\tau e^{\boldsymbol{A}\sigma}\, d\sigma\, d\tau\, \boldsymbol{g}, \tag{58}$$

where

$$\boldsymbol{r}_0 = \boldsymbol{r}|_{t=0} \text{ and } \boldsymbol{v}_0 = \boldsymbol{v}|_{t=0}.$$

---

[24] See A. Sommerfeld, *Lectures on Theoretical Physics*, Vol. I (Mechanics), § 30.

[25] Here the symbol $\times$ denotes the vector product.

Substituting for e^{At} the series

$$E + A\frac{t}{1!} + A^2\frac{t^2}{2!} + \cdots$$

and replacing A by its expression from (56), we have:

$$r = r_0 + v_0 t + \frac{1}{2} g t^2 - \omega \times \left(v_0 t^2 + \frac{1}{3} g t^3\right) + \omega \times \left[\omega \times \left(\frac{2}{3} v_0 t^3 + \frac{1}{6} g t^4\right)\right] + \cdots .$$

Considering that the angular velocity ω is small (for the earth, $\omega \approx 7.3 \times 10^{-5}\ \mathrm{sec}^{-1}$), we neglect the terms containing the second and higher powers of ω; for the additional displacement of the point due to the motion of the earth we then obtain the approximate formula

$$d = -\omega \times \left(v_0 t^2 + \frac{1}{3} g t^3\right).$$

Returning to the exact solution (58), let us compute e^{At}. As a preliminary we establish that the minimal polynomial of the operator A has the form

$$\psi(\lambda) = \lambda(\lambda^2 + 4\omega^2).$$

For we find from (56)

$$A^2 x = 4\omega \times (\omega \times x) = 4(\omega x)\omega = 4\omega^2 x,$$
$$A^3 x = -2\omega \times A^2 x = 8\omega^2(\omega \times x).$$

Hence and from (56) it follows that the operators E, A, A^2 are linearly independent and that

$$A^3 + 4\omega^2 A = O.$$

The minimal polynomial $\psi(\lambda)$ has the simple roots $0, 2\omega i, -2\omega i$. The Lagrange interpolation formula for e^{At} has the form

$$1 + \frac{\sin 2\omega t}{2\omega}\lambda + \frac{1 - \cos 2\omega t}{4\omega^2}\lambda^2.$$

Then

$$e^{At} = E + \frac{\sin 2\omega t}{2\omega} A + \frac{1 - \cos 2\omega t}{4\omega^2} A^2.$$

Substituting this expression for e^{At} in (58) and replacing the operator A by its expression from (56), we find

$$r = r_0 + v_0 t + g\frac{t^2}{2} - \omega \times \left(\frac{1 - \cos 2\omega t}{2\omega^2} v_0 + \frac{2\omega t - \sin 2\omega t}{4\omega^3} g\right) +$$
$$+ \omega \times \left[\omega \times \left(\frac{2\omega t - \sin 2\omega t}{2\omega^3} v_0 + \frac{-1 + 2\omega^2 t^2 + \cos 2\omega t}{4\omega^4} g\right)\right]. \qquad (59)$$

Let us consider the special case $\boldsymbol{v}_0 = \boldsymbol{o}$. When we expand the triple vector product we obtain:

$$\boldsymbol{r} = \boldsymbol{r}_0 + \boldsymbol{g}\frac{t^2}{2} + \frac{2\omega t - \sin 2\omega t}{4\omega^3}(\boldsymbol{g} \times \boldsymbol{\omega}) + \frac{\cos 2\omega t - 1 + 2\omega^2 t^2}{4\omega^3}(\boldsymbol{g} \sin\varphi\, \boldsymbol{\omega} - \omega \boldsymbol{g}),$$

where φ is the geographical latitude of the point whose motion we are considering. The term

$$\frac{2\omega t - \sin 2\omega t}{4\omega^3}(\boldsymbol{g} \times \boldsymbol{\omega})$$

represents the eastward displacement perpendicular to the plane of the meridian, and the last term on the right-hand side of the last formula gives the displacement in the meridian plane perpendicular to, and away from, the earth's axis.

4. Suppose now that the following system of linear differential equations of the second order is given:

$$\left.\begin{array}{l} \dfrac{d^2x_1}{dt^2} + a_{11}x_1 + a_{12}x_2 + \cdots + a_{1n}x_n = 0 \\ \dfrac{d^2x_2}{dt^2} + a_{21}x_1 + a_{22}x_2 + \cdots + a_{2n}x_n = 0 \\ \cdots\cdots\cdots\cdots\cdots\cdots\cdots\cdots \\ \dfrac{d^2x_n}{dt^2} + a_{n1}x_1 + a_{n2}x_2 + \cdots + a_{nn}x_n = 0\,, \end{array}\right\} \tag{60}$$

where the a_{ik} $(i, k = 1, 2, \ldots, n)$ are constant coefficients. Introducing again the column $x = (x_1, x_2, \ldots, x_n)$ and the square matrix $A = \| a_{ik} \|_1^n$, we rewrite (60) in matrix form

$$\frac{d^2x}{dt^2} + Ax = O. \tag{60'}$$

We consider, to begin with, the case in which $|A| \neq 0$. If $n = 1$, i.e., if x and A are scalars and $A \neq 0$, the general solution of the equation (60) can be written in the form

$$x = \cos(\sqrt{A}\,t)\,x_0 + (\sqrt{A})^{-1} \sin(\sqrt{A}\,t)\,\dot{x}_0 \tag{61}$$

where $x_0 = x\,|_{t=0}$ and $\dot{x}_0 = \left.\dfrac{dx}{dt}\right|_{t=0}$.

By direct verification we see that (61) is a solution of (60) for arbitrary n, where x is a column and A a non-singular square matrix.[26] Here we use the formulas

---

[26] By $\sqrt{A}$ we mean a matrix whose square is equal to A. $\sqrt{A}$, we know, exists when $|A| \neq 0$ (see p. 114).

$$\left.\begin{aligned}\cos(\sqrt{A}\,t) &= E - \frac{1}{2!}At^2 + \frac{1}{4!}A^2t^4 - \cdots,\\ (\sqrt{A})^{-1}\sin(\sqrt{A}\,t) &= Et - \frac{1}{3!}At^3 + \frac{1}{5!}A^2t^5 - \cdots.\end{aligned}\right\} \qquad (62)$$

Formula (61) comprises all solutions of the system (60) or (60′), as the initial values x_0 and $\dot{x}_0$ may be chosen arbitrarily.

The right-hand sides of the formulas (62) have a meaning even when $|A| = 0$. Therefore (61) is the general solution of the given system of differential equations also when $|A| = 0$, provided only that the functions $\cos(\sqrt{A}t)$ and $(\sqrt{A})^{-1}\sin(\sqrt{A}t)$, which are part of this expression, are interpreted as the right-hand sides of the formulas (62).

We leave it to the reader to verify that the general solution of the inhomogeneous system

$$\frac{d^2x}{dt^2} + Ax = f(t) \qquad (63)$$

satisfying the initial conditions $x|_{t=0} = x_0$ and $\left.\frac{dx}{dt}\right|_{t=0} = \dot{x}_0$ can be written in the form

$$x = \cos(\sqrt{A}\,t)\,x_0 + (\sqrt{A})^{-1}\sin(\sqrt{A}\,t)\,\dot{x}_0 + \\ + (\sqrt{A})^{-1}\int_0^t \sin[\sqrt{A}\,(t-\tau)]\,f(\tau)\,d\tau. \qquad (64)$$

If $t = t_0$ is taken as the initial time, then in (61) and (64) $\cos(\sqrt{A}t)$ and $\sin(\sqrt{A}t)$ must be replaced by $\cos(\sqrt{A}\,(t-t_0))$ and $\sin(\sqrt{A}\,(t-t_0))$, and $\int_0^t$ by $\int_{t_0}^t$.

In the special case

$$f(t) = h\sin(pt + \alpha)$$

(h is a constant column, and p and α are numbers), (64) can be replaced by:

$$x = \cos(\sqrt{A}\,t)\,c + (\sqrt{A})^{-1}\sin(\sqrt{A}\,t)\,d + (A - p^2E)^{-1}h\sin(pt + \alpha),$$

where c and d are columns with arbitrary constant elements. This formula has meaning when p^2 is not a characteristic value of the matrix A ($|A - p^2E| \neq 0$).

§ 6. Stability of Motion in the Case of a Linear System

1. Let $x_1, x_2, \ldots, x_n$ be parameters that characterize the displacement of 'perturbed' motion of a given mechanical system from an original motion,[27] and suppose that these parameters satisfy a system of differential equations of the first order:

$$\frac{dx_i}{dt} = f_i(x_1, x_2, \ldots, x_n, t) \qquad (i = 1, 2, \ldots, n); \tag{65}$$

the independent variable t in these equations is the time, and the right-hand sides $f_i(x_1, x_2, \ldots, x_n, t)$ are continuous functions of the variables $x_1, \ldots, x_n$ in some domain containing the point $x_1 = 0, x_2 = 0, \ldots, x_n = 0)$ for all $t \geqq t_0$ (t_0 is the initial time).

We now introduce the definition of stability of motion according to Lyapunov.[28]

The motion to be investigated is called *stable* if for every $\varepsilon > 0$ we can find a $\delta > 0$ such that for arbitrary initial values of the parameters $x_{10}, x_{20}, \ldots, x_{n0}$ (for $t = t_0$) with moduli less than δ the parameters $x_1, x_2, \ldots, x_n$ remain of moduli less than ε for the whole time of the motion ($t \geqq t_0$), i.e., if for every $\varepsilon > 0$ we can find a $\delta > 0$ such that from

$$|x_{i0}| < \delta \qquad (i = 1, 2, \ldots, n) \tag{66}$$

it follows that

$$|x_i(t)| < \varepsilon \qquad (t \geqq t_0). \tag{67}$$

If, in addition, for some $\delta > 0$ we always have $\lim_{t \to +\infty} x_i(t) = 0$ $(i = 1, 2, \ldots, n)$ as long as $|x_{i0}| < \delta$ $(i = 1, 2, \ldots, n)$, then the motion is called *asymptotically stable.*

We now consider a linear system, i.e., that special case when (65) is a system of linear homogeneous differential equations

$$\frac{dx_i}{dt} = \sum_{k=1}^{n} p_{ik}(t)\, x_k, \tag{68}$$

where the $p_{ik}(t)$ are continuous functions for $t \geqq t_0$ $(i, k = 1, 2, \ldots, n)$.

In matrix form the system (68) can be written as follows:

---

[27] In these parameters, the motion to be studied is characterized by constant zero values $x_1 = 0, x_2 = 0, \ldots, x_n = 0$. Therefore in the mathematical treatment of the problem we speak of the 'stability' of the zero solution of the system (65) of differential equations.

[28] See [14], p. 13; [9], pp. 10-11; or [36], pp. 11-12. See also [3].

$$\frac{dx}{dt} = P(t)\,x, \tag{68'}$$

where x is the column matrix with the elements $x_1, x_2, \ldots, x_n$ and $P(t) = \| p_{ik}(t) \|_1^n$ is the coefficient matrix.

We denote by

$$q_{1j}(t),\ q_{2j}(t),\ \ldots,\ q_{nj}(t) \qquad (j = 1, 2, \ldots, n) \tag{69}$$

n linearly independent solutions of (68).[29] The matrix $Q(t) = \| q_{ij} \|_1^n$ whose columns are these solutions is called an *integral matrix* of the system (68).

Every solution of the system of linear homogeneous differential equations is obtained as a linear combination of n linearly independent solutions with constant coefficients:

$$x_i = \sum_{j=1}^{n} c_j q_{ij}(t) \qquad (t = 1, 2, \ldots, n),$$

or in matrix form,

$$x = Q(t)\,c, \tag{70}$$

where c is the column matrix whose elements are arbitrary constants $c_1, c_2, \ldots, c_n$.

We now choose the special integral matrix for which

$$Q(t_0) = E; \tag{71}$$

in other words, in the choice of n linearly independent solutions of (69) we shall start from the following special initial conditions:[30]

$$q_{ij}(t_0) = \delta_{ij} = \begin{cases} 0 & (i \neq j), \\ 1 & (i = j) \end{cases} \qquad (i, j = 1, 2, \ldots, n).$$

Then setting $t = t_0$ in (70), we find from (71):

$$x_0 = c,$$

and therefore formula (70) assumes the form

$$x = Q(t)\,x_0 \tag{72}$$

or, in expanded form,

$$x_i = \sum_{j=1}^{n} q_{ij}(t)\,x_{j0} \qquad (i = 1, 2, \ldots, n). \tag{72'}$$

---

[29] Here the second subscript j denotes the number of the solution.

[30] Arbitrary initial conditions determine uniquely a certain solution of a given system.

We consider three cases:

1. $Q(t)$ is a *bounded matrix* in the interval $(t_0, +\infty)$, i.e., there exists a number M such that

$$|q_{ij}(t)| \leqq M \qquad (t \geqq t_0;\ i, j = 1, 2, \ldots, n).$$

In this case it follows from (72′) that

$$|x_i(t)| \leqq nM \max |x_{j0}|.$$

The condition of stability is satisfied. (It is sufficient to take $\delta < \frac{\varepsilon}{nM}$ in (66) and (67).) *The motion characterized by the zero solution* $x_1 = 0$, $x_2 = 0, \ldots, x_n = 0$ *is stable.*

2. $\lim_{t \to +\infty} Q(t) = O$. In this case the matrix $Q(t)$ is bounded in the interval $(t_0, +\infty)$ and therefore, as we have already explained, the motion is stable. Moreover, it follows from (72) that

$$\lim_{t \to +\infty} x(t) = o.$$

for every x_0. *The motion is asymptotically stable.*

3. $Q(t)$ is an *unbounded matrix* in the interval $(t_0, +\infty)$. This means that at least one of the functions $q_{ij}(t)$, say $q_{hk}(t)$, is not bounded in the interval. We take the initial conditions $x_{10} = 0$, $x_{20} = 0, \ldots,$ $x_{k-1,0} = 0$, $x_{k0} \neq 0$, $x_{k+1,0} = 0, \ldots, x_{n0} = 0$. Then

$$x_h(t) = q_{hk}(t)\, x_{k0}.$$

However small in modulus x_{k0} may be, the function $x_h(t)$ is unbounded. The condition (67) is not satisfied for any δ. *The motion is unstable.*

2. We now consider the special case where the coefficients in the system (68) are constants:

$$P(t) = P = \text{const.} \tag{73}$$

We have then (see § 5)

$$x = e^{P(t - t_0)} x_0. \tag{74}$$

Comparing (74) with (72), we find that in this case

$$Q(t) = e^{P(t - t_0)}. \tag{75}$$

We denote by

$$\psi(\lambda) = (\lambda - \lambda_1)^{m_1} (\lambda - \lambda_2)^{m_2} \cdots (\lambda - \lambda_s)^{m_s}$$

the minimal polynomial of the coefficient matrix P.

For the investigation of the integral matrix (75) we apply formula (17) on p. 104. In this case $f(\lambda) = e^{\lambda(t-t_0)}$ (t is regarded as a parameter), $f^{(j)}(\lambda_k) = (t-t_0)^j e^{\lambda_k(t-t_0)}$. Formula (17) yields

$$e^{P(t-t_0)} = \sum_{k=1}^{s} [Z_{k1} + Z_{k2}(t-t_0) + \cdots + Z_{km_k}(t-t_0)^{m_k-1}] e^{\lambda_k(t-t_0)}. \qquad (76)$$

We consider three cases:

1. $\operatorname{Re}\lambda_k \leqq 0$ $(k = 1, 2, \ldots, s)$; and moreover, for all λ_k with $\operatorname{Re}\lambda_k = 0$ the corresponding $m_k = 1$ (i.e., pure imaginary characteristic values are simple roots of the minimal polynomial).
2. $\operatorname{Re}\lambda_k < 0$ $(k = 1, 2, \ldots, s)$.
3. For some k we have $\operatorname{Re}\lambda_k > 0$; or $\operatorname{Re}\lambda_k = 0$, but $m_k > 1$.

From the formula (76) it follows that in the first case the matrix $Q(t) = e^{P(t-t_0)}$ is bounded in the interval $(t_0, +\infty)$, in the second case $\lim_{t\to+\infty} e^{P(t-t_0)} = 0$, and in the third case the matrix $e^{P(t-t_0)}$ is not bounded in the interval $(t_0, +\infty)$.[31]

Therefore in the first case the motion $(x_1 = 0, x_2 = 0, \ldots, x_n = 0)$ is stable, in the second case it is asymptotically stable, and in the third case it is unstable.

---

[31] Special consideration is only required in the case when in (76) for $e^{P(t-t_0)}$ there occur several terms of maximal growth (for $t \to +\infty$), i.e., with maximal $\operatorname{Re}\lambda_k = \alpha_0$ and (for the given $\operatorname{Re}\lambda_k = \alpha_0$) maximal value $m_k = m_0$. The expression (76) can be represented in the form

$$e^{P(t-t_0)} = e^{\alpha_0(t-t_0)}(t-t_0)^{m_0-1}\left[\sum_{j=1}^{r} Z_{kjm_0} e^{i\beta_j(t-t_0)} + (*)\right],$$

where $\beta_1, \beta_2, \ldots, \beta_r$ are distinct real numbers and $(*)$ denotes a matrix that tends to zero as $t \to +\infty$. From this representation it follows that the matrix $e^{P(t-t_0)}$ is not bounded for $a_0 + m_0 - 1 > 0$, because the matrix $\sum_{j=1}^{r} Z_{kjm_0} e^{i\beta_j(t-t_0)}$ cannot converge for $t \to +\infty$. We can see this by showing that

$$f(t) = \sum_{j=1}^{r} c_j e^{i\beta_j t},$$

where c_j are complex numbers and β_j real and distinct numbers, can converge to zero for $t \to +\infty$ only when $f(t) \equiv 0$. But, in fact, it follows from $\lim_{t\to+\infty} f(t) = 0$ that

$$\sum_{j=1}^{r} |c_j|^2 = \lim_{T\to+\infty} \frac{1}{T}\int_0^T |f(t)|^2\, dt = 0$$

and therefore

$$c_1 = c_2 = \cdots = c_n = 0.$$

The results of the investigation may be formulated in the form of the following theorem:[32]

THEOREM 3: *The zero solution of the linear system* (68) *for* $P = \text{const.}$ *is stable in the sense of Lyapunov if*

1) *the real parts of all the characteristic values of* P *are negative or zero,*

2) *those characteristic values whose real part is zero, i.e., the pure imaginary characteristic values (if any such exist), are simple roots of the minimal polynomial of* P*;*

and it is unstable if at least one of the conditions 1), 2) *is violated.*

The zero solution of the linear system (68) *is asymptotically stable if and only if all the characteristic values of* P *have negative real parts.*

The considerations above enable us to make a statement about the nature of the integral matrix $e^{P(t-t_0)}$ in the general case of arbitrary characteristic values of the constant matrix P.

THEOREM 4: *The integral matrix* $e^{P(t-t_0)}$ *of the linear system* (68) *for* $P = \text{const.}$ *is always representable in the form*

$$e^{P(t-t_0)} = Z_-(t) + Z_0 + Z_+(t),$$

where

1) $\lim_{t \to +\infty} Z_-(t) = O$,

2) Z_0 *is either constant or is a bounded matrix in the interval* $(t_0, +\infty)$ *that does not have a limit for* $t \to +\infty$,

3) $Z_+(t) = O$ *or* $Z_+(t)$ *is an unbounded matrix in the interval* $(t_0, +\infty)$.

Proof. On the right-hand side of (76) we divide all the summands into three groups. We denote by $Z\ (t)$ the sum of all the terms containing the factors $e^{\lambda_k(t-t_0)}$, with $\operatorname{Re} \lambda_k < 0$. We denote by Z_0 the sum of all those matrices Z_{k1} for which $\operatorname{Re} \lambda_k = 0$. We denote by $Z_+(t)$ the sum of all the remaining terms. It is easy to see that $Z_-(t)$, $Z_0(t)$, and $Z_+(t)$ have the properties 1), 2), 3) of the theorem.

---

[32] On the question of sharpening the criteria of stability and instability for quasi-linear systems (i.e., of non-linear systems that become linear after neglecting the non-linear terms), see further Chapter XIV, § 3.

CHAPTER VI

EQUIVALENT TRANSFORMATIONS OF POLYNOMIAL MATRICES. ANALYTIC THEORY OF ELEMENTARY DIVISORS

The first three sections of this chapter deal with the theory of equivalent polynomial matrices. On the basis of this, we shall develop, in the next three sections, the analytical theory of elementary divisors, i.e., the theory of the reduction of a constant (non-polynomial) square matrix A to a normal form $\tilde{A}$ $(A = T\tilde{A}T^{-1})$. In the last two sections of the chapter two methods for the construction of the transforming matrix T will be given.

§ 1. Elementary Transformations of a Polynomial Matrix

1. Definition 1: *A polynomial matrix, or λ-matrix, is a rectangular matrix $A(\lambda)$ whose elements are polynomials in λ:*

$$A(\lambda) = \| a_{ik}(\lambda) \| = \| a_{ik}^{(0)}\lambda^l + a_{ik}^{(1)}\lambda^{l-1} + \cdots + a_{ik}^{(l)} \| \qquad (i = 1, 2, \ldots, m;\ k = 1, 2, \ldots, n);$$

here l is the largest of the degrees of the polynomials $a_{ik}(\lambda)$.

Setting

$$A_j = \| a_{ik}^{(j)} \| \qquad (i = 1, 2, \ldots, m;\ k = 1, 2, \ldots, n;\ j = 0, 1, \ldots, l),$$

we may represent the polynomial matrix $A(\lambda)$ in the form of a matrix polynomial in λ, i.e., in the form of a polynomial in λ with matrix coefficients:

$$A(\lambda) = A_0\lambda^l + A_1\lambda^{l-1} + \cdots + A_{l-1}\lambda + A_l .$$

We introduce the following *elementary operations* on a polynomial matrix $A(\lambda)$:

1. Multiplication of any row, for example the i-th, by a number $c \neq 0$.

2. Addition to any row, for example the i-th, of any other row, for example the j-th, multiplied by any arbitrary polynomial $b(\lambda)$.

3. Interchange of any two rows, for example the i-th and the j-th.

We leave it to the reader to verify that the operations 1., 2., 3. are equivalent to a multiplication of the polynomial matrix $A(\lambda)$ on the *left* by the following square matrices of order m, respectively:[1]

$$
\left.
\begin{array}{l}
S' = \left\Vert \begin{array}{ccccc} 1 & \cdots & \overset{(i)}{\cdot} & \cdots & 0 \\ \vdots & \ddots & \vdots & & \vdots \\ \cdot & & c & & \cdot \\ \vdots & & & \ddots & \vdots \\ 0 & \cdots & \cdots & \cdots & 1 \end{array} \right\Vert, \quad
S'' = \left\Vert \begin{array}{ccccccc} 1 & \cdots & \overset{(i)}{\cdot} & \cdots & \overset{(j)}{\cdot} & \cdots & 0 \\ \vdots & \ddots & \vdots & & \vdots & & \vdots \\ \cdot & & 1 & \cdots & b(\lambda) & \cdots & \cdot \\ \vdots & & & \ddots & & & \vdots \\ \vdots & & & & \ddots & & \vdots \\ 0 & \cdots & \cdots & \cdots & \cdots & \cdots & 1 \end{array} \right\Vert, \\[2ex]
\qquad S''' = \left\Vert \begin{array}{ccccccc} 1 & & \overset{(i)}{\cdot} & & \overset{(j)}{\cdot} & & 0 \\ & \ddots & \vdots & & \vdots & & \\ \cdots & & 0 & \cdots & 1 & & \cdots \\ & & \vdots & \ddots & \vdots & & \\ \cdots & & 1 & \cdots & 0 & & \cdots \\ & & \vdots & & \vdots & \ddots & \\ 0 & & \cdot & & \cdot & & 1 \end{array} \right\Vert ;
\end{array}
\right\} \qquad (1)
$$

in other words, as the result of applying the operations 1., 2., 3. the matrix $A(\lambda)$ is transformed into $S' \cdot A(\lambda)$, $S'' \cdot A(\lambda)$, and $S''' \cdot A(\lambda)$, respectively. The operations of type 1., 2., 3. are therefore called *left elementary* operations.

In the same way we define the *right elementary* operations on a polynomial matrix (these are performed not on the rows, but on the columns);[2] the matrices (of order n) corresponding to them are:

---

[1] In the matrices (1) all the elements that are not shown are 1 on the main diagonal and 0 elsewhere.

[2] See footnote 1.

$$T' = \begin{Vmatrix} 1 & \cdots & \cdots & \cdots & 0 \\ \vdots & \ddots & & & \vdots \\ \vdots & & c & \cdots & \vdots \\ \vdots & & & \ddots & \vdots \\ 0 & \cdots & \cdots & \cdots & 1 \end{Vmatrix} \begin{matrix} \\ \\ (i), \\ \\ \\ \end{matrix} \qquad T'' = \begin{Vmatrix} 1 & \cdots & \cdots & \cdots & \cdots & 0 \\ \vdots & \ddots & & & & \vdots \\ \vdots & & 1 & \cdots & \cdots & \vdots \\ \vdots & & \vdots & \ddots & & \vdots \\ \vdots & & b(\lambda) & \cdots & \cdots & \vdots \\ \vdots & & & & \ddots & \vdots \\ 0 & \cdots & \cdots & \cdots & \cdots & 1 \end{Vmatrix} \begin{matrix} \\ \\ (i) \\ \\ (j) \\ \\ \\ \end{matrix},$$

$$T''' = \begin{Vmatrix} 1 & \cdots & \cdots & \cdots & \cdots & 0 \\ \vdots & \ddots & & & & \vdots \\ \vdots & & 0 & \cdots & 1 \;\cdots & \vdots \\ \vdots & & \vdots & \ddots & \vdots & \vdots \\ \vdots & & 1 & \cdots & 0 & \vdots \\ \vdots & & & & \ddots & \vdots \\ 0 & \cdots & \cdots & \cdots & \cdots & 1 \end{Vmatrix} \begin{matrix} \\ \\ (i) \\ \\ (j) \\ \\ \\ \end{matrix}.$$

The result of applying a right elementary operation is equivalent to multiplying the matrix $A(\lambda)$ on the *right* by the corresponding matrix T.

Note that T' and T''' coincide with S' and S''' and that T'' coincides with S'' when the indices i and j are interchanged in these matrices. The matrices of type S', S'', S''' (or, what is the same, T', T'', T''') will be called *elementary matrices*.

.The determinant of every elementary matrix does not depend on λ and is different from zero. Therefore each left (right) elementary operation has an inverse operation which is also a left (right) elementary operation.[3]

DEFINITION 2: *Two polynomial matrices* $A(\lambda)$ *and* $B(\lambda)$ *are called* 1) *left-equivalent,* 2) *right-equivalent,* 3) *equivalent if one of them can be obtained from the other by means of* 1) *left-elementary,* 2) *right elementary,* 3) *left and right elementary operations, respectively.*[4]

---

[3] It follows from this that if a matrix $B(\lambda)$ is obtained from $A(\lambda)$ by means of left (right; left and right) elementary operations, then $A(\lambda)$ can, conversely, be obtained from $B(\lambda)$ by means of elementary operations of the same type. The left elementary operations form a group, as do the right elementary operations.

[4] From the definition it follows that only matrices of the same dimensions can be left-equivalent, right-equivalent, or simply equivalent.

Let $B(\lambda)$ be obtained from $A(\lambda)$ by means of the left elementary operations corresponding to $S_1, S_2, \ldots, S_p$. Then

$$B(\lambda) = S_p S_{p-1} \cdots S_1 A(\lambda). \tag{2}$$

Denoting the product $S_p S_{p-1} \cdots S_1$ by $P(\lambda)$, we write (2) in the form

$$B(\lambda) = P(\lambda) A(\lambda), \tag{3}$$

where $P(\lambda)$, like each of the matrices $S_1, S_2, \ldots, S_p$ has a *constant*[5] non-zero determinant.

In the next section we shall prove that every square λ-matrix $P(\lambda)$ with a constant non-zero determinant can be represented in the form of a product of elementary matrices. Therefore (3) is equivalent to (2) and signifies left equivalence of the matrices $A(\lambda)$ and $B(\lambda)$.

In the case of right equivalence of the polynomial matrices $A(\lambda)$ and $B(\lambda)$ we shall have instead of (3) the equation

$$B(\lambda) = A(\lambda) Q(\lambda) \tag{3'}$$

and in the case of (two-sided) equivalence the equation

$$B(\lambda) = P(\lambda) A(\lambda) Q(\lambda). \tag{3''}$$

Here again, $P(\lambda)$ and $Q(\lambda)$ are matrices with non-zero determinants, independent of λ.

Thus, Definition 2 can be replaced by an equivalent definition.

DEFINITION 2′: *Two rectangular λ-matrices $A(\lambda)$ and $B(\lambda)$ are called* 1) *left-equivalent,* 2) *right-equivalent,* 3) *equivalent if*

1) $B(\lambda) = P(\lambda)A(\lambda)$,
2) $B(\lambda) = A(\lambda)Q(\lambda)$,
3) $B(\lambda) = P(\lambda)A(\lambda)Q(\lambda)$,

respectively, where $P(\lambda)$ and $Q(\lambda)$ are polynomial square matrices with constant non-zero determinants.

2. All the concepts introduced above are illustrated in the following important example.

We consider a system of m linear homogeneous differential equations of order l with constant coefficients, where $x_1, x_2, \ldots, x_n$ are n unknown functions of the independent variable t:

$$\left.\begin{array}{l} a_{11}(D)x_1 + a_{12}(D)x_2 + \cdots + a_{1n}(D)x_n = 0 \\ a_{21}(D)x_1 + a_{22}(D)x_2 + \cdots + a_{2n}(D)x_n = 0 \\ \dots\dots\dots\dots\dots\dots\dots\dots\dots \\ a_{m1}(D)x_1 + a_{m2}(D)x_2 + \cdots + a_{mn}(D)x_n = 0; \end{array}\right\} \tag{4}$$

[5] I.e., independent of λ.

here

$$a_{ik}(D)=a_{ik}^{(0)}D^l+a_{ik}^{(1)}D^{l-1}+\cdots+a_{ik}^{(l)} \qquad (i=1,2,\ldots,m;\ k=1,2,\ldots,n)$$

is a polynomial in D with constant coefficients; $D=\frac{d}{dt}$ is the differential operator.

The matrix of operator coefficients

$$A(D)=\|a_{ik}(D)\| \qquad (i=1,2,\ldots,m;\ k=1,2,\ldots,n)$$

is a polynomial matrix, or D-matrix.

Clearly, the left elementary operation 1. on the matrix $A(D)$ signifies term-by-term multiplication of the i-th differential equation of the system by the number $c\neq 0$. The left elementary operation 2. signifies the term-by-term addition to the i-th equation of the j-th equation which has previously been subjected to the differential operator $b(D)$. The left elementary operation 3. signifies an interchange of the i-th and j-th equation.

Thus, if we replace in (4) the matrix $A(D)$ of operator coefficients by a left-equivalent matrix $B(D)$, we obtain a reduced system of equations. Since, conversely, by the same reasoning, the original system is a consequence of the new system, the two systems of equations are equivalent.[6]

It is not difficult in this example to interpret the right elementary operations as well. The first of them signifies the introduction of a new unknown function $\bar{x}_i=\frac{1}{c}x_i$ for the unknown function x_i; the second signifies the introduction of a new unknown function $\bar{x}_j=x_j+b(D)x_i$ (instead of x_j); the third signifies the interchange of the terms in the equations that contain x_i and x_j (i.e., $\bar{x}_i=x_j$, $\bar{x}_j=x_i$).

§ 2. Canonical Form of a λ-Matrix

1. To begin with, we shall examine what comparatively simple form we can obtain for a rectangular polynomial matrix $A(\lambda)$ by means of left elementary operations only.

Let us assume that the first column of $A(\lambda)$ contains elements not identically equal to zero. Among them we choose a polynomial of least degree and by a permutation of the rows we make it into the element $a_{11}(\lambda)$. Then we divide $a_{i1}(\lambda)$ by $a_{11}(\lambda)$; we denote quotient and remainder by $q_{i1}(\lambda)$ and $r_{i1}(\lambda)$ $(i=2,\ldots,m)$:

[6] Here it is assumed that the unknown functions $x_1, x_2, \ldots, x_n$ are such that their derivatives of all orders, as far as they occur in the transformations, exist. With this restriction, two systems of equations with left-equivalent matrices $A(D)$ and $B(D)$ have the same solutions.

$$a_{i1}(\lambda)=a_{11}(\lambda)q_{i1}(\lambda)+r_{i1}(\lambda) \qquad (i=2,\ldots,m).$$

Now we subtract from the i-th row the first row multiplied by $q_{i1}(\lambda)$ $(i=2,\ldots,m)$. If not all the remainders $r_{i1}(\lambda)$ are identically equal to zero, then we choose one of them that is not equal to zero and is of least degree and put it into the place of $a_{11}(\lambda)$ by a permutation of the rows. As the result of all these operations, the degree of the polynomial $a_{11}(\lambda)$ is reduced.

Now we repeat this process. Since the degree of the polynomial $a_{11}(\lambda)$ is finite, this must come to an end at some stage—i.e., at this stage all the elements $a_{21}(\lambda), a_{31}(\lambda), \ldots, a_{m1}(\lambda)$ turn out to be identically equal to zero.

Next we take the element $a_{22}(\lambda)$ and apply the same procedure to the rows numbered 2, 3, ..., m, achieving $a_{32}(\lambda)=\ldots=a_{m2}(\lambda)=0$. Continuing still further, we finally reduce the matrix $A(\lambda)$ to the following form:

$$\left\|\begin{matrix} b_{11}(\lambda) & b_{12}(\lambda) & \ldots & b_{1m}(\lambda) & \ldots & b_{1n}(\lambda) \\ 0 & b_{22}(\lambda) & \ldots & b_{2m}(\lambda) & \ldots & b_{2n}(\lambda) \\ \cdot & \cdot & \cdot & \cdot & \cdot & \cdot \\ 0 & 0 & \ldots & b_{mm}(\lambda) & \ldots & b_{mn}(\lambda) \end{matrix}\right\|, \quad \left\|\begin{matrix} b_{11}(\lambda) & b_{12}(\lambda) & \ldots & b_{1n}(\lambda) \\ 0 & b_{22}(\lambda) & \ldots & b_{2n}(\lambda) \\ \cdot & \cdot & \cdot & \cdot \\ 0 & 0 & \ldots & b_{nn}(\lambda) \\ 0 & 0 & \ldots & 0 \\ \cdot & \cdot & \cdot & \cdot \\ 0 & 0 & \ldots & 0 \end{matrix}\right\|. \tag{5}$$

$$(m \leqq n) \qquad\qquad (m \geqq n)$$

If the polynomial $b_{22}(\lambda)$ is not identically equal to zero, then by applying a left elementary operation of the second type we can make the degree of the element $b_{12}(\lambda)$ less than the degree of $b_{22}(\lambda)$ (if $b_{22}(\lambda)$ is of degree zero, then $b_{12}(\lambda)$ becomes identically equal to zero). In the same way, if $b_{33}(\lambda)\not\equiv 0$, then by left elementary operations of the second type we make the degrees of the elements $b_{13}(\lambda)$, $b_{23}(\lambda)$ less than the degree of $b_{33}(\lambda)$ without changing the elements $b_{12}(\lambda)$, etc.

We have established the following theorem:

THEOREM 1: *An arbitrary rectangular polynomial matrix of dimension $m\times n$ can always be brought into the form* (5) *by means of left elementary operations, where the polynomials $b_{1k}(\lambda), b_{2k}(\lambda), \ldots, b_{k-1,k}(\lambda)$ are of degree less than that of $b_{kk}(\lambda)$, provided $b_{kk}(\lambda)\not\equiv 0$, and are all identically equal to zero if $b_{kk}(\lambda)=$* const. $\neq 0$ $(k=2,3,\ldots,\min(m,n))$.

Similarly, we prove

THEOREM 2: *An arbitrary rectangular polynomial matrix of dimension $m\times n$ can always be brought into the form*

$$\left\| \begin{matrix} c_{11}(\lambda) & 0 & \dots & 0 & 0 \dots 0 \\ c_{21}(\lambda) & c_{22}(\lambda) & \dots & 0 & 0 \dots 0 \\ \dots & \dots & \dots & \dots & \dots \\ c_{m1}(\lambda) & c_{m2}(\lambda) & \dots & c_{mm}(\lambda) & 0 \dots 0 \end{matrix} \right\|, \quad \left\| \begin{matrix} c_{11}(\lambda) & 0 & \dots & 0 \\ c_{21}(\lambda) & c_{22}(\lambda) & \dots & 0 \\ \dots & \dots & \dots & \dots \\ c_{n1}(\lambda) & c_{n2}(\lambda) & \dots & c_{nn}(\lambda) \\ \dots & \dots & \dots & \dots \\ c_{m1}(\lambda) & c_{m2}(\lambda) & \dots & c_{mn}(\lambda) \end{matrix} \right\| \tag{6}$$

$$(m \leqq n) \qquad\qquad (m \geqq n)$$

by means of right elementary operations, where the polynomials $c_{k1}(\lambda)$, $c_{k2}(\lambda), \dots, c_{k,k-1}(\lambda)$ *are of degree less than that of* $c_{kk}(\lambda)$, *provided* $c_{kk}(\lambda) \not\equiv 0$, *and all are identically equal to zero if* $c_{kk}(\lambda) = \text{const.} \neq 0$ $(k = 2, 3, \dots, \min(m, n))$.

2. From Theorems 1 and 2 we deduce the corollary:

COROLLARY: *If the determinant of a square polynomial matrix* $P(\lambda)$ *does not depend on* λ *and is different from zero, then the matrix can be represented in the form of a product of a finite number of elementary matrices.*

For by Theorem 1 the matrix $P(\lambda)$ can be brought into the form

$$\left\| \begin{matrix} b_{11}(\lambda) & b_{12}(\lambda) & \dots & b_{1n}(\lambda) \\ 0 & b_{22}(\lambda) & \dots & b_{2n}(\lambda) \\ \dots & \dots & \dots & \dots \\ 0 & 0 & \dots & b_{nn}(\lambda) \end{matrix} \right\| \tag{7}$$

by left elementary operations, where n is the order of $P(\lambda)$. Since in the application of elementary operations to a square polynomial matrix the determinant of the matrix is only multiplied by constant non-zero factors, the determinant of the matrix (7), like that of $P(\lambda)$, does not depend on λ and is different from 0, i.e.,

$$b_{11}(\lambda)\, b_{22}(\lambda) \cdots b_{nn}(\lambda) = \text{const.} \neq 0.$$

Hence

$$b_{kk}(\lambda) = \text{const.} \neq 0 \quad (k = 1, 2, \dots, n).$$

But then, also by Theorem 1, the matrix (7) has the diagonal form $\| b_k \delta_{ik} \|_1^n$ and can therefore be reduced to the unit matrix E by means of left elementary operations of type 1. But then, conversely, the unit matrix E can be transformed into $P(\lambda)$ by means of the left elementary operations whose matrices are $S_1, S_2, \dots, S_p$. Therefore

$$P(\lambda) = S_p S_{p-1} \cdots S_1 E = S_p S_{p-1} \cdots S_1 .$$

As we pointed out on p. 133, from this corollary there follows the equivalence of the two Definitions 2 and 2′ of equivalence of polynomial matrices.

3. Let us return to our example of the system of differential equations (4). We apply Theorem 1 to the matrix $\| a_{ik}(D) \|$ of operator coefficients. As we have shown on p. 135, the system (4) is then replaced by an equivalent system

$$\left.\begin{aligned} b_{11}(D)\,x_1 + b_{12}(D)\,x_2 + \cdots + b_{1s}(D)\,x_s &= -b_{1,s+1}(D)\,x_{s+1} - \cdots - b_{1n}(D)\,x_n, \\ b_{22}(D)\,x_2 + \cdots + b_{2s}(D)\,x_s &= -b_{2,s+1}(D)\,x_{s+1} - \cdots - b_{2n}(D)\,x_n, \\ \cdots\cdots\cdots\cdots\cdots\cdots\cdots\cdots & \\ b_{ss}(D)\,x_s &= -b_{s,s+1}(D)\,x_{s+1} - \cdots - b_{sn}(D)\,x_n \end{aligned}\right\} \quad (4')$$

where $s = \min(m, n)$. In this system we may choose the functions $x_{s+1}, \ldots, x_n$ arbitrarily, after which the functions $x_s, x_{s-1}, \ldots, x_1$ can be determined successively; however, at each stage of this process only one differential equation with one unknown function has to be integrated.

4. We now pass on to establishing the 'canonical' form into which a rectangular matrix $A(\lambda)$ can be brought by applying to it both left and right elementary operations.

Among all the elements $a_{ik}(\lambda)$ of $A(\lambda)$ that are not identically equal to zero we choose one which has the least degree in λ and by suitable permutations of the rows and columns we make this element into $a_{11}(\lambda)$. Then we find the quotients and remainders of the polynomials $a_{i1}(\lambda)$ and $a_{1k}(\lambda)$ on division by $a_{11}(\lambda)$:

$$a_{i1}(\lambda) = a_{11}(\lambda)\,q_{i1}(\lambda) + r_{i1}(\lambda), \quad a_{1k}(\lambda) = a_{11}(\lambda)\,q_{1k}(\lambda) + r_{1k}(\lambda)$$
$$(i = 2, 3, \ldots, m;\ k = 2, 3, \ldots, n).$$

If at least one of the remainders $r_{i1}(\lambda)$, $r_{1k}(\lambda)$ $(i = 2, \ldots, m;\ k = 2, \ldots, n)$, for example $r_{1k}(\lambda)$, is not identically equal to zero, then by subtracting from the k-th column the first column multiplied by $q_{1k}(\lambda)$, we replace $a_{1k}(\lambda)$ by the remainder $r_{1k}(\lambda)$, which is of smaller degree than $a_{11}(\lambda)$. Then we can again reduce the degree of the element in the top left corner of the matrix by putting in its place an element of smaller degree in λ.

But if all the remainders $r_{21}(\lambda), \ldots, r_{m1}(\lambda)$; $r_{12}(\lambda), \ldots, r_{1n}(\lambda)$ are identically equal to zero, then by subtracting from the i-th row the first multiplied by $q_{i1}(\lambda)$ $(i = 2, \ldots, m)$ and from the k-th column the first multiplied by $q_{1k}(\lambda)$ $(k = 2, \ldots, n)$, we reduce our polynomial matrix to the form

$$\left\| \begin{matrix} a_{11}(\lambda) & 0 & \ldots & 0 \\ 0 & a_{22}(\lambda) & \ldots & a_{2n}(\lambda) \\ \cdots & \cdots & \cdots & \cdots \\ 0 & a_{m2}(\lambda) & \ldots & a_{mn}(\lambda) \end{matrix} \right\|.$$

If at least one of the elements $a_{ik}(\lambda)$ $(i=2,\ldots,m;\ k=2,\ldots,n)$ is not divisible without remainder by $a_{11}(\lambda)$, then by adding to the first column that column which contains such an element we arrive at the preceding case and can therefore again replace the element $a_{11}(\lambda)$ by a polynomial of smaller degree.

Since the original element $a_{11}(\lambda)$ had a definite degree and since the process of reducing this degree cannot be continued indefinitely, we must, after a finite number of elementary operations, obtain a matrix of the form

$$\left\|\begin{matrix} a_1(\lambda) & 0 & \ldots & 0 \\ 0 & b_{22}(\lambda) & \ldots & b_{2n}(\lambda) \\ \cdot & \cdot & \cdot & \cdot \\ 0 & b_{m2}(\lambda) & \ldots & b_{mn}(\lambda) \end{matrix}\right\|, \tag{8}$$

in which all the elements $b_{ik}(\lambda)$ are divisible without remainder by $a_1(\lambda)$. If among these elements $b_{ik}(\lambda)$ there is one not identically equal to zero, then continuing the same reduction process on the rows numbered $2,\ldots,m$ and the columns $2,\ldots,n$, we reduce the matrix (8) to the form

$$\left\|\begin{matrix} a_1(\lambda) & 0 & 0 & \ldots & 0 \\ 0 & a_2(\lambda) & 0 & \ldots & 0 \\ 0 & 0 & c_{33}(\lambda) & \ldots & c_{3n}(\lambda) \\ \cdot & \cdot & \cdot & \cdot & \cdot \\ 0 & 0 & c_{m3}(\lambda) & \ldots & c_{mn}(\lambda) \end{matrix}\right\|,$$

where $a_2(\lambda)$ is divisible without remainder by $a_1(\lambda)$ and all the polynomials $c_{ik}(\lambda)$ are divisible without remainder by $a_2(\lambda)$. Continuing the process further, we finally arrive at a matrix of the form

$$\left\|\begin{matrix} a_1(\lambda) & 0 & \ldots & 0 & 0\ldots 0 \\ 0 & a_2(\lambda) & \ldots & 0 & 0\ldots 0 \\ \cdot & \cdot & \cdot & \cdot & \cdot \\ 0 & 0 & \ldots & a_s(\lambda) & 0\ldots 0 \\ 0 & 0 & \ldots & 0 & 0\ldots 0 \\ \cdot & \cdot & \cdot & \cdot & \cdot \\ 0 & 0 & \ldots & 0 & 0\ldots 0 \end{matrix}\right\|, \tag{9}$$

where the polynomials $a_1(\lambda)$, $a_2(\lambda)$, $\ldots$, $a_s(\lambda)$ $(s \leqq \min(m, n))$ are not identically equal to zero and each is divisible by the preceding one.

By multiplying the first s rows by suitable non-zero numerical factors, we can arrange that the highest coefficients of the polynomials $a_1(\lambda)$, $a_2(\lambda),\ldots, a_s(\lambda)$ are equal to 1.

DEFINITION 3: *A rectangular polynomial matrix is called a canonical diagonal matrix if it is of the form* (9), *where* 1) *the polynomials* $a_1(\lambda)$, $a_2(\lambda), \ldots, a_s(\lambda)$ *are not identically equal to zero and* 2) *each of the polynomials* $a_2(\lambda), \ldots, a_s(\lambda)$ *is divisible by the preceding. Moreover, it is assumed that the highest coefficients of all the polynomials* $a_1(\lambda), a_2(\lambda), \ldots,$ $a_s(\lambda)$ *are equal to* 1.

Thus, we have proved that: *An arbitrary rectangular polynomial matrix* $A(\lambda)$ *is equivalent to a canonical diagonal matrix.* In the next section we shall prove that: *The polynomials* $a_1(\lambda), a_2(\lambda), \ldots, a_s(\lambda)$ *are uniquely determined by the given matrix* $A(\lambda)$; and we shall set up formulas that connect these polynomials with the elements of $A(\lambda)$.

§ 3. Invariant Polynomials and Elementary Divisors of a Polynomial Matrix

1. We introduce the concept of invariant polynomials of a λ-matrix $A(\lambda)$.

Let $A(\lambda)$ be a polynomial matrix of rank r, i.e., the matrix has minors of order r not identically equal to zero, but all the minors of order greater than r are identically equal to zero in λ. We denote by $D_j(\lambda)$ the greatest common divisor of all the minors of order j in $A(\lambda)$ $(j=1,2,\ldots,r)$.[7] Then it is easy to see that in the series

$$D_r(\lambda), \quad D_{r-1}(\lambda), \ldots, D_1(\lambda), \quad D_0(\lambda) \equiv 1$$

each polynomial is divisible by the succeeding one.[8] The corresponding quotients will be denoted by $i_1(\lambda), i_2(\lambda), \ldots, i_r(\lambda)$:

$$i_1(\lambda)=\frac{D_r(\lambda)}{D_{r-1}(\lambda)}, \quad i_2(\lambda)=\frac{D_{r-1}(\lambda)}{D_{r-2}(\lambda)}, \ldots, i_r(\lambda)=\frac{D_1(\lambda)}{D_0(\lambda)}=D_1(\lambda). \tag{10}$$

DEFINITION 4: *The polynomials* $i_1(\lambda), i_2(\lambda), \ldots, i_r(\lambda)$ *defined by* (10) *are called the invariant polynomials of the rectangular matrix* $A(\lambda)$.

The term 'invariant polynomial' is explained by the following arguments. Let $A(\lambda)$ and $B(\lambda)$ be two equivalent polynomial matrices. Then they are obtained from one another by means of elementary operations. But an easy verification shows immediately that the elementary operations

[7] We take the highest coefficient in $D_j(\lambda)$ to be 1 $(j=1,2,\ldots,r)$.

[8] If we apply the Bézout decomposition with respect to the elements of any row to an arbitrary minor of order j, then every term in the decomposition is divisible by $D_{j-1}(\lambda)$; therefore every minor of order j, and hence $D_j(\lambda)$, is divisible by $D_{j-1}(\lambda)$ $(j=2,3,\ldots,r)$.

change neither the rank of $A(\lambda)$ nor the polynomials $D_1(\lambda)$, $D_2(\lambda)$, $\dots$, $D_r(\lambda)$. For when we apply to the identity (3″) the formula that expresses a minor of a product of matrices by the minors of the factors (see p. 12), we obtain for an arbitrary minor of $B(\lambda)$ the expression

$$B\begin{pmatrix} j_1 & j_2 \dots j_p \\ k_1 & k_2 \dots k_p \end{pmatrix}; \lambda\Big) = \sum_{\substack{1 \leqq \alpha_1 < \alpha_2 < \cdots < \alpha_p \leqq m \\ 1 \leqq \beta_1 < \beta_2 < \cdots < \beta_p \leqq n}} P\begin{pmatrix} j_1 & j_2 \dots j_p \\ \alpha_1 & \alpha_2 \dots \alpha_p \end{pmatrix} A\left(\begin{matrix} \alpha_1 & \alpha_2 \dots \alpha_p \\ \beta_1 & \beta_2 \dots \beta_p \end{matrix}; \lambda\right) Q\begin{pmatrix} \beta_1 & \beta_2 \dots \beta_p \\ k_1 & k_2 \dots k_p \end{pmatrix}$$

$$(p = 1, 2, \dots, \min(m, n)).$$

Hence it follows that all the minors of order greater than r of the matrix $B(\lambda)$ are zero, so that we have for the rank r^* of $B(\lambda)$:

$$r^* \leqq r.$$

Moreover, it follows from the same formula that $D_p^*(\lambda)$, the greatest common divisor of all the minors of order p of $B(\lambda)$, is divisible by $D_p(\lambda)$ $(p = 1, 2, \dots, \min(m, n))$. But the matrices $A(\lambda)$ and $B(\lambda)$ can exchange roles. Therefore $r \leqq r^*$ and $D_p(\lambda)$ is divisible by $D_p^*(\lambda)$ $(p = 1, 2, \dots, \min(m, n))$. Hence[9]

$$r = r^*, \quad D_1^*(\lambda) = D_1(\lambda), \quad D_2^*(\lambda) = D_2(\lambda), \dots, D_r^*(\lambda) = D_r(\lambda).$$

Since elementary operations do not change the polynomials $D_1(\lambda)$, $D_2(\lambda)$, $\dots$, $D_r(\lambda)$, they also leave the polynomials $i_1(\lambda)$, $i_2(\lambda)$, $\dots$, $i_r(\lambda)$ defined by (10) unchanged.

Thus, the polynomials $i_1(\lambda)$, $i_2(\lambda)$, $\dots$, $i_r(\lambda)$ remain *invariant* on transition from one matrix to another equivalent one.

If the polynomial matrix has the canonical diagonal form (9), then it is easy to see that for this matrix

$$D_1(\lambda) = a_1(\lambda), \quad D_2(\lambda) = a_1(\lambda)\, a_2(\lambda), \quad \dots, \quad D_r(\lambda) = a_1(\lambda)\, a_2(\lambda) \cdots a_r(\lambda).$$

But then, by (10), the diagonal polynomials in (9) $a_1(\lambda)$, $a_2(\lambda)$, $\dots$, $a_r(\lambda)$ coincide with the invariant polynomials

$$i_1(\lambda) = a_r(\lambda), \; i_2(\lambda) = a_{r-1}(\lambda), \; \dots, \; i_r(\lambda) = a_1(\lambda). \tag{11}$$

Here $i_1(\lambda)$, $i_2(\lambda)$, $\dots$, $i_r(\lambda)$ are at the same time the invariant polynomials of the original matrix $A(\lambda)$, because it is equivalent to (9).

The results obtained can be stated in the form of the following theorem.

---

[9] The highest coefficients in $D_p(\lambda)$ and $D_p^*(\lambda)$ $(p = 1, 2, \dots, r)$ are 1.

Theorem 3: *The rectangular polynomial matrix $A(\lambda)$ is always equivalent to a canonical diagonal matrix*

$$\left\| \begin{matrix} i_r(\lambda) & 0 & \dots & 0 & 0 & \dots & 0 \\ 0 & i_{r-1}(\lambda) & \dots & 0 & 0 & \dots & 0 \\ \cdot & \cdot & \cdot & \cdot & \cdot & \cdot & \cdot \\ 0 & 0 & \dots & i_1(\lambda) & 0 & \dots & 0 \\ 0 & 0 & & 0 & 0 & \dots & 0 \\ \cdot & \cdot & \cdot & \cdot & \cdot & \cdot & \cdot \\ 0 & 0 & & 0 & 0 & \dots & 0 \end{matrix} \right\| . \tag{12}$$

Moreover, r must here be the rank of $A(\lambda)$ and $i_1(\lambda), i_2(\lambda), \dots, i_r(\lambda)$ the invariant polynomials of $A(\lambda)$ defined by (10).

Corollary 1: *Two rectangular matrices of the same dimension $A(\lambda)$ and $B(\lambda)$ are equivalent if and only if they have the same invariant polynomials.*

The sufficiency of the condition was explained above. The necessity follows from the fact that two polynomial matrices having the same invariant polynomials are equivalent to one and the same canonical diagonal matrix and, therefore, to each other. Thus: *The invariant polynomials form a complete system of invariants of a λ-matrix.*

Corollary 2: *In the sequence of invariant polynomials*

$$i_1(\lambda)=\frac{D_r(\lambda)}{D_{r-1}(\lambda)},\quad i_2(\lambda)=\frac{D_{r-1}(\lambda)}{D_{r-2}(\lambda)},\quad \dots,\quad i_r(\lambda)=\frac{D_1(\lambda)}{D_0(\lambda)} \qquad (D_0(\lambda)\equiv 1) \tag{13}$$

every polynomial from the second onwards divides the preceding one.

This statement does not follow immediately from (13). It does follow from the fact that the polynomials $i_1(\lambda), i_2(\lambda), \dots, i_r(\lambda)$ coincide with the polynomials $a_r(\lambda), a_{r-1}(\lambda), \dots, a_1(\lambda)$ of the canonical diagonal matrix (9).

2. We now indicate a method of computing the invariant polynomials of a quasi-diagonal λ-matrix if the invariant polynomials of the matrices in the diagonal blocks are known.

Theorem 4: *If in a quasi-diagonal rectangular matrix*

$$C(\lambda)=\left\| \begin{matrix} A(\lambda) & O \\ O & B(\lambda) \end{matrix} \right\|$$

every invariant polynomial of $A(\lambda)$ divides every invariant polynomial of $B(\lambda)$, then the set of invariant polynomials of $C(\lambda)$ is the union of the invariant polynomials of $A(\lambda)$ and $B(\lambda)$.

Proof. We denote by $i_1'(\lambda), i_2'(\lambda), \ldots, i_r'(\lambda)$ and $i_1''(\lambda), i_2''(\lambda), \ldots, i_q''(\lambda)$, respectively, the invariant polynomials of the λ-matrices $A(\lambda)$ and $B(\lambda)$. Then[10]

$$A(\lambda) \sim \{i_r'(\lambda), \ldots, i_1'(\lambda), 0, \ldots, 0\}, \quad B(\lambda) \sim \{i_q''(\lambda), \ldots, i_1''(\lambda), 0, \ldots, 0\}$$

and therefore

$$C(\lambda) \sim \{i_r'(\lambda), \ldots, i_1'(\lambda), \quad i_q''(\lambda), \ldots, i_1''(\lambda), 0, \ldots, 0\}. \tag{14}$$

The λ-matrix on the right-hand side of this relation is of canonical diagonal form. By Theorem 3 the diagonal elements of this matrix that are not identically equal to zero then form a complete system of invariants of the polynomial matrix $C(\lambda)$. This proves the theorem.

In order to determine the invariant polynomials of $C(\lambda)$ in the general case of arbitrary invariant polynomials of $A(\lambda)$ and $B(\lambda)$ we make use of the important concept of elementary divisors.

We decompose the invariant polynomials $i_1(\lambda), i_2(\lambda), \ldots, i_r(\lambda)$ into irreducible factors over the given number field F:[11]

$$\begin{aligned} i_1(\lambda) &= [\varphi_1(\lambda)]^{c_1} [\varphi_2(\lambda)]^{c_2} \cdots [\varphi_s(\lambda)]^{c_s}, \\ i_2(\lambda) &= [\varphi_1(\lambda)]^{d_1} [\varphi_2(\lambda)]^{d_2} \cdots [\varphi_s(\lambda)]^{d_s}, \\ &\cdots\cdots\cdots\cdots \\ i_r(\lambda) &= [\varphi_1(\lambda)]^{l_1} [\varphi_2(\lambda)]^{l_2} \cdots [\varphi_s(\lambda)]^{l_s}. \end{aligned} \quad \begin{pmatrix} c_k \geqq d_k \geqq \cdots \geqq l_k \geqq 0; \\ k = 1, 2, \ldots, s \end{pmatrix} \tag{15}$$

Here $\varphi_1(\lambda), \varphi_2(\lambda), \ldots, \varphi_s(\lambda)$ are all the distinct factors irreducible over F (and with highest coefficient 1) that occur in $i_1(\lambda), i_2(\lambda), \ldots, i_r(\lambda)$.

DEFINITION 5: *All the powers among $[\varphi_1(\lambda)]^{c_1}, \ldots, [\varphi_s(\lambda)]^{l_s}$ in* (15), *as far as they are distinct from* 1, *are called the elementary divisors of the matrix $A(\lambda)$ in the field* F.[12]

THEOREM 5: *The set of elementary divisors of the rectangular quasi-diagonal matrix*

$$C(\lambda) = \begin{Vmatrix} A(\lambda) & O \\ O & B(\lambda) \end{Vmatrix}$$

is always obtained by combining the elementary divisors of $A(\lambda)$ with those of $B(\lambda)$.

[10] The symbol $\sim$ denotes here the equivalence of matrices; and braces $\{\ \}$, a diagonal rectangular matrix of the form (12).

[11] Some of the exponents $c_k, d_k, \ldots, l_k$ $(k = 1, 2, \ldots, s)$ may be equal to zero.

[12] The formulas (15) enable us to define not only the elementary divisors of $A(\lambda)$ in the field F in terms of the invariant polynomials but also, conversely, the invariant polynomials in terms of the elementary divisors.

Proof. We decompose the invariant polynomials of $A(\lambda)$ and $B(\lambda)$ into irreducible factors over F:[13]

$$
\begin{aligned}
i_1'(\lambda) &= [\varphi_1(\lambda)]^{c_1'}[\varphi_2(\lambda)]^{c_2'}\cdots[\varphi_s(\lambda)]^{c_s'}, & i_1''(\lambda) &= [\varphi_1(\lambda)]^{c_1''}[\varphi_2(\lambda)]^{c_2''}\cdots[\varphi_s(\lambda)]^{c_s''},\\
i_2'(\lambda) &= [\varphi_1(\lambda)]^{d_1'}[\varphi_2(\lambda)]^{d_2'}\cdots[\varphi_s(\lambda)]^{d_s'}, & i_2''(\lambda) &= [\varphi_1(\lambda)]^{d_1''}[\varphi_2(\lambda)]^{d_2''}\cdots[\varphi_s(\lambda)]^{d_s''},\\
&\cdots\cdots & &\cdots\cdots\\
i_r'(\lambda) &= [\varphi_1(\lambda)]^{h_1'}[\varphi_2(\lambda)]^{h_2'}\cdots[\varphi_s(\lambda)]^{h_s'}, & i_q''(\lambda) &= [\varphi_1(\lambda)]^{g_1''}[\varphi_2(\lambda)]^{g_2''}\cdots[\varphi_s(\lambda)]^{g_s''}.
\end{aligned}
$$

We denote by

$$c_1 \geqq d_1 \geqq \cdots \geqq l_1 > 0, \tag{16}$$

all the non-zero numbers among $c_1', d_1', \ldots, h_1', c_1'', d_1'', \ldots, g_1''$.

Then the matrix $C(\lambda)$ is equivalent to the matrix (14), and by a permutation of rows and of columns the latter can be brought into 'diagonal' form

$$\{[\varphi_1(\lambda)]^{c_1}\cdot(*),\ [\varphi_1(\lambda)]^{d_1}\cdot(*),\ \ldots,\ [\varphi_1(\lambda)]^{l_1}\cdot(*),\ (**),\ \ldots,\ (**)\} \tag{17}$$

where we have denoted by $(*)$ polynomials that are prime to $\varphi_1(\lambda)$ and by $(**)$ polynomials that are either prime to $\varphi_1(\lambda)$ or identically equal to zero. From the form of the matrix (17) we deduce immediately the following decomposition of the polynomials $D_r(\lambda)$, $D_{r-1}(\lambda)$, ... and $i_1(\lambda)$, $i_2(\lambda)$, ... of the matrix $C(\lambda)$:

$$D_r(\lambda) = [\varphi_1(\lambda)]^{c_1+d_1+\cdots+l_1}\cdot(*),\quad D_{r-1}(\lambda) = [\varphi_1(\lambda)]^{d_1+\cdots+l_1}\cdot(*),\ \ldots,$$
$$i_1(\lambda) = [\varphi_1(\lambda)]^{c_1}(*),\quad i_2(\lambda) = [\varphi_1(\lambda)]^{d_1}(*),\ \ldots.$$

Hence it follows that $[\varphi_1(\lambda)]^{c_1}$, $[\varphi_1(\lambda)]^{d_1}, \ldots, [\varphi_1(\lambda)]^{l_1}$, i.e., all the powers

$$[\varphi_1(\lambda)]^{c_1'},\ \ldots,\ [\varphi_1(\lambda)]^{h_1'},\ [\varphi_1(\lambda)]^{c_1''},\ \ldots,\ [\varphi_1(\lambda)]^{g_1''},$$

as far as they are distinct from 1, are elementary divisors of $C(\lambda)$.

The elementary divisors of $C(\lambda)$ that are powers of $\varphi_2(\lambda)$ are determined similarly, etc. This completes the proof of the theorem.

Note. The theory of equivalence for integral matrices (i.e., matrices whose elements are integers) can be constructed along similar lines. Here in 1., 2. (see pp. 130-31) $c = \pm 1$, $b(\lambda)$ is to be replaced by an integer, and in (3), (3′), (3″), in place of $P(\lambda)$ and $Q(\lambda)$ there are integral matrices with determinants equal to ± 1.

---

[13] If any irreducible polynomial $\varphi_k(\lambda)$ occurs as a factor in some invariant polynomials, but not in others, then in the latter we write $\varphi_k(\lambda)$ with a zero exponent.

3. Suppose now that $A = \| a_{ik} \|_1^n$ is a matrix with elements in the field F. We form its characteristic matrix

$$\lambda E - A = \begin{Vmatrix} \lambda - a_{11} & -a_{12} & \dots & -a_{1n} \\ -a_{21} & \lambda - a_{22} & \dots & -a_{2n} \\ \cdot & \cdot & \cdot & \cdot \\ -a_{n1} & -a_{n2} & \dots & \lambda - a_{nn} \end{Vmatrix}. \tag{18}$$

The characteristic matrix is a λ-matrix of rank n. Its invariant polynomials

$$i_1(\lambda) = \frac{D_n(\lambda)}{D_{n-1}(\lambda)}, \quad i_2(\lambda) = \frac{D_{n-1}(\lambda)}{D_{n-2}(\lambda)}, \quad \dots, \quad i_n(\lambda) = \frac{D_1(\lambda)}{D_0(\lambda)} \quad (D_0(\lambda) \equiv 1), \tag{19}$$

are called the *invariant polynomials of the matrix* A and the corresponding elementary divisors in F are called the *elementary divisors of the matrix* A *in the field* F. A knowledge of the invariant polynomials (and, hence, of the elementary divisors) of A enables us to investigate its structure. Therefore practical methods of computing the invariant polynomials of a matrix are of interest. The formulas (19) give an algorithm for computing these polynomials, but for large n this algorithm is very cumbrous.

Theorem 3 gives another method of computing invariant polynomials, based on the reduction of the characteristic matrix (18) to canonical diagonal form by means of elementary operations.

Example:

$$A = \begin{Vmatrix} 3 & 1 & 0 & 0 \\ -4 & -1 & 0 & 0 \\ 6 & 1 & 2 & 1 \\ -14 & -5 & -1 & 0 \end{Vmatrix}, \qquad \lambda E - A = \begin{Vmatrix} \lambda - 3 & -1 & 0 & 0 \\ 4 & \lambda + 1 & 0 & 0 \\ -6 & -1 & \lambda - 2 & -1 \\ 14 & 5 & 1 & \lambda \end{Vmatrix}.$$

In the characteristic matrix $\lambda E - A$ we add to the fourth row the third multiplied by λ:

$$\begin{Vmatrix} \lambda - 3 & -1 & 0 & 0 \\ 4 & \lambda + 1 & 0 & 0 \\ -6 & -1 & \lambda - 2 & -1 \\ 14 - 6\lambda & 5 - \lambda & \lambda^2 - 2\lambda + 1 & 0 \end{Vmatrix}.$$

Now adding to the first three columns the fourth, multiplied by -6, -1, and $\lambda - 2$, respectively, we obtain

$$\begin{Vmatrix} \lambda - 3 & -1 & 0 & 0 \\ 4 & \lambda + 1 & 0 & 0 \\ 0 & 0 & 0 & -1 \\ 14 - 6\lambda & 5 - \lambda & \lambda^2 - 2\lambda + 1 & 0 \end{Vmatrix}.$$

We add to the first column the second multiplied by $\lambda - 3$:

$$\left\| \begin{matrix} 0 & -1 & 0 & 0 \\ \lambda^2 - 2\lambda + 1 & \lambda + 1 & 0 & 0 \\ 0 & 0 & 0 & -1 \\ -\lambda^2 + 2\lambda - 1 & 5 - \lambda & \lambda^2 - 2\lambda + 1 & 0 \end{matrix} \right\| .$$

To the second and fourth rows we add the first multiplied by $\lambda + 1$ and $5 - \lambda$, respectively; we obtain

$$\left\| \begin{matrix} 0 & -1 & 0 & 0 \\ \lambda^2 - 2\lambda + 1 & 0 & 0 & 0 \\ 0 & 0 & 0 & -1 \\ -\lambda^2 + 2\lambda - 1 & 0 & \lambda^2 - 2\lambda + 1 & 0 \end{matrix} \right\| .$$

To the second row we add the fourth; then we multiply the first and third rows by -1. After permuting some rows and columns we obtain:

$$\left\| \begin{matrix} 1 & 0 & 0 & 0 \\ 0 & 1 & 0 & 0 \\ 0 & 0 & (\lambda - 1)^2 & 0 \\ 0 & 0 & 0 & (\lambda - 1)^2 \end{matrix} \right\| .$$

The matrix has two elementary divisors $(\lambda - 1)^2$ and $(\lambda - 1)^2$.

§ 4. Equivalence of Linear Binomials

1. In the preceding sections we have considered rectangular λ-matrices. In the present section we consider two square λ-matrices $A(\lambda)$ and $B(\lambda)$ of order n in which all the elements are of degree not higher than 1 in λ. These polynomial matrices may be represented in the form of matrix binomials:

$$A(\lambda) = A_0\lambda + A_1, \quad B(\lambda) = B_0\lambda + B_1 .$$

We shall assume that these binomials are of degree 1 and regular, i.e., that $| A_0 | \neq 0$, $| B_0 | \neq 0$ (see p. 76).

The following theorem gives a criterion for the equivalence of such binomials:

THEOREM 6: *If two regular binomials of the first degree* $A_0\lambda + A_1$ and $B_0\lambda + B_1$ *are equivalent, then they are strictly equivalent, i.e., in the identity*

$$B_0\lambda + B_1 = P(\lambda)(A_0\lambda + A_1)Q(\lambda) \tag{20}$$

the matrices $P(\lambda)$ *and* $Q(\lambda)$*—with constant non-zero determinants—can be replaced by constant non-singular matrices* P *and* Q:[14]

$$B_0\lambda + B_1 = P(A_0\lambda + A_1)Q . \tag{21}$$

---

[14] The identity (21) is equivalent to the two matrix equations: $B_0 = PA_0Q$ and $B_1 = PA_1Q$.

Proof. Since the determinant of $P(\lambda)$ does not depend on λ and is different from zero,[15] the inverse matrix $M(\lambda) = P^{-1}(\lambda)$ is also a polynomial matrix. With the help of this matrix we write (20) in the form

$$M(\lambda)(B_0\lambda + B_1) = (A_0\lambda + A_1)Q(\lambda). \tag{22}$$

Regarding $M(\lambda)$ and $Q(\lambda)$ as matrix polynomials, we divide $M(\lambda)$ on the left by $A_0\lambda + A_1$ and $Q(\lambda)$ on the right by $B_0\lambda + B_1$:

$$M(\lambda) = (A_0\lambda + A_1)S(\lambda) + M, \tag{23}$$

$$Q(\lambda) = T(\lambda)(B_0\lambda + B_1) + Q; \tag{24}$$

here M and Q are constant square matrices (independent of λ) of order n. We substitute these expressions for $M(\lambda)$ and $Q(\lambda)$ in (22). After a few small transformations, we obtain

$$(A_0\lambda + A_1)[T(\lambda) - S(\lambda)](B_0\lambda + B_1) = M(B_0\lambda + B_1) - (A_0\lambda + A_1)Q. \tag{25}$$

The difference in the brackets must be identically equal to zero; for otherwise the product on the left-hand side of (25) would be of degree $\geqq 2$, while the polynomial on the right-hand side of the equation is of degree not higher than 1. Therefore

$$S(\lambda) = T(\lambda); \tag{26}$$

But then we obtain from (25):

$$M(B_0\lambda + B_1) = (A_0\lambda + A_1)Q. \tag{27}$$

We shall now show that M is a non-singular matrix. For this purpose we divide $P(\lambda)$ on the left by $B_0\lambda + B_1$:

$$P(\lambda) = (B_0\lambda + B_1)U(\lambda) + P. \tag{28}$$

From (22), (23), and (28) we deduce:

$$\begin{aligned} E = M(\lambda)P(\lambda) &= M(\lambda)(B_0\lambda + B_1)U(\lambda) + M(\lambda)P \\ &= (A_0\lambda + A_1)Q(\lambda)U(\lambda) + (A_0\lambda + A_1)S(\lambda)P + MP \\ &= (A_0\lambda + A_1)[Q(\lambda)U(\lambda) + S(\lambda)P] + MP. \end{aligned} \tag{29}$$

---

[15] The equivalence of the binomials $A_0\lambda + A_1$ and $B_0\lambda + B_1$ means that an identity (20) exists in which $|P(\lambda)| = \text{const.} \neq 0$ and $|Q(\lambda)| = \text{const.} \neq 0$. However, in this case the last relations follow from (20) itself. For the determinants of regular binomials of the first degree are of degree n:

$$|A_0\lambda + A_1| = |A_0|\lambda^n + \ldots,\ |B_0\lambda + B_1| = |B_0|\lambda^n + \ldots;\ |A_0| \neq 0,\ |B_0| \neq 0.$$

Therefore it follows from

$$|B_0\lambda + B_1| = |P(\lambda)|\,|A_0\lambda + A_1|\,|Q(\lambda)|$$

that

$$|P(\lambda)| = \text{const.} \neq 0,\ |Q(\lambda)| = \text{const.} \neq 0.$$

Since the last term of this chain of equations must be of degree zero in λ (because it is equal to E), the expression in brackets must be identically equal to zero. But then from (29)

$$MP = E, \tag{30}$$

so that $|M| \neq 0$ and $M^{-1} = P$.

Multiplying both sides of (27) on the left by P, we obtain:

$$B_0\lambda + B_1 = P(A_0\lambda + A_1)Q.$$

The fact that P is non-singular follows from (30). That P and Q are non-singular also follows directly from (21), since this identity implies

$$B_0 = PA_0Q$$

and therefore

$$|P||A_0||Q| = |B_0| \neq 0.$$

This completes the proof of the theorem.

Note. From the proof it follows (see (24) and (28)) that the constant matrices P and Q by which we have replaced the λ-matrices $P(\lambda)$ and $Q(\lambda)$ in (20) can be taken as the left and right remainders, respectively, of $P(\lambda)$ and $Q(\lambda)$ on division by $B_0\lambda + B_1$.

§ 5. A Criterion for Similarity of Matrices

1. Let $A \; \| a_{ik} \|_1^n$ be a matrix with numerical elements from the field F. Its characteristic matrix $\lambda E - A$ is a λ-matrix of rank n and therefore has n invariant polynomials (see § 3)

$$i_1(\lambda), i_2(\lambda), \ldots, i_n(\lambda).$$

The following theorem shows that these invariant polynomials determine the original matrix A to within similarity transformations.

THEOREM 7: *Two matrices $A = \| a_{ik} \|_1^n$ and $B = \| b_{ik} \|_1^n$ are similar ($B = T^{-1}AT$) if and only if they have the same invariant polynomials or, what is the same, the same elementary divisors in the field* F.

Proof. The condition is *necessary*. For if the matrices A and B are similar, then there exists a non-singular matrix T such that

$$B = T^{-1}AT.$$

Hence

$$\lambda E - B = T^{-1}(\lambda E - A)T.$$

This equation shows that the characteristic matrices $\lambda E - A$ and $\lambda E - B$ are equivalent and therefore have the same invariant polynomials.

The condition is *sufficient*. Suppose that the characteristic matrices $\lambda E - A$ and $\lambda E - B$ have the same invariant polynomials. Then these λ-matrices are equivalent (see Corollary 1 to Theorem 3) and there exist, in consequence, two polynomial matrices $P(\lambda)$ and $Q(\lambda)$ such that

$$\lambda E - B = P(\lambda)(\lambda E - A)Q(\lambda). \tag{31}$$

Applying Theorem 6 to the matrix binomials $\lambda E - A$ and $\lambda E - B$, we may replace in (31) the λ-matrices $P(\lambda)$ and $Q(\lambda)$ by constant matrices P and Q:

$$\lambda E - B = P(\lambda E - A)Q; \tag{32}$$

moreover, P and Q may be taken (see the Note on p. 147) as the left remainder and the right remainder, respectively, of $P(\lambda)$ and $Q(\lambda)$ on division by $\lambda E - B$, i.e., by the Generalized Bézout Theorem, we may set:[16]

$$P = \widehat{P}(B),\ Q = Q(B) \tag{33}$$

Equating coefficients of the powers of λ on both sides of (32), we obtain:

$$B = PAQ, \quad E = PQ,$$

i.e.,

$$B = T^{-1}AT,$$

where

$$T = Q = P^{-1}$$

This proves the theorem.

2. *Note.* We have incidentally established the following result, which we state separately:

SUPPLEMENT TO THEOREM 7. *If $A = \| a_{ik} \|_1^n$ and $B = \| b_{ik} \|_1^n$ are two similar matrices,*

$$B = T^{-1}AT, \tag{34}$$

then we can choose as the transforming matrix T the matrix

$$T = Q(B) = [\widehat{P}(B)]^{-1}, \tag{35}$$

where $P(\lambda)$ and $Q(\lambda)$ are polynomial matrices in the identity

$$\lambda E - B = P(\lambda)(\lambda E - A)Q(\lambda)$$

which connects the equivalent characteristic matrices $\lambda E - A$ and $\lambda E - B$; in (35) $Q(B)$ denotes the right value of the matrix polynomial $Q(\lambda)$, and $\widehat{P}(B)$ the left value of $P(\lambda)$, when the argument is replaced by B.

---

[16] We recall that $\widehat{P}(B)$ is the left value of the polynomial $P(\lambda)$ and $Q(B)$ the right value of $Q(\lambda)$, when λ is replaced by B (see p. 81).

§ 6. The Normal Forms of a Matrix

1. Let

$$g(\lambda) = \lambda^m + \alpha_1\lambda^{m-1} + \cdots + \alpha_{m-1}\lambda + \alpha_m$$

be a polynomial with coefficients in F.

We consider the square matrix of order m

$$L = \left\| \begin{matrix} 0 & 0 & . & . & . & 0 & -\alpha_m \\ 1 & 0 & . & . & . & 0 & -\alpha_{m-1} \\ 0 & 1 & . & . & . & 0 & -\alpha_{m-2} \\ . & . & . & . & . & . & . \\ 0 & 0 & . & . & . & 1 & -\alpha_1 \end{matrix} \right\| . \tag{36}$$

It is not difficult to verify that $g(\lambda)$ is the characteristic polynomial of L:

$$|\lambda E - L| = \left| \begin{matrix} \lambda & 0 & 0 & . & . & . & 0 & \alpha_m \\ -1 & \lambda & 0 & . & . & . & 0 & \alpha_{m-1} \\ 0 & -1 & \lambda & . & . & . & 0 & \alpha_{m-2} \\ . & . & . & . & . & . & . & . \\ 0 & 0 & 0 & . & . & . & -1 & \alpha_1 + \lambda \end{matrix} \right| = g(\lambda).$$

On the other hand, the minor of the element α_m in the characteristic determinant is equal to ± 1. Therefore $D_{m-1}(\lambda) = 1$ and $i_1(\lambda) = \frac{D_m(\lambda)}{D_{m-1}(\lambda)} = D_m(\lambda) = g(\lambda)$, $i_2(\lambda) = \ldots = i_m(\lambda) = 1$.

Thus, L has a single invariant polynomial different from 1, namely $g(\lambda)$.

We shall call L the *companion matrix* of the polynomial $g(\lambda)$.

Let $A = \| a_{ik} \|_1^n$ be a matrix with the invariant polynomials

$$i_1(\lambda), i_2(\lambda), \ldots, i_t(\lambda), i_{t+1}(\lambda) = 1, \ldots, i_n(\lambda) = 1. \tag{37}$$

Here the polynomials $i_1(\lambda), i_2(\lambda), \ldots, i_t(\lambda)$ have positive degrees and, from the second onwards, each divides the preceding one. We denote the companion matrices of these polynomials by $L_1, L_2, \ldots, L_t$.

Then the quasi-diagonal matrix of order n

$$L_{\mathrm{I}} = \{ L_1, L_2, \ldots, L_t \} \tag{38}$$

has the polynomials (37) as its invariant polynomials (see Theorem 4 on p. 141). Since the matrices A and L_{I} have the same invariant polynomials, they are similar, i.e., there always exists a non-singular matrix U ($|U| \neq 0$) such that

$$A = UL_{\mathrm{I}}U^{-1}. \tag{I}$$

The matrix L_{I} is called the *first natural normal form* of the matrix A. This normal form is characterized by: 1) the quasi-diagonal form (38), 2) the special structure of the diagonal blocks (36), and 3) the additional condition: in the sequence of characteristic polynomials of the diagonal blocks every polynomial from the second onwards divides the preceding one.[17]

2. We now denote by

$$\chi_1(\lambda), \chi_2(\lambda), \ldots, \chi_u(\lambda) \tag{39}$$

the elementary divisors of $A = \| a_{ik} \|_1^n$ in the number field F. The corresponding companion matrices will be denoted by

$$L^{(1)}, L^{(2)}, \ldots, L^{(u)}.$$

Since $\chi_j(\lambda)$ is the only elementary divisor of $L^{(j)}$ $(j = 1, 2, \ldots, u)$,[18] the quasi-diagonal matrix

$$L_{\mathrm{II}} = \{L^{(1)}, L^{(2)}, \ldots, L^{(u)}\} \tag{40}$$

has, by Theorem 5, the polynomials (39) as its elementary divisors.

The matrices A and L_{II} have the same elementary divisors in F. Therefore the matrices are similar, i.e., there always exists a non-singular matrix V ($|V| \neq 0$) such that

$$A = VL_{\mathrm{II}}V^{-1}. \tag{II}$$

The matrix L_{II} is called the *second natural normal form* of the matrix A. This normal form is characterized by: 1) the quasi-diagonal form (40), 2) the special structure of the diagonal blocks (36), and 3) the additional condition: the characteristic polynomial of each diagonal block is a power of an irreducible polynomial over F.

Note. The elementary divisors of a matrix A, in contrast to the invariant polynomials, are essentially connected with the given number field F. If we choose instead of the original field F another number field (which also contains the elements of the given matrix A), then the elementary divisors may change. Together with the elementary divisors, the second natural normal form of a matrix also changes.

---

[17] From the conditions 1), 2), 3) it follows automatically that the characteristic polynomials of the diagonal blocks in L_{I} are the invariant polynomials of the matrix L_{I} and, hence, of A.

[18] $\chi_j(\lambda)$ is the only invariant polynomial of $L^{(j)}$ and is at the same time a power of a polynomial irreducible over F.

Suppose, for example, that $A = \| a_{ik} \|_1^n$ is a matrix with real elements. The characteristic polynomial of the matrix then has real coefficients. But this polynomial may have complex roots. If F is the field of real numbers, then among the elementary divisors there may also be powers of irreducible quadratic trinomials with real coefficients. If F is the field of complex numbers, then every elementary divisor has the form $(\lambda - \lambda_0)^p$.

3. Let us assume now that the number field F contains not only the elements of A, but also the characteristic values of the matrix.[19] Then the elementary divisors of A have the form[20]

$$(\lambda - \lambda_1)^{p_1},\ (\lambda - \lambda_2)^{p_2},\ \ldots,\ (\lambda - \lambda_u)^{p_u} \qquad (p_1 + p_2 + \cdots + p_u = n). \tag{41}$$

We consider one of these elementary divisors:

$$(\lambda - \lambda_0)^p$$

and associate with it the following matrix of order p:

$$\left\| \begin{matrix} \lambda_0 & 1 & 0 & \ldots & 0 \\ 0 & \lambda_0 & 1 & \ldots & 0 \\ \cdot & \cdot & \cdot & \cdot & \cdot \\ 0 & 0 & 0 & \ldots & 1 \\ 0 & 0 & 0 & \ldots & \lambda_0 \end{matrix} \right\| = \lambda_0 E^{(p)} + H^{(p)}. \tag{42}$$

It is easy to verify that this matrix has only the one elementary divisor $(\lambda - \lambda_0)^p$. The matrix (42) will be called the *Jordan block* corresponding to the elementary divisor $(\lambda - \lambda_0)^p$.

The Jordan blocks corresponding to the elementary divisors (41) will be denoted by

$$J_1,\ J_2,\ \ldots,\ J_u.$$

Then the quasi-diagonal matrix

$$J = \{ J_1,\ J_2,\ \ldots,\ J_u \}$$

has the powers (41) as its elementary divisors.

The matrix J can also be written in the form

$$J = \{ \lambda_1 E_1 + H_1,\ \lambda_2 E_2 + H_2,\ \ldots,\ \lambda_u E_u + H_u \};$$

where

$$E_k = E^{(p_k)},\ H_k = H^{(p_k)} \qquad (k = 1, 2, \ldots, u).$$

---

[19] This always holds for an arbitrary matrix A if F is the field of complex numbers.

[20] Among the numbers $\lambda_1, \lambda_2, \ldots, \lambda_u$ there may be some that are equal.

Since the matrices A and J have the same elementary divisors, they are similar, i.e., there exists a non-singular matrix $T(|T| \neq 0)$ such that

$$A = TJT^{-1} = T\{\lambda_1 E_1 + H_1,\ \lambda_2 E_2 + H_2,\ \ldots,\ \lambda_u E_u + H_u\} T^{-1}. \quad \text{(III)}$$

The matrix J is called the *Jordan normal form* or simply *Jordan form* of A. The Jordan normal form is characterized by its quasi-diagonal form and by the special structure (42) of the diagonal blocks.

The following scheme describes the Jordan matrix J for the elementary divisors $(\lambda - \lambda_1)^2$, $(\lambda - \lambda_2)^3$, $\lambda - \lambda_3$, $(\lambda - \lambda_4)^2$:

$$J = \left\| \begin{array}{cc:ccc:c:cc} \lambda_1 & 1 & 0 & 0 & 0 & 0 & 0 & 0 \\ 0 & \lambda_1 & 0 & 0 & 0 & 0 & 0 & 0 \\ \hdashline 0 & 0 & \lambda_2 & 1 & 0 & 0 & 0 & 0 \\ 0 & 0 & 0 & \lambda_2 & 1 & 0 & 0 & 0 \\ 0 & 0 & 0 & 0 & \lambda_2 & 0 & 0 & 0 \\ \hdashline 0 & 0 & 0 & 0 & 0 & \lambda_3 & 0 & 0 \\ \hdashline 0 & 0 & 0 & 0 & 0 & 0 & \lambda_4 & 1 \\ 0 & 0 & 0 & 0 & 0 & 0 & 0 & \lambda_4 \end{array} \right\|. \quad (43)$$

If (and only if) all the elementary divisors of a matrix A are of the first degree, the Jordan form is a diagonal matrix, and in this case we have:

$$A = T\{\lambda_1, \lambda_2, \ldots, \lambda_n\} T^{-1}. \quad (44)$$

Thus: *A matrix A has simple structure* (see Chapter III, § 8) *if and only if all its elementary divisors are of the first degree.*[21]

Instead of the Jordan block (42) sometimes the *'lower' Jordan block* of order p is used:

$$\left\| \begin{array}{cccccc} \lambda_0 & 0 & \cdots & & 0 & 0 \\ 1 & \lambda_0 & \cdots & & 0 & 0 \\ 0 & \ddots & \ddots & & \vdots & \vdots \\ \vdots & & \ddots & \ddots & \vdots & \vdots \\ \vdots & & & \ddots & \lambda_0 & 0 \\ 0 & \cdots & & 0 & 1 & \lambda_0 \end{array} \right\| = \lambda_0 E^{(p)} + F^{(p)}.$$

This matrix also has the single elementary divisor $(\lambda - \lambda_0)^p$ only. To the elementary divisors (41) there corresponds the *lower Jordan matrix*.[22]

---

[21] The elementary divisors of degree 1 are often called 'linear' or 'simple' elementary divisors.

[22] The matrix J is often called the *upper Jordan matrix*, in contrast to the lower Jordan matrix $J_{(1)}$.

$$J_{(1)} = \{ \lambda_1 E_1 + F_1, \lambda_2 E_2 + F_2, \ldots, \lambda_u E_u + F_u \}$$
$$(E_k = E^{(p_k)}, \; F_k = F^{(p_k)}; \; k = 1, 2, \ldots, u).$$

An arbitrary matrix A having the elementary divisors (41) is always similar to $J_{(1)}$, i.e., there exists a non-singular matrix T_1 ($|T_1| \neq 0$) such that

$$A = T_1 J_{(1)} T_1^{-1} = T_1 \{ \lambda_1 E_1 + F_1, \lambda_2 E_2 + F_2, \ldots, \lambda_u E_u + F_u \} T_1^{-1}. \quad \text{(IV)}$$

We also note that if $\lambda_0 \neq 0$, each of the two matrices

$$\lambda_0 (E^{(p)} + H^{(p)}), \qquad \lambda_0 (E^{(p)} + F^{(p)})$$

has only the single elementary divisor $(\lambda - \lambda_0)^p$. Therefore for a *non-singular* matrix A having the elementary divisors (41) we have, apart from (III) and (IV), the representations

$$A = T_2 \{ \lambda_1 (E_1 + H_1), \lambda_2 (E_2 + H_2), \ldots, \lambda_u (E_u + H_u) \} T_2^{-1}, \quad \text{(V)}$$
$$A = T_3 \{ \lambda_1 (E_1 + F_1), \lambda_2 (E_2 + F_2), \ldots, \lambda_u (E_u + F_u) \} T_3^{-1}. \quad \text{(VI)}$$

§ 7. The Elementary Divisors of the Matrix $f(A)$

1. In this section we consider the following problem:

Given the elementary divisors (in the field of complex numbers) of a matrix $A = \| a_{ik} \|_1^n$ and given a function $f(\lambda)$ defined on the spectrum of A, to determine the elementary divisors (in the field of complex numbers) of the matrix $f(A)$.

The matrix $f(A)$ does not alter if we replace the function $f(\lambda)$ by a polynomial that assumes on the spectrum of A the same values as $f(\lambda)$ (see Chapter V, § 1). Without loss of generality we may therefore assume in what follows that $f(\lambda)$ is a polynomial.

We denote by

$$(\lambda - \lambda_1)^{p_1}, \; (\lambda - \lambda_2)^{p_2}, \ldots, (\lambda - \lambda_u)^{p_u}$$

the elementary divisors of A.[23] Thus A is similar to the Jordan matrix

$$A = TJT^{-1},$$

and so

$$f(A) = Tf(J)T^{-1}.$$

[23] Among the numbers $\lambda_1, \lambda_2, \ldots, \lambda_u$ there may be some that are equal.

Moreover,

$$J = \{ J_1, J_2, \ldots, J_u \}, \quad J_i = \lambda_i E^{(p_i)} + H^{(p_i)} \quad (i = 1, 2, \ldots, u)$$

and

$$f(J) = \{ f(J_1), f(J_2), \ldots, f(J_u) \}, \tag{45}$$

where (see Example 2 on p. 100)

$$f(J_i) = \begin{Vmatrix} f(\lambda_i) & \frac{f'(\lambda_i)}{1!} & \cdots & \frac{f^{(p_i-1)}(\lambda_i)}{(p_i-1)!} \\ 0 & f(\lambda_i) & \ddots & \vdots \\ \vdots & & \ddots & \frac{f'(\lambda_i)}{1!} \\ 0 & 0 & \cdots & f(\lambda_i) \end{Vmatrix}. \tag{46}$$

Since the similar matrices $f(A)$ and $f(J)$ have the same elementary divisors, we shall from now on consider $f(J)$ instead of $f(A)$.

2. Let us determine the defect[24] d of $f(A)$ or, what is the same, of $f(J)$. The defect of a quasi-diagonal matrix is equal to the sum of the defects of the various diagonal blocks and the defect of $f(J_i)$ (see 46)) is equal to the smaller of the numbers k_i and p_i, where k_i is the multiplicity of λ_i as a root of $f(\lambda)$,[25] so that

$$f(\lambda_i) = f'(\lambda_i) = \cdots = f^{(k_i-1)}(\lambda_i) = 0, \quad f^{(k_i)}(\lambda_i) \neq 0 \qquad (i = 1, 2, \ldots, u).$$

We have thus arrived at the following theorem:

THEOREM 8: *The defect of the matrix $f(A)$, where A has the elementary divisors*

$$(\lambda - \lambda_1)^{p_1}, (\lambda - \lambda_2)^{p_2}, \ldots, (\lambda - \lambda_u)^{p_u} \tag{47}$$

is given by the formula

$$d = \sum_{i=1}^{u} \min(k_i, p_i); \tag{48}$$

---

[24] $d = n - r$, where r is the rank of $f(A)$. If the elementary divisors of a matrix are known, then the defect of the matrix is determined as the number of elementary divisors corresponding to the characteristic value 0, i.e., as the number of elementary divisors of the form λ^{ω}.

[25] k_i may be equal to zero; in that case $f(\lambda_i) \neq 0$.

here k_i is the multiplicity of λ_i as root of $f(\lambda)$ $(i=1, 2, \ldots, u)$.[26]

As an application of this theorem we shall determine all the elementary divisors of an arbitrary matrix $A = \| a_{ik} \|_1^n$ that corresponds to a characteristic value λ_0:

$$\underbrace{\lambda-\lambda_0, \ldots, \lambda-\lambda_0}_{g_1}; \quad \underbrace{(\lambda-\lambda_0)^2, \ldots, (\lambda-\lambda_0)^2}_{g_2}; \quad \ldots; \quad \underbrace{(\lambda-\lambda_0)^m, \ldots, (\lambda-\lambda_0)^m}_{g_m},$$

where $g_i \geqq 0$ $(i=1, 2, \ldots, m-1)$, $g_m > 0$, provided the defects

$$d_1, d_2, \ldots, d_m$$

of the matrices

$$A-\lambda_0 E, (A-\lambda_0 E)^2, \ldots, (A-\lambda_0 E)^m$$

are given.

For this purpose we note that $(A-\lambda_0 E)^j = f_j(A)$, where $f_j(\lambda) = (\lambda-\lambda_0)^j$ $(j=1, 2, \ldots, m)$. In order to determine the defect of $(A-\lambda_0 E)^j$ we have, therefore, to set $k_i = j$ in (48) for the elementary divisors corresponding to the characteristic value λ_0 and $k_i = 0$ for all the other terms $(j=1, 2, \ldots, m)$. Thus we obtain the formulas

$$\left.\begin{array}{l} g_1 + g_2 + g_3 + \cdots + g_m = d_1, \\ g_1 + 2g_2 + 2g_3 + \cdots + 2g_m = d_2, \\ g_1 + 2g_2 + 3g_3 + \cdots + 3g_m = d_3, \\ \cdots\cdots\cdots\cdots\cdots\cdots\cdots \\ g_1 + 2g_2 + 3g_3 + \cdots + mg_m = d_m. \end{array}\right\} \tag{49}$$

Hence[27]

$$g_j = 2d_j - d_{j-1} - d_{j+1} \qquad (j=1, 2, \ldots, m;\ d_0 = 0,\ d_{m+1} = d_m). \tag{50}$$

3. Let us return to the basic problem of determining the elementary divisors of the matrix $f(A)$. As we have mentioned above, the elementary divisors of $f(A)$ coincide with those of $f(J)$ and the elementary divisors of a quasi-diagonal matrix coincide with those of the diagonal blocks (see Theorem 5). Therefore the problem reduces to finding the elementary divisors of a matrix C of regular triangular form:

---

[26] In the general case, where $f(\lambda)$ is not a polynomial, then min (k_i, p_i) in (48) has to be interpreted as the number p_i if

$$f(\lambda_i) = f'(\lambda_i) = \ldots = f^{(p_i)}(\lambda_i) = 0$$

and as the number $k_i \leqq p_i$ if

$$f(\lambda_i) = f'(\lambda_i) = \ldots = f^{(k_i-1)}(\lambda_i) = 0,$$
$$f^{(k_i)}(\lambda_i) \neq 0$$

$(i=1, 2, \ldots, u)$.

[27] The number m is characterized by the fact that $d_{m-1} < d_m = d_{m+j}$ $(j=1, 2, \ldots)$.

$$C = \sum_{k=0}^{p-1} a_k H^k = \begin{Vmatrix} a_0 & a_1 & \cdots & a_{p-1} \\ 0 & a_0 & \ddots & \vdots \\ \vdots & & \ddots & a_1 \\ 0 & 0 & \cdots & a_0 \end{Vmatrix}. \tag{51}$$

We consider separately two cases:

1. $a_1 \neq 0$. The characteristic polynomial of C is obviously equal to

$$D_p(\lambda) = (\lambda - a_0)^p.$$

Since $D_{p-1}(\lambda)$ divides $D_p(\lambda)$ without remainder, we have

$$D_{p-1}(\lambda) = (\lambda - a_0)^g \quad (g \leqq p).$$

Here $D_{p-1}(\lambda)$ denotes the greatest common divisor of the minors of order $p-1$ in the characteristic matrix

$$\lambda E - C = \begin{Vmatrix} \lambda - a_0 & -a_1 & \cdots & -a_{p-1} \\ 0 & \lambda - a_0 & \ddots & \vdots \\ \vdots & & \ddots & -a_1 \\ \overset{+}{0} & 0 & \cdots & \lambda - a_0 \end{Vmatrix}.$$

It is easy to see that when the minor of the zero element marked by '+' is expanded, every term contains at least one factor $\lambda - a_0$, except the product of the elements on the main diagonal, which is $(-a_1)^{p-1}$ and is therefore in our case different from zero. But since $D_{p-1}(\lambda)$ must be a power of $\lambda - a_0$, we see that $g = 0$. But then it follows from

$$D_p(\lambda) = (\lambda - a_0)^p, \quad D_{p-1}(\lambda) = 1$$

that C has only the one elementary divisor $(\lambda - a_0)^p$.

2. $a_1 = \ldots = a_{k-1} = 0$, $a_k \neq 0$. In this case,

$$C = a_0 E + a_k H^k + \cdots + a_{p-1} H^{p-1}.$$

Therefore for the positive integer j the defect of the matrix

$$(C - a_0 E)^j = a_k^j H^{kj} + \cdots$$

is given by

$$d_j = \begin{cases} kj, & \text{when} \quad kj \leqq p, \\ p, & \text{when} \quad kj > p. \end{cases}$$

We set

$$p = qk + h \quad (0 \leqq h < k). \tag{52}$$

Then[28]

$$d_1 = k,\ d_2 = 2k,\ \ldots,\ d_q = qk,\ d_{q+1} = p. \tag{53}$$

Therefore we have by (50)

$$g_1 = \cdots = g_{q-1} = 0, \quad g_q = k - h,\ g_{q+1} = h.$$

Thus, the matrix C has the elementary divisors

$$\underbrace{(\lambda - a_0)^{q+1}, \ldots, (\lambda - a_0)^{q+1}}_{h}, \quad \underbrace{(\lambda - a_0)^{q}, \ldots, (\lambda - a_0)^{q}}_{k-h}, \tag{54}$$

where the integers $q > 0$ and $h \geqq 0$ are determined by (52).

4. Now we are in a position to ascertain what elementary divisors the matrix $f(J)$ has (see (45) and (46)). To each elementary divisor of A

$$(\lambda - \lambda_0)^p$$

there corresponds in $f(J)$ the diagonal cell

$$f(\lambda_0 E + H) = \sum_{i=0}^{p-1} \frac{f^{(i)}(\lambda_0)}{i!} H^i = \left\| \begin{matrix} f(\lambda_0) & \frac{f'(\lambda_0)}{1!} & \cdots & \frac{f^{(p-1)}(\lambda_0)}{(p-1)!} \\ 0 & f(\lambda_0) & \ddots & \vdots \\ \vdots & \vdots & \ddots & \vdots \\ \vdots & \vdots & \ddots & \frac{f'(\lambda_0)}{1!} \\ 0 & 0 & \cdots & f(\lambda_0) \end{matrix} \right\|. \tag{55}$$

Clearly the problem reduces to finding the elementary divisors of a cell of the form (55). But the matrix (55) is of the regular triangular form (51), where

$$a_0 = f(\lambda_0),\ a_1 = f'(\lambda_0),\ a_2 = \frac{f''(\lambda_0)}{2!},\ \ldots.$$

Thus we arrive at the theorem:

---

[28] In this case the number $q + 1$ plays the role of m in (49) and (50).

THEOREM 9: *The elementary divisors of the matrix $f(A)$ are obtained from those of A in the following way: To an elementary divisor*

$$(\lambda - \lambda_0)^p \tag{56}$$

of A for $p=1$ or for $p > 1$ and $f'(\lambda_0) \neq 0$ there corresponds a single elementary divisor

$$(\lambda - f(\lambda_0))^p \tag{57}$$

of $f(A)$; for $p > 1$ and $f'(\lambda_0) = \dots = f^{(k-1)}(\lambda_0) = 0$, $f^{(k)}(\lambda_0) \neq 0$ $(k < p)$ to the elementary divisor (56) *of A there correspond the following elementary divisors of $f(A)$*:

$$\underbrace{(\lambda - f(\lambda_0))^{q+1}, \dots, (\lambda - f(\lambda_0))^{q+1}}_{h}, \quad \underbrace{(\lambda - f(\lambda_0))^{q}, \dots, (\lambda - f(\lambda_0))^{q}}_{k-h}, \tag{58}$$

where

$$p = qk + h, \quad 0 \leqq q, \quad 0 \leqq h < k;$$

finally, for $p > 1$, $f'(\lambda_0) = \dots = f^{(p-1)}(\lambda_0) = 0$, to the elementary divisor (56) *there correspond p elementary divisors of the first degree of $f(A)$*:[29]

$$\lambda - f(\lambda_0), \dots, \lambda - f(\lambda_0). \tag{59}$$

We note the following special cases of this theorem.

1. *If $\lambda_1, \lambda_2, \dots, \lambda_n$ are the characteristic values of A, then $f(\lambda_1), f(\lambda_2), \dots, f(\lambda_n)$ are the characteristic values of $f(A)$.* (In both sequences each characteristic value is repeated as often as its multiplicity as a root of the characteristic equation indicates.)[30]

2. *If the derivative $f'(\lambda)$ is not zero on the spectrum of A,*[31] *then in going from A to $f(A)$ the elementary divisors are not 'split up' i.e., if A has the elementary divisors*

$$(\lambda - \lambda_1)^{p_1}, (\lambda - \lambda_2)^{p_2}, \dots, (\lambda - \lambda_n)^{p_n},$$

then $f(A)$ has the elementary divisors

$$(\lambda - f(\lambda_1))^{p_1}, (\lambda - f(\lambda_2))^{p_2}, \dots, (\lambda - f(\lambda_n))^{p_n}.$$

---

29 (57) is obtained from (58) by setting $k=1$; (59) is obtained from (58) by setting $k=p$ or $k > p$.

30 Statement 1. was established separately in Chapter IV, p. 84.

31 I.e., $f'(\lambda_i) \neq 0$ for those λ_i that are multiple roots of the minimal polynomial.

§ 8. A General Method of Constructing the Transforming Matrix

In many problems in the theory of matrices and its applications it is sufficient to know the normal form into which a given matrix $A = \| a_{ik} \|_1^n$ can be carried by similarity transformations. The normal form is completely determined by the invariant polynomials of the characteristic matrix $\lambda E - A$. To find the latter, we can use the defining formulas (see (10) on p. 139) or the reduction of the characteristic matrix $\lambda E - A$ to canonical diagonal form by elementary transformations.

In some problems, however, it is necessary to know not only the normal form $\tilde{A}$ of the given matrix A, but also a non-singular transforming matrix T.

1. An immediate method of determining T consists in the following. The equation

$$A = T\tilde{A}T^{-1}$$

can be written as:

$$AT - T\tilde{A} = O.$$

This matrix equation in T is equivalent to a system of n^2 linear homogeneous equations in the n^2 unknown elements of T. The determination of a transforming matrix reduces to the solution of this system of n^2 equations. Moreover, we have to choose from the set of all solutions one for which $|T| \neq 0$. The existence of such a solution is certain, since A and $\tilde{A}$ have the same invariant polynomials.[32]

Note that whereas the normal form is uniquely determined by the matrix A,[33] for the transforming matrix T we always have an innumerable set of values that are given by

$$T = UT_1, \tag{60}$$

where T_1 is one of the transforming matrices and U is an arbitrary matrix that is permutable with A.[34]

---

[32] From this fact follows the similarity of $\tilde{A}$ and A.

[33] This statement is unconditionally true as regards the first natural normal form. As far as the second normal form or the Jordan normal form is concerned, they are uniquely determined to within the order of the diagonal blocks.

[34] The formula (60) may be replaced by

$$T = T_1V_1,$$

where V is an arbitrary matrix permutable with $\tilde{A}$.

The method proposed above for determining a transforming matrix T is simple enough in concept but of little use in practice, since it requires a great many computations (even for $n=4$ we have to solve 16 linear equations).

2. We proceed to explain a more efficient method of constructing the transforming matrix T. This method is based on the Supplement to Theorem 7 (p. 148). According to this, we can choose as the transforming matrix

$$T=Q(\tilde{A}), \tag{61}$$

provided

$$\lambda E-\tilde{A}=P(\lambda)(\lambda E-A)Q(\lambda).$$

The latter equation expresses the equivalence of the characteristic matrices $\lambda E-A$ and $\lambda E-\tilde{A}$. Here $P(\lambda)$ and $Q(\lambda)$ are polynomial matrices with constant non-zero determinants.

For the actual process of finding $Q(\lambda)$ we reduce the two λ-matrices $\lambda E-A$ and $\lambda E-\tilde{A}$ to canonical form by means of the corresponding elementary transformations

$$\{i_n(\lambda), i_{n-1}(\lambda), \ldots, i_1(\lambda)\}=P_1(\lambda)(\lambda E-A)Q_1(\lambda) \tag{62}$$

$$\{i_n(\lambda), i_{n-1}(\lambda), \ldots, i_1(\lambda)\}=P_2(\lambda)(\lambda E-\tilde{A})Q_2(\lambda) \tag{63}$$

where

$$Q_1(\lambda)=T_1T_2, \ldots, T_{p_1}, \quad Q_2(\lambda)=T_1^*T_2^*\ldots T_{p_2}^*, \tag{64}$$

and where $T_1, \ldots, T_{p_1}, T_1^*, \ldots, T_{p_2}^*$ are the elementary matrices corresponding to the elementary operations on the columns of the λ-matrices $\lambda E-A$ and $\lambda E-\tilde{A}$. From (62), (63), and (64) it follows that

$$\lambda E-\tilde{A}=P(\lambda)(\lambda E-A)Q(\lambda),$$

where

$$Q(\lambda)=Q_1(\lambda)Q_2^{-1}(\lambda)=T_1T_2\cdots T_{p_1}T_{p_2}^{*-1}T_{p_2-1}^{*-1}\cdots T_1^{*-1}. \tag{65}$$

We can compute the matrix $Q(\lambda)$ by applying successively to the columns of the unit matrix E the elementary operations with the matrices $T_1, \ldots, T_{p_1}, T_{p_2}^{*-1}, \ldots, T_1^{*-1}$. After this (in accordance with (61)) we replace the argument λ in $Q(\lambda)$ by the matrix $\tilde{A}$.

Example.

$$A=\left\|\begin{array}{rrrr} 1 & -1 & 1 & -1 \\ -3 & 3 & -5 & 4 \\ 8 & -4 & 3 & -4 \\ 15 & -10 & 11 & -11 \end{array}\right\|.$$

Let us introduce a symbolic notation for the right elementary operations and the corresponding matrices (see pp. 130-131):

$$T' = [(c)\,i],\quad T'' = [i + (b(\lambda))\,j],\quad T''' = [ij].$$

In transforming the characteristic matrix $\lambda E - A$ into normal diagonal form we shall at the same time keep a record of the elementary right operations to be performed, i.e., the operations on the columns:

$$\lambda E - A = \begin{Vmatrix} \lambda - 1 & 1 & -1 & 1 \\ 3 & \lambda - 3 & 5 & -4 \\ -8 & 4 & \lambda - 3 & 4 \\ -15 & 10 & -11 & \lambda + 11 \end{Vmatrix},\quad \begin{Vmatrix} 0 & 0 & 0 & 1 \\ 4\lambda - 1 & \lambda + 1 & 1 & -4 \\ -4\lambda - 4 & 0 & \lambda + 1 & 4 \\ -\lambda^2 - 10\lambda - 4 & -\lambda - 1 & \lambda & \lambda + 11 \end{Vmatrix},$$

$$\begin{Vmatrix} 1 & 0 & 0 & 0 \\ -4 & \lambda + 1 & 1 & 4\lambda - 1 \\ 4 & 0 & \lambda + 1 & -4\lambda - 4 \\ \lambda + 11 & -\lambda - 1 & \lambda & -\lambda^2 - 10\lambda - 4 \end{Vmatrix};$$

$$\begin{Vmatrix} 1 & 0 & 0 & 0 \\ 0 & \lambda + 1 & 1 & 4\lambda - 1 \\ 0 & 0 & \lambda + 1 & -4\lambda - 4 \\ 0 & -\lambda - 1 & \lambda & -\lambda^2 - 10\lambda - 4 \end{Vmatrix},\quad \begin{Vmatrix} 1 & 0 & 0 & 0 \\ 0 & 0 & 1 & 0 \\ 0 & -\lambda^2 - 2\lambda - 1 & \lambda + 1 & -4\lambda^2 - 7\lambda - 3 \\ 0 & -\lambda^2 - 2\lambda - 1 & \lambda & -5\lambda^2 - 9\lambda - 4 \end{Vmatrix},$$

$$\begin{Vmatrix} 1 & 0 & 0 & 0 \\ 0 & 1 & 0 & 0 \\ 0 & \lambda + 1 & -\lambda^2 - 2\lambda - 1 & -4\lambda^2 - 7\lambda - 3 \\ 0 & \lambda & -\lambda^2 - 2\lambda - 1 & -5\lambda^2 - 9\lambda - 4 \end{Vmatrix};$$

$$\begin{Vmatrix} 1 & 0 & 0 & 0 \\ 0 & 1 & 0 & 0 \\ 0 & 0 & \lambda^2 + 2\lambda + 1 & 4\lambda^2 + 7\lambda + 3 \\ 0 & 0 & \lambda^2 + 2\lambda + 1 & 5\lambda^2 + 9\lambda + 4 \end{Vmatrix},\quad \begin{Vmatrix} 1 & 0 & 0 & 0 \\ 0 & 1 & 0 & 0 \\ 0 & 0 & \lambda^2 + 2\lambda + 1 & -\lambda^2 - 3\lambda - 2 \\ 0 & 0 & \lambda^2 + 2\lambda + 1 & -\lambda - 1 \end{Vmatrix},$$

$$\begin{Vmatrix} 1 & 0 & 0 & 0 \\ 0 & 1 & 0 & 0 \\ 0 & 0 & -\lambda^2 - 2\lambda - 1 & \lambda + 1 \\ 0 & 0 & \lambda^2 + 2\lambda + 1 & -\lambda^2 - 3\lambda - 2 \end{Vmatrix};$$

$$\begin{Vmatrix} 1 & 0 & 0 & 0 \\ 0 & 1 & 0 & 0 \\ 0 & 0 & \lambda + 1 & -\lambda^2 - 2\lambda - 1 \\ 0 & 0 & -\lambda^2 - 3\lambda - 2 & \lambda^2 + 2\lambda + 1 \end{Vmatrix},\quad \begin{Vmatrix} 1 & 0 & 0 & 0 \\ 0 & 1 & 0 & 0 \\ 0 & 0 & \lambda + 1 & 0 \\ 0 & 0 & -\lambda^2 - 3\lambda - 2 & -(\lambda + 1)^3 \end{Vmatrix},$$

$$\begin{Vmatrix} 1 & 0 & 0 & 0 \\ 0 & 1 & 0 & 0 \\ 0 & 0 & \lambda + 1 & 0 \\ 0 & 0 & 0 & (\lambda + 1)^3 \end{Vmatrix}.$$

Here

$$Q_1(\lambda) = [1 + (1 - \lambda)\,4]\,[2 - 4]\,[3 + 4]\,[14]\,[2 - (\lambda + 1)\,3]\,[4 + (1 - 4\lambda)\,3]\,[23] \times \times [4 - (5)\,3]\,[43]\,[4 + (\lambda + 1)\,3].$$

We have found the invariant polynomials $(\lambda+1)^3$, $(\lambda+1)$, 1, and 1 of A. The matrix has two elementary divisors, $(\lambda+1)^3$ and $(\lambda+1)$. Therefore the Jordan normal form is

$$J=\begin{Vmatrix} -1 & 1 & 0 & 0 \\ 0 & -1 & 1 & 0 \\ 0 & 0 & -1 & 0 \\ 0 & 0 & 0 & -1 \end{Vmatrix}.$$

By elementary operations we bring the matrix $\lambda E - J$ into normal diagonal form

$$\lambda E - J=\begin{Vmatrix} \lambda+1 & -1 & 0 & 0 \\ 0 & \lambda+1 & -1 & 0 \\ 0 & 0 & \lambda+1 & 0 \\ 0 & 0 & 0 & \lambda+1 \end{Vmatrix}, \begin{Vmatrix} \lambda+1 & -1 & 0 & 0 \\ 0 & 0 & -1 & 0 \\ 0 & (\lambda+1)^2 & \lambda+1 & 0 \\ 0 & 0 & 0 & \lambda+1 \end{Vmatrix},$$

$$\begin{Vmatrix} 0 & -1 & 0 & 0 \\ 0 & 0 & -1 & 0 \\ (\lambda+1)^3 & (\lambda+1)^2 & \lambda+1 & 0 \\ 0 & 0 & 0 & \lambda+1 \end{Vmatrix};$$

$$\begin{Vmatrix} 0 & 1 & 0 & 0 \\ 0 & 0 & 1 & 0 \\ (\lambda+1)^3 & 0 & 0 & 0 \\ 0 & 0 & 0 & \lambda+1 \end{Vmatrix}, \begin{Vmatrix} 1 & 0 & 0 & 0 \\ 0 & 1 & 0 & 0 \\ 0 & 0 & 0 & (\lambda+1)^3 \\ 0 & 0 & \lambda+1 & 0 \end{Vmatrix},$$

$$\begin{Vmatrix} 1 & 0 & 0 & 0 \\ 0 & 1 & 0 & 0 \\ 0 & 0 & \lambda+1 & 0 \\ 0 & 0 & 0 & (\lambda+1)^3 \end{Vmatrix}.$$

Here

$$Q_2(\lambda)=[2+(\lambda+1)\,3]\ [1+(\lambda+1)\,2]\ [12]\ [23]\ [34].$$

Therefore

$$\begin{aligned} Q(\lambda) &= Q_1(\lambda)\,Q_2^{-1}(\lambda) \\ &= [1+(1-\lambda)4]\ [2-4]\ [3+4]\ [14]\ [2-(\lambda+1)3]\ [4+(1-4\lambda)3]\ [23]\ [4-(5)\,3]\ \times \\ &\quad \times\ [43]\ [4+(\lambda+1)\,3]\ [34]\ [23]\ [12]\ [1-(\lambda+1)\,2]\ [2-(\lambda+1)\,3]. \end{aligned}$$

We apply these elementary operations successively to the unit matrix E:

$$E=\begin{Vmatrix} 1 & 0 & 0 & 0 \\ 0 & 1 & 0 & 0 \\ 0 & 0 & 1 & 0 \\ 0 & 0 & 0 & 1 \end{Vmatrix}, \begin{Vmatrix} 1 & 0 & 0 & 0 \\ 0 & 1 & 0 & 0 \\ 0 & 0 & 1 & 0 \\ 1-\lambda & -1 & 1 & 1 \end{Vmatrix},$$

$$\begin{Vmatrix} 0 & 0 & 0 & 1 \\ 0 & 1 & 0 & 0 \\ 0 & 0 & 1 & 0 \\ 1 & -1 & 1 & 1-\lambda \end{Vmatrix}, \begin{Vmatrix} 0 & 0 & 0 & 1 \\ 0 & 1 & 0 & 0 \\ 0 & -\lambda-1 & 1 & 0 \\ 1 & -\lambda-2 & 1 & 1-\lambda \end{Vmatrix};$$

$$\begin{Vmatrix} 0 & 0 & 0 & 1 \\ 0 & 1 & 0 & 0 \\ 0 & -\lambda-1 & 1 & 1-4\lambda \\ 1 & -\lambda-2 & 1 & 2-5\lambda \end{Vmatrix}, \quad \begin{Vmatrix} 0 & 0 & 0 & 1 \\ 0 & 0 & 1 & 0 \\ 0 & 1 & -\lambda-1 & 1-4\lambda \\ 1 & 1 & -\lambda-2 & 2-5\lambda \end{Vmatrix}, \quad \begin{Vmatrix} 0 & 0 & 0 & 1 \\ 0 & 0 & 1 & -5 \\ 0 & 1 & -\lambda-1 & \lambda+6 \\ 1 & 1 & -\lambda-2 & 12 \end{Vmatrix};$$

$$\begin{Vmatrix} 0 & 0 & 1 & 0 \\ 0 & 0 & -5 & 1 \\ 0 & 1 & \lambda+6 & -\lambda-1 \\ 1 & 1 & 12 & -\lambda-2 \end{Vmatrix}, \quad \begin{Vmatrix} 0 & 0 & 1 & \lambda+1 \\ 0 & 0 & -5 & -5\lambda-4 \\ 0 & 1 & \lambda+6 & \lambda^2+6\lambda+5 \\ 1 & 1 & 12 & 11\lambda+10 \end{Vmatrix}, \quad \begin{Vmatrix} \lambda+1 & 0 & 0 & 1 \\ -5\lambda-4 & 0 & 0 & -5 \\ \lambda^2+6\lambda+5 & 0 & 1 & \lambda+6 \\ 11\lambda+10 & 1 & 1 & 12 \end{Vmatrix},$$

$$\begin{Vmatrix} \lambda+1 & 0 & 0 & 1 \\ -5\lambda-4 & 0 & 0 & -5 \\ \lambda^2+6\lambda+5 & -\lambda-1 & 1 & \lambda+6 \\ 10\lambda+9 & -\lambda & 1 & 12 \end{Vmatrix}.$$

Thus

$$Q(\lambda) = \begin{Vmatrix} \lambda+1 & 0 & 0 & 1 \\ -5\lambda-4 & 0 & 0 & -5 \\ \lambda^2+6\lambda+5 & -\lambda-1 & 1 & \lambda+6 \\ 10\lambda+9 & -\lambda & 1 & 12 \end{Vmatrix}$$

$$= \begin{Vmatrix} 0 & 0 & 0 & 0 \\ 0 & 0 & 0 & 0 \\ 1 & 0 & 0 & 0 \\ 0 & 0 & 0 & 0 \end{Vmatrix} \lambda^2 + \begin{Vmatrix} 1 & 0 & 0 & 0 \\ -5 & 0 & 0 & 0 \\ 6 & -1 & 0 & 1 \\ 10 & -1 & 0 & 0 \end{Vmatrix} \lambda + \begin{Vmatrix} 1 & 0 & 0 & 1 \\ -4 & 0 & 0 & -5 \\ 5 & -1 & 1 & 6 \\ 9 & 0 & 1 & 12 \end{Vmatrix}.$$

Observing that

$$J^2 = \begin{Vmatrix} 1 & -2 & 1 & 0 \\ 0 & 1 & -2 & 0 \\ 0 & 0 & 1 & 0 \\ 0 & 0 & 0 & 1 \end{Vmatrix},$$

we have

$$T = Q(J) = \begin{Vmatrix} 0 & 0 & 0 & 0 \\ 0 & 0 & 0 & 0 \\ 1 & 0 & 0 & 0 \\ 0 & 0 & 0 & 0 \end{Vmatrix} \cdot \begin{Vmatrix} 1 & -2 & 1 & 0 \\ 0 & 1 & -2 & 0 \\ 0 & 0 & 1 & 0 \\ 0 & 0 & 0 & 1 \end{Vmatrix} +$$

$$+ \begin{Vmatrix} 1 & 0 & 0 & 0 \\ -5 & 0 & 0 & 0 \\ 6 & -1 & 0 & 1 \\ 12 & -1 & 0 & 0 \end{Vmatrix} \cdot \begin{Vmatrix} -1 & 1 & 0 & 0 \\ 0 & -1 & 1 & 0 \\ 0 & 0 & -1 & 0 \\ 0 & 0 & 0 & -1 \end{Vmatrix} + \begin{Vmatrix} 1 & 0 & 0 & 1 \\ -4 & 0 & 0 & -5 \\ 5 & -1 & 1 & 6 \\ 9 & 0 & 1 & 12 \end{Vmatrix}$$

$$= \begin{Vmatrix} 0 & 1 & 0 & 1 \\ 1 & -5 & 0 & -5 \\ 0 & 4 & 1 & 5 \\ -1 & 11 & 0 & 12 \end{Vmatrix}.$$

Check:

$$AT = \begin{Vmatrix} 0 & -1 & 1 & -1 \\ -1 & 6 & -5 & 5 \\ 0 & -4 & 3 & -5 \\ 1 & -12 & 11 & -12 \end{Vmatrix}, \quad TJ = \begin{Vmatrix} 0 & -1 & 1 & -1 \\ -1 & 6 & -5 & 5 \\ 0 & -4 & 3 & -5 \\ 1 & -12 & 11 & -12 \end{Vmatrix},$$

i.e., $AT = TJ$.

$$|T| = \begin{vmatrix} 0 & 1 & 0 & 1 \\ 1 & -5 & 0 & -5 \\ 0 & 4 & 1 & 5 \\ -1 & 11 & 0 & 12 \end{vmatrix} = -1 \neq 0.$$

Therefore

$$A = TJT^{-1}.$$

§ 9. Another Method of Constructing a Transforming Matrix

1. We shall now explain another method of constructing a transforming matrix which often leads to fewer computations than the method of the preceding section. However, we shall apply this second method only when the Jordan normal form and thus the elementary divisors

$$(\lambda - \lambda_1)^{p_1}, \ (\lambda - \lambda_2)^{p_2}, \ldots \tag{66}$$

of the given matrix A are known.

Let $A = TJT^{-1}$, where

$$J = \{\lambda_1 E^{(p_1)} + H^{(p_1)}, \lambda_2 E^{(p_2)} + H^{(p_2)}, \ldots\} = \begin{Vmatrix} \overbrace{\begin{matrix} \lambda_1 & 1 & \cdots & 0 \\ \cdot & \cdot & \cdot & \cdot \\ \cdot & & \cdot & \cdot \\ \cdot & & & 1 \\ 0 & \cdots & \cdot & \lambda_1 \end{matrix}}^{p_1} & & \\ & \overbrace{\begin{matrix} \lambda_2 & 1 & \cdots & 0 \\ \cdot & \cdot & \cdot & \cdot \\ \cdot & & \cdot & \cdot \\ \cdot & & & 1 \\ 0 & \cdots & \cdot & \lambda_2 \end{matrix}}^{p_2} & \\ & & \ddots \end{Vmatrix}.$$

Then denoting the k-th column of T by t_k $(k = 1, 2, \ldots, n)$, we replace the matrix equation

$$AT = TJ$$

by the equivalent system of equations

$$At_1 = \lambda_1 t_1,\ At_2 = \lambda_1 t_2 + t_1, \ldots, A\, t_{p_1} = \lambda_1 t_{p_1} + t_{p_1-1} \tag{67}$$

$$At_{p_1+1} = \lambda_2 t_{p_1+1},\ At_{p_1+2} = \lambda_2 t_{p_1+2} + t_{p_1+1}, \ldots, At_{p_1+p_2} = \lambda_2 t_{p_1+p_2} + t_{p_1+p_2-1} \tag{68}$$

$$\cdots\cdots\cdots\cdots\cdots\cdots\cdots\cdots\cdots$$

which we rewrite as follows:

$$(A - \lambda_1 E)\, t_1 = O,\ (A - \lambda_1 E)\, t_2 = t_1, \ldots, (A - \lambda_1 E)\, t_{p_1} = t_{p_1-1} \tag{67'}$$

$$(A - \lambda_2 E)\, t_{p_1+1} = O,\ (A - \lambda_2 E)\, t_{p_1+2} = t_{p_1+1}, \ldots, (A - \lambda_2 E)\, t_{p_1+p_2} = t_{p_1+p_2-1} \tag{68'}$$

$$\cdots\cdots\cdots\cdots\cdots\cdots\cdots\cdots\cdots$$

Thus, all the columns of T are split into 'Jordan chains' of columns: $[t_1, t_2, \ldots, t_{p_1}]$, $[t_{p_1+1}, t_{p_1+2}, \ldots, t_{p_1+p_2}]$, ...

To every Jordan block of J (or, what is the same, to every elementary divisor (66)) there corresponds its Jordan chain of columns. Each Jordan chain of columns is characterized by a system of equations of type (67), (68), etc.

The task of finding a transforming matrix T reduces to that of finding the Jordan chains that would give in all n linearly independent columns.

We shall show that these Jordan chains of columns can be determined by means of the reduced adjoint matrix $C(\lambda)$ (see Chapter IV, § 6).

For the matrix $C(\lambda)$ we have the identity

$$(\lambda E - A)\, C(\lambda) = \psi(\lambda)\, E. \tag{69}$$

where $\psi(\lambda)$ is the minimal polynomial of A.

Let

$$\psi(\lambda) = (\lambda - \lambda_0)^m \chi(\lambda) \qquad (\chi(\lambda_0) \neq 0).$$

We differentiate the identity (69) term by term $m - 1$ times:

$$\left.\begin{aligned} (\lambda E - A)\, C'(\lambda) + C(\lambda) &= \psi'(\lambda)\, E \\ (\lambda E - A)\, C''(\lambda) + 2C'(\lambda) &= \psi''(\lambda)\, E \\ &\cdots\cdots\cdots \\ (\lambda E - A)\, C^{(m-1)}(\lambda) + (m-1)\, C^{(m-2)}(\lambda) &= \psi^{(m-1)}(\lambda)\, E. \end{aligned}\right\} \tag{70}$$

Substituting λ_0 for λ in (69) and (70) and observing that the right-hand sides are zero, we obtain

$$(A - \lambda_0 E)\, C = O,\ (A - \lambda_0 E)\, D = C,\ (A - \lambda_0 E)\, F = D, \ldots, (A - \lambda_0 E) K = G; \tag{71}$$

where

$$\left.\begin{aligned} C = C(\lambda_0),\ D = \frac{1}{1!} C'(\lambda_0),\ F = \frac{1}{2!} C''(\lambda_0), \ldots, G = \frac{1}{(m-2)!} C^{(m-2)}(\lambda_0) \\ K = \frac{1}{(m-1)!} C^{(m-1)}(\lambda_0). \end{aligned}\right\} \tag{72}$$

In (71) we replace the matrices (72) by their k-th columns ($k = 1, 2, \ldots, n$). We obtain:

$$(A - \lambda_0 E) C_k = o, \ (A - \lambda_0 E) D_k = C_k, \ldots, (A - \lambda_0 E) K_k = G_k \qquad (73)$$
$$(k = 1, 2, \ldots, n).$$

Since $C = C(\lambda_0) \neq O$,[35] we can find a k ($\leqq n$) such that

$$C_k \neq o. \qquad (74)$$

Then the m columns

$$C_k, D_k, F_k, \ldots, G_k, K_k \qquad (75)$$

are linearly independent. For let

$$\gamma C_k + \delta D_k + \ldots + \varkappa K_k = o. \qquad (76)$$

Multiplying both sides of (76) successively by $A - \lambda_0 E, \ldots, (A - \lambda_0 E)^{m-1}$, we obtain

$$\delta C_k + \ldots + \varkappa G_k = o, \ldots, \varkappa C_k = o. \qquad (77)$$

From (76) and (77) we find by (74):

$$\gamma = \delta = \ldots = \varkappa = 0.$$

Since the linearly independent columns (75) satisfy the system of equations (73), they form a Jordan chain of vectors corresponding to the elementary divisor $(\lambda - \lambda_0)^m$ (compare (73) with (67′)).

If $C_k = o$ for some k, but $D_k \neq o$, then the columns $D_k, \ldots, G_k, K_k$ form a Jordan chain of $m - 1$ vectors, etc.

2. We shall now show first of all how to construct a transforming matrix T in the case where the elementary divisors of A are pairwise co-prime:

$$(\lambda - \lambda_1)^{m_1}, \ldots, (\lambda - \lambda_s)^{m_s},$$
$$(\lambda_i \neq \lambda_j \text{ for } i \neq j;\ i, j = 1, 2, \ldots, s).$$

With the elementary divisor $(\lambda - \lambda_j)^{m_j}$ we associate the Jordan chain of columns

$$C^{(j)}, D^{(j)}, \ldots, G^{(j)}, K^{(j)},$$

constructed as indicated above. Then

$$(A - \lambda_j E) C^{(j)} = o, \ (A - \lambda_j E) D^{(j)} = C^{(j)}, \ldots, (A - \lambda_j E) K^{(j)} = G^{(j)}. \qquad (78)$$

When we give to j the values $1, 2, \ldots, s$, we obtain s Jordan chains containing n columns in all. These columns are linearly independent.

[35] From $C(\lambda_0) = O$ it would follow that all the elements of $C(\lambda)$ have a common divisor of positive degree, in contradiction to the definition of $C(\lambda)$.

For, suppose that

$$\sum_{j=1}^{s} [\gamma_j C^{(j)} + \delta_j D^{(j)} + \dots + \varkappa_j K^{(j)}] = o. \tag{79}$$

We multiply both sides of (79) on the left by

$$(A-\lambda_1 E)^{m_1} \cdots (A-\lambda_{j-1}E)^{m_{j-1}} (A-\lambda_j E)^{m_j - 1} (A-\lambda_{j+1}E)^{m_{j+1}} \cdots (A-\lambda_s E)^{m_s} \tag{80}$$

and obtain

$$\varkappa_j = 0 .$$

Replacing $m_j - 1$ successively by $m_j - 2$, $m_j - 3$, ... in (80), we find:

$$\gamma_j = \delta_j = \dots = \varkappa_j = 0 \qquad (j = 1, 2, \dots, s),$$

and this is what we had to prove.

We define the matrix T by the formula

$$T = (C^{(1)}, D^{(1)}, \dots, K^{(1)}; C^{(2)}, D^{(2)}, \dots, K^{(2)}; \dots; C^{(s)}, D^{(s)}, \dots, K^{(s)}). \tag{81}$$

Example.

$$\begin{array}{c} \\ A = \end{array} \begin{array}{cccc} \mathit{8} & \mathit{3} & \mathit{-10} & \mathit{-3} \\ \left\| \begin{array}{cccc} 3 & -1 & -4 & 2 \\ 2 & 3 & -2 & -4 \\ 2 & -1 & -3 & 2 \\ 1 & 2 & -1 & -3 \end{array} \right\| \\ \dots & \dots & \dots & \dots \\ 3 & 2 & 2 & 1 \\ \dots & \dots & \dots & \dots \\ 1 & 4 & 0 & 2 \end{array} ;$$

$\psi(\lambda) = \Delta(\lambda) = (\lambda - 1)^2 (\lambda + 1)^2 = \lambda^4 - 2\lambda^2 + 1$.
elementary divisors: $(\lambda - 1)^2, (\lambda + 1)^2$,

$$\Psi(\lambda, \mu) = \frac{\psi(\mu) - \psi(\lambda)}{\mu - \lambda} = \mu^3 + \lambda\mu^2 + (\lambda^2 - 2)\mu + \lambda^3 - 2\lambda,$$

$$C(\lambda) = \Psi(\lambda E, A) = A^3 + \lambda A^2 + (\lambda^2 - 2) A + (\lambda^3 - 2\lambda) E .$$

We make up the first column $C_1(\lambda)$:

$$C_1(\lambda) = [A^3]_1 + \lambda [A^2]_1 + (\lambda^2 - 2) A_1 + (\lambda^3 - 2\lambda) E_1 .$$

For the computation of the first column of A^2 we multiply all the rows of A into the first column of A. We obtain:[36] $[A^2]_1 = (1, 4, 0, 2)$. Multiplying all the rows of A into this column, we find: $[A^3]_1 = (3, 6, 2, 3)$.

Therefore

$$C_1(\lambda) = \left\| \begin{array}{c} 3 \\ 6 \\ 2 \\ 3 \end{array} \right\| + \lambda \left\| \begin{array}{c} 1 \\ 4 \\ 0 \\ 2 \end{array} \right\| + (\lambda^2 - 2) \left\| \begin{array}{c} 3 \\ 2 \\ 2 \\ 1 \end{array} \right\| + (\lambda^3 - 2\lambda) \left\| \begin{array}{c} 1 \\ 0 \\ 0 \\ 0 \end{array} \right\| = \left\| \begin{array}{c} \lambda^3 + 3\lambda^2 - \lambda - 3 \\ 2\lambda^2 + 4\lambda + 2 \\ 2\lambda^2 - 2 \\ \lambda^2 + 2\lambda + 1 \end{array} \right\| .$$

[36] The columns into which we multiply the rows are written underneath the rows of A. The elements of the row of column-sums are set up in italics, for checking.

Hence $C_1(1) = (0, 8, 0, 4)$ and $C_1'(1) = (8, 8, 4, 4)$. As $C_1(-1) = (0, 0, 0, 0)$, we pass on to the second column and, proceeding as before, we find: $C_2(-1) = (-4, 0, -4, 0)$ and $C_2'(-1) = (4, -4, 4, -4)$. We set up the matrix:

$$(C_1(1), C_1'(1); C_2(-1), C_2'(-1)) = \left\| \begin{matrix} 0 & 8 & -4 & 4 \\ 8 & 8 & 0 & -4 \\ 0 & 4 & -4 & 4 \\ 4 & 4 & 0 & -4 \end{matrix} \right\|.$$

We cancel[37] 4 in the first two columns and -4 in the last two columns.

$$T = \left\| \begin{matrix} 0 & 2 & 1 & -1 \\ 2 & 2 & 0 & 1 \\ 0 & 1 & 1 & -1 \\ 1 & 1 & 0 & 1 \end{matrix} \right\|.$$

We leave it to the reader to verify that

$$AT = T \cdot \left\| \begin{matrix} 1 & 1 & 0 & 0 \\ 0 & 1 & 0 & 0 \\ 0 & 0 & -1 & 1 \\ 0 & 0 & 0 & -1 \end{matrix} \right\|$$

3. Coming now to the general case, we shall investigate the Jordan chains of vectors corresponding to a characteristic value λ_0 for which there are p elementary divisors $(\lambda - \lambda_0)^m$, q elementary divisors $(\lambda - \lambda_0)^{m-1}$, r elementary divisors $(\lambda - \lambda_0)^{m-2}$, etc.

As a preliminary to this, we establish some properties of the matrices

$$C = C(\lambda_0),\ D = C'(\lambda_0),\ F = \frac{1}{2!} C''(\lambda_0),\ \ldots,\ K = \frac{1}{(m-1)!} C^{(m-1)}(\lambda_0). \tag{82}$$

1. *The matrices* (82) *can be represented in the form of polynomials in* A:

$$C = h_1(A),\ D = h_2(A),\ \ldots,\ K = h_m(A), \tag{83}$$

where

$$h_i(\lambda) = \frac{\psi(\lambda)}{(\lambda - \lambda_0)^i} \quad (i = 1, 2, \ldots, m). \tag{84}$$

For

$$C(\lambda) = \Psi(\lambda E, A),$$

where

$$\Psi(\lambda, \mu) = \frac{\psi(\mu) - \psi(\lambda)}{\mu - \lambda}.$$

---

[37] A Jordan chain remains a Jordan chain when all its columns are multiplied by a number $c \neq 0$.

Therefore

$$\frac{1}{k!} C^{(k)}(\lambda_0) = \frac{1}{k!} \Psi^{(k)}(\lambda_0 E, A), \tag{85}$$

where

$$\begin{aligned}\frac{1}{k!} \psi^{(k)}(\lambda_0, \mu) &= \frac{1}{k!} \left[\frac{\partial^k}{\partial \lambda^k} \Psi(\lambda, \mu) \right]_{\lambda=\lambda_0} \\ &= \frac{1}{k!} \left[\frac{\partial^k}{\partial \lambda^k} \frac{\psi(\mu)}{\mu - \lambda} \right]_{\lambda=\lambda_0} = \frac{\psi(\mu)}{(\mu - \lambda_0)^{k+1}} = h_{k+1}(\mu).\end{aligned} \tag{86}$$

(83) follows from (82), (85), and (86).

2. *The matrices* (82) *have the ranks*

$$p,\ 2p + q,\ 3p + 2q + r,\ \ldots.$$

This property of the matrices (82) follows immediately from 1. and Theorem 8 (§ 7), if we equate the rank to $n - d$ and use formula (48) for the defect of a function on A (p. 154).

3. *In the sequence of matrices* (82) *every column of each matrix is a linear combination of the columns of every following matrix.*

Let us take two matrices $h_i(A)$ and $h_k(A)$ in (82) (see 1.). Suppose that $i < k$. Then it follows from (84) that:

$$h_i(A) = h_k(A)\,(A - \lambda_0 E)^{k-i}.$$

Hence the j-th column y_j ($j = 1, 2, \ldots, n$) of $h_i(A)$ is expressed linearly by the columns $z_1, z_2, \ldots, z_n$ of $h_k(A)$:

$$y_j = \sum_{g=1}^{n} \alpha_g z_g,$$

where $\alpha_1, \alpha_2, \ldots, \alpha_n$ are the elements of the j-th column of $(A - \lambda_0 E)^{k-i}$.

4. *Without changing the basic formulas* (71) *we may replace any column in C by an arbitrary linear combination of all the columns, provided we make the corresponding replacements in $D, \ldots, K$.*

We now proceed to the construction of the Jordan chains of columns for the elementary divisors

$$\underbrace{(\lambda - \lambda_0)^m, \ldots, (\lambda - \lambda_0)^m}_{p};\ \underbrace{(\lambda - \lambda_0)^{m-1}, \ldots, (\lambda - \lambda_0)^{m-1}}_{q};\ \ldots.$$

Using the properties 2. and 4., we transform the matrix C into the form

$$C = (C_1, C_2, \ldots, C_p;\ o, o, \ldots, o); \tag{87}$$

where the columns $C_1, C_2, \ldots, C_p$ are linearly independent. Now

$$D = (D_1, D_2, \ldots, D_p;\ D_{p+1}, \ldots, D_n).$$

By 3., for every i $(1 \leqq i \leqq p)$ C_i is a linear combination of the columns $D_1, D_2, \ldots, D_n$:

$$C_i = \alpha_1 D_1 + \cdots + \alpha_p D_p + \alpha_{p+1} D_{p+1} + \cdots + \alpha_n D_n. \tag{88}$$

We multiply both sides of this equation by $A - \lambda_0 E$. Observing (see (73)) that

$$(A - \lambda_0 E)\, C_i = o \quad (i = 1, 2, \ldots, p), \quad (A - \lambda_0 E)\, D_j = C_j \quad (j = 1, 2, \ldots, n),$$

we obtain by (87)

$$o = \alpha_1 C_1 + \alpha_2 C_2 + \cdots + \alpha_p C_p;$$

hence in (88)

$$\alpha_1 = \cdots = \alpha_p = 0.$$

Therefore the columns $C_1, C_2, \ldots, C_p$ are linearly independent combinations of the columns $D_{p+1}, \ldots, D_n$. Therefore by 4. and 2., we can, without changing the matrix C, take the columns $C_1, \ldots, C_p$ instead of $D_{p+1}, \ldots, D_{2p}$ and zeros instead of $D_{2p+q+1}, \ldots, D_n$.

Then the matrix D assumes the form

$$D = (D_1, \ldots, D_p;\ C_1, C_2, \ldots, C_p;\ D_{2p+1}, \ldots, D_{2p+q};\ o, o, \ldots, o). \tag{89}$$

In the same way, preserving the forms (87) and (89) of the matrices C and D, we can represent the next matrix F in the form

$$\left.\begin{aligned} F = (&F_1, \ldots, F_p;\ D_1, \ldots, D_p;\ F_{2p+1}, \ldots, F_{2p+q};\ C_1, \ldots, C_p; \\ &D_{2p+1}, \ldots, D_{2p+q}, F_{3p+2q+1}, \ldots, F_{3p+2q+r}, o, \ldots, o), \end{aligned}\right\} \tag{90}$$

etc.

Formulas (73) gives us the Jordan chains

$$\left.\begin{aligned} &\underbrace{\overbrace{(C_1, D_1, \ldots, K_1)}^{m}, \ldots, \overbrace{(C_p, D_p, \ldots, K_p)}^{m};}_{p} \\ &\underbrace{\overbrace{(D_{2p+1}, F_{2p+1} \ \ldots, K_{2p+1},)}^{m-1} \ldots, \overbrace{(D_{2p+q}, F_{2p+q}, \ldots, K_{2p+q})}^{m-1};}_{q} \ \ldots \end{aligned}\right\} \tag{91}$$

These Jordan chains are linearly independent. For all the columns C_i in (91) are linearly independent, because they form p linearly independent columns of C. All the columns C_i, D_j in (91) are independent, because they form $2p + q$ independent columns in D, etc.; finally, *all the columns in* (91) *are independent, because they form* $n_0 = mp + (m-1)q + \ldots$ *independent columns in* K. The number of columns in (91) is equal to the sum of the exponents of the elementary divisors corresponding to the given characteristic value λ_0.

Suppose that the matrix $A = \| a_{ik} \|_1^n$ has s distinct characteristic values λ_j

$$(j = 1, 2, \ldots, s;$$

$$\Delta(\lambda) = (\lambda - \lambda_1)^{n_1} (\lambda - \lambda_2)^{n_2} \cdots (\lambda - \lambda_s)^{n_s}$$

$$\psi(\lambda) = (\lambda - \lambda_1)^{m_1} (\lambda - \lambda_2)^{m_2} \cdots (\lambda - \lambda_s)^{m_s}).$$

For each characteristic value λ_j we form its system of independent Jordan chains (91); the number of columns in this system is equal to n_j ($j = 1, 2, 3, \ldots, s$). All the chains so obtained contain $n = n_1 + n_2 + \ldots + n_s$ columns.

These n *columns are linearly independent and form one of the required transforming matrices* T.

The proof of the linear independence of these n columns proceeds as follows.

Every linear combination of these n columns can be represented in the form

$$\sum_{j=1}^{s} H_j = o, \tag{92}$$

where H_j is a linear combination of columns in the Jordan chains (91) corresponding to the characteristic value λ_j ($j = 1, 2, \ldots, s$). But every column in the Jordan chain corresponding to the characteristic value λ_j satisfies the equation

$$(A - \lambda_j E)^{m_j} x = o.$$

Therefore

$$(A - \lambda_j E)^{m_j} H_j = o. \tag{93}$$

We take a fixed number j ($1 \leqq j \leqq s$) and construct the Lagrange-Sylvester interpolation polynomial $r(\lambda)$ (See Chapter V, §§ 1, 2) with the following values on the spectrum of the matrix:

$$r(\lambda_i) = r'(\lambda_i) = \ldots = r^{(m_i - 1)}(\lambda_i) = 0 \text{ for } i \neq j$$

and

$$r(\lambda_j) = 1, \; r'(\lambda_j) = \cdots = r^{(m_j - 1)}(\lambda_j) = 0.$$

Then, for every $i \neq j$, $r(\lambda)$ is divisible by $(\lambda - \lambda_1)^{m_i}$ without remainder; therefore by (93),

$$r(A) H_i = o \qquad (i \neq j). \tag{94}$$

In exactly the same way, the difference $r(\lambda) - 1$ is divisible by $(\lambda - \lambda_j)^{m_j}$ without remainder; therefore

$$r(A) H_j = H_j. \tag{95}$$

Multiplying both sides of (92) by $r(A)$, we obtain from (94) and (95):

$$H_j = o.$$

This is valid for every $j = 1, 2, \ldots, s$. But H_j is a linear combination of independent columns corresponding to one and the same characteristic value λ_j $(j = 1, 2, \ldots, s)$. Therefore all the coefficients in the linear combination H_j $(j = 1, 2, \ldots, s)$, and hence all the coefficients in (92), are equal to zero.

Note. Let us point out some transformations on the columns of the matrix T under which it is transformed into the same Jordan form (with the same arrangement of the Jordan diagonal blocks):

I. *Multiplication of all the columns of an arbitrary Jordan chain by a non-zero number.*

II. *Addition to each column (beginning with the second) of a Jordan chain of the preceding column of the same chain, multiplied by one and the same arbitrary number.*

III. *Addition to all the columns of a Jordan chain of the corresponding columns of another chain containing the same or a larger number of columns and corresponding to the same characteristic value.*

Example 1.

$$A = \begin{Vmatrix} 1 & 0 & 0 & 1 & -1 \\ 0 & 1 & -2 & 3 & -3 \\ 0 & 0 & -1 & 2 & -2 \\ 1 & -1 & 1 & 0 & 1 \\ 1 & -1 & 1 & -1 & 2 \end{Vmatrix}; \qquad \begin{gathered} \Delta(\lambda) = (\lambda - 1)^4 (\lambda + 1), \\ \psi(\lambda) = (\lambda - 1)^2 (\lambda + 1) = \lambda^3 - \lambda^2 - \lambda + 1. \\ \text{Elementary divisors of the matrix } A \\ (\lambda - 1)^2,\ (\lambda - 1)^2,\ \lambda + 1. \end{gathered}$$

$$J = \begin{Vmatrix} 1 & 1 & 0 & 0 & 0 \\ 0 & 1 & 0 & 0 & 0 \\ 0 & 0 & 1 & 1 & 0 \\ 0 & 0 & 0 & 1 & 0 \\ 0 & 0 & 0 & 0 & -1 \end{Vmatrix},$$

$$\Psi(\lambda, \mu) = \frac{\psi(\mu) - \psi(\lambda)}{\mu - \lambda} = \mu^2 + (\lambda - 1)\mu + \lambda^2 - \lambda - 1,$$
$$C(\lambda) = \Psi(\lambda E, A) = A^2 + (\lambda - 1) A + (\lambda^2 - \lambda - 1) E.$$

Let us compute successively the column of A^2 and the corresponding columns of $C(\lambda)$, $C(1)$, $C'(\lambda)$, $C'(1)$, $C(-1)$. We must obtain two linearly independent columns of $C(1)$ and one non-zero column of $C(-1)$.

$$C(\lambda)=\begin{Vmatrix}1&0&0&2&*\\0&1&0&2&*\\0&0&1&0&*\\2&-2&2&-1&*\\2&-2&2&-2&*\end{Vmatrix}+(\lambda-1)\begin{Vmatrix}1&0&0&1&*\\0&1&-2&3&*\\0&0&-1&2&*\\1&-1&1&0&*\\1&-1&1&-1&*\end{Vmatrix}+(\lambda^2-\lambda-1)\begin{Vmatrix}1&0&0&0&0\\0&1&0&0&0\\0&0&1&0&0\\0&0&0&1&0\\0&0&0&0&1\end{Vmatrix}$$

$$C(+1)=\begin{Vmatrix}0&0&0&2&*\\0&0&0&2&*\\0&0&0&0&*\\2&-2&2&-2&*\\2&-2&2&-2&*\end{Vmatrix},\quad C'(\lambda)=\begin{Vmatrix}1&0&0&1&*\\0&1&-2&3&*\\0&0&-1&2&*\\1&-1&1&0&*\\1&-1&1&-1&*\end{Vmatrix}+(2\lambda-1)\begin{Vmatrix}1&0&0&0&0\\0&1&0&0&0\\0&0&1&0&0\\0&0&0&1&0\\0&0&0&0&1\end{Vmatrix},$$

$$C'(+1)=\begin{Vmatrix}2&*&*&1&*\\0&*&*&3&*\\0&*&*&2&*\\1&*&*&1&*\\1&*&*&-1&*\end{Vmatrix},\quad C(-1)=\begin{Vmatrix}0&0&0&*&*\\0&0&4&*&*\\0&0&4&*&*\\0&0&0&*&*\\0&0&0&*&*\end{Vmatrix}.$$

Therefore[38]

$$T=(C_1(+1),C_1'(+1),C_4(+1),C_4'(+1),C_3(-1))=\begin{Vmatrix}0&2&2&1&0\\0&0&2&3&4\\0&0&0&2&4\\2&1&-2&1&0\\2&1&-2&-1&0\end{Vmatrix}.$$

The matrix T can be simplified a little. We

1) Divide the fifth column by 4;
2) Add the first column to the third and the second to the fourth;
3) Subtract the third column from the fourth;
4) Divide the first and second columns by 2;
5) Subtract the first column, multiplied by ½, from the second.

Then we obtain the matrix

$$T_1=\begin{Vmatrix}0&1&2&1&0\\0&0&2&1&1\\0&0&0&2&1\\1&0&0&2&0\\1&0&0&0&0\end{Vmatrix}.$$

We leave it to the reader to verify that $AT_1=T_1J$ and $|T_1|\neq 0$.

Example 2.

$$A=\begin{Vmatrix}1&-1&1&-1\\-3&3&-5&4\\8&-4&3&-4\\15&-10&11&-11\end{Vmatrix};\qquad \begin{aligned}\Delta(\lambda)&=(\lambda+1)^4,\\ \psi(\lambda)&=(\lambda+1)^3.\end{aligned}$$

Elementary divisors: $(\lambda+1)^3$, $\lambda+1$.

[38] Here the subscript denotes the number of the column; for example, $C_3(-1)$ denotes the third column of $C(-1)$.

$$J = \begin{Vmatrix} -1 & 1 & 0 & 0 \\ 0 & -1 & 1 & 0 \\ 0 & 0 & -1 & 0 \\ 0 & 0 & 0 & -1 \end{Vmatrix}.$$

We form the polynomials

$$h_1(\lambda) = \frac{\psi(\lambda)}{\lambda + 1} = (\lambda + 1)^2, \; h_2(\lambda) = \frac{\psi(\lambda)}{(\lambda + 1)^2} = \lambda + 1, \; h_3(\lambda) = \frac{\psi(\lambda)}{(\lambda + 1)^3} = 1$$

and the matrices[39]

$$C = h_1(A) = (A + E)^2, \; D = h_2(A) = A + E, \; F = E:$$

$$C = \begin{Vmatrix} 0 & 0 & 0 & 0 \\ 2 & -1 & 1 & -1 \\ 0 & 0 & 0 & 0 \\ -2 & 1 & -1 & 1 \end{Vmatrix}, \; D = \begin{Vmatrix} 2 & -1 & 1 & -1 \\ -3 & 4 & -5 & 4 \\ 8 & -4 & 4 & -4 \\ 15 & -10 & 11 & -10 \end{Vmatrix}, \; F = \begin{Vmatrix} 1 & 0 & 0 & 0 \\ 0 & 1 & 0 & 0 \\ 0 & 0 & 1 & 0 \\ 0 & 0 & 0 & 1 \end{Vmatrix}.$$

For the first three columns of T we take the third column of these matrices: $T = (C_3, D_3, F_3, *)$. In the matrices C, D, F, we subtract twice the third column from the first and we add the third column to the second and to the fourth. We obtain

$$\tilde{C} = \begin{Vmatrix} 0 & 0 & 0 & 0 \\ 0 & 0 & 1 & 0 \\ 0 & 0 & 0 & 0 \\ 0 & 0 & -1 & 0 \end{Vmatrix}, \; \tilde{D} = \begin{Vmatrix} 0 & 0 & 1 & 0 \\ 7 & -1 & -5 & -1 \\ 0 & 0 & 4 & 0 \\ -7 & 1 & 11 & 1 \end{Vmatrix}, \; \tilde{F} = \begin{Vmatrix} 1 & 0 & 0 & 0 \\ 0 & 1 & 0 & 0 \\ -2 & 1 & 1 & 1 \\ 0 & 0 & 0 & 1 \end{Vmatrix}.$$

In the matrices $\tilde{D}$, $\tilde{F}$, we add the fourth column, multiplied by 7, to the first and subtract the fourth column from the second. We obtain

$$\tilde{C} = \begin{Vmatrix} 0 & 0 & 0 & 0 \\ 0 & 0 & 1 & 0 \\ 0 & 0 & 0 & 0 \\ 0 & 0 & -1 & 0 \end{Vmatrix}, \; \tilde{\tilde{D}} = \begin{Vmatrix} 0 & 0 & 1 & 0 \\ 0 & 0 & -5 & -1 \\ 0 & 0 & 4 & 0 \\ 0 & 0 & 11 & 1 \end{Vmatrix}, \; \tilde{\tilde{F}} = \begin{Vmatrix} 1 & 0 & 0 & 0 \\ 0 & 1 & 0 & 0 \\ 5 & 0 & 1 & 1 \\ 7 & -1 & 0 & 1 \end{Vmatrix}.$$

For the last column of T we take the first column of $\tilde{\tilde{F}}$. Then we have

$$T = (C_3, D_3, F_3, \tilde{\tilde{F}}_1) = \begin{Vmatrix} 0 & 1 & 0 & 1 \\ 1 & -5 & 0 & 0 \\ 0 & 4 & 1 & 5 \\ -1 & 11 & 0 & 7 \end{Vmatrix}.$$

As a check, we can verify that $AT = TJ$ and that $|T| \neq 0$.

CHAPTER VII

THE STRUCTURE OF A LINEAR OPERATOR IN AN n-DIMENSIONAL SPACE

(*Geometrical Theory of Elementary Divisors*)

The analytic theory of elementary divisors expounded in the preceding chapter has enabled us to determine for every square matrix a similar matrix having 'normal' or 'canonical' form. On the other hand, we have seen in Chapter III that the behaviour of a linear operator in an n-dimensional space with respect to various bases is given by means of a class of similar matrices. The existence of a matrix of normal form in such a class is closely connected with important and deep properties of a linear operator in an n-dimensional space. The study of these properties is the object of the present chapter. The investigation of the structure of a linear operator will lead us, independently of the contents of the preceding chapter, to the theory of transformations of a matrix to a normal form. Therefore the contents of this chapter may be called *the geometrical theory of elementary divisors.*[1]

§ 1. The Minimal Polynomial of a Vector and a Space (with Respect to a Given Linear Operator)

1. We consider an n-dimensional vector space $\boldsymbol{R}$ over the field F and a linear operator $\boldsymbol{A}$ in this space.

Let $\boldsymbol{x}$ be an arbitrary vector of $\boldsymbol{R}$. We form the sequence of vectors

$$\boldsymbol{x},\ \boldsymbol{A}\boldsymbol{x},\ \boldsymbol{A}^2\boldsymbol{x},\ \ldots. \tag{1}$$

Since the space is finite-dimensional, there is an integer p $(1 \leqq p \leqq n)$ such that the vectors $\boldsymbol{x}, \boldsymbol{A}\boldsymbol{x}, \ldots, \boldsymbol{A}^{p-1}\boldsymbol{x}$ are linearly independent, while $\boldsymbol{A}^p\boldsymbol{x}$ is a linear combination of these vectors with coefficients in F:

---

[1] The account of the geometric theory of elementary divisors to be given here is based on our paper [167]. For other geometrical constructions of the theory of elementary divisors, see [22], §§ 96-99 and also [53].

$$A^p x = -\gamma_1 A^{p-1}x - \gamma_2 A^{p-2}x - \cdots - \gamma_p x. \tag{2}$$

We form the monic polynomial $\varphi(\lambda) = \lambda^p + \gamma_1 \lambda^{p-1} + \cdots + \gamma_{p-1}\lambda + \gamma_p$. (A *monic* polynomial is a polynomial in which the coefficient of the highest power of the variable is unity.) Then (2) can be written:

$$\varphi(A)x = o. \tag{3}$$

Every polynomial $\varphi(\lambda)$ for which (3) holds will be called an *annihilating polynomial for the vector* x.[2] But it is easy to see that of all the monic annihilating polynomials of x the one we have constructed is of least degree. This polynomial will be called the *minimal annihilating polynomial of* x or simply the *minimal polynomial of* x.

Note that every annihilating polynomial $\tilde{\varphi}(\lambda)$ of x is divisible by the minimal polynomial $\varphi(\lambda)$.

For let

$$\tilde{\varphi}(\lambda) = \varphi(\lambda)\varkappa(\lambda) + \varrho(\lambda),$$

where $\varkappa(\lambda)$, $\varrho(\lambda)$ are quotient and remainder on dividing $\tilde{\varphi}(\lambda)$ by $\varphi(\lambda)$. Then

$$\tilde{\varphi}(A)x = \varkappa(A)\varphi(A)x + \varrho(A)x = \varrho(A)x$$

and therefore $\varrho(A)x = o$. But the degree of $\varrho(\lambda)$ is less than that of the minimal polynomial $\varphi(\lambda)$. Hence $\varrho(\lambda) \equiv 0$.

From what we have proved it follows, in particular, that every vector x has only one minimal polynomial.

2. We choose a basis $e_1, e_2, \ldots, e_n$ in R. We denote by $\varphi_1(\lambda), \varphi_2(\lambda), \ldots, \varphi_n(\lambda)$ the minimal polynomials of the basis vectors $e_1, e_2, \ldots, e_n$ and by $\psi(\lambda)$ the least common multiple of these polynomials ($\psi(\lambda)$ is taken with highest coefficient 1). Then $\psi(\lambda)$ is an annihilating polynomial for all the basis vectors $e_1, e_2, \ldots, e_n$. Since every vector $x \in R$ is representable in the form $x = x_1 e_1 + x_2 e_2 + \cdots + x_n e_n$, we have

$$\psi(A)x = x_1\psi(A)e_1 + x_2\psi(A)e_2 + \cdots + x_n\psi(A)e_n = o,$$

i.e.,

$$\psi(A) = O. \tag{4}$$

The polynomial $\psi(\lambda)$ is called *an annihilating polynomial for the whole space* R. Let $\tilde{\psi}(\lambda)$ be an arbitrary annihilating polynomial for the whole space R. Then $\tilde{\psi}(\lambda)$ is an annihilating polynomial for the basis vectors

[2] Of course, the phrase 'with respect to the given operator A' is tacitly understood. For the sake of brevity, this circumstance is not mentioned in the definition, because throughout this entire chapter we shall deal with a single operator A.

$e_1, e_2, \ldots, e_n$. Therefore $\tilde{\psi}(\lambda)$ must be a common multiple of the minimal polynomials $\varphi_1(\lambda), \varphi_2(\lambda), \ldots, \varphi_n(\lambda)$ of these vectors and must therefore be divisible without remainder by their *least* common multiple $\psi(\lambda)$. Hence it follows that, of all the annihilating polynomials for the whole space $\boldsymbol{R}$, the one we have constructed, $\psi(\lambda)$, has the least degree and it is monic. This polynomial is uniquely determined by the space $\boldsymbol{R}$ and the operator $\boldsymbol{A}$ and is called the *minimal polynomial of the space* $\boldsymbol{R}$.[3] The uniqueness of the minimal polynomial of the space $\boldsymbol{R}$ follows from the statement proved above: *every annihilating polynomial* $\tilde{\psi}(\lambda)$ *of the space* $\boldsymbol{R}$ *is divisible by the minimal polynomial* $\psi(\lambda)$. Although the construction of the minimal polynomial $\psi(\lambda)$ was associated with a definite basis $e_1, e_2, \ldots, e_n$, the polynomial $\psi(\lambda)$ itself does not depend on the choice of this basis (this follows from the uniqueness of the minimal polynomial for the space $\boldsymbol{R}$).

Finally we mention that the minimal polynomial of the space $\boldsymbol{R}$ annihilates every vector $\boldsymbol{x}$ of $\boldsymbol{R}$ so that *the minimal polynomial of the space is divisible by the minimal polynomial of every vector in the space.*

§ 2. Decomposition into Invariant Subspaces with Co-Prime Minimal Polynomials

1. If some collection of vectors $\boldsymbol{R}'$ forming part of $\boldsymbol{R}$ has the property that the sum of any two vectors of $\boldsymbol{R}'$ and the product of any vector of $\boldsymbol{R}'$ by a number $\alpha \in F$ always belongs to $\boldsymbol{R}'$, then that manifold $\boldsymbol{R}'$ is itself a vector space, a *subspace* of $\boldsymbol{R}$.

If two subspaces $\boldsymbol{R}'$ and $\boldsymbol{R}''$ of $\boldsymbol{R}$ are given and if it is known that

1. $\boldsymbol{R}'$ and $\boldsymbol{R}''$ have no vector in common except the null vector, and
2. every vector $\boldsymbol{x}$ of $\boldsymbol{R}$ can be represented in the form of a sum

$$\boldsymbol{x} = \boldsymbol{x}' + \boldsymbol{x}'' \quad (\boldsymbol{x}' \in \boldsymbol{R}', \boldsymbol{x}'' \in \boldsymbol{R}''), \tag{5}$$

then we shall say that the space $\boldsymbol{R}$ is *decomposed* into the two subspaces $\boldsymbol{R}'$ and $\boldsymbol{R}''$ and shall write:

$$\boldsymbol{R} = \boldsymbol{R}' + \boldsymbol{R}'' \tag{6}$$

Note that the condition 1. implies the uniqueness of the representation (5). For if for a certain vector $\boldsymbol{x}$ we had two distinct representations in the form of a sum of terms from $\boldsymbol{R}'$ and $\boldsymbol{R}''$, (5) and

$$\boldsymbol{x} = \tilde{\boldsymbol{x}}' + \tilde{\boldsymbol{x}}'' \quad (\tilde{\boldsymbol{x}}' \in \boldsymbol{R}', \tilde{\boldsymbol{x}}'' \in \boldsymbol{R}'') \tag{7}$$

then, subtracting (7) from (5) term by term, we would obtain:

[3] If in some basis $e_1, e_2, \ldots, e_n$ a matrix $A = \| a_{ik} \|_1^n$ then the annihilating or minimal polynomial of the space $\boldsymbol{R}$ (with respect to $\boldsymbol{A}$) is the annihilating or minimal polynomial of the matrix A, and vice versa. Compare with Chapter IV, § 6.

$$\boldsymbol{x}' - \tilde{\boldsymbol{x}}' = \tilde{\boldsymbol{x}}'' - \boldsymbol{x}'',$$

i.e., equality of the non-null vectors $\boldsymbol{x}' - \tilde{\boldsymbol{x}}' \in \boldsymbol{R}'$ and $\tilde{\boldsymbol{x}}'' - \boldsymbol{x}'' \in \boldsymbol{R}''$, which, by 1., is impossible.

Thus, condition 1. may be replaced by the requirement that the representation (5) be unique. In this form, the definition of decomposition immediately extends to an arbitrary number of subspaces.

Let

$$\boldsymbol{R} = \boldsymbol{R}' + \boldsymbol{R}''$$

and let $\boldsymbol{e}_1', \boldsymbol{e}_2', \ldots, \boldsymbol{e}_{n'}'$ and $\boldsymbol{e}_1'', \boldsymbol{e}_2'', \ldots, \boldsymbol{e}_{n''}''$ be bases of $\boldsymbol{R}'$ and $\boldsymbol{R}''$, respectively. Then the reader can easily prove that all these $n' + n''$ vectors are linearly independent and form a basis of $\boldsymbol{R}$, so that a basis of the whole space is formed from bases of the subspaces. It follows, in particular, that $n = n' + n''$.

Example 1. Suppose that in a three-dimensional space three directions, not parallel to one and the same plane, are given. Since every vector in the space can be split, uniquely, into components in these three directions, we have

$$\boldsymbol{R} = \boldsymbol{R}' + \boldsymbol{R}'' + \boldsymbol{R}''',$$

where $\boldsymbol{R}$ is the set of all the vectors of one space, $\boldsymbol{R}'$ the set of all vectors parallel to the first direction, $\boldsymbol{R}''$ to the second, and $\boldsymbol{R}'''$ to the third. In this case, $n = 3$ and $n' = n'' = n''' = 1$.

Example 2. Suppose that in a three-dimensional space a plane and a line intersecting the plane are given. Then

$$\boldsymbol{R} = \boldsymbol{R}' + \boldsymbol{R}'',$$

where $\boldsymbol{R}$ is the set of all vectors of our space, $\boldsymbol{R}'$ the set of all vectors parallel to the given plane, and $\boldsymbol{R}''$ the set of all vectors parallel to the given line. In this example, $n = 3$, $n' = 2$, $n'' = 1$.

2. A subspace $\boldsymbol{R}' \subset \boldsymbol{R}$ is called *invariant* with respect to the operator $\boldsymbol{A}$ if $\boldsymbol{A}\boldsymbol{R}' \subset \boldsymbol{R}'$, i.e. if $\boldsymbol{x} \in \boldsymbol{R}'$ implies $\boldsymbol{A}\boldsymbol{x} \in \boldsymbol{R}'$. In other words, the operator $\boldsymbol{A}$ carries a vector of an invariant subspace into a vector of the same subspace.

In what follows we shall carry out a decomposition of the whole space into subspaces invariant with respect to $\boldsymbol{A}$. The decomposition reduces the study of the behavior of an operator in the whole space to the study of its behavior in the various component subspaces.

We shall now prove the following theorem:

THEOREM 1 (First Theorem on the Decomposition of a Space into Invariant Subspaces): *If for a given operator* $\boldsymbol{A}$ *the minimal polynomial* $\psi(\lambda)$ *of the space is represented over* F *in the form of a product of two co-prime polynomials* $\psi_1(\lambda)$ *and* $\psi_2(\lambda)$ *(with highest coefficients* 1)

$$\psi(\lambda) = \psi_1(\lambda)\,\psi_2(\lambda), \tag{8}$$

then the whole space $\boldsymbol{R}$ *splits into two invariant subspaces* $\boldsymbol{I}_1$ *and* $\boldsymbol{I}_2$

$$\boldsymbol{R} = \boldsymbol{I}_1 + \boldsymbol{I}_2, \tag{9}$$

whose minimal polynomials are $\psi_1(\lambda)$ *and* $\psi_2(\lambda)$, *respectively.*

Proof. We denote by $\boldsymbol{I}_1$ the set of all vectors $\boldsymbol{x} \in \boldsymbol{R}$ satisfying the equation $\psi_1(\boldsymbol{A})\,\boldsymbol{x} = \boldsymbol{o}$. $\boldsymbol{I}_2$ is similarly defined by the equation $\psi_2(\boldsymbol{A})\,\boldsymbol{x} = \boldsymbol{o}$. $\boldsymbol{I}_1$ and $\boldsymbol{I}_2$ so defined are subspaces of $\boldsymbol{R}$.

Since $\psi_1(\lambda)$ and $\psi_2(\lambda)$ are co-prime, it follows that there exist polynomials $\chi_1(\lambda)$ and $\chi_2(\lambda)$ (with coefficients in F) such that

$$1 = \psi_1(\lambda)\,\chi_1(\lambda) + \psi_2(\lambda)\,\chi_2(\lambda). \tag{10}$$

Now let $\boldsymbol{x}$ be an arbitrary vector of $\boldsymbol{R}$. In (10) we replace λ by $\boldsymbol{A}$ and we apply both sides of the operator equation so obtained to the vector $\boldsymbol{x}$:

$$\boldsymbol{x} = \psi_1(\boldsymbol{A})\,\chi_1(\boldsymbol{A})\,\boldsymbol{x} + \psi_2(\boldsymbol{A})\,\chi_2(\boldsymbol{A})\,\boldsymbol{x}, \tag{11}$$

i.e.,

$$\boldsymbol{x} = \boldsymbol{x}' + \boldsymbol{x}'', \tag{12}$$

where

$$\boldsymbol{x}' = \psi_2(\boldsymbol{A})\,\chi_2(\boldsymbol{A})\,\boldsymbol{x}, \quad \boldsymbol{x}'' = \psi_1(\boldsymbol{A})\,\chi_1(\boldsymbol{A})\,\boldsymbol{x}. \tag{13}$$

Furthermore,

$$\psi_1(\boldsymbol{A})\,\boldsymbol{x}' = \psi(\boldsymbol{A})\,\chi_2(\boldsymbol{A})\,\boldsymbol{x} = \boldsymbol{o}, \quad \psi_2(\boldsymbol{A})\,\boldsymbol{x}'' = \psi(\boldsymbol{A})\,\chi_1(\boldsymbol{A})\,\boldsymbol{x} = \boldsymbol{o},$$

i.e.,

$$\boldsymbol{x}' \in \boldsymbol{I}_1, \text{ and } \boldsymbol{x}'' \in \boldsymbol{I}_2.$$

$\boldsymbol{I}_1$ and $\boldsymbol{I}_2$ have only the null vector in common. For if $\boldsymbol{x}_0 \in \boldsymbol{I}_1$ and $\boldsymbol{x}_0 \in \boldsymbol{I}_2$, i.e., $\psi_1(\boldsymbol{A})\,\boldsymbol{x}_0 = \boldsymbol{o}$ and $\psi_2(\boldsymbol{A})\,\boldsymbol{x}_0 = \boldsymbol{o}$, then by (11)

$$\boldsymbol{x}_0 = \chi_1(\boldsymbol{A})\,\psi_1(\boldsymbol{A})\,\boldsymbol{x}_0 + \chi_2(\boldsymbol{A})\,\psi_2(\boldsymbol{A})\,\boldsymbol{x}_0 = \boldsymbol{o}.$$

Thus we have proved that $\boldsymbol{R} = \boldsymbol{I}_1 + \boldsymbol{I}_2$.

Now suppose that $\boldsymbol{x} \in \boldsymbol{I}_1$. Then $\psi_1(\boldsymbol{A})\boldsymbol{x} = \boldsymbol{o}$. Multiplying both sides of this equation by $\boldsymbol{A}$ and reversing the order of $\boldsymbol{A}$ and $\psi_1(\boldsymbol{A})$, we obtain $\psi_1(\boldsymbol{A})\boldsymbol{A}\boldsymbol{x} = \boldsymbol{o}$, i.e., $\boldsymbol{A}\boldsymbol{x} \in \boldsymbol{I}_1$. This proves that the subspace $\boldsymbol{I}_1$ is invariant with respect to $\boldsymbol{A}$. The invariance of the subspace $\boldsymbol{I}_2$ is proved similarly.

We shall now show that $\psi_1(\lambda)$ is the minimal polynomial of $\boldsymbol{I}_1$. Let $\tilde{\psi}_1(\lambda)$ be an arbitrary annihilating polynomial for $\boldsymbol{I}_1$, and $\boldsymbol{x}$ an arbitrary vector of $\boldsymbol{R}$. Using the decomposition (12) already established, we write:

$$\tilde{\psi}_1(\boldsymbol{A})\psi_2(\boldsymbol{A})\boldsymbol{x} = \psi_2(\boldsymbol{A})\tilde{\psi}_1(\boldsymbol{A})\boldsymbol{x}' + \tilde{\psi}_1(\boldsymbol{A})\psi_2(\boldsymbol{A})\boldsymbol{x}'' = \boldsymbol{o}.$$

Since $\boldsymbol{x}$ is an arbitrary vector of $\boldsymbol{R}$, it follows that the product $\tilde{\psi}_1(\lambda)\psi_2(\lambda)$ is an annihilating polynomial for $\boldsymbol{R}$ and is therefore divisible by $\psi(\lambda) = \psi_1(\lambda)\psi_2(\lambda)$ without remainder; in other words, $\tilde{\psi}_1(\lambda)$ is divisible by $\psi_1(\lambda)$. But $\tilde{\psi}_1(\lambda)$ is an arbitrary annihilating polynomial for $\boldsymbol{I}_1$ and $\psi_1(\lambda)$ is a particular one of the annihilating polynomials (by the definition of $\boldsymbol{I}_1$). Hence $\psi_1(\lambda)$ is the minimal polynomial of $\boldsymbol{I}_1$. In exactly the same way it is shown that $\psi_2(\lambda)$ is the minimal polynomial for the invariant subspace $\boldsymbol{I}_2$.

This completes the proof of the theorem.

Let us decompose $\psi(\lambda)$ into irreducible factors over F:

$$\psi(\lambda) = [\varphi_1(\lambda)]^{c_1}[\varphi_2(\lambda)]^{c_2}\cdots[\varphi_s(\lambda)]^{c_s} \tag{14}$$

(here $\varphi_1(\lambda), \varphi_2(\lambda), \ldots, \varphi_s(\lambda)$ are distinct irreducible polynomials over F with highest coefficient 1). Then by the theorem we have

$$\boldsymbol{R} = \boldsymbol{I}_1 + \boldsymbol{I}_2 + \ldots + \boldsymbol{I}_s, \tag{15}$$

where $\boldsymbol{I}_k$ is an invariant subspace with the minimal polynomial $[\varphi_k(\lambda)]^{c_k}$ $(k = 1, 2, \ldots, s)$.

Thus, the theorem reduces the study of the behaviour of a linear operator in an arbitrary space to the study of the behaviour of this operator in a space where the minimal polynomial is a power of an irreducible polynomial over F. We shall take advantage of this to prove the following important theorem:

THEOREM 2: *In a vector space there always exists a vector whose minimal polynomial coincides with the minimal polynomial of the whole space.*

We consider first the special case where the minimal polynomial of the space $\boldsymbol{R}$ is a power of an irreducible polynomial $\varphi(\lambda)$:

$$\psi(\lambda) = [\varphi(\lambda)]^l.$$

In $\boldsymbol{R}$ we choose a basis $\boldsymbol{e}_1, \boldsymbol{e}_2, \ldots, \boldsymbol{e}_n$. The minimal polynomial of $\boldsymbol{e}_1$ is a divisor of $\psi(\lambda)$ and is therefore representable in the form $[\varphi(\lambda)]^{l_i}$, where $l_i \leqq l$ $(i=1, 2, \ldots, n)$.

But the minimal polynomial of the space is the least common multiple of the minimal polynomials of the basis vectors, so that $\psi(\lambda)$ is the largest of the powers $[\varphi(\lambda)]^{l_i}$ $(i=1, 2, \ldots, n)$. In other words, $\psi(\lambda)$ coincides with the minimal polynomial of one of the basis vectors $\boldsymbol{e}_1, \boldsymbol{e}_2, \ldots, \boldsymbol{e}_n$.

Turning now to the general case, we prove the following preliminary lemma:

LEMMA: *If the minimal polynomials of the vectors* $\boldsymbol{e}'$ *and* $\boldsymbol{e}''$ *are co-prime, then the minimal polynomial of the sum vector* $\boldsymbol{e}' + \boldsymbol{e}''$ *is equal to the product of the minimal polynomials of the constituent vectors.*

Proof. Let $\chi_1(\lambda)$ and $\chi_2(\lambda)$ be the minimal polynomials of the vectors $\boldsymbol{e}'$ and $\boldsymbol{e}''$. By assumption, $\chi_1(\lambda)$ and $\chi_2(\lambda)$ are co-prime. Let $\chi(\lambda)$ be an arbitrary annihilating polynomial of the vector $\boldsymbol{e} = \boldsymbol{e}' + \boldsymbol{e}''$. Then

$$\chi_2(\boldsymbol{A})\,\chi(\boldsymbol{A})\,\boldsymbol{e}' = \chi_2(\boldsymbol{A})\,\chi(\boldsymbol{A})\,\boldsymbol{e} - \chi(\boldsymbol{A})\,\chi_2(\boldsymbol{A})\,\boldsymbol{e}'' = \boldsymbol{o},$$

i.e., $\chi_2(\lambda)\chi(\lambda)$ is an annihilating polynomial of $\boldsymbol{e}'$. Therefore $\chi_2(\lambda)\chi(\lambda)$ is divisible by $\chi_1(\lambda)$, and since $\chi_1(\lambda)$ and $\chi_2(\lambda)$ are co-prime, $\chi(\lambda)$ is divisible by $\chi_1(\lambda)$. It is proved similarly that $\chi(\lambda)$ is divisible by $\chi_2(\lambda)$. But $\chi_1(\lambda)$ and $\chi_2(\lambda)$ are co-prime. Therefore $\chi(\lambda)$ is divisible by the product $\chi_1(\lambda)\,\chi_2(\lambda)$. Thus, every annihilating polynomial of the vector $\boldsymbol{e}$ is divisible by $\chi_1(\lambda)\,\chi_2(\lambda)$. Therefore $\chi_1(\lambda)\chi_2(\lambda)$ is the minimal polynomial of the vector $\boldsymbol{e} = \boldsymbol{e}' + \boldsymbol{e}''$.

We now return to Theorem 2. For the proof in the general case we use the decomposition (15). Since the minimal polynomials of the subspaces $\boldsymbol{I}_1, \boldsymbol{I}_2, \ldots, \boldsymbol{I}_s$ are powers of irreducible polynomials, our assertion is already proved for these subspaces. Therefore there exist vectors $\boldsymbol{e}' \in \boldsymbol{I}_1, \boldsymbol{e}'' \in \boldsymbol{I}_2, \ldots, \boldsymbol{e}^{(s)} \in \boldsymbol{I}_s$ whose minimal polynomials are $[\varphi_1(\lambda)]^{c_1}$, $[\varphi_2(\lambda)]^{c_2}, \ldots, [\varphi_s(\lambda)]^{c_s}$, respectively. By the lemma, the minimal polynomial of the vector $\boldsymbol{e} = \boldsymbol{e}' + \boldsymbol{e}'' + \cdots + \boldsymbol{e}^{(s)}$ is equal to the product

$$[\varphi_1(\lambda)]^{c_1}\,[\varphi_2(\lambda)]^{c_2}\cdots[\varphi_s(\lambda)]^{c_s},$$

i.e., to the minimal polynomial of the space $\boldsymbol{R}$.

§ 3. Congruence. Factor Space

1. Suppose given a subspace $\boldsymbol{I} \subset \boldsymbol{R}$. We shall say that two vectors $\boldsymbol{x}, \boldsymbol{y}$ of $\boldsymbol{R}$ are *congruent* modulo $\boldsymbol{I}$ and shall write $\boldsymbol{x} \equiv \boldsymbol{y} \pmod{\boldsymbol{I}}$ if and only if $\boldsymbol{y} - \boldsymbol{x} \in \boldsymbol{I}$. It is easy to verify that the concept of congruence so introduced has the following properties:

For all $\boldsymbol{x}, \boldsymbol{y}, \boldsymbol{z} \in \boldsymbol{R}$

1. $\boldsymbol{x} \equiv \boldsymbol{x} \pmod{\boldsymbol{I}}$ (reflexivity of congruence).
2. From $\boldsymbol{x} \equiv \boldsymbol{y} \pmod{\boldsymbol{I}}$ it follows that $\boldsymbol{y} \equiv \boldsymbol{x} \pmod{\boldsymbol{I}}$ (symmetry of congruence).
3. From $\boldsymbol{x} \equiv \boldsymbol{y} \pmod{\boldsymbol{I}}$ and $\boldsymbol{y} \equiv \boldsymbol{z} \pmod{\boldsymbol{I}}$ it follows that $\boldsymbol{x} \equiv \boldsymbol{z} \pmod{\boldsymbol{I}}$ (transitivity of congruence).

The presence of these three properties enables us to make use of congruence to divide all the vectors of the space into classes, by assigning vectors that are pairwise congruent (mod $\boldsymbol{I}$) to the same class (vectors of distinct classes are incongruent (mod $\boldsymbol{I}$)). The class containing the vector $\boldsymbol{x}$ will be denoted by $\bar{\boldsymbol{x}}$.[5] The subspace $\boldsymbol{I}$ is one of these classes, namely $\bar{\boldsymbol{o}}$. Note that to every congruence $\boldsymbol{x} \equiv \boldsymbol{y} \pmod{\boldsymbol{I}}$ there corresponds the equality[6] of the associated classes: $\bar{\boldsymbol{x}} = \bar{\boldsymbol{y}}$.

It is elementary to prove that congruences may be added term by term and multiplied by a number of F:

1. From
$$\boldsymbol{x} \equiv \boldsymbol{x}' \text{ and } \boldsymbol{y} \equiv \boldsymbol{y}' \qquad \pmod{\boldsymbol{I}}$$
it follows that
$$\boldsymbol{x} + \boldsymbol{y} \equiv \boldsymbol{x}' + \boldsymbol{y}' \qquad \pmod{\boldsymbol{I}}.$$

2. From
$$\boldsymbol{x} \equiv \boldsymbol{x}' \qquad \pmod{\boldsymbol{I}}$$
it follows that
$$\alpha\boldsymbol{x} \equiv \alpha\boldsymbol{x}' \qquad \pmod{\boldsymbol{I}} \ (\alpha \in \text{F}).$$

These properties of congruence show that the operations of addition and multiplication by a number of F do not 'break up' the classes. If we take two classes $\bar{\boldsymbol{x}}$ and $\bar{\boldsymbol{y}}$ and add elements $\boldsymbol{x}, \boldsymbol{x}', \ldots$ of the first class to arbitrary elements $\boldsymbol{y}, \boldsymbol{y}', \ldots$ of the second class, then all the sums so obtained belong to one and the same class, which we call the *sum* of the classes $\bar{\boldsymbol{x}}$ and $\bar{\boldsymbol{y}}$ and denote by $\bar{\boldsymbol{x}} + \bar{\boldsymbol{y}}$. Similarly, if all the vectors $\boldsymbol{x}, \boldsymbol{x}', \ldots$ of the class $\boldsymbol{x}$ are multiplied by a number $\alpha \in \text{F}$, then the products belong to one class, which we denote by $\alpha\bar{\boldsymbol{x}}$.

Thus, in the manifold $\bar{\boldsymbol{R}}$ of all classes $\bar{\boldsymbol{x}}, \bar{\boldsymbol{y}}, \ldots$ two operations are introduced: 'addition' and 'multiplication by a number of F.' It is easy to verify that these operations have the properties set forth in the definition of a vector space (Chapter III, § 1). Therefore $\bar{\boldsymbol{R}}$, as well as $\boldsymbol{R}$, is a vector

[5] Since each class contains an infinite set of vectors, there is, by this condition, an infinite number of ways of designating the class.

[6] That is, identity.

space over the field F. We shall say that $\bar{\boldsymbol{R}}$ is a *factor space* of $\boldsymbol{R}$. If $n, m, \bar{n}$ are the dimensions of the spaces $\boldsymbol{R}, \boldsymbol{I}, \bar{\boldsymbol{R}}$, respectively, then $\bar{n} = n - m$.

2. All the concepts introduced in this section can be illustrated very well by the following example.

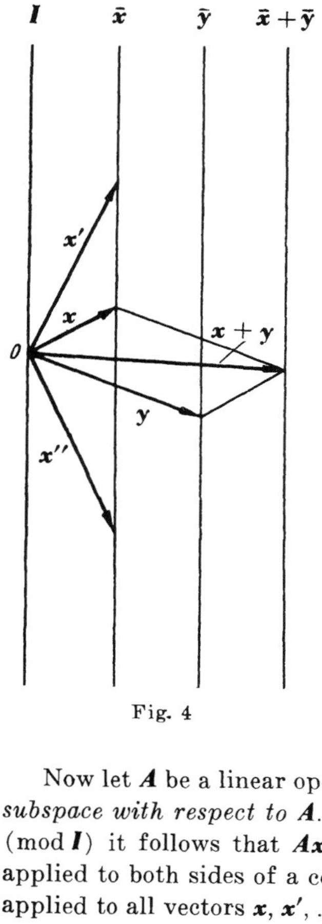

Fig. 4

Example. Let $\boldsymbol{R}$ be the set of all vectors of a three-dimensional space and F the field of real numbers. For greater clarity, we shall represent vectors in the form of directed segments beginning at a point O. Let $\boldsymbol{I}$ be a straight line passing through O (more accurately: the set of vectors that lie along some line passing through O; Fig. 4.).

The congruence $\boldsymbol{x} \equiv \boldsymbol{x}' \pmod{\boldsymbol{I}}$ signifies that the vectors $\boldsymbol{x}$ and $\boldsymbol{x}'$ differ by a vector of $\boldsymbol{I}$, i.e., the segment containing the end-points of $\boldsymbol{x}$ and $\boldsymbol{x}'$ is parallel to $\boldsymbol{I}$. Therefore the class $\bar{\boldsymbol{x}}$ is represented by the line passing through the end-point of $\boldsymbol{x}$ and parallel to $\boldsymbol{I}$ (more accurately: by the 'bundle' of vectors starting from O whose end-points lie on that line). 'Bundles' may be added and multiplied by a real number (by adding and multiplying the vectors that occur in the bundles). These 'bundles' are also the elements of the factor space $\bar{\boldsymbol{R}}$. In this example, $n = 3$, $m = 1$, $\bar{n} = 2$.

We obtain another example by taking for $\boldsymbol{I}$ a plane passing through O. In this example, $n = 3$, $m = 2$, $\bar{n} = 1$.

Now let $\boldsymbol{A}$ be a linear operator in $\boldsymbol{R}$. Let us assume that $\boldsymbol{I}$ *is an invariant subspace with respect to* $\boldsymbol{A}$. The reader will easily prove that from $\boldsymbol{x} \equiv \boldsymbol{x}'$ $\pmod{\boldsymbol{I}}$ it follows that $\boldsymbol{Ax} \equiv \boldsymbol{Ax}' \pmod{\boldsymbol{I}}$, so that the operator $\boldsymbol{A}$ can be applied to both sides of a congruence. In other words, if the operator $\boldsymbol{A}$ is applied to all vectors $\boldsymbol{x}, \boldsymbol{x}', \ldots$ of a class $\bar{\boldsymbol{x}}$, then the vectors $\boldsymbol{Ax}, \boldsymbol{Ax}', \ldots$ also belong to one class, which we denote by $\boldsymbol{A}\bar{\boldsymbol{x}}$. The linear operator $\boldsymbol{A}$ carries classes into classes and is, thus, a linear operator in $\bar{\boldsymbol{R}}$.

We shall say that the vectors $\boldsymbol{x}_1, \boldsymbol{x}_2, \ldots, \boldsymbol{x}_p$ are *linearly dependent modulo* $\boldsymbol{I}$ if there exist numbers $\alpha_1, \alpha_2, \ldots, \alpha_p$ in F, not all equal to zero, such that

$$\alpha_1\boldsymbol{x}_1 + \alpha_2\boldsymbol{x}_2 + \cdots + \alpha_p\boldsymbol{x}_p \equiv \boldsymbol{o} \pmod{\boldsymbol{I}}. \tag{16}$$

Note that not only the concept of linear dependence of vectors, but also all the concepts, statements, and reasonings, in the preceding sections of this chapter can be repeated word for word with the symbol '$=$' replaced throughout by the symbol '$\equiv$ (mod $\boldsymbol{I}$),' where $\boldsymbol{I}$ is some fixed subspace invariant with respect to $\boldsymbol{A}$.

Thus, we can introduce the concepts of an annihilating polynomial and of the minimal polynomial of a vector or a space (mod $\boldsymbol{I}$). All these concepts will be called 'relative,' in contrast to the 'absolute' concepts that were introduced earlier (and that hold for the symbol '$=$').

The reader should observe that *the relative minimal polynomial (of a vector or a space) is a divisor of the absolute one.* For example, let $\sigma_1(\lambda)$ be the relative minimal polynomial of a vector $\boldsymbol{x}$ and $\sigma(\lambda)$ the corresponding absolute minimal polynomial.

Then

$$\sigma(\boldsymbol{A})\boldsymbol{x} = \boldsymbol{o},$$

and hence it follows that also

$$\sigma(\boldsymbol{A})\boldsymbol{x} \equiv \boldsymbol{o} \quad (\text{mod } \boldsymbol{I}).$$

Therefore $\sigma(\lambda)$ is a relative annihilating polynomial of $\boldsymbol{x}$ and as such is divisible by the relative minimal polynomial $\sigma_1(\lambda)$.

Side by side with the 'absolute' statements of the preceding sections we have 'relative' statements. For example, we have the statement: 'In every space there always exists a vector whose relative minimal polynomial coincides with the relative minimal polynomial of the whole space.'

The truth of all 'relative' statements depends on the fact that by operating with congruences modulo $\boldsymbol{I}$ we deal essentially with equalities—however not in the space $\boldsymbol{R}$, but in the space $\overline{\boldsymbol{R}}$.

§ 4. Decomposition of a Space into Cyclic Invariant Subspaces

1. Let $\sigma(\lambda) = \lambda^p + \alpha_1 \lambda^{p-1} + \cdots + \alpha_{p-1}\lambda + \alpha_p$ be the minimal polynomial of a vector $\boldsymbol{e}$. Then the vectors

$$\boldsymbol{e}, \boldsymbol{A}\boldsymbol{e}, \ldots, \boldsymbol{A}^{p-1}\boldsymbol{e} \tag{17}$$

are linearly independent, and

$$\boldsymbol{A}^p\boldsymbol{e} = -\alpha_p\boldsymbol{e} - \alpha_{p-1}\boldsymbol{A}\boldsymbol{e} - \cdots - \alpha_1\boldsymbol{A}^{p-1}\boldsymbol{e}. \tag{18}$$

The vectors (17) form a basis of a p-dimensional subspace $\boldsymbol{I}$. We shall call this subspace *cyclic* in view of the special character of the basis (17) and of (18).[7] The operator $\boldsymbol{A}$ carries the first vector of (17) into the second, the second into the third, etc. The last basis vector is carried by $\boldsymbol{A}$ into a linear combination of the basis vectors in accordance with (18). Thus, $\boldsymbol{A}$ carries every basis vector into a vector of $\boldsymbol{I}$ and hence an arbitrary vector of $\boldsymbol{I}$ into another vector of $\boldsymbol{I}$. In other words, *a cyclic subspace is always invariant with respect to* $\boldsymbol{A}$.

Every vector $\boldsymbol{x} \in \boldsymbol{I}$ is representable in the form of a linear combination of the basis vectors (17), i.e., in the form

$$\boldsymbol{x} = \chi(\boldsymbol{A})\,\boldsymbol{e}, \tag{19}$$

where $\chi(\lambda)$ is a polynomial in λ of degree $\leqq p-1$ with coefficients in F. By forming all possible polynomials $\chi(\lambda)$ of degree $\leqq p-1$ with coefficients in F we obtain all the vectors of $\boldsymbol{I}$, each once only, i.e., for only one polynomial $\chi(\lambda)$. In view of the basis (17) or the formula (19) we shall say that the vector $\boldsymbol{e}$ *generates* the subspace.

Note that *the minimal polynomial of the generating vector* $\boldsymbol{e}$ *is also the minimal polynomial of the whole subspace* $\boldsymbol{I}$.

2. We are now ready to establish the fundamental proposition of the whole theory, according to which the space $\boldsymbol{R}$ splits into cyclic subspaces.

Let $\psi_1(\lambda) = \psi(\lambda) = \lambda^m + \alpha_1\lambda^{m-1} + \cdots + \alpha_m$ be the minimal polynomial of the space $\boldsymbol{R}$. Then there exists a vector $\boldsymbol{e}$ in the space for which this polynomial is minimal (Theorem 2, p. 180). Let $\boldsymbol{I}_1$ denote the cyclic subspace with the basis

$$\boldsymbol{e},\ \boldsymbol{A}\boldsymbol{e},\ \ldots,\ \boldsymbol{A}^{m-1}\boldsymbol{e}. \tag{20}$$

If $n = m$, then $\boldsymbol{R} = \boldsymbol{I}_1$. Suppose that $n > m$ and that the polynomial

$$\psi_2(\lambda) = \lambda^p + \beta_1\lambda^{p-1} + \cdots + \beta_p$$

is the minimal polynomial of $\boldsymbol{R}$ (mod $\boldsymbol{I}_1$). By the remark at the end of § 3, $\psi_2(\lambda)$ is a divisor of $\psi_1(\lambda)$, i.e., there exists a polynomial $\varkappa(\lambda)$ such that

$$\psi_1(\lambda) = \psi_2(\lambda)\,\varkappa(\lambda). \tag{21}$$

---

[7] It would be more accurate to call this subspace: cyclic with respect to the linear operator $\boldsymbol{A}$. But since the whole theory is built up with reference to a single operator $\boldsymbol{A}$, the words 'with respect to the linear operator $\boldsymbol{A}$' are omitted for the sake of brevity (see the similar remark in footnote 2, p. 176).

Moreover, in $\boldsymbol{R}$ there exists a vector $\boldsymbol{g}^*$ whose relative minimal polynomial is $\psi_2(\lambda)$. Then

$$\psi_2(\boldsymbol{A})\,\boldsymbol{g}^* \equiv \boldsymbol{o} \qquad (\mathrm{mod}\ \boldsymbol{I}_1), \tag{22}$$

i.e., there exists a polynomial $\chi(\lambda)$ of degree $\leqq m-1$ such that

$$\psi_2(\boldsymbol{A})\,\boldsymbol{g}^* = \chi(\boldsymbol{A})\,\boldsymbol{e}. \tag{23}$$

We apply the operator $\varkappa(\boldsymbol{A})$ to both sides of the equation. Then by (21) we obtain on the left $\psi_1(\boldsymbol{A})\boldsymbol{g}^*$, i.e. zero, because $\psi_1(\lambda)$ is the absolute minimal polynomial of the space; therefore

$$\varkappa(\boldsymbol{A})\,\chi(\boldsymbol{A})\,\boldsymbol{e} = \boldsymbol{o}.$$

This equation shows that the product $\varkappa(\lambda)\,\chi(\lambda)$ is an annihilating polynomial of the vector $\boldsymbol{e}$ and is therefore divisible by the minimal polynomial $\psi_1(\lambda) = \varkappa(\lambda)\,\psi_2(\lambda)$, so that $\chi(\lambda)$ is divisible by $\psi_2(\lambda)$:

$$\chi(\lambda) = \varkappa_1(\lambda)\,\psi_2(\lambda), \tag{24}$$

where $\varkappa_1(\lambda)$ is a polynomial. Using this decomposition of $\chi(\lambda)$, we may rewrite (23) as follows:

$$\psi_2(\boldsymbol{A})\,[\boldsymbol{g}^* - \varkappa_1(\boldsymbol{A})\,\boldsymbol{e}] = \boldsymbol{o}. \tag{25}$$

We now introduce the vector

$$\boldsymbol{g} = \boldsymbol{g}^* - \varkappa_1(\boldsymbol{A})\,\boldsymbol{e}. \tag{26}$$

Then (25) can be written as follows:

$$\psi_2(\boldsymbol{A})\,\boldsymbol{g} = \boldsymbol{o}. \tag{27}$$

The last equation shows that $\psi_2(\lambda)$ is an absolute annihilating polynomial of the vector $\boldsymbol{g}$ and is therefore divisible by the absolute minimal polynomial of $\boldsymbol{g}$. On the other hand, we have from (26):

$$\boldsymbol{g} \equiv \boldsymbol{g}^* \quad (\mathrm{mod}\ \boldsymbol{I}_1). \tag{28}$$

Hence $\psi_2(\lambda)$, being the relative minimal polynomial of $\boldsymbol{g}^*$, is the same for $\boldsymbol{g}$ as well. Comparing the last two statements, we deduce that $\psi_2(\lambda)$ is simultaneously the relative and the absolute minimal polynomial of $\boldsymbol{g}$.

From the fact that $\psi_2(\lambda)$ is the absolute minimal polynomial of $\boldsymbol{g}$ it follows that the subspace $\boldsymbol{I}_2$ with the basis

$$\boldsymbol{g},\ \boldsymbol{A}\boldsymbol{g},\ \ldots,\ \boldsymbol{A}^{p-1}\boldsymbol{g} \tag{29}$$

is cyclic.

From the fact that $\psi_2(\lambda)$ is the relative minimal polynomial of $\boldsymbol{g}$ (mod $\boldsymbol{I}_1$) it follows that the vectors (29) are linearly independent (mod $\boldsymbol{I}_1$), i.e., no linear combination with coefficients not all zero can be equal to a linear combination of the vectors (20). Since the latter are themselves linearly independent, our last statement asserts the linear independence of the $m + p$ vectors

$$\boldsymbol{e},\ \boldsymbol{A}\boldsymbol{e}, \ldots, \boldsymbol{A}^{m-1}\boldsymbol{e};\ \boldsymbol{g},\ \boldsymbol{A}\boldsymbol{g}, \ldots, \boldsymbol{A}^{p-1}\boldsymbol{g}. \tag{30}$$

The vectors (30) form a basis of the invariant subspace $\boldsymbol{I}_1 + \boldsymbol{I}_2$ of dimension $m + p$.

If $n = m + p$, then $\boldsymbol{R} = \boldsymbol{I}_1 + \boldsymbol{I}_2$. If $n > m + p$, we consider $\boldsymbol{R}$ (mod $\boldsymbol{I}_1 + \boldsymbol{I}_2$) and continue our process of separating cyclic subspaces. Since the whole space $\boldsymbol{R}$ is of finite dimension n, this process must come to an end with some subspace $\boldsymbol{I}_t$, where $t \leqq n$.

We have arrived at the following theorem:

THEOREM 3 (Second Theorem on the Decomposition of a Space into Invariant Subspaces): *Relative to a given linear operator* $\boldsymbol{A}$ *the space can always be split into cyclic subspaces* $\boldsymbol{I}_1, \boldsymbol{I}_2, \ldots, \boldsymbol{I}_t$ *with minimal polynomials* $\psi_1(\lambda), \psi_2(\lambda), \ldots, \psi_t(\lambda)$

$$\boldsymbol{R} = \boldsymbol{I}_1 + \boldsymbol{I}_2 + \ldots + \boldsymbol{I}_t \tag{31}$$

such that $\psi_1(\lambda)$ *coincides with the minimal polynomial* $\psi(\lambda)$ *of the whole space and that each* $\psi_i(\lambda)$ *is divisible by* $\psi_{i+1}(\lambda)$ $(i = 1, 2, \ldots, t-1)$.

3. We now mention some properties of cyclic spaces. Let $\boldsymbol{R}$ be a cyclic n-dimensional space and $\psi(\lambda) = \lambda^m + \ldots$ its minimal polynomial. Then it follows from the definition of a cyclic space that $m = n$. Conversely, suppose that $\boldsymbol{R}$ is an arbitrary space and that it is known that $m = n$. Applying the proof of the decomposition theorem, we represent $\boldsymbol{R}$ in the form (31). But the dimension of the cyclic subspace $\boldsymbol{I}_1$ is m, because its minimal polynomial coincides with the minimal polynomial of the whole space. Since $m = n$ by assumption, we have $\boldsymbol{R} = \boldsymbol{I}_1$, i.e., $\boldsymbol{R}$ is a cyclic space.

Thus we have established the following *criterion for cyclicity of a space*:

THEOREM 4: *A space is cyclic if and only if its dimension is equal to the degree of its minimal polynomial.*

Next, suppose that we have a decomposition of a cyclic space $\boldsymbol{R}$ into two invariant subspaces $\boldsymbol{I}_1$ and $\boldsymbol{I}_2$:

$$\boldsymbol{R} = \boldsymbol{I}_1 + \boldsymbol{I}_2. \tag{32}$$

We denote the dimensions of $\boldsymbol{R}$, $\boldsymbol{I}_1$, and $\boldsymbol{I}_2$ by n, n_1, and n_2, their minimal polynomials by $\psi(\lambda)$, $\psi_1(\lambda)$, and $\psi_2(\lambda)$, and the degrees of these minimal polynomials by m, m_1, and m_2, respectively. Then

$$m_1 \leqq n_1, \quad m_2 \leqq n_2. \tag{33}$$

We add these inequalities term by term:

$$m_1 + m_2 \leqq n_1 + n_2. \tag{34}$$

Since $\psi(\lambda)$ is the least common multiple of $\psi_1(\lambda)$ and $\psi_2(\lambda)$, we have

$$m \leqq m_1 + m_2. \tag{35}$$

Moreover, it follows from (32) that

$$n = n_1 + n_2. \tag{36}$$

(34), (35), and (36) give us a chain of relations

$$m \leqq m_1 + m_2 \leqq n_1 + n_2 = n. \tag{37}$$

But since the space $\boldsymbol{R}$ is cyclic, the extreme numbers of this chain, m and n, are equal. Therefore we have equality in the middle terms, i.e.,

$$m = m_1 + m_2 = n_1 + n_2.$$

From the fact that $m = m_1 + m_2$ we deduce that $\psi_1(\lambda)$ and $\psi_2(\lambda)$ are co-prime.

Bearing (33) in mind, we find from $m_1 + m_2 = n_1 + n_2$ that

$$m_1 = n_1 \quad \text{and} \quad m_2 = n_2. \tag{38}$$

These equations mean that the subspaces $\boldsymbol{I}_1$ and $\boldsymbol{I}_2$ are cyclic.

Thus we have arrived at the following proposition:

THEOREM 5: *A cyclic space can only split into invariant subspaces that* 1. *are also cyclic and* 2. *have co-prime minimal polynomials.*

The same arguments (in the opposite order) show that Theorem 5 has a converse:

THEOREM 6: *If a space is split into invariant subspaces that* 1. *are cyclic and* 2. *have co-prime minimal polynomials, then the space itself is cyclic.*

Suppose now that $\boldsymbol{R}$ is a cyclic space and that its minimal polynomial is a power of an irreducible polynomial over F: $\psi(\lambda) = [\varphi(\lambda)]^c$. In this case, the minimal polynomial of every invariant subspace of $\boldsymbol{R}$ must also be a power of this irreducible polynomial $\varphi(\lambda)$. Therefore the minimal polynomials of any two invariant subspaces cannot be co-prime. But then, by what we have proved, $\boldsymbol{R}$ cannot split into invariant subspaces.

Suppose, conversely, that some space $\boldsymbol{R}$ is known not to split into invariant subspaces. Then $\boldsymbol{R}$ is a cyclic space, for otherwise, by the second decomposition theorem, it could be split into cyclic subspaces; moreover, the minimal polynomial of $\boldsymbol{R}$ must be a power of an irreducible polynomial, because otherwise $\boldsymbol{R}$ could be split into invariant subspaces, by the first decomposition theorem.

Thus we have reached the following conclusion:

THEOREM 7: *A space does not split into invariant subspaces if and only if* 1. *it is cyclic and* 2. *its minimal polynomial is a power of an irreducible polynomial over* F.

We now return to the decomposition (31) and split the minimal polynomials $\psi_1(\lambda), \psi_2(\lambda), \ldots, \psi_t(\lambda)$ of the cyclic subspaces $\boldsymbol{I}_1, \boldsymbol{I}_2, \ldots, \boldsymbol{I}_t$ into irreducible factors over F:

$$\left.\begin{array}{l}\psi_1(\lambda)=[\varphi_1(\lambda)]^{c_1}[\varphi_2(\lambda)]^{c_2}\cdots[\varphi_s(\lambda)]^{c_s},\\ \psi_2(\lambda)=[\varphi_1(\lambda)]^{d_1}[\varphi_2(\lambda)]^{d_2}\cdots[\varphi_s(\lambda)]^{d_s},\\ \cdots\cdots\cdots\cdots\cdots\cdots\\ \psi_t(\lambda)=[\varphi_1(\lambda)]^{l_1}[\varphi_2(\lambda)]^{l_2}\cdots[\varphi_s(\gamma)]^{l_s}\end{array}\right\} \qquad (39)$$
$$(c_k \geqq d_k \geqq \ldots \geqq l_k \geqq 0;^{8}\ k=1,,2,\ldots,s).$$

To $\boldsymbol{I}_1$ we apply the first decomposition theorem. Then we obtain

$$\boldsymbol{I}_1 = \boldsymbol{I}_1' + \boldsymbol{I}_1'' + \cdots + \boldsymbol{I}_1^{(s)};$$

where $\boldsymbol{I}_1', \boldsymbol{I}_1'', \ldots, \boldsymbol{I}_1^{(s)}$ are cyclic subspaces with the minimal polynomials $[\varphi_1(\lambda)]^{c_1}, [\varphi_2(\lambda)]^{c_2}, \ldots, [\varphi_s(\lambda)]^{c_s}$. Similarly we decompose the spaces $\boldsymbol{I}_2, \ldots, \boldsymbol{I}_t$. In this way we obtain a decomposition of the whole space $\boldsymbol{R}$ into cyclic subspaces with the minimal polynomials $[\varphi_k(\lambda)]^{ck}, [\varphi_k(\lambda)]^{dk}, \ldots, [\varphi_k(\lambda)]^{lk}$ $(k=1,2,\ldots,s)$. (Here we neglect the powers whose exponents are zero.) From Theorem 7 it follows that these cyclic subspaces are indecomposable (into invariant subspaces). We have thus arrived at the following theorem:

THEOREM 8 (Third Theorem on the Decomposition of a Space into Invariant Subspaces): *A space can always be split into cyclic invariant subspaces*

$$\boldsymbol{R} = \boldsymbol{I}' + \boldsymbol{I}'' + \cdots + \boldsymbol{I}^{(u)}. \qquad (40)$$

such that the minimal polynomial of each of these cyclic subspaces is a power of an irreducible polynomial.

This theorem gives the decomposition of a space into indecomposable invariant subspaces.

[8] Some of the exponents $d_k, \ldots, l_k$ for $k > 1$ may be equal to zero.

Note. Theorem 8 (the third decomposition theorem) has been proved by applying the first two decomposition theorems. But it can also be obtained by other means, namely, as an immediate (and almost trivial) corollary of Theorem 7.

For if the space $\boldsymbol{R}$ splits at all, then it can always be split into indecomposable invariant subspaces:

$$\boldsymbol{R} = \boldsymbol{I}' + \boldsymbol{I}'' + \dots + \boldsymbol{I}^{(u)}. \tag{40}$$

By Theorem 7, each of the constituent subspaces is cyclic and has as its minimal polynomial a power of an irreducible polynomial over F.

§ 5. The Normal Form of a Matrix

1. Let $\boldsymbol{I}_1$ be an m-dimensional invariant subspace of $\boldsymbol{R}$. In $\boldsymbol{I}_1$ we take an arbitrary basis $\boldsymbol{e}_1, \boldsymbol{e}_2, \dots, \boldsymbol{e}_m$ and complement it to a basis

$$\boldsymbol{e}_1, \boldsymbol{e}_2, \dots, \boldsymbol{e}_m, \boldsymbol{e}_{m+1}, \dots, \boldsymbol{e}_n$$

of $\boldsymbol{R}$. Let us see what the matrix A of the operator $\boldsymbol{A}$ looks like in this basis. We remind the reader that the k-th column of A consists of the coordinates of the vector $\boldsymbol{A}\boldsymbol{e}_k$ $(k = 1, 2, \dots, n)$. For $k \leqq m$ the vector $\boldsymbol{A}\boldsymbol{e}_k \in \boldsymbol{I}_1$ (by the invariance of $\boldsymbol{I}_1$) and the last $n - m$ coordinates of $\boldsymbol{A}\boldsymbol{e}_k$ are zero. Therefore A has the following form

$$A = \begin{Vmatrix} \overbrace{A_1}^{m} & \overbrace{A_3}^{n-m} \\ O & A_2 \end{Vmatrix} \begin{matrix} \}\, m \\ \}\, n-m \end{matrix}, \tag{41}$$

where A_1 and A_2 are square matrices of orders m and $n - m$, respectively, and A_3 is a rectangular matrix. The fact that the fourth 'block' is zero expresses the invariance of the subspace $\boldsymbol{I}_1$. The matrix A_1 gives the operator $\boldsymbol{A}$ in $\boldsymbol{I}_1$ (with respect to the basis $\boldsymbol{e}_1, \boldsymbol{e}_2, \dots, \boldsymbol{e}_m$).

Let us assume now that $\boldsymbol{e}_{m+1}, \dots, \boldsymbol{e}_n$ is the basis of some invariant subspace $\boldsymbol{I}_2$, so that $\boldsymbol{R} = \boldsymbol{I}_1 + \boldsymbol{I}_2$ and a basis of the whole space is formed from the two parts that are the bases of the invariant subspaces $\boldsymbol{I}_1$ and $\boldsymbol{I}_2$. Then obviously the block A_3 in (41) is also equal to zero and the matrix A has the quasi-diagonal form

$$A = \left\| \begin{matrix} A_1 & O \\ O & A_2 \end{matrix} \right\| = \{ A_1, A_2 \}, \tag{42}$$

where A_1 and A_2 are, respectively, square matrices of orders m and $n - m$ which give the operator in the subspaces $\boldsymbol{I}_1$ and $\boldsymbol{I}_2$ (with respect to the bases $\boldsymbol{e}_1, \boldsymbol{e}_2, \ldots, \boldsymbol{e}_m$ and $\boldsymbol{e}_{m+1}, \ldots, \boldsymbol{e}_n$). It is not difficult to see that, conversely, to a quasi-diagonal form of the matrix there always corresponds a decomposition of the space into invariant subspaces (and the basis of the whole space is formed from the bases of these subspaces).

2. By the second decomposition theorem, we can split the whole space $\boldsymbol{R}$ into cyclic subspaces $\boldsymbol{I}_1, \boldsymbol{I}_2, \ldots, \boldsymbol{I}_t$:

$$\boldsymbol{R} = \boldsymbol{I}_1 + \boldsymbol{I}_2 + \cdots + \boldsymbol{I}_t. \tag{43}$$

In the sequence of minimal polynomials of these subspaces $\psi_1(\lambda), \psi_2(\lambda), \ldots, \psi_t(\lambda)$ each factor is a divisor of the proceeding one (from which it follows automatically that the first polynomial is the minimal polynomial of the whole space).

Let

$$\begin{aligned} \psi_1(\lambda) &= \lambda^m + \alpha_1 \lambda^{m-1} + \cdots + \alpha_m, \\ \psi_2(\lambda) &= \lambda^p + \beta_1 \lambda^{p-1} + \cdots + \beta_p, \\ &\cdots\cdots\cdots\cdots \\ \psi_t(\lambda) &= \lambda^v + \varepsilon_1 \lambda^{v-1} + \cdots + \varepsilon_v. \end{aligned} \qquad (m \geq p \geq \cdots \geq v). \tag{44}$$

We denote by $\boldsymbol{e}, \boldsymbol{g}, \ldots, \boldsymbol{l}$ generating vectors of the subspaces $\boldsymbol{I}_1, \boldsymbol{I}_2, \ldots, \boldsymbol{I}_t$ and we form a basis of the whole space from the following bases of the cyclic subspaces:

$$\boldsymbol{e}, A\boldsymbol{e}, \ldots, A^{m-1}\boldsymbol{e}; \boldsymbol{g}, A\boldsymbol{g}, \ldots, A^{p-1}\boldsymbol{g}; \ldots; \boldsymbol{l}, A\boldsymbol{l}, \ldots, A^{v-1}\boldsymbol{l}. \tag{45}$$

Let us see what the matrix L_{I} corresponding to $\boldsymbol{A}$ in this basis looks like.

As we have explained at the beginning of this section, the matrix L_{I} must have quasi-diagonal form

$$L_{\mathrm{I}} = \begin{pmatrix} L_1 & O & \ldots & O \\ O & L_2 & \ldots & O \\ \cdot & \cdot & \cdot & \cdot \\ O & O & \ldots & L_t \end{pmatrix}. \tag{46}$$

The matrix L_1 corresponds to the operator $\boldsymbol{A}$ in $\boldsymbol{I}_1$ with respect to the basis $\boldsymbol{e}_1 = \boldsymbol{e}, \boldsymbol{e}_2 = A\boldsymbol{e}, \ldots, \boldsymbol{e}_m = A^{m-1}\boldsymbol{e}$. By applying the rule for the formation of the matrix for a given operator in a given basis (Chapter III, p. 67), we find

$$L_1 = \left\| \begin{array}{cccccc} 0 & 0 & \ldots & 0 & -\alpha_m \\ 1 & 0 & \ldots & 0 & -\alpha_{m-1} \\ 0 & 1 & \ddots & & \vdots \\ \vdots & & \ddots & \ddots & \vdots \\ \vdots & & & 0 & -\alpha_2 \\ 0 & 0 & \ldots & 1 & -\alpha_1 \end{array} \right\| . \tag{47}$$

Similarly

$$L_2 = \left\| \begin{array}{cccccc} 0 & 0 & \ldots & 0 & -\beta_p \\ 1 & 0 & \ldots & 0 & -\beta_{p-1} \\ 0 & 1 & \ddots & & \vdots \\ \vdots & & \ddots & \ddots & \vdots \\ \vdots & & & 0 & -\beta_2 \\ 0 & 0 & \ldots & 1 & -\beta_1 \end{array} \right\| . \tag{48}$$

Computing the characteristic polynomials of the matrices $L_1, L_2, \ldots, L_t$, we find:

$$|\lambda E - L_1| = \psi_1(\lambda), \; |\lambda E - L_2| = \psi_2(\lambda), \; \ldots, \; |\lambda E - L_t| = \psi_t(\lambda)$$

(for cyclic subspaces the characteristic polynomial of an operator $\boldsymbol{A}$ coincides with the minimal polynomial of the subspace relative to this operator).

The matrix L_{I} corresponds to the operator $\boldsymbol{A}$ in the 'canonical' basis (45). If A is the matrix corresponding to $\boldsymbol{A}$ in an arbitrary basis, then A is similar to L_{I}, i.e., there exists a non-singular matrix T such that

$$A = T L_{\text{I}} T^{-1}. \tag{49}$$

Of the matrix L_{I} we shall say that it has the *first natural normal form.* This form is characterized by

1) The quasi-diagonal form;
2) The special structure of the diagonal blocks (47), (48), etc.
3) The additional condition: the characteristic polynomial of each diagonal block is divisible by the characteristic polynomial of the following block.

If we start not from the second, but from the third decomposition theorem, then in exactly the same way we would obtain a matrix L_{II} corresponding to the operator $\boldsymbol{A}$ in the appropriate basis—a matrix having the *second natural normal form*, which is characterized by

1) The quasi-diagonal form

$$L_{\text{II}} = \{ L^{(1)}, L^{(2)}, \ldots, L^{(u)} \};$$

2) The special structure of the diagonal blocks (47), (48), etc.;
3) The additional condition: the characteristic polynomial of each block is a power of an irreducible polynomial over F.

3. In the following section we shall show that in the class of similar matrices corresponding to one and the same operator there is one and only one matrix having the first normal form,[9] and one and only one [10] having the second normal form. Moreover, we shall give an algorithm for the computation of the polynomials $\psi_1(\lambda), \psi_2(\lambda), \dots, \psi_t(\lambda)$ from the elements of the matrix A. Knowledge of these polynomials enables us to write out all the elements of the matrices L_{I} and L_{II} similar to A and having the first and second natural normal forms, respectively.

§ 6. Invariant Polynomials. Elementary Divisors

1. We[11] denote by $D_p(\lambda)$ the greatest common divisor of all the minors of order p of the characteristic matrix $A_\lambda = \lambda E - A$ $(p = 1, 2, \dots, n)$.[12] Since in the sequence

$$D_n(\lambda),\ D_{n-1}(\lambda),\ \dots,\ D_1(\lambda),$$

each polynomial is divisible by the following, the formulas

$$i_1(\lambda) = \frac{D_n(\lambda)}{D_{n-1}(\lambda)},\ i_2(\lambda) = \frac{D_{n-1}(\lambda)}{D_{n-2}(\lambda)},\ \dots,\ i_n(\lambda) = \frac{D_1(\lambda)}{D_0(\lambda)} \quad (D_0(\lambda) \equiv 1) \tag{50}$$

define n polynomials whose product is equal to the characteristic polynomial

$$\Delta(\lambda) = |\lambda E - A| = D_n(\lambda) = i_1(\lambda)\, i_2(\lambda) \cdots i_n(\lambda). \tag{51}$$

We split the polynomials $i_p(\lambda)$ $(p = 1, 2, \dots, n)$ into irreducible factors over F:

$$i_p(\lambda) = [\varphi_1(\lambda)]^{\gamma_p} [\varphi_2(\lambda)]^{\delta_p} \cdots \quad (p = 1, 2, \dots, n); \tag{52}$$

where $\varphi_1(\lambda),\ \varphi_2(\lambda), \dots$ are distinct irreducible polynomials over F.

---

[9] This does not mean that there exists only one canonical basis of the form (45). There may be many canonical bases, but to all of them there corresponds one and the same matrix L_{I}.

[10] To within the order of the diagonal blocks.

[11] In subsection 1. of the present section we repeat the basic concepts of Chapter VI, § 3 for the characteristic matrix that were there established for an arbitrary polynomial matrix.

[12] We always take the highest coefficient of the greatest common divisor as 1.

The polynomials $i_1(\lambda), i_2(\lambda), \ldots, i_n(\lambda)$ are called the *invariant polynomials*, and all the non-constant powers among $[\varphi_1(\lambda)]^{\gamma_p}, [\varphi_2(\lambda)]^{\delta_p}, \ldots$ are called the *elementary divisors*, of the characteristic matrix $A_\lambda = \lambda E - A$ or, simply, of A.

The product of all the elementary divisors, like the product of all the invariant polynomials, is equal to the characteristic polynomial $\Delta(\lambda) = |\lambda E - A|$.

The name 'invariant polynomial' is justified by the fact that two similar matrices A and $\tilde{A}$,

$$\tilde{A} = T^{-1} AT, \tag{53}$$

always have identical invariant polynomials

$$i_p(\lambda) = \tilde{i}_p(\lambda) \quad (p = 1, 2, \ldots, n). \tag{54}$$

For it follows from (53) that

$$\tilde{A}_\lambda = \lambda E - \tilde{A} = T^{-1}(\lambda E - A)\, T = T^{-1} A_\lambda T. \tag{55}$$

Hence (see Chapter I, § 2) we obtain a relation between the minors of the similar matrices A_λ and $\tilde{A}_\lambda$:

$$\begin{aligned} &\tilde{A}_\lambda \begin{pmatrix} i_1 & i_2 & \ldots & i_p \\ k_1 & k_2 & \ldots & k_p \end{pmatrix} \\ &= \sum_{\substack{\alpha_1 < \alpha_2 < \cdots < \alpha_p \\ \beta_1 < \beta_2 < \cdots < \beta_p}} T^{-1} \begin{pmatrix} i_1 & i_2 & \ldots & i_p \\ \alpha_1 & \alpha_2 & \ldots & \alpha_p \end{pmatrix} A_\lambda \begin{pmatrix} \alpha_1 & \alpha_2 & \ldots & \alpha_p \\ \beta_1 & \beta_2 & \ldots & \beta_p \end{pmatrix} T \begin{pmatrix} \beta_1 & \beta_2 & \ldots & \beta_p \\ k_1 & k_2 & \ldots & k_p \end{pmatrix} \\ &\qquad (p = 1, 2, \ldots, n). \end{aligned} \tag{56}$$

This equation shows that every common divisor of all the minors of order p of A_λ is a common divisor of all the minors of order p of $\tilde{A}_\lambda$, and vice versa (since A and $\tilde{A}$ can interchange places). Hence it follows that $D_p(\lambda) = \tilde{D}_p(\lambda)$ $(p = 1, 2, \ldots, n)$ and that (54) holds.

Since all the matrices representing a given operator $\boldsymbol{A}$ in various bases are similar and therefore have the same invariant polynomials and the same elementary divisors, we can speak of the invariant polynomials and the elementary divisors of an operator $\boldsymbol{A}$.

2. We choose now for $\tilde{A}$ the matrix $L_{\rm I}$ having the first natural normal form and we compute the invariant polynomials of A starting from the form of the matrix $\tilde{A}_\lambda = \lambda E - \tilde{A}$ (in (57) this matrix is written out for the case $m = 5$, $p = 4$, $q = 4$, $r = 3$):

$$\left\| \begin{array}{ccccc|cccc|cccc|ccc}
\lambda & 0 & 0 & 0 & \alpha_5 & 0 & 0 & 0 & 0 & 0 & 0 & 0 & 0 & 0 & 0 & 0 \\
-1 & \lambda & 0 & 0 & \alpha_4 & 0 & 0 & 0 & 0 & 0 & 0 & 0 & 0 & 0 & 0 & 0 \\
0 & -1 & \lambda & 0 & \alpha_3 & 0 & 0 & 0 & 0 & 0 & 0 & 0 & 0 & 0 & 0 & 0 \\
0 & 0 & -1 & \lambda & \alpha_2 & 0 & 0 & 0 & 0 & 0 & 0 & 0 & 0 & 0 & 0 & 0 \\
0 & 0 & 0 & -1 & \alpha_1+\lambda & 0 & 0 & 0 & 0 & 0 & 0 & 0 & 0 & 0 & 0 & 0 \\
\hline
0 & 0 & 0 & 0 & 0 & \lambda & 0 & 0 & \beta_4 & 0 & 0 & 0 & 0 & 0 & 0 & 0 \\
0 & 0 & 0 & 0 & 0 & -1 & \lambda & 0 & \beta_3 & 0 & 0 & 0 & 0 & 0 & 0 & 0 \\
0 & 0 & 0 & 0 & 0 & 0 & -1 & \lambda & \beta_2 & 0 & 0 & 0 & 0 & 0 & 0 & 0 \\
0 & 0 & 0 & 0 & 0 & 0 & 0 & -1 & \beta_1+\lambda & 0 & 0 & 0 & 0 & 0 & 0 & 0 \\
\hline
0 & 0 & 0 & 0 & 0 & 0 & 0 & 0 & 0 & \lambda & 0 & 0 & \gamma_4 & 0 & 0 & 0 \\
0 & 0 & 0 & 0 & 0 & 0 & 0 & 0 & 0 & -1 & \lambda & 0 & \gamma_3 & 0 & 0 & 0 \\
0 & 0 & 0 & 0 & 0 & 0 & 0 & 0 & 0 & 0 & -1 & \lambda & \gamma_2 & 0 & 0 & 0 \\
0 & 0 & 0 & 0 & 0 & 0 & 0 & 0 & 0 & 0 & 0 & -1 & \gamma_1+\lambda & 0 & 0 & 0 \\
\hline
0 & 0 & 0 & 0 & 0 & 0 & 0 & 0 & 0 & 0 & 0 & 0 & 0 & \lambda & 0 & \varepsilon_3 \\
0 & 0 & 0 & 0 & 0 & 0 & 0 & 0 & 0 & 0 & 0 & 0 & 0 & -1 & \lambda & \varepsilon_2 \\
0 & 0 & 0 & 0 & 0 & 0 & 0 & 0 & 0 & 0 & 0 & 0 & 0 & 0 & -1 & \varepsilon_1+\lambda
\end{array} \right\| . \quad (57)$$

Using Laplace's Theorem, we find

$$D_n(\lambda) = |\lambda E - \widetilde{A}| = |\lambda E - L_1|\,|\lambda E - L_2| \cdots |\lambda E - L_t| = \psi_1(\lambda)\psi_2(\lambda)\cdots\psi_t(\lambda). \quad (58)$$

Now let us find $D_{n-1}(\lambda)$. We consider the minor of the element a_m. This minor, apart from a factor ± 1, is equal to

$$|\lambda E - L_2| \cdots |\lambda E - L_t| = \psi_2(\lambda) \cdots \psi_t(\lambda). \quad (59)$$

We shall show that this minor of order $n-1$ is a divisor of all the other minors of order $n-1$, so that

$$D_{n-1}(\lambda) = \psi_2(\lambda) \cdots \psi_t(\lambda). \quad (60)$$

For this purpose we first take the minor of an element outside the diagonal blocks and show that it vanishes. To obtain this minor we have to suppress one row and one column in the matrix (57). The lines crossed out in this case intersect two distinct diagonal blocks, so that in each of these blocks only one line is crossed out. Suppose, for example, that in the j-th diagonal block one of the rows is crossed out. In the minor we take that vertical strip which contains this diagonal block. In this strip, which has s columns, all the rows except $s-1$ rows consist entirely of zeros (we have denoted the order of A_j by s). Expanding the determinant of order $n-1$ by Laplace's Theorem with respect to the minors of order s in this strip, we see that it is equal to zero.

Now we take the minor of an element inside one of the diagonal blocks. In this case the lines crossed out 'mutilate' only one of the diagonal blocks, say the j-th, and the matrix of the minor is again quasi-diagonal. Therefore the minor is equal to

$$\psi_1(\lambda) \cdots \psi_{j-1}(\lambda)\, \psi_{j+1}(\lambda) \cdots \psi_t(\lambda)\, \chi(\lambda), \tag{61}$$

where $\chi(\lambda)$ is the determinant of the 'mutilated' j-th diagonal block. Since $\psi_i(\lambda)$ is divisible by $\psi_{i+1}(\lambda)$ $(i=1, 2, \ldots, t-1)$, the product (61) is divisible by (59). Thus, equation (60) can be regarded as proved. By similar arguments we obtain:

$$\left.\begin{aligned} D_{n-2}(\lambda) &= \psi_3(\lambda) \cdots \psi_t(\lambda), \\ &\cdots\cdots\cdots\cdots \\ D_{n-t+1}(\lambda) &= \psi_t(\lambda), \\ D_{n-t}(\lambda) &= \cdots = D_1(\lambda) = 1. \end{aligned}\right\} \tag{62}$$

From (58), (60), and (62) we find:

$$\begin{aligned} &\psi_1(\lambda) = \frac{D_n(\lambda)}{D_{n-1}(\lambda)} = i_1(\lambda), \quad \psi_2(\lambda) = \frac{D_{n-1}(\lambda)}{D_{n-2}(\lambda)} = i_2(\lambda), \quad \ldots, \\ &\psi_t(\lambda) = \frac{D_{n-t+1}(\lambda)}{D_{n-t}(\lambda)} = i_t(\lambda), \\ &\qquad i_{t+1}(\lambda) = \cdots = i_n(\lambda) = 1. \end{aligned} \tag{63}$$

The formulas (63) show that the polynomials $\psi_1(\lambda)$, $\psi_2(\lambda), \ldots, \psi_t(\lambda)$ coincide with the invariant polynomials, other than 1, of the operator $\boldsymbol{A}$ (or the corresponding matrix A).

Let us give three equivalent formulations of the results obtained:

THEOREM 9 (More precise form of the Second Decomposition Theorem): *If $\boldsymbol{A}$ is a linear operator in $\boldsymbol{R}$, then the space $\boldsymbol{R}$ can be decomposed into cyclic subspaces*

$$\boldsymbol{R} = \boldsymbol{I}_1 + \boldsymbol{I}_2 + \ldots + \boldsymbol{I}_t$$

such that in the sequence of minimal polynomials $\psi_1(\lambda)$, $\psi_2(\lambda), \ldots, \psi_t(\lambda)$ of the subspaces $\boldsymbol{I}_1, \boldsymbol{I}_2, \ldots, \boldsymbol{I}_t$, each is divisible by the following. The polynomials $\psi_1(\lambda)$, $\psi_2(\lambda), \ldots, \psi_t(\lambda)$ are uniquely determined: they coincide with the invariant polynomials, other than 1, of the operator $\boldsymbol{A}$.

THEOREM 9′: *For every linear operator $\boldsymbol{A}$ in $\boldsymbol{R}$ there exists a basis in which the matrix L_{I} that gives the operator is of the first natural normal form. This matrix is uniquely determined when the operator $\boldsymbol{A}$ is given: the characteristic polynomials of the diagonal blocks of L_{I} are the invariant polynomials of $\boldsymbol{A}$.*

THEOREM 9″: *In every class of similar matrices (with elements in* F*) there exists one and only one matrix L_{I} having the first natural normal form. The characteristic polynomials of the diagonal blocks of L_{I} coincide with the invariant polynomials (other than 1) of every matrix of that class.*

On p. 194 we established that two similar matrices have the same invariant polynomials. Now suppose, conversely, that two matrices A and B with elements in F are known to have the same invariant polynomials. Since the matrix L_{I} is uniquely determined when these polynomials are given, the two matrices A and B are similar to one and the same matrix L_{I} and, therefore, to each other. We thus arrive at the following proposition:

THEOREM 10: *Two matrices with elements in* F *are similar if and only if they have the same invariant polynomials.*[13]

3. The characteristic polynomial $\Delta(\lambda)$ of the operator $\boldsymbol{A}$ coincides with $D_n(\lambda)$, and hence with the product of all invariant polynomials:

$$\Delta(\lambda) = \psi_1(\lambda)\,\psi_2(\lambda)\cdots\psi_t(\lambda). \tag{64}$$

But $\psi_1(\lambda)$ is the minimal polynomial of the whole space with respect to $\boldsymbol{A}$; hence $\psi_1(\boldsymbol{A}) = \boldsymbol{O}$ and by (64)

$$\Delta(\boldsymbol{A}) = \boldsymbol{O}. \tag{65}$$

Thus we have incidentally obtained the Hamilton-Cayley Theorem (see Chapter IV, § 4):

Every linear operator (every square matrix) satisfies its characteristic equation.

In § 4 by splitting the polynomials $\psi_1(\lambda), \psi_2(\lambda), \ldots, \psi_t(\lambda)$ into irreducible factors over F:

$$\begin{aligned} \psi_1(\lambda) &= [\varphi_1(\lambda)]^{c_1}\,[\varphi_2(\lambda)]^{c_2}\cdots[\varphi_s(\lambda)]^{c_s}, \\ \psi_2(\lambda) &= [\varphi_1(\lambda)]^{d_1}\,[\varphi_2(\lambda)]^{d_2}\cdots[\varphi_s(\lambda)]^{d_s}, \quad \begin{pmatrix} c_k \geqq d_k \geqq \cdots \geqq l_k, \\ k = 1, 2, \ldots, s \end{pmatrix}, \\ \cdots&\cdots\cdots\cdots\cdots\cdots\cdots \\ \psi_t(\lambda) &= [\varphi_1(\lambda)]^{l_1}\,[\varphi_2(\lambda)]^{l_2}\cdots[\varphi_s(\lambda)]^{l_s} \end{aligned} \tag{66}$$

we were led to the third decomposition theorem. To each power with non-zero exponent on the right-hand sides of (66) there corresponds an invariant subspace in this decomposition.

By (63) all the powers, other than 1, among $[\varphi_k(\lambda)]^{c_k}, \ldots, [\varphi_k(\lambda)]^{l_k}$ ($k = 1, 2, \ldots, s$) are the elementary divisors of $\boldsymbol{A}$ (or A) in the field F (see p. 194).

Thus we arrive at the following more precise statement of the third decomposition theorem:

---

[13] Or (what is the same) the same elementary divisors in the field F.

THEOREM 11: *If $\boldsymbol{A}$ is a linear operator in a vector space $\boldsymbol{R}$ over a field* F, *then $\boldsymbol{R}$ can be split into cyclic subspaces whose minimal polynomials are the elementary divisors of $\boldsymbol{A}$ in* F.

Let

$$\boldsymbol{R} = \boldsymbol{I}' + \boldsymbol{I}'' + \dots + \boldsymbol{I}^{(u)} \tag{67}$$

be such a decomposition. We denote by $\boldsymbol{e}', \boldsymbol{e}'', \dots, \boldsymbol{e}^{(u)}$ generating vectors of the subspaces $\boldsymbol{I}', \boldsymbol{I}'', \dots, \boldsymbol{I}^{(u)}$ and from the 'cyclic' bases of these subspaces we form a basis of the whole space

$$\boldsymbol{e}', \boldsymbol{Ae}', \dots; \ \boldsymbol{e}'', \boldsymbol{Ae}'', \dots; \ \boldsymbol{e}^{(u)}, \boldsymbol{Ae}^{(u)}, \dots . \tag{68}$$

It is easy to see that the matrix L_{II} corresponding to the operator $\boldsymbol{A}$ in the basis (68) has quasi-diagonal form, like L_{I}:

$$L_{\mathrm{II}} = \{L_1, L_2, \dots, L_u\}. \tag{69}$$

The diagonal blocks $L_1, L_2, \dots, L_u$ are of the same structure as the blocks (47) and (48) of L_{I}. However, the characteristic polynomials of these diagonal blocks are not the invariant polynomials, but the elementary divisors of $\boldsymbol{A}$. The matrix L_{II} has the second natural normal form (see § 5).

We have arrived at another formulation of Theorem 11:

THEOREM 11′: *For every linear operator $\boldsymbol{A}$ in $\boldsymbol{R}$ (over the field* F*) there exists a basis in which the matrix L_{II} giving the operator is of the second natural normal form; the characteristic polynomials of the diagonal blocks are the elementary divisors of $\boldsymbol{A}$ in* F.

This theorem also admits a formulation in terms of matrices:

THEOREM 11″: *A matrix A with elements in the field* F *is always similar to a matrix L_{II} having the second natural normal form in which the characteristic polynomials of the diagonal blocks are the elementary divisors of A.*

Theorem 11 and the associated Theorems 11′ and 11″ have, in a certain sense, a converse.

Let

$$\boldsymbol{R} = \boldsymbol{I}' + \boldsymbol{I}'' + \dots + \boldsymbol{I}^{(u)}$$

be an arbitrary decomposition of a space $\boldsymbol{R}$ into indecomposable invariant subspaces. Then by Theorem 7 the subspaces $\boldsymbol{I}', \boldsymbol{I}'', \dots, \boldsymbol{I}^{(u)}$ are cyclic and their minimal polynomials are powers of irreducible polynomials over F. We may write these powers, after adding powers with zero exponent if necessary, in the form[14]

[14] At least one of the numbers $l_1, l_2, \dots, l_t$ is positive.

$$\begin{array}{l} [\varphi_1(\lambda)]^{c_1}, [\varphi_2(\lambda)]^{c_2}, \ldots, [\varphi_s(\lambda)]^{c_s}, \\ [\varphi_1(\lambda)]^{d_1}, [\varphi_2(\lambda)]^{d_2}, \ldots, [\varphi_s(\lambda)]^{d_s}, \\ \cdots\cdots\cdots\cdots\cdots\cdots \\ [\varphi_1(\lambda)]^{l_1}, [\varphi_2(\lambda)]^{l_2}, \ldots, [\varphi_s(\lambda)]^{l_s}. \end{array} \quad \begin{pmatrix} c_k \geqq d_k \geqq \cdots \geqq l_k \geqq 0, \\ k=1, 2, \ldots, s \end{pmatrix}. \tag{70}$$

We denote the sum of the subspaces whose minimal polynomials are in the first row by $\boldsymbol{I}_1$. Similarly, we introduce $\boldsymbol{I}_2, \ldots, \boldsymbol{I}_t$ (t is the number of rows in (70)). By Theorem 6, the subspaces $\boldsymbol{I}_1, \boldsymbol{I}_2, \ldots, \boldsymbol{I}_t$ are cyclic and their minimal polynomials $\psi_1(\lambda), \psi_2(\lambda), \ldots, \psi_t(\lambda)$ are determined by the formulas (66). Here in the sequence $\psi_1(\lambda), \psi_2(\lambda), \ldots, \psi_t(\lambda)$ each polynomial is divisible by the following. But then Theorem 9 is immediately applicable to the decomposition

$$\boldsymbol{R} = \boldsymbol{I}_1 + \boldsymbol{I}_2 + \ldots + \boldsymbol{I}_t .$$

By this theorem

$$\psi_p(\lambda) = i_p(\lambda) \quad (p = 1, 2, \ldots, n),$$

and therefore, by (66), all the powers (70) with non-zero exponent are the elementary divisors of $\boldsymbol{A}$ in the field F. Thus we have the following theorem:

THEOREM 12: *If the vector space $\boldsymbol{R}$ (over the field F) is split in any way into decomposable invariant subspaces (with respect to an operator $\boldsymbol{A}$), then the minimal polynomials of these subspaces are all the elementary divisors of $\boldsymbol{A}$ in F.*

There is an equivalent formulation in terms of matrices:

THEOREM 12′: *In each class of similar matrices (with elements in F) there exists only one matrix (to within the order of the diagonal blocks) having the second normal form L_{II}; the characteristic polynomials of its diagonal blocks are the elementary divisors of every matrix of the given class.*

Suppose that the space $\boldsymbol{R}$ is split into two invariant subspaces (with respect to an operator $\boldsymbol{A}$)

$$\boldsymbol{R} = \boldsymbol{I}_1 + \boldsymbol{I}_2.$$

When we split $\boldsymbol{I}_1$ and $\boldsymbol{I}_2$ into indecomposable subspaces, we obtain at the same time a decomposition of the whole space $\boldsymbol{R}$ into indecomposable subspaces. Hence, bearing Theorem 12 in mind, we obtain:

THEOREM 13: *If the space $\boldsymbol{R}$ is split into invariant subspaces with respect to an operator $\boldsymbol{A}$, then the elementary divisors of $\boldsymbol{A}$ in each of these invariant subspaces, taken in their totality, form a complete system of elementary divisors of $\boldsymbol{A}$ in $\boldsymbol{R}$.*

This theorem has the following matrix form:

THEOREM 13′: *A complete system of elementary divisors in* F *of a quasi-diagonal matrix is obtained as the union of the elementary divisors of the diagonal blocks.*

Theorem 13′ is often used for the actual process of finding the elementary divisors of a matrix.

§ 7. The Jordan Normal Form of a Matrix

1. Suppose that all the roots of the characteristic polynomial $\Delta(\lambda)$ of an operator $\boldsymbol{A}$ belong to the field F. This will hold true, in particular, if F is the field of all complex numbers.

In this case, the decomposition of the invariant polynomials into elementary divisors in F will look as follows:

$$\begin{aligned} i_1(\lambda) &= (\lambda-\lambda_1)^{c_1}\,(\lambda-\lambda_2)^{c_2}\cdots(\lambda-\lambda_s)^{c_s},\\ i_2(\lambda) &= (\lambda-\lambda_1)^{d_1}\,(\lambda-\lambda_2)^{d_2}\cdots(\lambda-\lambda_s)^{d_s},\\ &\cdots\cdots\cdots\cdots\cdots\\ i_t(\lambda) &= (\lambda-\lambda_1)^{l_1}\,(\lambda-\lambda_2)^{l_2}\cdots(\lambda-\lambda_s)^{l_s}. \end{aligned} \quad \begin{pmatrix} c_k \geqq d_k \geqq \cdots \geqq l_k \geqq 0, \\ c_k > 0;\ k=1,2,\ldots,s \end{pmatrix} \tag{71}$$

Since the product of all the invariant polynomials is equal to the characteristic polynomial $\Delta(\lambda)$, $\lambda_1, \lambda_2, \ldots, \lambda_s$ in (71) are all the distinct roots of $\Delta(\lambda)$.

We take an arbitrary elementary divisor

$$(\lambda-\lambda_0)^p; \tag{72}$$

here λ_0 is one of the numbers $\lambda_1, \lambda_2, \ldots, \lambda_s$ and p is one of the (non-zero) exponents $c_k, d_k, \ldots, l_k$ $(k=1, 2, \ldots, s)$.

To this elementary divisor there corresponds in (67) a definite cyclic subspace $\boldsymbol{I}$, generated by a vector which we denote by $\boldsymbol{e}$. For this vector $(\lambda-\lambda_0)^p$ is the minimal polynomial.

We consider the vectors

$$\boldsymbol{e}_1 = (\boldsymbol{A}-\lambda_0\boldsymbol{E})^{p-1}\boldsymbol{e},\ \ \boldsymbol{e}_2 = (\boldsymbol{A}-\lambda_0\boldsymbol{E})^{p-2}\boldsymbol{e},\ \ldots,\ \boldsymbol{e}_p = \boldsymbol{e}. \tag{73}$$

The vectors $\boldsymbol{e}_1, \boldsymbol{e}_2, \ldots, \boldsymbol{e}_p$ are linearly independent, since otherwise there would be an annihilating polynomial for $\boldsymbol{e}$ of degree less than p, which is impossible. Now we note that

$$(\boldsymbol{A}-\lambda_0\boldsymbol{E})\,\boldsymbol{e}_1 = \boldsymbol{o},\ (\boldsymbol{A}-\lambda_0\boldsymbol{E})\,\boldsymbol{e}_2 = \boldsymbol{e}_1,\ \ldots,\ (\boldsymbol{A}-\lambda_0\boldsymbol{E})\,\boldsymbol{e}_p = \boldsymbol{e}_{p-1} \tag{74}$$

or

$$\boldsymbol{A}\boldsymbol{e}_1 = \lambda_0\boldsymbol{e}_1,\ \boldsymbol{A}\boldsymbol{e}_2 = \lambda_0\boldsymbol{e}_2 + \boldsymbol{e}_1,\ \ldots,\ \boldsymbol{A}\boldsymbol{e}_p = \lambda_0\boldsymbol{e}_p + \boldsymbol{e}_{p-1}. \tag{75}$$

With the help of (75) we can easily write down the matrix corresponding to $\boldsymbol{A}$ in $\boldsymbol{I}$ for the basis (73). This matrix looks as follows:

$$\begin{Vmatrix} \lambda_0 & 1 & 0 & \dots & 0 \\ 0 & \lambda_0 & 1 & \dots & 0 \\ \cdot & & \cdot & \cdot & \cdot \\ \cdot & & & \cdot & \cdot \\ \cdot & & & \cdot & 1 \\ 0 & 0 & 0 & \dots & \lambda_0 \end{Vmatrix} = \lambda_0 E^{(p)} + H^{(p)}, \tag{76}$$

where $E^{(p)}$ is the unit matrix of order p and $H^{(p)}$ the matrix of order p which has 1's along the first superdiagonal and 0's everywhere else.

Linearly independent vectors $\boldsymbol{e}_1, \boldsymbol{e}_2, \dots, \boldsymbol{e}_p$ for which (75) holds form a so-called *Jordan chain of vectors* in $\boldsymbol{I}$. From Jordan chains connected with each subspace $\boldsymbol{I}', \boldsymbol{I}'', \dots, \boldsymbol{I}^{(u)}$ we form a *Jordan basis* of $\boldsymbol{R}$. If we now denote the minimal polynomials of these subspaces, i.e., the elementary divisors of $\boldsymbol{A}$, by

$$(\lambda - \lambda_1)^{p_1}, (\lambda - \lambda_2)^{p_2}, \dots, (\lambda - \lambda_u)^{p_u} \tag{77}$$

(the numbers $\lambda_1, \lambda_2, \dots, \lambda_u$ need not all be distinct), then the matrix J corresponding to $\boldsymbol{A}$ in a Jordan basis has the following quasi-diagonal form:

$$J = \{ \lambda_1 E^{(p_1)} + H^{(p_1)}, \ \lambda_2 E^{(p_2)} + H^{(p_2)}, \dots, \lambda_u E^{(p_u)} + H^{(p_u)} \}. \tag{78}$$

We shall say of the matrix J that it is of *Jordan normal form* or simply *Jordan form*. The matrix J can be written down at once when the elementary divisors of $\boldsymbol{A}$ in the field F containing all the characteristic roots of the equation $\Delta(\lambda) = 0$ are known.

Every matrix A is similar to a matrix J of Jordan normal form, i.e., for an arbitrary matrix A there always exists a non-singular matrix T $(|T| \neq 0)$ such that

$$A = TJT^{-1}.$$

If all the elementary divisors of $\boldsymbol{A}$ are of the first degree (and in that case only), the Jordan form is a diagonal matrix and we have:

$$A = T \{ \lambda_1, \lambda_2, \dots, \lambda_n \} T^{-1}.$$

Thus: *A linear operator $\boldsymbol{A}$ has simple structure* (see Chapter III, § 8) *if and only if all the elementary divisors of $\boldsymbol{A}$ are linear.*

Let us number the vectors $\boldsymbol{e}_1, \boldsymbol{e}_2, \dots, \boldsymbol{e}_p$ defined by (70) in the reverse order:

$$\boldsymbol{g}_1 = \boldsymbol{e}_p = \boldsymbol{e}, \ \boldsymbol{g}_2 = \boldsymbol{e}_{p-1} = (\boldsymbol{A} - \lambda_0 \boldsymbol{E})\boldsymbol{e}, \dots, \boldsymbol{g}_p = \boldsymbol{e}_1 = (\boldsymbol{A} - \lambda_0 \boldsymbol{E})^{p-1}\boldsymbol{e}. \tag{79}$$

Then

$$(\boldsymbol{A}-\lambda_0\boldsymbol{E})\,\boldsymbol{g}_1=\boldsymbol{g}_2,\quad (\boldsymbol{A}-\lambda_0\boldsymbol{E})\,\boldsymbol{g}_2=\boldsymbol{g}_3,\ \ldots,\ (\boldsymbol{A}-\lambda_0\boldsymbol{E})\,\boldsymbol{g}_p=\boldsymbol{o};$$

hence

$$\boldsymbol{A}\boldsymbol{g}_1=\lambda_0\boldsymbol{g}_1+\boldsymbol{g}_2,\quad \boldsymbol{A}\boldsymbol{g}_2=\lambda_0\boldsymbol{g}_2+\boldsymbol{g}_3,\ \ldots,\ \boldsymbol{A}\boldsymbol{g}_p=\lambda_0\boldsymbol{g}_p.$$

The vectors (79) form a basis in the cyclic invariant subspace $\boldsymbol{I}$ that corresponds in (67) to the elementary divisor $(\lambda-\lambda_0)^p$.

In this basis, as is easy to see, to the operator $\boldsymbol{A}$ there corresponds the matrix

$$\left\|\begin{matrix} \lambda_0 & 0 & 0 & \ldots & 0 \\ 1 & \lambda_0 & 0 & \ldots & 0 \\ 0 & 1 & \lambda_0 & & \cdot \\ \cdot & \cdot & \cdot & \cdot & \cdot \\ \cdot & & \cdot & \cdot & \cdot \\ 0 & 0 & \ldots & 1 & \lambda_0 \end{matrix}\right\| = \lambda_0 E^{(p)} + F^{(p)}.$$

We shall say of the vectors (79) that they form a *lower Jordan chain* of vectors. If we take a lower Jordan chain of vectors in each subspace $\boldsymbol{I}'$, $\boldsymbol{I}''$, $\ldots$, $\boldsymbol{I}^{(u)}$ of (67), we can form from these chains a *lower Jordan basis* in which to the operator $\boldsymbol{A}$ there corresponds the quasi-diagonal matrix

$$J_1=\{\lambda_1 E^{(p_1)}+F^{(p_1)},\ \lambda_2 E^{(p_2)}+F^{(p_2)},\ \ldots,\ \lambda_u E^{(p_u)}+F^{(p_u)}\}. \tag{80}$$

We shall say of the matrix J_1 that it is of *lower Jordan form.* In contrast to (80), we shall sometimes call (78) an *upper Jordan matrix.*

Thus: *Every matrix A is similar to an upper and to a lower Jordan matrix.*

§ 8. Krylov's Method of Transforming the Secular Equation

1. When a matrix $A=\|a_{ik}\|_1^n$ is given, then its characteristic (secular) equation can be written in the form

$$\Delta(\lambda)\equiv(-1)^n\begin{vmatrix} a_{11}-\lambda & a_{12} & \ldots & a_{1n} \\ a_{21} & a_{22}-\lambda & \ldots & a_{2n} \\ \cdot & \cdot & \cdot & \cdot \\ a_{n1} & a_{n2} & \ldots & a_{nn}-\lambda \end{vmatrix}=0. \tag{81}$$

On the left-hand side of this equation is the characteristic polynomial $\Delta(\lambda)$ of degree n. For the direct computation of the coefficients of this polynomial it is necessary to expand the characteristic determinant

$| A - \lambda E |$; and for large n this involves very cumbersome computational work, because λ occurs in the diagonal elements of the determinant.[15]

In 1937, A. N. Krylov [251] proposed a transformation of the characteristic determinant as a result of which λ occurs only in the elements of one column (or row).

Krylov's transformation simplifies the computation of the coefficients of the characteristic equation considerably.[16]

In this section we shall give an algebraic method of transforming the characteristic equation which differs somewhat from Krylov's own method.[17]

We consider an n-dimensional vector space $\boldsymbol{R}$ with basis $\boldsymbol{e}_1, \boldsymbol{e}_2, \ldots, \boldsymbol{e}_n$ and the linear operator $\boldsymbol{A}$ in $\boldsymbol{R}$ determined by a given matrix $A = \| a_{ik} \|_1^n$ in this basis. We take an arbitrary vector $\boldsymbol{x} \neq \boldsymbol{o}$ in $\boldsymbol{R}$ and form the sequence of vectors

$$\boldsymbol{x}, \ \boldsymbol{A}\boldsymbol{x}, \ \boldsymbol{A}^2\boldsymbol{x}, \ \ldots \ . \tag{82}$$

Suppose that the first p vectors $\boldsymbol{x}, \boldsymbol{A}\boldsymbol{x}, \ldots, \boldsymbol{A}^{p-1}\boldsymbol{x}$ of this sequence are linearly independent and that the $(p+1)$-st vector $\boldsymbol{A}^p\boldsymbol{x}$ is a linear combination of these p vectors:

$$\boldsymbol{A}^p\boldsymbol{x} = -\alpha_p\boldsymbol{x} - \alpha_{p-1}\boldsymbol{A}\boldsymbol{x} - \cdots - \alpha_1\boldsymbol{A}^{p-1}\boldsymbol{x} \tag{83}$$

or

$$\varphi(\boldsymbol{A})\boldsymbol{x} = \boldsymbol{o}, \tag{84}$$

where

$$\varphi(\lambda) = \lambda^p + \alpha_1\lambda^{p-1} + \cdots + \alpha_p. \tag{85}$$

All the further vectors in (82) can also be expressed linearly by the first p vectors of the sequence.[18] Thus, in (82) there are p linearly independent

---

[15] We recall that the coefficient of λ^k in $\Delta(\lambda)$ is equal (apart from the sign) to the sum of all the principal minors of order $n-k$ in A $(k = 1, 2, \ldots, n)$. Thus, even for $n = 6$, the direct determination of the coefficient of λ in $\Delta(\lambda)$ would require the computation of six determinants of order 5; that of λ^2 would require fifteen determinants of order 4; etc.

[16] The algebraic analysis of Krylov's method of transforming the secular equation is contained in a number of papers [268], [269], [211], [168], and [149].

[17] Krylov arrived at his method of transformation by starting from a system of n linear differential equations with constant coefficients. Krylov's approach in algebraic form can be found, for example, in [268] and [168] and in § 21 of the book [25].

[18] When we apply the operator $\boldsymbol{A}$ to both sides of (83) we express $\boldsymbol{A}^{p+1}\boldsymbol{x}$ linearly in terms of $\boldsymbol{A}\boldsymbol{x}, \ldots, \boldsymbol{A}^{p-1}\boldsymbol{x}, \boldsymbol{A}^p\boldsymbol{x}$. But $\boldsymbol{A}^p\boldsymbol{x}$, by (83), is expressed linearly in terms of $\boldsymbol{x}, \boldsymbol{A}\boldsymbol{x}, \ldots, \boldsymbol{A}^{p-1}\boldsymbol{x}$. Hence we obtain a similar expression for $\boldsymbol{A}^{p+1}\boldsymbol{x}$. By applying the operator $\boldsymbol{A}$ to the expression thus obtained for $\boldsymbol{A}^{p+1}\boldsymbol{x}$, we express $\boldsymbol{A}^{p+2}\boldsymbol{x}$ in terms of $\boldsymbol{x}, \boldsymbol{A}\boldsymbol{x}, \ldots, \boldsymbol{A}^{p+1}\boldsymbol{x}$, etc.

vectors and this maximal number of linearly independent vectors in (82) is always realized by the *first* p vectors.

The polynomial $\varphi(\lambda)$ is the minimal (annihilating) polynomial of the vector $\boldsymbol{x}$ with respect to the operator $\boldsymbol{A}$ (see § 1). *The method of Krylov consists in an effective determination of the minimal polynomial* $\varphi(\lambda)$ *of* $\boldsymbol{x}$.

We consider separately two cases: the *regular* case, where $p=n$; and the *singular* case, where $p<n$.

The polynomial $\varphi(\lambda)$ is a divisor of the minimal polynomial $\psi(\lambda)$ of the whole space $\boldsymbol{R}$,[19] and $\psi(\lambda)$ in turn is a divisor of the characteristic polynomial $\Delta(\lambda)$. Therefore $\varphi(\lambda)$ is always a divisor of $\Delta(\lambda)$.

In the regular case, $\varphi(\lambda)$ and $\Delta(\lambda)$ are of the same degree and, since their highest coefficients are equal, they coincide. Thus, *in the regular case*

$$\Delta(\lambda) \equiv \psi(\lambda) \equiv \varphi(\lambda),$$

and therefore *in the regular case Krylov's method is a method of computing the coefficients of the characteristic polynomial* $\Delta(\lambda)$.

In the singular case, as we shall see later, Krylov's method does not enable us to determine $\Delta(\lambda)$, and in this case it only determines the divisor $\varphi(\lambda)$ of $\Delta(\lambda)$.

In explaining Krylov's transformation, we shall denote the coordinates of $\boldsymbol{x}$ in the given basis $\boldsymbol{e}_1, \boldsymbol{e}_2, \ldots, \boldsymbol{e}_n$ by $a, b, \ldots, l$, and the coordinates of the vector $\boldsymbol{A}^k\boldsymbol{x}$ by $a_k, b_k, \ldots, l_k$ $(k=1, 2, \ldots, n)$.

2. *Regular case*: $p=n$. In this case, the vectors $\boldsymbol{x}, \boldsymbol{A}\boldsymbol{x}, \ldots, \boldsymbol{A}^{n-1}\boldsymbol{x}$ are linearly independent and the equations (83), (84), and (85) assume the form

$$\boldsymbol{A}^n\boldsymbol{x} = -\alpha_n\boldsymbol{x} - \alpha_{n-1}\boldsymbol{A}\boldsymbol{x} - \cdots - \alpha_1\boldsymbol{A}^{n-1}\boldsymbol{x} \tag{86}$$

or

$$\Delta(\boldsymbol{A})\boldsymbol{x} = \boldsymbol{o}, \tag{87}$$

where

$$\Delta(\lambda) = \lambda^n + \alpha_1\lambda^{n-1} + \cdots + \alpha_{n-1}\lambda + \alpha_n. \tag{88}$$

The condition of linear independence of the vectors $\boldsymbol{x}, \boldsymbol{A}\boldsymbol{x}, \ldots, \boldsymbol{A}^{n-1}\boldsymbol{x}$ may be written analytically as follows (see Chapter III, § 1):

$$M = \begin{vmatrix} a & b & \ldots & l \\ a_1 & b_1 & \ldots & l_1 \\ \cdot & \cdot & \cdots & \cdot \\ a_{n-1} & b_{n-1} & \ldots & l_{n-1} \end{vmatrix} \neq 0. \tag{89}$$

We consider the matrix formed from the coordinate vectors $\boldsymbol{x}, \boldsymbol{A}\boldsymbol{x}, \ldots, \boldsymbol{A}^n\boldsymbol{x}$:

[19] $\psi(\lambda)$ is the minimal polynomial of A.

$$\left\| \begin{matrix} a & b & \dots & l \\ a_1 & b_1 & \dots & l_1 \\ \dots & \dots & \dots & \dots \\ a_{n-1} & b_{n-1} & \dots & l_{n-1} \\ a_n & b_n & & l_n \end{matrix} \right\| . \tag{90}$$

In the regular case the rank of this matrix is n. The first n rows of the matrix are linearly independent, and the last, $(n+1)$-st, row is a linear combination of the preceding n.

We obtain the dependence between the rows of (90) when we replace the vector equation (86) by the equivalent system of n scalar equations

$$\left. \begin{aligned} -\alpha_n a - \alpha_{n-1}a_1 - \cdots - \alpha_1 a_{n-1} &= a_n \\ -\alpha_n b - \alpha_{n-1}b_1 - \cdots - \alpha_1 b_{n-1} &= b_n \\ \dots\dots\dots\dots\dots\dots\dots\dots \\ -\alpha_n l - \alpha_{n-1}l_1 - \cdots - \alpha_1 l_{n-1} &= l_n . \end{aligned} \right\} . \tag{91}$$

From this system of n linear equations we may determine the unknown coefficients $\alpha_1, \alpha_2, \dots, \alpha_n$ uniquely,[20] and substitute their values in (88). This elimination of $\alpha_1, \alpha_2, \dots, \alpha_n$ from (88) and (91) can be performed symmetrically. For this purpose we rewrite (88) and (91) as follows:

$$\left. \begin{aligned} a\alpha_n + a_1\alpha_{n-1} + \cdots + a_{n-1}\alpha_1 + a_n\alpha_0 &= 0 \\ b\alpha_n + b_1\alpha_{n-1} + \cdots + b_{n-1}\alpha_1 + b_n\alpha_0 &= 0 \\ \dots\dots\dots\dots\dots\dots\dots\dots \\ l\alpha_n + l_1\alpha_{n-1} + \cdots + l_{n-1}\alpha_1 + l_n\alpha_0 &= 0 \\ 1\alpha_n + \lambda\alpha_{n-1} + \cdots + \lambda^{n-1}\alpha_1 + [\lambda^n - \Delta(\lambda)]\alpha_0 &= 0 \end{aligned} \right\} \quad (\alpha_0 = 1).$$

Since this system of $n+1$ equations in the $n+1$ unknown $\alpha_0, \alpha_2, \dots, \alpha_n$ has a non-zero solution $(a_0 = 1)$, its determinant must vanish:

$$\left| \begin{matrix} a & a_1 & \dots & a_{n-1} & a_n \\ b & b_1 & \dots & b_{n-1} & b_n \\ \dots & \dots & \dots & \dots & \dots \\ l & l_1 & \dots & l_{n-1} & l_n \\ 1 & \lambda & \dots & \lambda^{n-1} & \lambda^n - \Delta(\lambda) \end{matrix} \right| = 0 . \tag{92}$$

Hence we determine $\Delta(\lambda)$ after a preliminary transposition of the determinant (92) with respect to the main diagonal:

[20] By (89), the determinant of this system is different from zero .

$$M\Delta(\lambda) = \begin{vmatrix} a & b & \dots & l & 1 \\ a_1 & b_1 & \dots & l_1 & \lambda \\ \dots & \dots & \dots & \dots & \dots \\ a_{n-1} & b_{n-1} & \dots & l_{n-1} & \lambda^{n-1} \\ a_n & b_n & \dots & l_n & \lambda^n \end{vmatrix}, \tag{93}$$

where the constant factor M is determined by (89) and differs from zero.

The identity (93) represents Krylov's transformation. In Krylov's determinant on the right-hand side of the identity, λ occurs only in the elements of the last column; the remaining elements of the determinant do not depend on λ.

Note. In the regular case, the whole space $\boldsymbol{R}$ is cyclic (with respect to $\boldsymbol{A}$). If we choose the vectors $\boldsymbol{x}, \boldsymbol{Ax}, \dots, \boldsymbol{A}^{n-1}\boldsymbol{x}$ as a basis, then in this basis the operator $\boldsymbol{A}$ corresponds to a matrix $\tilde{A}$ having the natural normal form

$$\tilde{A} = \left\| \begin{matrix} 0 & 0 & \dots & 0 & -\alpha_n \\ 1 & 0 & \dots & 0 & -\alpha_{n-1} \\ \cdot & \ddots & & & \vdots \\ \cdot & & \ddots & & \cdot \\ \cdot & & & 0 & -\alpha_2 \\ 0 & & \dots & 1 & -\alpha_1 \end{matrix} \right\|. \tag{94}$$

The transition from the original basis $\boldsymbol{e}_1, \boldsymbol{e}_2, \dots, \boldsymbol{e}_n$ to the basis $\boldsymbol{x}, \boldsymbol{Ax}, \dots, \boldsymbol{A}^{n-1}\boldsymbol{x}$ is accomplished by means of the non-singular transforming matrix

$$T = \left\| \begin{matrix} a & a_1 & \dots & a_{n-1} \\ b & b_1 & \dots & b_{n-1} \\ \dots & \dots & \dots & \dots \\ l & l_1 & \dots & l_{n-1} \end{matrix} \right\|. \tag{95}$$

and then

$$A = T\tilde{A}T^{-1}. \tag{96}$$

3. *Singular case*: $p < n$. In this case, the vectors $\boldsymbol{x}, \boldsymbol{Ax}, \dots, \boldsymbol{A}^{n-1}\boldsymbol{x}$ are linearly dependent, so that

$$M = \begin{vmatrix} a & b & \dots & l \\ a_1 & b_1 & \dots & l_1 \\ \dots & \dots & \dots & \dots \\ a_{n-1} & b_{n-1} & \dots & l_{n-1} \end{vmatrix} = 0.$$

Now (93) had been deduced under the assumption $M \neq 0$. But both sides of this equation are rational integral functions of λ and of the parameters $a, b, \ldots, l$.[21] Therefore it follows by a 'continuity' argument that (93) also holds for $M = 0$. But then, when Krylov's determinant is expanded, all the coefficients turn out to be zero. Thus in the singular case $(p < n)$ the formula (93) goes over into the trivial identity $0 = 0$.

Let us consider the matrix formed from the coordinates of the vectors $\boldsymbol{x}, \boldsymbol{A}\boldsymbol{x}, \ldots, \boldsymbol{A}^p\boldsymbol{x}$

$$\left\| \begin{matrix} a & b & \ldots & l \\ a_1 & b_1 & \ldots & l_1 \\ \cdot & \cdot & \cdot & \cdot \\ a_{p-1} & b_{p-1} & \ldots & l_{p-1} \\ a_p & b_p & \ldots & l_p \end{matrix} \right\|. \tag{97}$$

This matrix is of rank p and the first p rows are linearly independent, but the last, $(p+1)$-st, row is a linear combination of the first p rows with the coefficients $-\alpha_p, -\alpha_{p-1}, \ldots, -\alpha_1$ (see (83)). From the n coordinates $a, b, \ldots, l$ we can choose p coordinates $c, f, \ldots, h$ such that the determinant formed from the coordinates of the vectors $\boldsymbol{x}, \boldsymbol{A}\boldsymbol{x}, \ldots, \boldsymbol{A}^{p-1}\boldsymbol{x}$ is different from zero:

$$M^* = \left| \begin{matrix} c & f & \ldots & h \\ c_1 & f_1 & \ldots & h_1 \\ \cdot & \cdot & \cdot & \cdot \\ c_{p-1} & f_{p-1} & \ldots & h_{p-1} \end{matrix} \right|. \tag{98}$$

Furthermore, it follows from (83) that:

$$\left. \begin{aligned} -\alpha_p c - \alpha_{p-1} c_1 - \cdots - \alpha_1 c_{p-1} &= c_p \\ -\alpha_p f - \alpha_{p-1} f_1 - \cdots - \alpha_1 f_{p-1} &= f_p \\ \cdots\cdots\cdots\cdots\cdots\cdots & \\ -\alpha_p h - \alpha_{p-1} h_1 - \cdots - \alpha_1 h_{p-1} &= h_p. \end{aligned} \right\} \tag{99}$$

From this system of equations the coefficients $\alpha_1, \alpha_2, \ldots, \alpha_p$ of the polynomial $\varphi(\lambda)$ (the minimal polynomial of $\boldsymbol{x}$) are uniquely determined. In exact analogy with the regular case (however, with the value n replaced by p and the letters $a, b, \ldots, l$ by $c, f, \ldots, h$), we may eliminate $\alpha_1, \alpha_2, \ldots, \alpha_p$ from (85) and (99) and obtain the following formula for $\varphi(\lambda)$:

---

[21] $a_i = a_{11}^{(i)} a + a_{12}^{(i)} b + \cdots + a_{1n}^{(i)} l, \quad b_i = a_{21}^{(i)} a + a_{22}^{(i)} b + \cdots + a_{2n}^{(i)} l$, etc. $(i = 1, 2, \ldots, n)$, where $a_{jk}^{(i)}$ $(j, k = 1, 2, \ldots, n)$ are the elements of A^i $(i = 1, 2, \ldots, n)$.

$$M^* \varphi(\lambda) = \begin{vmatrix} c & f & \dots & h & 1 \\ c_1 & f_1 & \dots & h_1 & \lambda \\ \cdot & \cdot & \cdot & \cdot & \cdot \\ c_{p-1} & f_{p-1} & \dots & h_{p-1} & \lambda^{p-1} \\ c_p & f_p & \dots & h_p & \lambda^p \end{vmatrix}. \tag{100}$$

4. Let us now clarify the problem: for what matrices $A = \| a_{ik} \|_1^n$ and for what choice of the original vector $\boldsymbol{x}$ or, what is the same, of the initial parameters $a, b, \dots, l$ the regular case holds.

We have seen that in the regular case

$$\Delta(\lambda) \equiv \psi(\lambda) \equiv \varphi(\lambda).$$

The fact that the characteristic polynomial $\Delta(\lambda)$ coincides with the minimal polynomial $\psi(\lambda)$ means that in the matrix $A = \| a_{ik} \|_1^n$ there are no two elementary divisors with one and the same characteristic value, i.e., all the elementary divisors are co-prime in pairs. In the case where A is a matrix of simple structure, this requirement is equivalent to the condition that the characteristic equation of A have no multiple roots.

The fact that the polynomials $\psi(\lambda)$ and $\varphi(\lambda)$ coincide means that for $\boldsymbol{x}$ we have chosen a vector that generates (by means of $\boldsymbol{A}$) the whole space $\boldsymbol{R}$. Such a vector always exists, by Theorem 2 of § 2.

But if the condition $\Delta(\lambda) \equiv \psi(\lambda)$ is not satisfied, then however we choose the vector $\boldsymbol{x} \neq \boldsymbol{o}$, we do not obtain $\Delta(\lambda)$, since the polynomial $\varphi(\lambda)$ obtained by Krylov's method is a divisor of $\psi(\lambda)$ which in this case does not coincide with $\Delta(\lambda)$ but is only a factor of it. By varying the vector $\boldsymbol{x}$ we may obtain for $\varphi(\lambda)$ every divisor of $\psi(\lambda)$.[22]

The results we have reached can be stated in the form of the following theorem:

THEOREM 14: *Krylov's transformation gives an expression for the characteristic polynomial $\Delta(\lambda)$ of the matrix $A = \| a_{ik} \|_1^n$ in the form of the determinant* (93) *if and only if two conditions are satisfied:*

1. *The elementary divisors of A are co-prime in pairs.*

2. *The initial parameters $a, b, \dots, l$ are the coordinates of a vector $\boldsymbol{x}$ that generates the whole n-dimensional space (by means of the operator $\boldsymbol{A}$ corresponding to the matrix A).*[23]

---

[22] See, for example, [168], p. 48.

[23] In analytical form, this condition means that the columns $\boldsymbol{x}, \boldsymbol{Ax}, \dots, \boldsymbol{A}^{n-1}\boldsymbol{x}$ are linearly independent, where $\boldsymbol{x} = (a, b, \dots, l)$.

In general, the Krylov transformation leads to some divisor $\varphi(\lambda)$ of the characteristic polynomial $\Delta(\lambda)$. This divisor $\varphi(\lambda)$ is the minimal polynomial of the vector $\boldsymbol{x}$ with the coordinates $a, b, \dots, l$ (where $a, b, \dots, l$ are the initial parameters in the Krylov transformation).

5. Let us show how to find the coordinates of a characteristic vector $\boldsymbol{y}$ for an arbitrary characteristic value λ_0 which is a root of the polynomial $\varphi(\lambda)$ obtained by Krylov's method.[24]

We shall seek a vector $\boldsymbol{y} \neq \boldsymbol{o}$ in the form

$$\boldsymbol{y} = \xi_1 \boldsymbol{x} + \xi_2 A\boldsymbol{x} + \cdots + \xi_p A^{p-1}\boldsymbol{x}. \tag{101}$$

Substituting this expression for $\boldsymbol{y}$ in the vector equation

$$A\boldsymbol{y} = \lambda_0 \boldsymbol{y}$$

and using (83), we obtain

$$\begin{aligned}\xi_1 A\boldsymbol{x} + \xi_2 A^2\boldsymbol{x} + \cdots + \xi_{p-1} A^{p-1}\boldsymbol{x} + \xi_p(-\alpha_p \boldsymbol{x} - \alpha_{p-1}A\boldsymbol{x} - \cdots - \alpha_1 A^{p-1}\boldsymbol{x})& \\ = \lambda_0(\xi_1\boldsymbol{x} + \xi_2 A\boldsymbol{x} + \cdots + \xi_p A^{p-1}\boldsymbol{x}).&\end{aligned} \tag{102}$$

Hence, among other things, it follows that $\xi_p \neq 0$, because the equation $\xi_p = 0$ would yield by (102) a linear dependence among the vectors $\boldsymbol{x}$, $A\boldsymbol{x}$, $\dots$, $A^{p-1}\boldsymbol{x}$. In what follows we set $\xi_p = 1$. Then we obtain from (102):

$$\left.\begin{aligned}\xi_p = 1,\ \xi_{p-1} = \lambda_0\xi_p + \alpha_1,\ \xi_{p-2} = \lambda_0\xi_{p-1} + \alpha_2, \dots, \xi_1 = \lambda_0\xi_2 + \alpha_{p-1},\\ 0 = \lambda_0\xi_1 + \alpha_p.\end{aligned}\right\} \tag{103}$$

The first of these equations determine for us in succession the values ξ_p, ξ_{p-1}, $\dots$, ξ_1 (the coordinates of $\boldsymbol{y}$ in the 'new' basis $\boldsymbol{x}$, $A\boldsymbol{x}$, $\dots$, $A^{p-1}\boldsymbol{x}$); the last equation is a consequence of the preceding ones and of the relation $\lambda_0^p + \alpha_1\lambda_0^{p-1} + \dots + \alpha_p = 0$.

The coordinates a', b', $\dots$, l' of the vector $\boldsymbol{y}$ in the original basis may be found from the following formulas, which follow from (101):

$$\left.\begin{aligned}a' &= \xi_1 a + \xi_2 a_1 + \cdots + \xi_p a_{p-1}\\ b' &= \xi_1 b + \xi_2 b_1 + \cdots + \xi_p b_{p-1}\\ &\dots\dots\dots\dots\dots\\ l' &= \xi_1 l + \xi_2 l_1 + \cdots + \xi_p l_{p-1}.\end{aligned}\right\} \tag{104}$$

Example 1.

We recommend to the reader the following scheme of computations.

[24] The following arguments hold both in the regular case $p = n$ and the singular case $p < n$.

Under the given matrix A we write the row of the coordinates of $\boldsymbol{x}$: $a, b, \dots, l$. These numbers are given arbitrarily (with only one condition: at least one is different from zero). Under the row $a, b, \dots, l$ we write the row $a_1, b_1, \dots, l_1$, i.e., the coordinates of the vector $\boldsymbol{Ax}$. The numbers $a_1, b_1, \dots, l_1$ are obtained by multiplying the row $a, b, \dots, l$ successively into the rows of the given matrix A. For example, $a_1 = a_{11}a + a_{12}b + \dots + {}_{1n}l$, $b_1 = a_{21}a + a_{22}b + \dots + a_{2n}l$, etc. Under the row $a_1, b_1, \dots, l_1$ we write the row $a_2, b_2, \dots, l_2$, etc. Each of the rows, beginning with the second, is determined by multiplying the preceding row successively into the rows of the given matrix.

Above the given matrix we write the sum row as a check.

	8	*3*	*−10*	*−3*		
$A=$	3	−1	−4	2		
	2	3	−2	−4		
	2	−1	−3	2		
	1	2	−1	−3		
$\boldsymbol{x} = \boldsymbol{e}_1 + \boldsymbol{e}_2$	1	1	0	0	−1	1
$\boldsymbol{Ax}$	2	5	1	3	−1	−1
$\boldsymbol{A}^2\boldsymbol{x}$	3	5	2	2	1	−1
$\boldsymbol{A}^3\boldsymbol{x}$	0	9	−1	5	1	1 .
$\boldsymbol{A}^4\boldsymbol{x}$	5	9	4	4		
$\boldsymbol{y}$	0	8	0	4		
	0	2	0	1		
$\boldsymbol{z}$	−4	0	−4	0		
	1	0	1	0		

The given case is regular, because

$$M = \begin{vmatrix} 1 & 1 & 0 & 0 \\ 2 & 5 & 1 & 3 \\ 3 & 5 & 2 & 2 \\ 0 & 9 & -1 & 5 \end{vmatrix} = -16 \neq 0.$$

Krylov's determinant has the form

$$-16\,\Delta(\lambda) = \begin{vmatrix} 1 & 1 & 0 & 0 & 1 \\ 2 & 5 & 1 & 3 & \lambda \\ 3 & 5 & 2 & 2 & \lambda^2 \\ 0 & 9 & -1 & 5 & \lambda^3 \\ 5 & 9 & 4 & 4 & \lambda^4 \end{vmatrix}.$$

Expanding this determinant and cancelling -16 we find:

$$\Delta(\lambda) = \lambda^4 - 2\lambda^2 + 1 = (\lambda - 1)^2 (\lambda + 1)^2.$$

We denote by

$$y = \xi_1 x + \xi_2 Ax + \xi_3 A^2 x + \xi_4 A^3 x$$

a characteristic vector of A corresponding to the characteristic value $\lambda_0 = 1$. We find the numbers $\xi_1, \xi_2, \xi_3, \xi_4$ by the formulas (103):

$$\xi_4 = 1, \ \xi_3 = 1 \cdot \lambda_0 + 0 = 1, \ \xi_2 = 1 \cdot \lambda_0 - 2 = -1, \ \xi_1 = -1 \cdot \lambda_0 + 0 = -1.$$

The control equation $-1 \cdot \lambda_0 + 1 = 0$ is, of course, satisfied.

We place the numbers $\xi_1, \xi_2, \xi_3, \xi_4$ in a vertical column parallel to the columns of x, Ax, A^2x, A^3x. Multiplying the column $\xi_1, \xi_2, \xi_3, \xi_4$ into the columns a_1, a_2, a_3, a_4, we obtain the first coordinate a' of the vector y in the original basis e_1, e_2, e_3, e_4; similarly we obtain b', c', d'. As coordinates of y we find (after cancelling by 4): 0, 2, 0, 1. Similarly, we determine the coordinates 1, 0, 1, 0 of a characteristic vector z for the characteristic value $\lambda_0 = -1$.

Furthermore, by (94) and (95),

$$A = T\tilde{A}T^{-1}$$

where

$$\tilde{A} = \left\| \begin{matrix} 0 & 0 & 0 & -1 \\ 1 & 0 & 0 & 0 \\ 0 & 1 & 0 & 2 \\ 0 & 0 & 1 & 0 \end{matrix} \right\|, \qquad T = \left\| \begin{matrix} 1 & 2 & 3 & 0 \\ 1 & 5 & 5 & 9 \\ 0 & 1 & 2 & -1 \\ 0 & 3 & 2 & 5 \end{matrix} \right\|.$$

Example 2. We consider the same matrix A, but as initial parameters we take the numbers $a = 1$, $b = 0$, $c = 0$, $d = 0$.

$$\begin{matrix} & \mathit{8} & \mathit{3} & \mathit{-10} & \mathit{-3} \\ A = & \left\| \begin{matrix} 3 & -1 & -4 & 2 \\ 2 & 3 & -2 & -4 \\ 2 & -1 & -3 & 2 \\ 1 & 2 & -1 & -3 \end{matrix} \right\| \end{matrix}$$

$x = e_1$	1	0	0	0
Ax	3	2	2	1
A^2x	1	4	0	2
A^3x	3	6	2	3

But in this case

$$M = \left| \begin{matrix} 1 & 0 & 0 & 0 \\ 3 & 2 & 2 & 1 \\ 1 & 4 & 0 & 2 \\ 3 & 6 & 2 & 3 \end{matrix} \right| = 0$$

and $p = 3$. We have a singular case to deal with.

Taking the first three coordinates of the vectors $\boldsymbol{x}$, $\boldsymbol{Ax}$, $\boldsymbol{A^2x}$, $\boldsymbol{A^3x}$, we write the Krylov determinant in the form

$$\begin{vmatrix} 1 & 0 & 0 & 1 \\ 3 & 2 & 2 & \lambda \\ 1 & 4 & 0 & \lambda^2 \\ 3 & 6 & 2 & \lambda^3 \end{vmatrix}.$$

Expanding this determinant and cancelling -8, we obtain:

$$\varphi(\lambda) = \lambda^3 - \lambda^2 - \lambda + 1 = (\lambda - 1)^2 (\lambda + 1).$$

Hence we find three characteristic values: $\lambda_1 = 1$, $\lambda_2 = 1$, $\lambda_3 = -1$. The fourth characteristic value can be obtained from the condition that the sum of all the characteristic values must be equal to the trace of the matrix. But $\operatorname{tr} A = 0$. Hence $\lambda_4 = -1$.

These examples show that in applying Krylov's method, when we write down successively the rows of the matrix

$$\left\| \begin{matrix} a & b & \dots & l \\ a_1 & b_1 & \dots & l_1 \\ a_2 & b_2 & \dots & l_2 \\ \dots & \dots & \dots & \dots \\ \dots & \dots & \dots & \dots \end{matrix} \right\| \tag{105}$$

it is necessary to watch the rank of the matrix obtained so that we stop after the first row (the $(p+1)$-st from above) that is a linear combination of the preceding ones. The determination of the rank is connected with the computation of certain determinants. Moreover, after obtaining Krylov's determinant in the form (93) or (100), in order to expand it with respect to the elements of the last column we have to compute a certain number of determinants of order $p-1$ (in the regular case, of order $n-1$).

Instead of expanding Krylov's determinant we can determine the coefficients $\alpha_1, \alpha_2, \dots$ directly from the system of equations (91) (or (99)) by applying any efficient method of solution to the system—for example, the elimination method. This method can be applied immediately to the matrix

$$\left\| \begin{matrix} a & b & \dots & l & 1 \\ a_1 & b_1 & \dots & l_1 & \lambda \\ a_2 & b_2 & \dots & l_2 & \lambda^2 \\ \dots & \dots & \dots & \dots & \dots \\ \dots & \dots & \dots & \dots & \dots \end{matrix} \right\| \tag{106}$$

by using it in parallel with the computation of the corresponding rows by Krylov's method. We shall then discover at once a row of the matrix (105) that depends on the preceding ones, without computing any determinant.

Let us explain this in some detail. In the first row of (106) we take an arbitrary element $c \neq 0$ and we use it to make the element c_1 under it into zero, by subtracting from the second row the first row multiplied by c_1/c. Next we take an element $f_1^* \neq 0$ in the second row and by means of c and f_1^* we make the elements c_2 and f_2 into zero, etc.[25] As a result of such a transformation, the element in the last column of (106) is replaced by a polynomial of degree k, $g_k(\lambda) = \lambda^k + \cdots$ $(k = 0, 1, 2, \ldots)$

Since under our transformation the rank of the matrix formed from the first k rows for any k and the first n columns of (106) does not change, the $(p+1)$-st row of the matrix must, after the transformation, have the form

$$0, 0, \ldots, 0, g_p(\lambda).$$

Our transformation does not change the value of the Krylov determinant

$$\begin{vmatrix} c & f & \ldots h & 1 \\ c_1 & f_1 & \ldots h_1 & \lambda \\ \cdot & \cdot & \cdot\;\cdot\;\cdot & \cdot \\ c_{p-1} & f_{p-1} & \ldots h_{p-1} & \lambda^{p-1} \\ c_p & f_p & \ldots h_p & \lambda^p \end{vmatrix} = M^*\varphi(\lambda).$$

Therefore

$$M^*\varphi(\lambda) = cf_1^* \cdots g_p(\lambda), \tag{107}$$

i.e.,[26] $g_p(\lambda)$ is the required polynomial $\varphi(\lambda) : g_p(\lambda) \equiv \varphi(\lambda)$.

We recommend the following simplification. After obtaining the k-th transformed row of (106)

$$a_{k-1}^*, b_{k-1}^*, \ldots, l_{k-1}^*, g_{k-1}(\lambda), \tag{108}$$

one should obtain the following $(k+1)$-st row by multiplying $a_{k-1}^*, b_{k-1}^*, \ldots, l_{k-1}^*$ (and not the original $a_{k-1}, b_{k-1}, \ldots, l_{k-1}$) into the rows of the given matrix.[27] Then we find the $(k+1)$-st row in the form

$$\tilde{a}_k, \tilde{b}_k, \ldots, \tilde{l}_k, \lambda g_{k-1}(\lambda),$$

and after subtracting the preceding rows, we obtain:

---

[25] The elements $c, f_1^*, \ldots$ must not belong to the last column containing the powers of λ.

[26] We recall that the highest coefficients of $\varphi(\lambda)$ and $g_p(\lambda)$ are 1.

[27] The simplification consists in the fact that in the row of (108) to be transformed $k - 1$ elements are equal to zero. Therefore it is simple to multiply such a row into the rows of A.

$$a_k^*,\ b_k^*,\ \ldots,\ l_k^*,\ g_k(\lambda).$$

The slight modification of Krylov's method that we have recommended (its combination with the elimination method) enables us to find at once the polynomial $\varphi(\lambda)$ that we are interested in (in the regular case, $\Delta(\lambda)$) without computing any determinants or solving any auxiliary system of equations.[28]

Example.

$$\begin{matrix} \mathit{4} & \mathit{4} & \mathit{1} & \mathit{5} & \mathit{0} \end{matrix}$$

$$A = \left\| \begin{matrix} 1 & 1 & -1 & 1 & 0 \\ 1 & 2 & -1 & 0 & 1 \\ -1 & 2 & 3 & -1 & 0 \\ 1 & -2 & 1 & 2 & -1 \\ 2 & 1 & -1 & 3 & 0 \end{matrix} \right\|$$

0	0	0	0	1	1
0	1	0	-1	0	λ
0	2	3	-4	-2	$\lambda^2 \quad [2-4\lambda]$
0	-2	3	0	0	$\lambda^2-4\lambda+2$
-5	-7	5	7	-5	$\lambda^3-4\lambda^2+2\lambda \quad [5+7\lambda]$
-5	0	5	0	0	$\lambda^3-4\lambda^2+9\lambda+5$
-10	-10	20	0	-15	$\lambda^4-4\lambda^3+9\lambda^2+5\lambda \quad [15-5(\lambda^2-4\lambda+2) -2(\lambda^3-4\lambda^2+9\lambda+5)]$
0	0	-5	0	0	$\lambda^4-6\lambda^3+12\lambda^2+7\lambda-5$
5	5	-15	-5	5	$\lambda^5-6\lambda^4+12\lambda^3+7\lambda^2-5\lambda \quad [-5-5\lambda+(\lambda^3-4\lambda^2+9\lambda+5)-2(\lambda^4-6\lambda^3+12\lambda^2+7\lambda-5)]$
0	0	0	0	0	$\underbrace{\lambda^5-8\lambda^4+25\lambda^3-21\lambda^2-15\lambda+10}_{\Delta(\lambda)}$

[28] Apart from the method of Krylov, we have acquainted the reader in Chapter IV with the method of D. K. Faddeev for the computation of the coefficients of the characteristic polynomial. Faddeev's method involves more computations than Krylov's but it is more general, being without singular cases. We also refer the reader to the very effective method of A. M. Danilevskiĭ [131]; see also the expository paper [376] and the book [15], § 24. See also [5] and [194].

CHAPTER VIII

MATRIX EQUATIONS

In this chapter we consider certain types of matrix equations that occur in various problems in the theory of matrices and its applications.

§ 1. The Equation $AX = XB$

1. Suppose that the equation

$$AX = XB \tag{1}$$

is given, where A and B are square matrices (in general of different orders)

$$A = \| a_{ij} \|_1^m, \qquad B = \| b_{kl} \|_1^n$$

and where X is an unknown rectangular matrix of dimension $m \times n$:

$$X = \| x_{jk} \| \qquad (j = 1, 2, \ldots, m; \quad k = 1, 2, \ldots, n).$$

We write down the elementary divisors of A and B (in the field of complex numbers):

$$(A): (\lambda - \lambda_1)^{p_1}, \quad (\lambda - \lambda_2)^{p_2}, \ \ldots, \ (\lambda - \lambda_u)^{p_u} \quad (p_1 + p_2 + \cdots + p_u = m),$$
$$(B): (\lambda - \mu_1)^{q_1}, \quad (\lambda - \mu_2)^{q_2}, \ \ldots, \ (\lambda - \mu_v)^{q_v} \quad (q_1 + q_2 + \cdots + q_v = n).$$

In accordance with these elementary divisors we reduce A and B to Jordan normal form

$$A = U\tilde{A}U^{-1}, \qquad B = V\tilde{B}V^{-1}, \tag{2}$$

where U and V are square non-singular matrices of orders m and n, respectively, and $\tilde{A}$ and $\tilde{B}$ are the Jordan matrices:

$$\begin{aligned} \tilde{A} &= \{ \lambda_1 E^{(p_1)} + H^{(p_1)}, \quad \lambda_2 E^{(p_2)} + H^{(p_2)}, \quad \ldots, \quad \lambda_u E^{(p_u)} + H^{(p_u)} \}, \\ \tilde{B} &= \{ \mu_1 E^{(q_1)} + H^{(q_1)}, \quad \mu_2 E^{(q_2)} + H^{(q_2)}, \quad \ldots, \quad \mu_v E^{(q_v)} + H^{(q_v)} \}. \end{aligned} \tag{3}$$

Replacing A and B in (1) by their expressions given in (2), we obtain:

$$U\tilde{A}U^{-1}X = XV\tilde{B}V^{-1}.$$

We multiply both sides of this equation on the left by U^{-1} and on the right by V:

$$\tilde{A}U^{-1}XV = U^{-1}XV\tilde{B}. \tag{4}$$

When we introduce in place of X a new unknown matrix $\tilde{X}$ (of the same dimension $m \times n$)

$$\tilde{X} = U^{-1}XV, \tag{5}$$

we can write equation (4) as follows:

$$\tilde{A}\tilde{X} = \tilde{X}\tilde{B}. \tag{6}$$

We have thus replaced the matrix equation (1) by the equation (6), of the same form, in which the given matrices have Jordan normal form.

We partition $\tilde{X}$ into blocks corresponding to the quasi-diagonal form of the matrices $\tilde{A}$ and $\tilde{B}$:

$$\tilde{X} = (X_{\alpha\beta}) \qquad (\alpha = 1, 2, \ldots, u;\quad \beta = 1, 2, \ldots, v)$$

(here $X_{\alpha\beta}$ is a rectangular matrix of dimension $p_\alpha \times q_\beta$ ($\alpha = 1, 2, \ldots, u$; $\beta = 1, 2, \ldots, v$)).

Using the rule for multiplying a partitioned matrix by a quasi-diagonal one (see p. 42), we carry out the multiplication of the matrices on the left-hand and right-hand sides of (6). Then this equation breaks up into uv matrix equations

$$[\lambda_\alpha E^{(p_\alpha)} + H^{(p_\alpha)}]\, X_{\alpha\beta} = X_{\alpha\beta}\, [\mu_\beta E^{(q_\beta)} + H^{(q_\beta)}]$$
$$(\alpha = 1, 2, \ldots, u;\quad \beta = 1, 2, \ldots, v),$$

which we rewrite as follows:

$$(\mu_\beta - \lambda_\alpha) X_{\alpha\beta} = H_\alpha X_{\alpha\beta} - X_{\alpha\beta} G_\beta \qquad (\alpha = 1, 2, \ldots, u;\quad \beta = 1, 2, \ldots, v); \tag{7}$$

we have used here the abbreviations

$$H_\alpha = H^{(p_\alpha)} \quad G_\beta = H^{(q_\beta)} \qquad (\alpha = 1, 2, \ldots, u;\quad \beta = 1, 2, \ldots, v). \tag{8}$$

Let us take one of the equations (7). Two cases can occur:

1. $\lambda_\alpha \neq \mu_\beta$. We iterate equation (7) $r - 1$ times:[1]

---

[1] We multiply both sides of (7) by $\mu_\beta - \lambda_\alpha$ and in each term of the right-hand side we replace $(\mu_\beta - \lambda_\alpha) X_{\alpha\beta}$ by $H_\alpha X_{\alpha\beta} - X_{\alpha\beta} G_\beta$. This process is repeated $r - 1$ times.

$$(\mu_\beta - \lambda_\alpha)^r X_{\alpha\beta} = \sum_{\sigma+\tau=r} (-1)^\tau \binom{r}{\tau} H_\alpha^\sigma X_{\alpha\beta} G_\beta^\tau . \tag{9}$$

Note that, by (8),

$$H_\alpha^{p_\alpha} = G_\beta^{q_\beta} = O. \tag{10}$$

If in (9) we take $r \geqq p_\alpha + q_\beta - 1$, then in each term of the sum on the right-hand side of (9) at least one of the relations

$$\sigma \geqq p_\alpha, \quad \tau \geqq q_\beta$$

is satisfied, so that by (10) either $H_\alpha^\sigma = O$ or $G_\beta^\tau = O$. Moreover, since in this case $\lambda_\alpha \neq \mu_\beta$, we find from (9):

$$X_{\alpha\beta} = O. \tag{11}$$

2. $\lambda_\alpha = \mu_\beta$. In this case equation (7) assumes the form

$$H_\alpha X_{\alpha\beta} = X_{\alpha\beta} G_\beta . \tag{12}$$

In the matrices H_α and G_β the elements of the first superdiagonal are equal to 1, and all the remaining elements are zero. Taking this specific structure of H_α and G_β into account and setting

$$X_{\alpha\beta} = \| \xi_{ik} \| \qquad (i = 1, 2, \ldots, p_\alpha; \quad k = 1, 2, \ldots, q_\beta),$$

we replace the matrix equation (12) by the following equivalent system of scalar equations:[2]

$$\xi_{i+1,k} = \xi_{i,k-1} \; (\xi_{i0} = \xi_{p_\alpha+1,k} = 0; \quad i = 1, 2, \ldots, p_\alpha; \quad k = 1, 2, \ldots, q_\beta). \tag{13}$$

The equations (13) have this meaning:

1) In the matrix $X_{\alpha\beta}$ the elements of every line parallel to the main diagonal are equal;

2) $\quad \xi_{21} = \xi_{31} = \cdots = \xi_{p_\alpha 1} = \xi_{p_\alpha 2} = \cdots = \xi_{p_\alpha, q_\beta - 1} = 0.$

Let $p_\alpha = q_\beta$. Then $X_{\alpha\beta}$ is a square matrix. From 1) and 2) it follows that in $X_{\alpha\beta}$ all the elements below the main diagonal are zero, all the elements in the main diagonal are equal to a certain number $c_{\alpha\beta}$, all the elements of the first superdiagonal are equal to a number $c'_{\alpha\beta}$, etc.; i.e.,

---

[2] From the structure of the matrices H_α and G_β it follows that the product $H_\alpha X_{\alpha\beta}$ is obtained from $X_{\alpha\beta}$ by shifting all the rows one place upwards and filling the last row with zeros; similarly, $X_{\alpha\beta} G_\beta$ is obtained from $X_{\alpha\beta}$ by shifting all the columns one place to the right and filling the first column with zeros (see Chapter I, p. 14). To simplify the notation we do not write the additional indices α, β in ξ_{ik}.

$$X_{\alpha\beta}=\left\| \begin{matrix} c_{\alpha\beta} & c'_{\alpha\beta} & \cdots & c_{\alpha\beta}^{(p_\alpha-1)} \\ 0 & c_{\alpha\beta} & \ddots & \vdots \\ \vdots & & \ddots & \vdots \\ \vdots & & \ddots & c'_{\alpha\beta} \\ 0 & \cdots & 0 & c_{\alpha\beta} \end{matrix} \right\| = T_{p_\alpha}; \qquad (14)$$

$$(p_\alpha=q_\beta)$$

here $c_{\alpha\beta}$, $c'_{\alpha\beta}$, ..., $c_{\alpha\beta}^{(p_\alpha-1)}$ are arbitrary parameters (the equations (12) do not impose any restrictions on the values of these parameters).

It is easy to see that for $p_\alpha < q_\beta$

$$X_{\alpha\beta}=(\overbrace{\;0\;}^{q_\beta-p_\alpha},\ T_{p_\alpha}) \qquad (15)$$

and for $p_\alpha > q_\beta$

$$X_{\alpha\beta}=\begin{pmatrix} T_{q_\beta} \\ 0 \end{pmatrix}\}\, p_\alpha-q_\beta. \qquad (16)$$

We shall say of the matrices (14), (15), and (16) that they have *regular* upper triangular form. The number of arbitrary parameters in $X_{\alpha\beta}$ is equal to the smaller of the numbers p_α and q_β. The scheme below shows the structure of the matrices $X_{\alpha\beta}$ for $\lambda_\alpha=\mu_\beta$ (the arbitrary parameters are here denoted by a, b, c, and d):

$$X_{\alpha\beta}=\left\| \begin{matrix} a & b & c & d \\ 0 & a & b & c \\ 0 & 0 & a & b \\ 0 & 0 & 0 & a \end{matrix} \right\|, \quad X_{\alpha\beta}=\left\| \begin{matrix} 0 & 0 & a & b & c \\ 0 & 0 & 0 & a & b \\ 0 & 0 & 0 & 0 & a \end{matrix} \right\|, \quad X_{\alpha\beta}=\left\| \begin{matrix} a & b & c \\ 0 & a & b \\ 0 & 0 & a \\ 0 & 0 & 0 \\ 0 & 0 & 0 \end{matrix} \right\|$$

$$(p_\alpha=q_\beta=4) \qquad (p_\alpha=3,\ q_\beta=5) \qquad (p_\alpha=5,\ q_\beta=3)$$

In order to subsume case 1 also in the count of arbitrary parameters in X, we denote by $d_{\alpha\beta}(\lambda)$ the greatest common divisor of the elementary divisors $(\lambda-\lambda_\alpha)^{p_\alpha}$ and $(\lambda-\mu_\beta)^{q_\beta}$ and by $\delta_{\alpha\beta}$ the degree of the polynomial $d_{\alpha\beta}(\lambda)$ $(\alpha=1, 2, \ldots, u;\ \beta=1, 2, \ldots, v)$. In case 1, we have $\delta_{\alpha\beta}=0$; in case 2, $\delta_{\alpha\beta}=\min(p_\alpha, q_\beta)$. Thus, in both cases the number of arbitrary parameters in $X_{\alpha\beta}$ is equal to $\delta_{\alpha\beta}$. The number of arbitrary parameters in $\tilde{X}$ is determined by the formula.

$$N=\sum_{\alpha=1}^{u}\sum_{\beta=1}^{v}\delta_{\alpha\beta}.$$

In what follows it will be convenient to denote the general solution of (6) by $X_{\tilde{A}\tilde{B}}$ (so far we have denoted it by $\tilde{X}$).

The results obtained in this section can be stated in the form of the following theorem:

THEOREM 1: *The general solution of the matrix equation*

$$AX = XB$$

where

$$A = \| a_{ik} \|_1^m = U\tilde{A}U^{-1} = U\{\lambda_1 E^{(p_1)} + H^{(p_1)}, \dots, \lambda_u E^{(p_u)} + H^{(p_u)}\} U^{-1},$$
$$B = \| b_{ik} \|_1^n = V\tilde{B}V^{-1} = V\{\mu_1 E^{(q_1)} + H^{(q_1)}, \dots, \mu_v E^{(q_v)} + H^{(q_v)}\} V^{-1}$$

is given by the formula

$$X = UX_{\tilde{A}\tilde{B}} V^{-1}. \tag{17}$$

Here $X_{\tilde{A}\tilde{B}}$ *is the general solution of the equation*

$$\tilde{A}\tilde{X} = \tilde{X}\tilde{B}$$

and has the following structure:

$X_{\tilde{A}\tilde{B}}$ *is decomposed into blocks*

$$X_{\tilde{A}\tilde{B}} = (\overbrace{X_{\alpha\beta}}^{q_\beta})\}\, p_\alpha \qquad (\alpha = 1, 2, \dots, u;\ \beta = 1, 2, \dots, v);$$

if $\lambda_\alpha \neq \mu_\beta$, *then the null matrix stands in the place* $X_{\alpha\beta}$, *but if* $\lambda_\alpha = \mu_\beta$, *then an arbitrary regular upper triangular matrix stands in the place* $X_{\alpha\beta}$.

$X_{\tilde{A}\tilde{B}}$, *and therefore also* X, *depends linearly on* N *arbitrary parameters* $c_1, c_2, \dots, c_N$

$$X = \sum_{j=1}^{N} c_j X_j, \tag{18}$$

where N *is determined by the formula*

$$N = \sum_{\alpha=1}^{u} \sum_{\beta=1}^{v} \delta_{\alpha\beta} \tag{19}$$

(*here* $\delta_{\alpha\beta}$ *denotes the degree of the greatest common divisor of* $(\lambda - \lambda_\alpha)^{p_\alpha}$ *and* $(\lambda - \mu_\beta)^{q_\beta}$).

Note that the matrices $X_1, X_2, \dots, X_N$ that occur in (18) are solutions of the original equation (1) (X_j is obtained from X by giving to the parameter c_j the value 1 and to the remaining parameters the value 0; $j = 1, ,2, \dots, N$). These solutions are linearly independent, since otherwise for certain values of the parameters $c_1, c_2, \dots, c_N$, not all zero, the matrix X, and therefore $X_{\tilde{A}\tilde{B}}$, would be the null matrix, which is impossible. Thus (18) shows that every solution of the original equation is a linear combination of N linearly independent solutions.

If the matrices A and B do not have common characteristic values (if the characteristic polynomials $|\lambda E - A|$ and $|\lambda E - B|$ are co-prime), *then* $N = \sum_{\alpha=1}^{u}\sum_{\beta=1}^{v} \delta_{\alpha\beta} = 0$, *and so* $X = O$, *i.e., in this case the equation* (1) *has only the trivial solution* $X = O$.

Note. Suppose that the elements of A and B belong to some number field F. Then we cannot say that the elements of U, V, and $X_{\tilde{A}\tilde{B}}$ that occur in (17) also belong to F. The elements of these matrices may be taken in an extension field F_1 which is obtained from F by adjoining the roots of the characteristic equations $|\lambda E - A| = 0$ and $|\lambda E - B| = 0$. We always have to deal with such an extension of the ground field when we use the reduction of given matrices to Jordan normal form.

However, the matrix equation (1) is equivalent to a system of mn linear homogeneous equations, where the unknowns are the elements x_{jk} ($j = 1, 2, 3, \ldots, m$; $k = 1, 2, \ldots, n$) of the required matrix X:

$$\sum_{j=1}^{m} a_{ij}x_{jk} = \sum_{h=1}^{n} x_{ih}b_{hk} \qquad (i = 1, 2, \ldots, m;\ k = 1, 2, \ldots, n). \tag{20}$$

What we have shown is that this system has N linearly independent solutions, where N is determined by (19). But it is well known that fundamental linearly independent solutions can be chosen in the ground field F to which the coefficients of (20) belong. Thus, in (18) the matrices $X_1, X_2, \ldots, X_N$ can be so chosen that their elements lie in F. If we then give to the arbitrary parameters in (18) all possible values in F, we obtain all the matrices X with elements in F that satisfy the equation (1).[3]

§ 2. The Special Case $A = B$. Commuting Matrices

1. Let us consider the special case of the equation (1):

$$AX = XA, \tag{21}$$

where $A = \| a_{ik} \|_1^n$ is a given matrix and $X = \| x_{ik} \|_1^n$ an unknown matrix. We have come to a problem of Frobenius: to determine all the matrices X that commute with a given matrix A.

We reduce A to Jordan normal form:

$$A = U\tilde{A}U^{-1} = U\{\lambda_1 E^{(p_1)} + H^{(p_1)}, \ldots, \lambda_u E^{(p_u)} + H^{(p_u)}\}U^{-1}. \tag{22}$$

---

[3] The matrices $A = \| a_{ij} \|_1^m$ and $B = \| b_{kl} \|_1^n$ determine a linear operator $\hat{F}(X) = AX - XB$ in the space of rectangular matrices X of dimension $m \times n$. A treatment of operators of this type is contained in the paper [179].

Then when we set in (17) $V = U$, $\tilde{B} = \tilde{A}$ and denote $X_{\tilde{A}\tilde{A}}$ simply by $X_{\tilde{A}}$, we obtain all solutions of (21), i.e., all matrices that commute with A, in the following form:

$$X = U X_{\tilde{A}} U^{-1}, \tag{23}$$

where $X_{\tilde{A}}$ denotes an arbitrary matrix permutable with $\tilde{A}$. As we have explained in the preceding section, $X_{\tilde{A}}$ is split into u^2 blocks

$$X_{\tilde{A}} = (X_{\alpha\beta})_1^u$$

corresponding to the splitting of the Jordan matrix $\tilde{A}$ into blocks; $X_{\alpha\beta}$ is either the null matrix or an arbitrary regular upper triangular matrix, depending on whether $\lambda_\alpha \neq \lambda_\beta$ or $\lambda_\alpha = \lambda_\beta$.

As an example, we write down the elements of $X_{\tilde{A}}$ in the case where A has the following elementary divisors:

$$(\lambda - \lambda_1)^4,\ (\lambda - \lambda_1)^3,\ (\lambda - \lambda_2)^2,\ \lambda - \lambda_2 \quad (\lambda_1 \neq \lambda_2).$$

In this case $X_{\tilde{A}}$ has the following form:

$$\left\| \begin{array}{cccc|ccc|cc|c} a & b & c & d & e & f & g & 0 & 0 & 0 \\ 0 & a & b & c & 0 & e & f & 0 & 0 & 0 \\ 0 & 0 & a & b & 0 & 0 & e & 0 & 0 & 0 \\ 0 & 0 & 0 & a & 0 & 0 & 0 & 0 & 0 & 0 \\ \hline 0 & h & k & l & m & p & q & 0 & 0 & 0 \\ 0 & 0 & h & k & 0 & m & p & 0 & 0 & 0 \\ 0 & 0 & 0 & h & 0 & 0 & m & 0 & 0 & 0 \\ \hline 0 & 0 & 0 & 0 & 0 & 0 & 0 & r & s & t \\ 0 & 0 & 0 & 0 & 0 & 0 & 0 & 0 & r & 0 \\ \hline 0 & 0 & 0 & 0 & 0 & 0 & 0 & 0 & w & z \end{array} \right\| \qquad (a,\ b,\ \ldots,\ z \text{ are arbitrary parameters}).$$

The number of parameters in $X_{\tilde{A}}$ is equal to N, where $N = \sum_{\alpha, \beta = 1}^{u} \delta_{\alpha\beta}$; here $\delta_{\alpha\beta}$ denotes the degree of the greatest common divisor of the polynomials $(\lambda - \lambda_\alpha)^{p_\alpha}$ and $(\lambda - \lambda_\beta)^{p_\beta}$.

Let us bring the invariant polynomials of A into the discussion: $i_1(\lambda)$, $i_2(\lambda)$, ..., $i_t(\lambda)$; $i_{t+1}(\lambda) = \ldots = i_n(\lambda) = 1$. We denote the degrees of these polynomials by $n_1 \geqq n_2 \geqq \ldots \geqq n_t > n_{t+1} = \ldots = 0$. Since each invariant polynomial is a product of certain co-prime elementary divisors, the formula for N can be written as follows:

$$N = \sum_{g,j=1}^{t} \varkappa_{gj}, \tag{24}$$

where $\varkappa_{gj}$ is the degree of the greatest common divisor of $i_g(\lambda)$ and $i_j(\lambda)$ $(g, j = 1, 2, \ldots, t)$. But the greatest common divisor of $i_g(\lambda)$ and $i_j(\lambda)$ is one of these polynomials and therefore $\varkappa_{gj} = \min(n_g, n_j)$. Hence we obtain:

$$N = n_1 + 3n_2 + \cdots + (2t - 1)\, n_t.$$

N is the number of linearly independent matrices that commute with A (we may assume that the elements of these matrices belong to the ground field F containing the elements of A; see the remark at the end of the preceding section). We have arrived at the following theorem:

THEOREM 2: *The number of linearly independent matrices that commute with the matrix $A = \| a_{ik} \|_1^n$ is given by the formula*

$$N = n_1 + 3n_2 + \ldots + (2t - 1)\, n_t. \tag{25}$$

where $n_1, n_2, \ldots, n_t$ are the degrees of the non-constant invariant polynomials $i_1(\lambda), i_2(\lambda), \ldots, i_t(\lambda)$ of A.

Note that

$$n = n_1 + n_2 + \ldots + n_t. \tag{26}$$

From (25) and (26) it follows that

$$N \geqq n, \tag{27}$$

where the equality sign holds if and only if $t = 1$, i.e., if all the elementary divisors of A are co-prime in pairs.

2. Let $g(\lambda)$ be an arbitrary polynomial in λ. Then $g(A)$ is permutable with A. There arises the converse question: when can every matrix that is permutable with A be expressed as a polynomial in A? Every matrix that commutes with A would then be a linear combination of the linearly independent matrices

$$E, A, A^2, \ldots, A^{n_1-1}.$$

Hence $N = n_1 \leqq n$; on comparing this with (27), we obtain: $N = n_1 = n$.

COROLLARY 1 TO THEOREM 2: *All the matrices that are permutable with A can be expressed as polynomials in A if and only if $n_1 = n$, i.e., if all the elementary divisors of A are co-prime in pairs.*

3. The polynomials in a matrix that commutes with A also commute with A. We raise the question: when can all the matrices that commute with A be expressed in the form of polynomials in one and the same matrix C? Let us consider the case in which they can be so expressed. Then since by the Hamilton-Cayley Theorem the matrix C satisfies its characteristic equation, every matrix that commutes with C must be expressible linearly by the matrices

$$E, C, C^2, \ldots, C^{n-1}.$$

Therefore in this case $N \leqq n$. Comparing this with (27), we find that $N=n$. Hence from (25) and (26) we also have $n_1=n$.

COROLLARY 2 TO THEOREM 2: *All the matrices that are permutable with A can be expressed in the form of polynomials in one and the same matrix C if and only if $n_1=n$, i.e. if and only if all the elementary divisors of $\lambda E - A$ are co-prime. In this case all the matrices that are permutable with A can be represented in the form of polynomials in A.*

4. We mention a very important property of permutable matrices.

THEOREM 3: *If two matrices $A=\| a_{ik} \|_1^n$ and $B=\| b_{ik} \|_1^n$ are permutable and if one of them, say A, has quasi-diagonal form*

$$A=\{\overbrace{A_1}^{s_1}, \overbrace{A_2}^{s_2}\}, \tag{28}$$

where the matrices A_1 and A_2 do not have characteristic values in common, then the other matrix also has the same quasi-diagonal form

$$B=\{\overbrace{B_1}^{s_1}, \overbrace{B_2}^{s_2}\}. \tag{29}$$

Proof. We split B into blocks corresponding to the quasi-diagonal form (28):

$$B=\begin{pmatrix} \overbrace{B_1}^{s_1} & \overbrace{X}^{s_2} \\ Y & B_2 \end{pmatrix}.$$

From the relation $AB=BA$ we obtain four matrix equations:

$$1.\ A_1B_1=B_1A_1,\quad 2.\ A_1X=XA_2,\quad 3.\ A_2Y=YA_1,\quad 4.\ A_2B_2=B_2A_2. \tag{30}$$

As we explained in § 1 (p. 220), the second and third of the equations in (30) only have the solutions $X=O$, $Y=O$, since A_1 and A_2 have no characteristic values in common. This proves our statement. The first and fourth of the equations in (30) express the permutability of A_1 and B_1 and of A_2 and B_2.

In geometrical language, this theorem runs as follows:

THEOREM 3′: *If*

$$\boldsymbol{R} = \boldsymbol{I}_1 + \boldsymbol{I}_2$$

is a decomposition of the whole space $\boldsymbol{R}$ *into invariant subspaces* $\boldsymbol{I}_1$ *and* $\boldsymbol{I}_2$ *with respect to an operator* $\boldsymbol{A}$ *and if the minimal polynomials of these subspaces (with respect to* $\boldsymbol{A}$*) are co-prime, then* $\boldsymbol{I}_1$ *and* $\boldsymbol{I}_2$ *are invariant with respect to any linear operator* $\boldsymbol{B}$ *that commutes with* $\boldsymbol{A}$.

Let us also give a geometrical proof of this statement. We denote by $\psi_1(\lambda)$ and $\psi_2(\lambda)$ the minimal polynomials of $\boldsymbol{I}_1$ and $\boldsymbol{I}_2$ with respect to $\boldsymbol{A}$. From the fact that they are co-prime it follows that all the vectors of $\boldsymbol{R}$ that satisfy the equation $\psi_1(\boldsymbol{A})\boldsymbol{x} = \boldsymbol{o}$ belong to $\boldsymbol{I}_1$ and all the vectors that satisfy $\psi_2(\boldsymbol{A})\boldsymbol{x} = \boldsymbol{o}$ belong to $\boldsymbol{I}_2$.[4] Let $\boldsymbol{x}_1 \in \boldsymbol{I}_1$. Then $\psi_1(\boldsymbol{A})\boldsymbol{x}_1 = \boldsymbol{o}$. The permutability of $\boldsymbol{A}$ and $\boldsymbol{B}$ implies that of $\psi_1(\boldsymbol{A})$ and $\boldsymbol{B}$, so that

$$\psi_1(\boldsymbol{A})\,\boldsymbol{B}\boldsymbol{x}_1 = \boldsymbol{B}\psi_1(\boldsymbol{A})\,\boldsymbol{x}_1 = \boldsymbol{o}\,,$$

i.e., $\boldsymbol{B}\boldsymbol{x}_1 \in \boldsymbol{I}_1$. The invariance of $\boldsymbol{I}_2$ with respect to $\boldsymbol{B}$ is proved similarly.

This theorem leads to a number of corollaries:

COROLLARY 1: *If the linear operators* $\boldsymbol{A}, \boldsymbol{B}, \ldots, \boldsymbol{L}$ *are pairwise permutable, then the whole space* $\boldsymbol{R}$ *can be split into subspaces invariant with respect to all the operators* $\boldsymbol{A}, \boldsymbol{B}, \ldots, \boldsymbol{L}$

$$\boldsymbol{R} = \boldsymbol{I}_1 + \boldsymbol{I}_2 + \ldots + \boldsymbol{I}_w$$

such that the minimal polynomial of each of these subspaces with respect to any one of the operators $\boldsymbol{A}, \boldsymbol{B}, \ldots, \boldsymbol{L}$ *is a power of an irreducible polynomial.*

As a special case of this we obtain:

COROLLARY 2: *If the linear operators* $\boldsymbol{A}, \boldsymbol{B}, \ldots, \boldsymbol{L}$ *are pairwise permutable and all the characteristic values of these operators belong to the ground field, then the whole space* $\boldsymbol{R}$ *can be split into subspaces* $\boldsymbol{I}_1, \boldsymbol{I}_2, \ldots, \boldsymbol{I}_w$*, invariant with respect to all the operators such that each operator* $\boldsymbol{A}, \boldsymbol{B}, \ldots, \boldsymbol{L}$ *has equal characteristic values in each of them.*

Finally, we mention a further special case of this statement:

COROLLARY 3: *If* $\boldsymbol{A}, \boldsymbol{B}, \ldots, \boldsymbol{L}$ *are pairwise permutable operators of simple structure* (see Chapter III, § 8), *then a basis of the space can be formed from common characteristic vectors of these operators.*

We also give the matrix form of the last statement:

Permutable matrices of simple structure can be brought into diagonal form simultaneously by a similarity transformation.

---

[4] See Theorem 1 of Chapter VII (p. 179).

§ 3. The Equation $AX - XB = C$

1. Suppose that the matrix equation

$$AX - XB = C \tag{31}$$

is given, where $A = \| a_{ij} \|_1^m$ and $B = \| b_{kl} \|_1^n$ are given square matrices of order m and n and where $C = \| c_{jk} \|$ and $X = \| x_{jk} \|$ are a given and an unknown rectangular matrix, respectively, of dimension $m \times n$. The equation (31) is equivalent to a system of mn scalar equations in the elements of X:

$$\sum_{j=1}^{m} a_{ij} x_{jk} - \sum_{l=1}^{n} x_{il} b_{lk} = c_{ik} \quad (i = 1, 2, \ldots, m;\ k = 1, 2, \ldots, n). \tag{31'}$$

The corresponding homogeneous system of equations

$$\sum_{j=1}^{m} a_{ij} x_{jk} - \sum_{l=1}^{n} x_{il} b_{lk} = 0 \quad (i = 1, 2, \ldots, m;\ k = 1, 2, \ldots, n),$$

can be written in matrix form as follows:

$$AX - XB = O. \tag{32}$$

Thus, if (32) only has the trivial solution $X = O$, then (31) has a unique solution. But we have established in § 1 that the only solution of (32) is the trivial one if and only if A and B do not have common characteristic values. Therefore, *if the matrices A and B do not have characteristic values in common, then* (31) *has a unique solution; but if the matrices A and B have characteristic values in common, then two cases may arise depending on the 'constant' term C: either the equation* (31) *is contradictory, or it has an infinite number of solutions given by the formula*

$$X = X_0 + X_1,$$

where X_0 is a fixed particular solution of (31) *and X_1 the general solution of the homogeneous equation* (32) (the structure of X_1 was described in § 1).

§ 4. The Scalar Equation $f(X) = O$

1. To begin with, let us consider the equation

$$g(X) = O, \tag{33}$$

where

$$g(\lambda) = (\lambda - \lambda_1)^{a_1} (\lambda - \lambda_2)^{a_2} \cdots (\lambda - \lambda_h)^{a_h}$$

is a given polynomial in the variable λ and X is an unknown square matrix of order n. Since the minimal polynomial of X, i.e., the first invariant polynomial $i_1(\lambda)$, must be a divisor of $g(\lambda)$, the elementary divisors of X must have the following form:

$$(\lambda-\lambda_{j_1})^{p_{j_1}},\ (\lambda-\lambda_{j_2})^{p_{j_2}},\ \ldots,\ (\lambda-\lambda_{j_v})^{p_{j_v}} \quad \begin{pmatrix} j_1, j_2, \ldots, j_v = 1, 2, \ldots, h, \\ p_{j_1} \leqq a_{j_1}, p_{j_2} \leqq a_{j_2}, \ldots, p_{j_v} \leqq a_{j_v}, \\ p_{j_1} + p_{j_2} + \cdots + p_{j_v} = n \end{pmatrix}$$

(among the indices $j_1, j_2, \ldots, j_v$ there may be some that are equal; n is the given order of the unknown matrix X).

We represent X in the form

$$X = T\{\lambda_{j_1}E^{(p_{j_1})} + H^{(p_{j_1})}, \ldots, \lambda_{j_v}E^{(p_{j_v})} + H^{(p_{j_v})}\}\, T^{-1}, \tag{34}$$

where T is an arbitrary non-singular matrix of order n. The set of solutions of the equation (33) with a given order of the unknown matrix splits, by formula (34), into a finite number of classes of similar matrices.

Example 1. Let the equation

$$X^m = O \tag{35}$$

be given.

If a certain power of a matrix is the null matrix, then the matrix is called *nilpotent*. The least exponent for which the power of the matrix is the null matrix is called the *index of nilpotency*.

Obviously, the solutions of (35) are all the nilpotent matrices with an index of nilpotency $\mu \leqq m$. The formula that comprises all the solutions of a given order n looks as follows (T is an arbitrary non-singular matrix):

$$X = T\{H^{(p_1)}, H^{(p_2)}, \ldots, H^{(p_v)}\}\, T^{-1} \quad \begin{pmatrix} p_1, p_2, \ldots, p_v \leqq m, \\ p_1 + p_2 + \cdots + p_v = n \end{pmatrix}. \tag{36}$$

Example 2. Let the equation

$$X^2 = X \tag{37}$$

be given.

A matrix satisfying this equation is called *idempotent*. The elementary divisors of an idempotent matrix can only be λ or $\lambda - 1$. Therefore an idempotent matrix can be described as a matrix of simple structure (i.e., reducible to diagonal form) with characteristic values 0 or 1. The formula comprising all the idempotent matrices of a given order n has the form

$$X = T\{\underbrace{1, 1, \ldots, 1, 0, \ldots, 0}_{n}\}T^{-1}, \tag{38}$$

where T is an arbitrary non-singular matrix of order n.

2. Let us now consider the more general equation

$$f(X)=O, \tag{39}$$

where $f(\lambda)$ is a regular function of λ in some domain G of the complex plane. We shall require of the unknown solution $X=\| x_{ik} \|_1^n$ that its characteristic values belong to G and that their multiplicities be as follows:

$$\begin{array}{ll} \text{Zeros:} & \lambda_1, \lambda_2, \ldots, \\ \text{Multiplicities:} & a_1, a_2, \ldots. \end{array}$$

As in the preceding case, every elementary divisor of X must have the form

$$(\lambda-\lambda_i)^{p_i} \quad (p_i \leqq a_i),$$

and therefore

$$X=T\{\lambda_{j_1}E^{(p_{j_1})}+H^{(p_{j_1})}, \ldots, \lambda_{j_\nu}E^{(p_{j_\nu})}+H^{(p_{j_\nu})}\}T^{-1} \tag{40}$$
$$(j_1, j_2, \ldots, j_\nu=1, 2, \ldots;\ p_{j_1} \leqq a_{j_1}, p_{j_2} \leqq a_{j_2}, \ldots, p_{j_\nu} \leqq a_{j_\nu};$$
$$p_{j_1}+p_{j_2}+\cdots+p_{j_\nu}=n)$$

(T is an arbitrary non-singular matrix).

§ 5. Matrix Polynomial Equations

1. Let us consider the equations

$$A_0X^m+A_1X^{m-1}+\cdots+A_m=O, \tag{41}$$

$$Y^mA_0+Y^{m-1}A_1+\cdots+A_m=O, \tag{42}$$

where $A_0, A_1, \ldots, A_m$ are given square matrices of order n and X, Y are unknown square matrices of the same order. The equation (33) investigated in the preceding section is a very special—one could almost say, trivial—case of (41) and (42) and is obtained by setting $A_i=\alpha_iE$, where α_i is a number and $i=1, 2, \ldots, m$.

The following theorem establishes a connection between (41), (42), and (33).

THEOREM 4: *Every solution of the matrix equation*

$$A_0X^m + A_1X^{m-1} + \cdots + A_m = O$$

satisfies the scalar equation

$$g(X) = O, \tag{43}$$

where

$$g(\lambda) \equiv |A_0\lambda^m + A_1\lambda^{m-1} + \cdots + A_m|. \tag{44}$$

The same scalar equation is satisfied by every solution Y of the matrix equation

$$Y^mA_0 + Y^{m-1}A_1 + \cdots + A_m = O.$$

Proof. We denote by $F(\lambda)$ the matrix polynomial

$$F(\lambda) = A_0\lambda^m + A_1\lambda^{m-1} + \cdots + A_m.$$

Then the equations (41) and (42) can be written as follows (see p. 81):

$$F(X) = O, \quad \hat{F}(Y) = O.$$

By the generalized Bézout Theorem (Chapter IV, § 3), if X and Y are solutions of these equations, the matrix polynomial $F(\lambda)$ is divisible on the right by $\lambda E - X$ and on the left by $\lambda E - Y$:

$$F(\lambda) = Q(\lambda)(\lambda E - X) = (\lambda E - Y)Q_1(\lambda).$$

Hence

$$g(\lambda) = |F(\lambda)| = |Q(\lambda)|\,\Delta(\lambda) = |Q_1(\lambda)|\,\Delta_1(\lambda). \tag{45}$$

where $\Delta(\lambda) = |\lambda E - X|$ and $\Delta_1(\lambda) = |\lambda E - Y|$ are the characteristic polynomials of X and Y. By the Hamilton-Cayley Theorem (Chapter IV, § 4),

$$\Delta(X) = O, \quad \Delta(Y) = O.$$

Therefore (45) implies that

$$g(X) = g(Y) = O,$$

and the theorem is proved.

Note that the Hamilton-Cayley Theorem is a special case of this theorem. For every square matrix A, when substituted for λ, satisfies the equation

$$\lambda E - A = O.$$

Therefore, by the theorem just proved,

$$\Delta(A) = O,$$

where $\Delta(\lambda) = |\lambda E - A|$.

2. Theorem 4 can be generalized as follows:

THEOREM 5:[5] *If $X_0, X_1, \ldots, X_m$ are pairwise permutable square matrices of order n that satisfy the matrix equation*

$$A_0X_0 + A_1X_1 + \cdots + A_mX_m = O \tag{46}$$

($A_0, A_1, \ldots, A_m$ are given square matrices of order n), then the same matrices $X_0, X_1, \ldots, X_m$ satisfy the scalar equation

$$g(X_0, X_1, \ldots, X_m) = O, \tag{47}$$

where

$$g(\xi_0, \xi_1, \ldots, \xi_m) = |A_0\xi_0 + A_1\xi_1 + \cdots + A_m\xi_m|. \tag{48}$$

Proof. We set[6]

$$F(\xi_0, \xi_1, \ldots, \xi_m) = \| f_{ik}(\xi_0, \xi_1, \ldots, \xi_m) \|_1^n = A_0\xi_0 + A_1\xi_1 + \cdots + A_m\xi_m.$$

$\xi_0, \xi_1, \ldots, \xi_m$ are scalar variables.

We denote by $\widehat{F}(\xi_0, \xi_1, \ldots, \xi_m) = \| \widehat{f_{ik}}(\xi_0, \xi_1, \ldots, \xi_m) \|_1^n$ the adjoint matrix of F ($\widehat{f_{ik}}$ is the algebraic complement of f_{ki} in the determinant $|F(\xi_0, \xi_1, \ldots, \xi_m)| = |f_{ik}|_1^n$ $(i, k = 1, 2, \ldots, n)$). Then every element $\widehat{f_{ik}}$ $(i, k = 1, 2, \ldots, n)$ of $\widehat{F}$ is a homogeneous polynomial in $\xi_0, \xi_1, \ldots, \xi_m$ of degree $n-1$, so that $\widehat{F}$ can be represented in the form

$$\widehat{F} = \sum_{j_0 + j_1 + \cdots + j_m = n-1} F_{j_0 j_1 \ldots j_m} \xi_0^{j_0} \xi_1^{j_1} \cdots \xi_m^{j_m},$$

where $F_{j_0 j_1 \ldots j_m}$ are certain constant matrices of order n.

From the definition of $\widehat{F}$ there follows the identity

$$\widehat{F}F = g(\xi_0, \xi_1, \ldots, \xi_m) E.$$

We write this in the following form:

$$\sum_{j_0 + j_1 + \cdots + j_{m-1} = n-1} F_{j_0 j_1 \ldots j_m} (A_0\xi_0 + A_1\xi_1 + \cdots + A_m\xi_m) \xi_0^{j_0} \xi_1^{j_1} \cdots \xi_m^{j_m} = g(\xi_0, \xi_1, \ldots, \xi_m) E. \tag{49}$$

The transition from the left-hand side of (49) to the right-hand side is accomplished by removing the parentheses and collecting similar terms. In this process *we have to permute the variables $\xi_0, \xi_1, \ldots, \xi_m$ among each other, but we do not have to permute the variables $\xi_0, \xi_1, \ldots, \xi_m$ with the matrix coefficients A_i and $F_{j_0 j_1 \ldots j_m}$*. Therefore the equation (49) is not violated when we substitute for the variables $\xi_0, \xi_1, \ldots, \xi_m$ the pairwise permutable matrices $X_0, X_1, \ldots, X_m$:

---

[5] See [318].

[6] The $f_{ik}(\xi_0, \xi_1, \ldots, \xi_m)$ are linear forms in $\xi_0, \xi_1, \ldots, \xi_m$ $(i, k = 1, 2, \ldots, n)$.

$$\sum_{j_0+j_1+\cdots+j_m=n-1} F_{j_0 j_1 \cdots j_m}(A_0X_0 + A_1X_1 + \cdots + A_mX_m)\, X_0^{j_0}X_1^{j_1}\cdots X_m^{j_m}$$
$$= g(X_0, X_1, \ldots, X_m). \qquad (50)$$

But, by assumption,

$$A_0X_0 + A_1X_1 + \cdots + A_mX_m = O.$$

Therefore we find from (50):

$$g(X_0, X_1, \ldots, X_m) = O,$$

and this is what we had to prove.

Note 1. Theorem 5 remains valid if (46) is replaced by

$$X_0A_0 + X_1A_1 + \cdots + X_mA_m = O. \qquad (51)$$

For we can apply Theorem 5 to the equation

$$A_0'X_0 + A_1'X_1 + \cdots + A_m'X_m = O$$

and then go over term by term to the transposed matrices.

Note 2. Theorem 4 is obtained as a special case of Theorem 5, when we take for $X_0, X_1, \ldots, X_m$

$$X^m, X^{m-1}, \ldots, X, E.$$

3. We have shown that every solution of (41) satisfies the scalar equation (of degree $\leqq mn$)

$$g(\lambda) = 0.$$

But the set of matrix solutions of this equation with a given order n splits into a finite number of classes of similar matrices (see § 4). Therefore all the solutions of (41) have to be looked for among the matrices of the form

$$T_iD_iT_i^{-1} \qquad (52)$$

(here D_i are well-defined matrices; if we wish, we may assume that the D_i have Jordan normal form. T_i are arbitrary non-singular matrices of order n; $i = 1, 2, \ldots, n$). In (41) we substitute for X the matrix (52) and choose T_i such that the equation (41) is satisfied. For each T_i we obtain a linear equation

$$A_0T_iD_i^m + A_1T_iD_i^{m-1} + \cdots + A_mT_i = O \qquad (i = 1, 2, \ldots, n). \qquad (53)$$

A natural method of finding solutions T_i of (53) is to replace the matrix equation by a system of linear homogeneous scalar equations in the elements of the required matrix T_i. Each non-singular solution T_i of (53), when substituted in (52), yields a solution of the given equation (41). Similar arguments may be applied to the equation (42).

In the following two sections we shall consider special cases of (41) connected with the extraction of m-th roots of a matrix.

§ 6. The Extraction of m-th Roots of a Non-Singular Matrix

1. In this section and the following, we deal with the equation

$$X^m = A, \tag{54}$$

where A is a given matrix and X an unknown matrix (both of order n) and m is a given positive integer.

In this section we consider the case $|A| \neq 0$ (A is non-singular). All the characteristic values of A are different from zero in this case (since $|A|$ is the product of these characteristic values).

We denote by

$$(\lambda - \lambda_1)^{p_1}, (\lambda - \lambda_2)^{p_2}, \ldots, (\lambda - \lambda_u)^{p_u} \tag{55}$$

the elementary divisors of A and reduce A to Jordan normal form :[7]

$$A = U\tilde{A}U^{-1} = U\{\lambda_1 E_1 + H_1, \ldots, \lambda_u E_u + H_u\} U^{-1}. \tag{56}$$

Since the characteristic values of the unknown matrix X, when raised to the m-th power, give the characteristic values of A, all the characteristic values of X are also different from zero. Therefore the derivative of $f(\lambda) = \lambda^m$ does not vanish on these characteristic values. But then (see Chapter VI, p. 158) the elementary divisors of X do not 'decompose' when X is raised to the m-th power. From what we have said, it follows that the elementary divisors of X are:

$$(\lambda - \xi_1)^{p_1}, (\lambda - \xi_2)^{p_2}, \ldots, (\lambda - \xi_u)^{p_u}. \tag{57}$$

where $\xi_j^m = \lambda_j$, i.e., ξ_j is one of the m-th roots of λ_j ($\xi_j = \sqrt[m]{\lambda_j}$; $j = 1, 2, \ldots, u$).

We now determine $\sqrt[m]{\lambda_j E_j + H_j}$ in the following way. In the λ-plane we take a circle, with center λ_j, not containing the origin. In this circle we have m distinct branches of the function $\sqrt[m]{\lambda}$. These branches can be distinguished from one another by the value they assume at the center λ_j of the circle. We denote by $\sqrt[m]{\lambda}$ that branch whose value at λ_j coincides with the characteristic value ξ_j of the unknown matrix X, and starting from this branch we define the matrix function $\sqrt[m]{\lambda_j E_j + H_j}$ by means of the series

[7] Here $E_j = E^{(p_j)}$ and $H_j = H^{(p_j)}$ ($j = 1, 2, \ldots, u$).

$$\sqrt[m]{\lambda_j E_j + H_j} = \lambda_j^{\frac{1}{m}} E_j + \frac{1}{m}\lambda_j^{\frac{1}{m}-1} H_j + \frac{1}{2!}\frac{1}{m}\left(\frac{1}{m}-1\right)\lambda_j^{\frac{1}{m}-2} H_j^2 + \cdots, \tag{58}$$

which breaks off.

Since the derivative of the function $\sqrt[m]{\lambda}$ at λ_j is not zero, the matrix (58) has only one elementary divisor $(\lambda - \xi_j)^{p_j}$, where $\xi_j = \sqrt[m]{\lambda}$ (here $j = 1, 2, 3, \ldots, u$). Hence it follows that the quasi-diagonal matrix

$$\{\sqrt[m]{\lambda_1 E_1 + H_1}, \sqrt[m]{\lambda_2 E_2 + H_2}, \ldots, \sqrt[m]{\lambda_u E_u + H_u}\}$$

has the elementary divisors (57), i.e., the same elementary divisors as the unknown matrix X. Therefore there exists a non-singular matrix T ($|T| \neq 0$) such that

$$X = T\,\{\sqrt[m]{\lambda_1 E_1 + H_1}, \sqrt[m]{\lambda_2 E_2 + H_2}, \ldots, \sqrt[m]{\lambda_u E_u + H_u}\}\,T^{-1}. \tag{59}$$

In order to determine T, we note that if on both sides of the identity

$$(\sqrt[m]{\lambda})^m = \lambda$$

we substitute the matrix $\lambda_j E_j + H_j$ ($j = 1, 2, \ldots, u$) in place of λ, we obtain:

$$(\sqrt[m]{\lambda_j E_j + H_j})^m = \lambda_j E_j + H_j \quad (j = 1, 2, \ldots, u).$$

Now from (54) and (59) it follows that

$$A = T\,\{\lambda_1 E_1 + H_1, \lambda_2 E_2 + H_2, \ldots, \lambda_u E_u + H_u\}\,T^{-1}. \tag{60}$$

Comparing (56) and (60) we find:

$$T = U X_{\tilde{A}}, \tag{61}$$

where $X_{\tilde{A}}$ is an arbitrary non-singular matrix permutable with $\tilde{A}$ (the structure of $X_{\tilde{A}}$ is described in detail in § 2).

When we substitute in (59) for T the expression $UX_{\tilde{A}}$ we obtain a formula that comprises all the solutions of the equation (54):

$$X = U X_{\tilde{A}}\,\{\sqrt[m]{\lambda_1 E_1 + H_1}, \sqrt[m]{\lambda_2 E_2 + H_2}, \ldots, \sqrt[m]{\lambda_u E_u + H_u}\}\,X_{\tilde{A}}^{-1} U^{-1}. \tag{62}$$

The multivalence of the right-hand side of this formula has a discrete as well as a continuous character: the discrete (in this case finite) character arises from the choice of the distinct branches of the function $\sqrt[m]{\lambda}$ in the various blocks of the quasi-diagonal matrix (for $\lambda_i = \lambda_k$ the branches of $\sqrt[m]{\lambda}$ in the j-th and k-th diagonal blocks may even be distinct); the continuous character arises from the arbitrary parameters contained in $X_{\tilde{A}}$.

All solutions of (54) will be called *m-th roots of* A and will be denoted by the many-valued symbol $\sqrt[m]{A}$. We point out that $\sqrt[m]{A}$ is, in general, not a function of the matrix A (i.e., is not representable in the form of a polynomial in A).

Note. If all the elementary divisors of A are co-prime in pairs, i.e., if the numbers $\lambda_1, \lambda_2, \ldots, \lambda_u$ are all distinct, then the matrix $X_{\tilde{A}}$ has quasi-diagonal form

$$X_{\tilde{A}} = \{X_1, X_2, \ldots, X_u\},$$

where X_j is permutable with $\lambda_j E_j + H_j$ and therefore permutable with every function of $\lambda_j E_j + H_j$ and, in particular, with $\sqrt[m]{\lambda_j E_j + H_j}$ $(j = 1, 2, \ldots, u)$. Therefore in this case (62) assumes the form

$$X = U\left\{\sqrt[m]{\lambda_1 E_1 + H_1}, \sqrt[m]{\lambda_2 E_2 + H_2}, \ldots, \sqrt[m]{\lambda_u E_u + H_u}\right\} U^{-1}.$$

Thus, if the elementary divisors of A are co-prime in pairs, then in the formula for $X = \sqrt[m]{A}$ only a discrete multivalence occurs. In this case every value of $\sqrt[m]{A}$ can be represented as a polynomial in A.

2. *Example.* Suppose it is required to find all square roots of

$$A = \begin{Vmatrix} 1 & 1 & 0 \\ 0 & 1 & 0 \\ 0 & 0 & 1 \end{Vmatrix},$$

i.e., all solutions of the equation

$$X^2 = A.$$

In this case A has already the Jordan normal form. Therefore in (62) we can set $A = \tilde{A}$, $U = E$. The matrix $X_{\tilde{A}}$ in this case looks as follows:

$$X_{\tilde{A}} = \begin{Vmatrix} a & b & c \\ 0 & a & 0 \\ 0 & d & e \end{Vmatrix},$$

where a, b, c, d, and e are arbitrary parameters.

The formula (62), which gives all the required solutions X, now assumes the following form:

$$X = \begin{Vmatrix} a & b & c \\ 0 & a & 0 \\ 0 & d & e \end{Vmatrix} \begin{Vmatrix} \varepsilon & \frac{\varepsilon}{2} & 0 \\ 0 & \varepsilon & 0 \\ 0 & 0 & \eta \end{Vmatrix} \begin{Vmatrix} a & b & c \\ 0 & a & 0 \\ 0 & d & e \end{Vmatrix}^{-1} \quad (\varepsilon^2 = \eta^2 = 1). \tag{63}$$

Without changing X we may multiply $X_{\tilde{A}}$ in (62) by a scalar so that $|X_{\tilde{A}}| = 1$. Then this leads to the equation $a^2 e = 1$; and hence $e = a^{-2}$.

Let us compute the elements of $X_{\tilde{A}}^{-1}$. For this purpose we write down the linear transformation with the matrix coefficients of $X_{\tilde{A}}$:

$$\begin{aligned} y_1 &= ax_1 + bx_2 + cx_3, \\ y_2 &= ax_2, \\ y_3 &= dx_2 + a^{-2}x_3. \end{aligned}$$

We solve this system of equations with respect to x_1, x_2, x_3. Then we obtain the transformation with the inverse matrix $X_{\tilde{A}}^{-1}$:

$$\begin{aligned} x_1 &= a^{-1}y_1 - (a^{-2}b - cd)\, y_2 - acy_3, \\ x_2 &= a^{-1}y_2, \\ x_3 &= -a\,dy_2 + a^2 y_3. \end{aligned}$$

Hence we find:

$$X_{\tilde{A}}^{-1} = \left\| \begin{matrix} a & b & c \\ 0 & a & 0 \\ 0 & d & a^{-2} \end{matrix} \right\|^{-1} = \left\| \begin{matrix} a^{-1} & cd - a^{-2}b & -ac \\ 0 & a^{-1} & 0 \\ 0 & -ad & a^2 \end{matrix} \right\|.$$

The formula (63) yields:

$$X = \left\| \begin{matrix} \varepsilon & (\varepsilon - \eta)\, a\, cd + \frac{\varepsilon}{2} & a^2 c\, (\eta - \varepsilon) \\ 0 & \varepsilon & 0 \\ 0 & (\varepsilon - \eta)\, da^{-1} & \eta \end{matrix} \right\|$$

$$= \left\| \begin{matrix} \varepsilon & (\varepsilon - \eta)\, vw + \frac{\varepsilon}{2} & (\eta - \varepsilon)\, v \\ 0 & \varepsilon & 0 \\ 0 & (\varepsilon - \eta)\, w & \eta \end{matrix} \right\| \quad (v = a^2 c,\ w = a^{-1} d). \tag{64}$$

The solution X depends on two arbitrary parameters u and w and two arbitrary signs ε and η.

§ 7. The Extraction of *m*-th Roots of a Singular Matrix

1. We pass on to the discussion of the case where $|A| = 0$ (A is a singular matrix).

As in the first case, we reduce A to the Jordan normal form:

$$A = U\{\lambda_1 E^{(p_1)} + H^{(p_1)}, \ldots, \lambda_u E^{(p_u)} + H^{(p_u)};\ H^{(q_1)}, H^{(q_2)}, \ldots, H^{(q_t)}\}\, U^{-1}; \tag{65}$$

here we have denoted by $(\lambda - \lambda_1)^{p_1}, \ldots, (\lambda - \lambda_u)^{p_u}$ the elementary divisors of A that correspond to non-zero characteristic values, and by $\lambda^{q_1}, \lambda^{q_2}, \ldots, \lambda^{q_t}$ the elementary divisors with characteristic value zero.

Then

$$A = U\{A_1, A_2\}\, U^{-1}, \tag{66}$$

where

$$A_1 = \{\lambda_1 E^{(p_1)} + H^{(p_1)}, \ldots, \lambda_u E^{(p_u)} + H^{(p_u)}\},\ A_2 = \{H^{(q_1)}, H^{(q_2)}, \ldots, H^{(q_t)}\}. \tag{67}$$

Note that A_1 is a non-singular matrix ($|A_1| \neq 0$) and A_2 a nilpotent matrix with index of nilpotency $\mu = \max(q_1, q_2, \ldots, q_t)$ ($A_2^{\mu} = O$).

The original equation (54) implies that A commutes with the unknown matrix X and therefore the similar matrices

$$U^{-1}AU = \{A_1, A_2\} \quad \text{and} \quad U^{-1}XU \tag{68}$$

also commute.

As we have shown in § 2 (Theorem 3), from the permutability of the matrices (68) and the fact that A_1 and A_2 do not have characteristic values in common, it follows that the second matrix in (68) has a corresponding quasi-diagonal form

$$U^{-1}XU = \{X_1, X_2\}. \tag{69}$$

When we replace the matrices A and X in (54) by the similar matrices

$$\{A_1, A_2\} \quad \text{and} \quad \{X_1, X_2\},$$

we replace (54) by two equations:

$$X_1^m = A_1, \tag{70}$$

$$X_2^m = A_2. \tag{71}$$

Since $|A_1| \neq 0$, the results of the preceding section are applicable to (70). Therefore we find X_1 by the formula (62):

$$X_1 = X_{A_1}\{\sqrt[m]{\lambda_1 E^{(p_1)} + H^{(p_1)}}, \ldots, \sqrt[m]{\lambda_u E^{(p_u)} + H^{(p_u)}}\}\, X_{A_1}^{-1}. \tag{72}$$

Thus it remains to consider the equation (71), i.e., to find all m-th roots of the nilpotent matrix A_2, which already has the Jordan normal form

$$A_2 = \{H^{(q_1)}, H^{(q_2)}, \ldots, H^{(q_t)}\}; \tag{73}$$

$\mu = \max(q_1, q_2, \ldots, q_t)$ is the index of nilpotency of A_2.

From $A_2^{\mu} = O$ and (71) we find

$$X_2^{m\mu} = O.$$

The last equation shows that the required matrix X_2 is also nilpotent with an index of nilpotency ν, where $m(\mu - 1) < \nu \leq m\mu$. We reduce X_2 to the Jordan form:

$$X_2 = T\{H^{(v_1)}, H^{(v_2)}, \ldots, H^{(v_s)}\}\, T^{-1} \tag{74}$$
$$(v_1, v_2, \ldots, v_s \leqq \nu).$$

Now we raise both sides of (74) to the m-th power. We obtain:

$$A_2 = X_2^m = T\{[H^{(v_1)}]^m, [H^{(v_2)}]^m, \ldots, [H^{(v_s)}]^m\}\, T^{-1}. \tag{75}$$

2. Let us now elarify the question of what elementary divisors the matrix $[H^{(v)}]^m$ has.[8] We denote by $\boldsymbol{H}$ the linear operator given by $H^{(v)}$ in a v-dimensional vector space with the basis $\boldsymbol{e}_1, \boldsymbol{e}_2, \ldots, \boldsymbol{e}_v$. Then from the form of the matrix $H^{(v)}$ (in $H^{(v)}$ all the elements of the first superdiagonal are equal to 1 and all the remaining elements are 0) it follows that

$$\boldsymbol{He}_1 = \boldsymbol{o},\ \boldsymbol{He}_2 = \boldsymbol{e}_1,\ \ldots,\ \boldsymbol{He}_v = \boldsymbol{e}_{v-1}. \tag{76}$$

These equations show that the vectors $\boldsymbol{e}_1, \boldsymbol{e}_2, \ldots, \boldsymbol{e}_v$ form a Jordan chain for $\boldsymbol{H}$, corresponding to the elementary divisor λ^v.

We write (76) as follows:

$$\boldsymbol{He}_j = \boldsymbol{e}_{j-1} \qquad (j = 1, 2, \ldots, v;\ \boldsymbol{e}_0 = \boldsymbol{o}).$$

Obviously,

$$\boldsymbol{H}^m \boldsymbol{e}_j = \boldsymbol{e}_{j-m} \qquad (j = 1, 2, \ldots, v;\ \boldsymbol{e}_0 = \boldsymbol{e}_{-1} = \cdots \boldsymbol{e}_{-m+1} = \boldsymbol{o}). \tag{77}$$

We express v in the form

$$v = km + r \qquad (r < m),$$

where k and r are non-negative integers. We arrange the basis vectors $\boldsymbol{e}_1, \boldsymbol{e}_2, \ldots, \boldsymbol{e}_v$ in the following way:

$$\begin{array}{llll} \boldsymbol{e}_1, & \boldsymbol{e}_2, & \ldots, & \boldsymbol{e}_m, \\ \boldsymbol{e}_{m+1}, & \boldsymbol{e}_{m+2}, & \ldots, & \boldsymbol{e}_{2m}, \\ \cdots & \cdots & \cdots & \cdots \\ \boldsymbol{e}_{(k-1)m+1}, & \boldsymbol{e}_{(k-1)m+2}, & \ldots, & \boldsymbol{e}_{km}, \\ \boldsymbol{e}_{km+1}, & \ldots, & \boldsymbol{e}_{km+r}. & \end{array} \tag{78}$$

This table has m columns: the first r columns contain $k+1$ vectors each, the remaining ones k vectors. The equation (77) shows that the vectors of each column form a Jordan chain with respect to the operator $\boldsymbol{H}^m$. If instead

[8] This question is answered by Theorem 9 of Chapter VI (p. 158). Here we are compelled to use another method of investigating the problem, because we have to find not only the elementary divisor of the matrix $[H^{(v)}]^m$, but also a matrix $p_{v,m}$ transforming $[H^{(v)}]^m$ into Jordan form.

of numbering the vectors (78) by rows we number them by columns, we obtain a new basis in which the matrix of the operator $\boldsymbol{H}^m$ has the following Jordan normal form:[9]

$$\{\underbrace{H^{(k+1)}, \ldots, H^{(k+1)}}_{r}, \underbrace{H^{(k)}, \ldots, H^{(k)}}_{m-r}\};$$

and therefore

$$[H^{(v)}]^m = P_{v,m} \{\underbrace{H^{(k+1)}, \ldots, H^{(k+1)}}_{r}, \underbrace{H^{(k)}, \ldots, H^{(k)}}_{m-r}\} P^{-1}_{v,m}, \tag{79}$$

where the matrix $P_{v,m}$ (describing the transition from the one basis to the other) has the following form (see Chapter III, § 4):

$$P_{v,m} = \left\| \begin{matrix} \overbrace{1 \quad 0 \ldots 0}^{m} & 0 & \ldots \\ 0 \quad 0 \ldots 0 & 1 & \ldots \\ \cdot \quad \cdot \quad \cdot \quad \cdot & \cdot & \cdot \\ \cdot \quad \cdot \quad \cdot \quad \cdot & \cdot & \cdot \\ 0 \quad 0 \ldots 0 & & \\ 0 \quad 1 \ldots 0 & & \\ \cdot \quad \cdot \quad \cdot \quad \cdot & \cdot & \cdot \end{matrix} \right\| \left. \vphantom{\begin{matrix} 1 \\ 0 \\ \cdot \\ \cdot \\ 0 \end{matrix}} \right\} k+1\,. \tag{80}$$

The matrix $H^{(v)}$ has the single elementary divisor λ^v. When $H^{(v)}$ is raised to the m-th power, this elementary divisor 'falls apart.' As (79) shows, $[H^{(v)}]^m$ has the elementary divisors:

$$\underbrace{\lambda^{k+1}, \ldots, \lambda^{k+1}}_{r}, \underbrace{\lambda^{k}, \ldots, \lambda^{k}}_{m-r}.$$

Turning now to (75), we set:

$$v_i = k_i m + r_i \quad (0 \leq r_i < m,\ k_i \geqq 0; \quad i = 1, 2, \ldots, s). \tag{81}$$

Then, by (79), equation (75) can be written as follows:

$$A_2 = X_2^m = TP\{\underbrace{H^{(k_1+1)}, \ldots, H^{(k_1+1)}}_{r_1}, \underbrace{H^{(k_1)}, \ldots, H^{(k_1)}}_{m-r_1}, \underbrace{H^{(k_2+1)}, \ldots, H^{(k_2+1)}}_{r_2}, H^{(k_2)}, \ldots\} P^{-1} T^{-1}. \tag{82}$$

where
$$P = \{P_{v_1,m}, P_{v_2,m}, \ldots, P_{v_s,m}\}$$

[9] In the case $k = 0$, the blocks $\underbrace{H^{(k)}, \ldots, H^{(k)}}_{m-r}$ are absent, and the matrix has the form $\{\underbrace{H^{(1)}, \ldots, H^{(1)}}_{r}\}$.

Comparing (82) with (73), we see that the blocks

$$H^{(k_1+1)}, \ldots, H^{(k_1+1)}, H^{(k_1)}, \ldots, H^{(k_1)}, H^{(k_2+1)}, \ldots, H^{(k_2+1)}, \ldots \tag{83}$$

must coincide, apart from the order, with the blocks

$$H^{(q_1)}, H^{(q_2)}, \ldots, H^{(q_t)}. \tag{84}$$

3. Let us call a system of elementary divisors $\lambda^{v_1}, \lambda^{v_2}, \ldots, \lambda^{v_s}$ *admissible* for X_2 if after raising of the matrix to the m-th power these elementary divisors split and generate the given system of elementary divisors of A_2: $\lambda^{q_1}, \lambda^{q_2}, \lambda^{q_3}, \ldots, \lambda^{q_t}$. The number of admissible systems of elementary divisors is always finite, because

$$\max(v_1, v_2, \ldots, v_s) \leqq m\mu, \quad v_1 + v_2 + \cdots + v_s = n_2 \tag{85}$$

(n_2 is the order of A_2).

In every concrete case the admissible systems of elementary divisors for X_2 can easily be determined by a finite number of trials.

Let us show that for each admissible system of elementary divisors $\lambda^{v_1}, \lambda^{v_2}, \ldots, \lambda^{v_s}$ form a corresponding solution of (71) and let us determine all these solutions. In this case there exists a transforming matrix Q such that

$$\{H^{(k_1+1)}, \ldots, H^{(k_1+1)}, H^{(k_1)}, \ldots, H^{(k_1)}, H^{(k_2+1)}, \ldots\} = Q^{-1}A_2Q. \tag{86}$$

The matrix Q describes the permutation of the blocks in the quasi-diagonal matrix that brings about the proper renumbering of the basis vectors. Therefore Q can be regarded as known. Using (86), we obtain from (82):

$$A_2 = TPQ^{-1}A_2QP^{-1}T^{-1}.$$

Hence

$$TPQ^{-1} = X_{A_2},$$

or

$$T = X_{A_2}QP^{-1}, \tag{87}$$

where X_{A_2} is an arbitrary matrix that commutes with A_2.

Substituting (87) for T in (74), we have

$$X_2 = X_{A_2}QP^{-1}\{H^{(v_1)}, H^{(v_2)}, \ldots, H^{(v_s)}\}PQ^{-1}X_{A_2}^{-1}. \tag{88}$$

From (69), (72), and (88) we obtain a general formula which comprises all the solutions:

$$X = U\{X_{A_1}, X_{A_2}QP^{-1}\}\{\sqrt[m]{\lambda_1 E^{(p_1)} + H^{(p_1)}}, \ldots, \sqrt[m]{\lambda_u E^{(p_u)} + H^{(p_u)}}, \\ H^{(v_1)}, \ldots, H^{(v_s)}\} \cdot \{X_{A_1}^{-1}, PQ^{-1}X_{A_2}^{-1}\}U^{-1}. \tag{89}$$

We draw the reader's attention to the fact that the m-th root of a singular matrix does not always exist. Its existence is bound up with the existence of a system of admissible elementary divisors for X_2.

It is easy to see, for example, that the equation

$$X^m = H^{(p)}$$

has no solution for $m > 1$, $p > 1$.

Example. Suppose it is required to extract the square root of

$$A = \begin{Vmatrix} 0 & 1 & 0 \\ 0 & 0 & 0 \\ 0 & 0 & 0 \end{Vmatrix},$$

i.e., to find all the solutions of the equation

$$X^2 = A.$$

In this case, $A = A_2$, $X = X_2$, $m = 2$, $t = 2$, $q_1 = 2$, and $q_2 = 1$. The matrix X can only have the one elementary divisor λ^3. Therefore $s = 1$, $v_1 = 3$, $k_1 = 1$, $r_1 = 1$ and (see (80))

$$P = P_{3,2} = \begin{Vmatrix} 1 & 0 & 0 \\ 0 & 0 & 1 \\ 0 & 1 & 0 \end{Vmatrix} = P^{-1}, Q = E.$$

Moreover, as in the example on page 233, in (88) we may set:

$$X_{A_2} = \begin{Vmatrix} a & b & c \\ 0 & a & 0 \\ 0 & d & a^{-2} \end{Vmatrix}, \quad X_{A_2}^{-1} = \begin{Vmatrix} a^{-1} & cd - a^{-2}b & -ac \\ 0 & a^{-1} & 0 \\ 0 & -ad & a^2 \end{Vmatrix}.$$

From this formula we obtain

$$X = X_2 = X_{A_2} P^{-1} H^{(3)} P X_{A_2}^{-1} = \begin{Vmatrix} 0 & \alpha & \beta \\ 0 & 0 & 0 \\ 0 & \beta^{-1} & 0 \end{Vmatrix},$$

where $\alpha = ca^{-1} - a^2 d$ and $\beta = a^3$ are arbitrary parameters.

§ 8. The Logarithm of a Matrix

1. We consider the matrix equation

$$e^X = A. \tag{90}$$

All the solutions of this equation are called (natural) *logarithms* of A and are denoted by $\ln A$.

The characteristic values λ_j of A are connected with the characteristic values ξ_j of X by the formula $\lambda_j = e^{\xi_j}$; therefore, if the equation (90) has a solution, then all the characteristic values of A are different from zero, and A is non-singular ($|A| \neq 0$). Thus, the condition $|A| \neq 0$ is necessary for the existence of solutions of the equation (90). Below, we shall see that this condition is also sufficient.

Suppose, then, that $|A| \neq 0$. We write down the elementary divisors of A:

$$(\lambda - \lambda_1)^{p_1}, (\lambda - \lambda_2)^{p_2}, \ldots, (\lambda - \lambda_u)^{p_u}$$

$$(\lambda_1 \lambda_2 \cdots \lambda^u \neq 0, \quad p_1 + p_2 + \cdots + p_u = n). \tag{91}$$

Corresponding to these elementary divisors we reduce A to the Jordan normal form:

$$\begin{aligned} A &= U\tilde{A}U^{-1} \\ &= U\{\lambda_1 E^{(p_1)} + H^{(p_1)}, \lambda_2 E^{(p_2)} + H^{(p_2)}, \ldots, \lambda_u E^{(p_u)} + H^{(p_u)}\} U^{-1}. \end{aligned} \tag{92}$$

Since the derivative of the function e^{ξ} is different from zero for all values of ξ, we know (see Chapter VI, p. 158) that in the transition from X to $A = e^X$ the elementary divisors do not split, so that X has the elementary divisors

$$(\lambda - \xi_1)^{p_1}, (\lambda - \xi_2)^{p_2}, \ldots, (\lambda - \xi_u)^{p_u}, \tag{93}$$

where $e^{\xi_j} = \lambda_j \quad (j = 1, 2, \ldots, u)$, i.e., ξ_j is one of the values of $\ln \lambda_j$ ($j = 1, 2, 3, \ldots, u$).

In the plane of the complex variable λ we draw a circle with center at λ_j and with radius less than $|\lambda_j|$ and we denote by $f_j(\lambda) = \ln \lambda$ that branch of the function $\ln \lambda$ in this circle which at λ_j assumes the value equal to the characteristic value ξ_j of X ($j = 1, 2, \ldots, u$). After this, we set:

$$\ln(\lambda_j E^{(p_j)} + H^{(p_j)}) = f_j(\lambda_j E^{(p_j)} + H^{(p_j)}) = \ln \lambda_j E^{(p_j)} + \lambda_j^{-1} H^{(p_j)} + \cdots. \tag{94}$$

Since the derivative of $\ln \lambda$ vanishes nowhere (in the finite part of the λ-plane), the matrix (94) has only the one elementary divisor $(\lambda - \xi_j)^{p_j}$. Therefore the quasi-diagonal matrix

$$\{\ln(\lambda_1 E^{(p_1)} + H^{(p_1)}), \ln(\lambda_2 E^{(p_2)} + H^{(p_2)}), \ldots, \ln(\lambda_u E^{(p_u)} + H^{(p_u)})\} \tag{95}$$

has the same elementary divisors as the unknown matrix X. Therefore there exists a matrix T ($|T| \neq 0$) such that

$$X = T\{\ln(\lambda_1 E^{(p_1)} + H^{(p_1)}), \ldots, \ln(\lambda_u E^{(p_u)} + H^{(p_u)})\} T^{-1}. \tag{96}$$

In order to determine T, we note that

$$A = e^X = T\{\lambda_1 E^{(p_1)} + H^{(p_1)}, \ldots, \lambda_u E^{(p_u)} + H^{(p_u)}\} T^{-1}. \tag{97}$$

Comparing (97) and (92), we find:

$$T = UX_{\tilde{A}}, \tag{98}$$

where $X_{\tilde{A}}$ is an arbitrary matrix that commutes with $\tilde{A}$. Substituting the expression for T from (98) into (96), we obtain a general formula that comprises all the logarithms of the matrix:

$$X = UX_{\tilde{A}}\{\ln(\lambda_1 E^{(p_1)} + H^{(p_1)}),$$
$$\ln(\lambda_2 E^{(p_2)} + H^{(p_2)}), \ldots, \ln(\lambda_u E^{(p_u)} + H^{(p_u)})\} X_{\tilde{A}}^{-1} U^{-1}. \tag{99}$$

Note. If all the elementary divisors of A are co-prime, then on the right-hand side of (99) the factors $X_{\tilde{A}}$ and $X_{\tilde{A}}^{-1}$ can be omitted (see a similar remark on p. 233).

CHAPTER IX

LINEAR OPERATORS IN A UNITARY SPACE

§ 1. General Considerations

In Chapters III and VII we studied linear operators in an arbitrary n-dimensional vector space. All the bases of such a space are of equal standing. To a given linear operator there corresponds in each basis a certain matrix. The matrices corresponding to one and the same operator in the various bases are similar. Thus, the study of linear operators in an n-dimensional vector space enables us to bring out those properties of matrices that are inherent in an entire class of similar matrices.

At the beginning of this chapter we shall introduce a metric into an n-dimensional space by assigning in a special way to each pair of vectors a certain number, the 'scalar product' of the two vectors. By means of the scalar product we shall define the 'length' of a vector and the cosine of the 'angle' between two vectors. This metrization leads to a unitary space if the ground field F is the field of all complex numbers and to a euclidean space if F is the field of all real numbers.

In the present chapter we shall study the properties of linear operators that are connected with the metric of the space. All the bases of the space are by no means of equal standing with respect to the metric. However, this does hold true of all *orthonormal* bases. The transition from one orthonormal basis to another in a unitary space is brought about by means of a special—namely, unitary—transformation (in a euclidean space, an orthogonal transformation). Therefore all the matrices that correspond to one and the same linear operator in two distinct bases of a unitary (euclidean) space are unitarily (orthogonally) similar. Thus, by studying linear operators in an n-dimensional metrized space we study the properties of matrices that remain invariant under transition from a given matrix to a unitarily—or orthogonally—similar one. This will lead in a natural way to the investigation of properties of special classes of matrices (normal, hermitian, unitary, symmetric, skew-symmetric, orthogonal matrices).

§ 2. Metrization of a Space

1. We consider a vector space $\boldsymbol{R}$ over the field of complex numbers. To every pair of vectors $\boldsymbol{x}$ and $\boldsymbol{y}$ of $\boldsymbol{R}$ given in a definite order let a certain complex number be assigned, the so-called *scalar product*, or *inner product*, of the vectors, denoted by $(\boldsymbol{xy})$ or $(\boldsymbol{x}, \boldsymbol{y})$. Suppose further that the 'scalar multiplication' has the following properties:

For arbitrary vectors $\boldsymbol{x}, \boldsymbol{y}, \boldsymbol{z}$ of $\boldsymbol{R}$ and an arbitrary complex number α, let[1]

$$\left.\begin{array}{ll} 1. & (\boldsymbol{xy}) = \overline{(\boldsymbol{yx})}, \\ 2. & (\alpha\boldsymbol{x}, \boldsymbol{y}) = \alpha(\boldsymbol{xy}), \\ 3. & (\boldsymbol{x}+\boldsymbol{y}, \boldsymbol{z}) = (\boldsymbol{xz}) + (\boldsymbol{yz}). \end{array}\right\} \quad (1)$$

Then we shall say that a *hermitian metric* is introduced in $\boldsymbol{R}$.

Note that 1., 2., and 3. have the following consequences for arbitrary $\boldsymbol{x}, \boldsymbol{y}, \boldsymbol{z}$ in $\boldsymbol{R}$:

$$\begin{array}{ll} 2'. & (\boldsymbol{x}, \alpha\boldsymbol{y}) = \bar{\alpha}(\boldsymbol{xy}), \\ 3'. & (\boldsymbol{x}, \boldsymbol{y}+\boldsymbol{z}) = (\boldsymbol{xy}) + (\boldsymbol{xz}). \end{array}$$

From 1. we deduce that for every vector $\boldsymbol{x}$ the scalar product $(\boldsymbol{x}\boldsymbol{x})$ is a real number. This number is called the *norm* of $\boldsymbol{x}$ and is denoted by $N\boldsymbol{x}$: $N\boldsymbol{x} = (\boldsymbol{x}, \boldsymbol{x})$.

If for every vector $\boldsymbol{x}$ of $\boldsymbol{R}$

$$4. \quad N\boldsymbol{x} = (\boldsymbol{x}\boldsymbol{x}) \geqq 0, \quad (2)$$

then the hermitian metric is called *positive semi-definite*. And if, moreover,

$$5. \quad N\boldsymbol{x} = (\boldsymbol{x}\boldsymbol{x}) > 0 \text{ for } \boldsymbol{x} \neq \boldsymbol{o}, \quad (3)$$

then the hermitian metric is called *positive definite*.

DEFINITION 1: *A vector space* $\boldsymbol{R}$ *with a positive-definite hermitian metric will be called a unitary space.*[2]

In this chapter we shall consider finite-dimensional unitary spaces.[3]

By the *length* of the vector $\boldsymbol{x}$ we mean[4] $+\sqrt{N\boldsymbol{x}} = +\sqrt{(\boldsymbol{x}, \boldsymbol{x})} = |\boldsymbol{x}|$. From 2. and 5. it follows that every vector other than the null vector has a positive

[1] A number with a bar over it denotes the complex conjugate of the number.

[2] The study of n-dimensional vector spaces with an arbitrary (not positive-definite) metric is taken up in the paper [319].

[3] In §§ 2-7 of this chapter, wherever it is not expressly stated that the space is finite-dimensional, all the arguments remain valid for infinite-dimensional spaces.

[4] The symbol $+\sqrt{\ }$ denotes the non-negative (arithmetical) value of the root.

length and that the null vector has length 0. A vector x is called *normalized* (or is said to be a *unit vector*) if $|x| = 1$. To normalize an arbitrary vector $x \neq o$ it is sufficient to multiply it by any complex number λ for which $|\lambda| = \frac{1}{|x|}$.

By analogy with the ordinary three-dimensional vector spaces, two vectors x and y are called *orthogonal* (in symbols: $x \perp y$) if $(xy) = 0$. In this case it follows from 1., 3., and 3′. that

$$\mathrm{N}(x+y) = (x+y, x+y) = (xx) + (yy) = \mathrm{N}x + \mathrm{N}y,$$

i.e. (the theorem of Pythagoras!),

$$|x+y|^2 = |x|^2 + |y|^2 \quad (x \perp y).$$

Let R be a unitary space of finite dimension n. We consider an arbitrary basis $e_1, e_2, \ldots, e_n$ of R. Let us denote by x_i and y_i $(i = 1, 2, \ldots, n)$ the coordinates of the vectors x and y in this basis:

$$x = \sum_{i=1}^{n} x_i e_i, \quad y = \sum_{i=1}^{n} y_i e_i.$$

Then by 2., 3., 2′., and 3′.,

$$(xy) = \sum_{i,k=1}^{n} h_{ik} x_i \bar{y}_k, \tag{4}$$

where

$$h_{ik} = (e_i e_k) \quad (i, k = 1, 2, \ldots, n). \tag{5}$$

In particular,

$$\mathrm{N}x = (xx) = \sum_{i,k=1}^{n} h_{ik} x_i \bar{x}_k. \tag{6}$$

From 1. and (5) we deduce

$$h_{ki} = \bar{h}_{ik} \quad (i, k = 1, 2, \ldots, n). \tag{7}$$

2. A form $\sum_{i,k=1}^{n} h_{ik} x_i \bar{x}_k$, where $h_{ki} = \bar{h}_{ik}$ $(i, k = 1, 2, \ldots, n)$ is called *hermitian*.[5] Thus, the norm of a vector, i.e., the square of its length, is a hermitian form in its coordinates. Hence the name 'hermitian metric.' The form on the right-hand side of (6) is, by 4., *non-negative*:

$$\sum_{i,k=1}^{n} h_{ik} x_i \bar{x}_k \geq 0 \tag{8}$$

for all values of the variables $x_1, x_2, \ldots, x_n$. By the additional condition 5., the form is in fact *positive definite*, i.e., the equality sign in (8) only holds when all the x_i are zero $(i = 1, 2, \ldots, n)$.

[5] In accordance with this, the expression on the right-hand side of (4) is called a hermitian bilinear form (in $x_1, x_2, \ldots, x_n$ and $y_1, y_2, \ldots, y_n$).

DEFINITION 2: *A system of vectors* $\mathbf{e}_1, \mathbf{e}_2, \ldots, \mathbf{e}_m$ *is called orthonormal if*

$$(\mathbf{e}_i\mathbf{e}_k) = \delta_{ik} = \begin{cases} 0, & \text{for } i \neq k, \\ 1, & \text{for } i = k \end{cases} \quad (i, k = 1, 2, \ldots, m). \tag{9}$$

When $m = n$, where n is the dimension of the space, we obtain an *orthonormal basis* of the space.

In § 7 we shall prove that *every n-dimensional space has an orthonormal basis.*

Let x_i and y_i $(i = 1, 2, \ldots, n)$ be the coordinates of $\mathbf{x}$ and $\mathbf{y}$ in an orthonormal basis. Then by (4), (5), and (9)

$$\left.\begin{aligned} (\mathbf{xy}) &= \sum_{i=1}^{n} x_i \bar{y}_i, \\ \mathrm{N}\mathbf{x} = (\mathbf{xx}) &= \sum_{i=1}^{n} |x_i|^2. \end{aligned}\right\} \tag{10}$$

Let us take an arbitrary fixed basis in an n-dimensional space $\boldsymbol{R}$. In this basis every metrization of the space is connected with a certain positive-definite hermitian form $\sum_{i,k=1}^{n} h_{ik} x_i \bar{x}_k$; and conversely, by (4), every such form determines a certain positive-definite hermitian metric in $\boldsymbol{R}$. However, *these metrics do not all give essentially different unitary n-dimensional spaces.* For let us take two such metrics with the respective scalar products $(\mathbf{xy})$ and $(\mathbf{xy})'$. We determine orthonormal bases in $\boldsymbol{R}$ with respect to these metrices: $\mathbf{e}_i$ and $\mathbf{e}_i'$ $(i = 1, 2, \ldots, n)$. Let the vector $\mathbf{x}$ in $\boldsymbol{R}$ be mapped onto the vector $\mathbf{x}'$ in $\boldsymbol{R}$, where $\mathbf{x}'$ is the vector whose coordinates in the basis $\mathbf{e}_i'$ are the same as the coordinates of $\mathbf{x}$ in the basis $\mathbf{e}_i$ $(i = 1, 2, \ldots, n)$. $(\mathbf{x} \to \mathbf{x}'.)$ This mapping is *affine.*[6] Moreover, by (10),

$$(\mathbf{xy}) = (\mathbf{x}'\mathbf{y}')'.$$

Therefore: *To within an affine transformation of the space all positive definite hermitian metrizations of an n-dimensional vector space coincide.*

If the field F is the field of real numbers, then a metric satisfying the postulates 1., 2., 3., 4., and 5. is called *euclidean.*

DEFINITION 3: *A vector space* $\boldsymbol{R}$ *over the field of real numbers with a positive euclidean metric is called a euclidean space.*

If x_i and y_i $(i = 1, 2, \ldots, n)$ are the coordinates of the vectors $\mathbf{x}$ and $\mathbf{y}$ in some basis $\mathbf{e}_1, \mathbf{e}_2, \ldots, \mathbf{e}_n$ of an n-dimensional euclidean space, then

---

[6] I.e., the operator A that maps the vector $\mathbf{x}$ of $\boldsymbol{R}$ onto the vector $\mathbf{x}'$ of $\boldsymbol{R}'$ is linear and non-singular.

$$(\boldsymbol{xy}) = \sum_{i,k=1}^{n} s_{ik} x_i y_k, \qquad \mathrm{N}\boldsymbol{x} = |\boldsymbol{x}|^2 = \sum_{i,k=1}^{n} s_{ik} x_i x_k.$$

Here $s_{ik} = s_{ki}$ $(i, k = 1, 2, \ldots, n)$ are real numbers.[7] The expression $\sum_{i,k=1}^{n} s_{ik} x_i x_k$ is called a *quadratic form* in $x_1, x_2, \ldots, x_n$. From the fact that the metric is positive definite it follows that the quadratic form $\sum_{i,k=1}^{n} s_{ik} x_i x_k$, which gives this metric analytically, is *positive definite*, i.e., $\sum_{i,k=1}^{n} s_{ik} x_i x_k > 0$ if $\sum_{i=1}^{n} x_i^2 > 0$.

In an orthonormal basis

$$(\boldsymbol{xy}) = \sum_{i=1}^{n} x_i y_i, \qquad \mathrm{N}\boldsymbol{x} = |\boldsymbol{x}|^2 = \sum_{i=1}^{n} x_i^2. \tag{11}$$

For $n = 3$ we obtain the well-known formulas for the scalar product of two vectors and for the square of the length of a vector in a three-dimensional euclidean space.

§ 3. Gram's Criterion for Linear Dependence of Vectors

1. Suppose that the vectors $\boldsymbol{x}_1, \boldsymbol{x}_2, \ldots, \boldsymbol{x}_m$ of a unitary or of a euclidean space $\boldsymbol{R}$ are linearly dependent, i.e., that there exist numbers[8] $c_1, c_2, \ldots, c_m$ not all zero, such that

$$c_1 \boldsymbol{x}_1 + c_2 \boldsymbol{x}_2 + \cdots + c_m \boldsymbol{x}_m = \boldsymbol{o}. \tag{12}$$

When we perform the scalar multiplication by $\boldsymbol{x}_1, \boldsymbol{x}_2, \ldots, \boldsymbol{x}_m$ in succession on both sides of this equation, we obtain

$$\left.\begin{array}{l} (\boldsymbol{x}_1\boldsymbol{x}_1)\,\bar{c}_1 + (\boldsymbol{x}_1\boldsymbol{x}_2)\,\bar{c}_2 + \cdots + (\boldsymbol{x}_1\boldsymbol{x}_m)\,\bar{c}_m = 0 \\ (\boldsymbol{x}_2\boldsymbol{x}_1)\,\bar{c}_1 + (\boldsymbol{x}_2\boldsymbol{x}_2)\,\bar{c}_2 + \cdots + (\boldsymbol{x}_2\boldsymbol{x}_m)\,\bar{c}_m = 0 \\ \cdots\cdots\cdots\cdots\cdots\cdots\cdots\cdots \\ (\boldsymbol{x}_m\boldsymbol{x}_1)\,\bar{c}_1 + (\boldsymbol{x}_m\boldsymbol{x}_2)\,\bar{c}_2 + \cdots + (\boldsymbol{x}_m\boldsymbol{x}_m)\,\bar{c}_m = 0. \end{array}\right\} \tag{13}$$

Regarding $\bar{c}_1, \bar{c}_2, \ldots, \bar{c}_m$ as a non-zero solution of the system (13) of linear homogeneous equations with the determinant

---

[7] $s_{ik} = (\boldsymbol{e}_i \boldsymbol{e}_k)$ $(i, k = 1, 2, \ldots, n)$.

[8] In the case of a euclidean space, $c_1, c_2, \ldots, c_m$ are real numbers.

$$G(\boldsymbol{x}_1, \boldsymbol{x}_2, \ldots, \boldsymbol{x}_m) = \begin{vmatrix} (\boldsymbol{x}_1\boldsymbol{x}_1) & (\boldsymbol{x}_1\boldsymbol{x}_2) & \ldots & (\boldsymbol{x}_1\boldsymbol{x}_m) \\ (\boldsymbol{x}_2\boldsymbol{x}_1) & (\boldsymbol{x}_2\boldsymbol{x}_2) & \ldots & (\boldsymbol{x}_2\boldsymbol{x}_m) \\ \cdot & \cdot & \cdot & \cdot \\ (\boldsymbol{x}_m\boldsymbol{x}_1) & (\boldsymbol{x}_m\boldsymbol{x}_2) & \ldots & (\boldsymbol{x}_m\boldsymbol{x}_m) \end{vmatrix}, \tag{14}$$

we conclude that this determinant must vanish:

$$G(\boldsymbol{x}_1, \boldsymbol{x}_2, \ldots, \boldsymbol{x}_m) = 0.$$

$G(\boldsymbol{x}_1, \boldsymbol{x}_2, \ldots, \boldsymbol{x}_m)$ is called the *Gramian* of the vectors $\boldsymbol{x}_1, \boldsymbol{x}_2, \ldots, \boldsymbol{x}_m$.

Suppose, conversely, that the Gramian (14) is zero. Then the system of equations (13) has a non-zero solution $\bar{c}_1, \bar{c}_2, \ldots, \bar{c}_m$. Equations (13) can be written as follows:

$$\left.\begin{array}{l} (\boldsymbol{x}_1,\ c_1\boldsymbol{x}_1 + c_2\boldsymbol{x}_2 + \cdots + c_m\boldsymbol{x}_m) = 0 \\ (\boldsymbol{x}_2,\ c_1\boldsymbol{x}_1 + c_2\boldsymbol{x}_2 + \cdots + c_m\boldsymbol{x}_m) = 0 \\ \cdot\ \cdot\ \cdot\ \cdot\ \cdot\ \cdot\ \cdot\ \cdot\ \cdot\ \cdot\ \cdot\ \cdot\ \cdot\ \cdot\ \cdot\ \cdot \\ (\boldsymbol{x}_m,\ c_1\boldsymbol{x}_1 + c_2\boldsymbol{x}_2 + \cdots + c_m\boldsymbol{x}_m) = 0. \end{array}\right\} \tag{13'}$$

Multiplying these equations by $c_1, c_2, \ldots, c_m$ respectively, and then adding, we obtain:

$$\mathrm{N}(c_1\boldsymbol{x}_1 + c_2\boldsymbol{x}_2 + \cdots + c_m\boldsymbol{x}_m) = 0;$$

and since the metric is positive definite

$$c_1\boldsymbol{x}_1 + c_2\boldsymbol{x}_2 + \cdots + c_m\boldsymbol{x}_m = \boldsymbol{o},$$

i.e., the vectors $\boldsymbol{x}_1, \boldsymbol{x}_2, \ldots, \boldsymbol{x}_m$ are linearly dependent.

Thus we have proved:

THEOREM 1: *The vectors $\boldsymbol{x}_1, \boldsymbol{x}_2, \ldots, \boldsymbol{x}_m$ are linearly independent if and only if their Gramian is not equal to zero.*

We note the following property of the Gramian:

If any principal minor of the Gramian is zero, then the Gramian is zero.

For a principal minor is the Gramian of part of the vectors. When this principal minor vanishes, it follows that these vectors are linearly dependent and then the whole system of vectors is dependent.

2. *Example.* Let $f_1(t), f_2(t), \ldots, f_n(t)$ be n complex functions of a real argument t, sectionally continuous in the closed interval $[\alpha, \beta]$. It is required to determine conditions under which they are linearly dependent. For this purpose, we introduce a positive-definite metric into the space of functions sectionally continuous in $[\alpha, \beta]$ by setting

$$(f, g) = \int_{\alpha}^{\beta} f(t)\overline{g(t)}\, dt.$$

Then Gram's criterion (Theorem 1) applied to the given function yields the required condition:

$$\begin{vmatrix} \int_{\alpha}^{\beta} f_1(t)\overline{f_1(t)}\,dt & \dots & \int_{\alpha}^{\beta} f_1(t)\overline{f_n(t)}\,dt \\ \cdot & \cdot\;\cdot\;\cdot & \cdot \\ \int_{\alpha}^{\beta} f_n(t)\overline{f_1(t)}\,dt & \dots & \int_{\alpha}^{\beta} f_n(t)\overline{f_n(t)}\,dt \end{vmatrix} = 0.$$

§ 4. Orthogonal Projection

1. Let $\boldsymbol{x}$ be an arbitrary vector in a unitary or euclidean space $\boldsymbol{R}$ and $\boldsymbol{S}$ an m-dimensional subspace with a basis $\boldsymbol{x}_1, \boldsymbol{x}_2, \ldots, \boldsymbol{x}_m$. We shall show that $\boldsymbol{x}$ can be represented (and moreover, represented uniquely) in the form

$$\boldsymbol{x} = \boldsymbol{x}_S + \boldsymbol{x}_N, \tag{15}$$

where

$$\boldsymbol{x}_S \in \boldsymbol{S} \text{ and } \boldsymbol{x}_N \perp \boldsymbol{S}$$

(the symbol $\perp$ denotes orthogonality of vectors; orthogonality to a subspace means orthogonality to every vector of the subspace); *$\boldsymbol{x}_S$ is the orthogonal projection of $\boldsymbol{x}$ onto $\boldsymbol{S}$, $\boldsymbol{x}_N$ the projecting vector.*

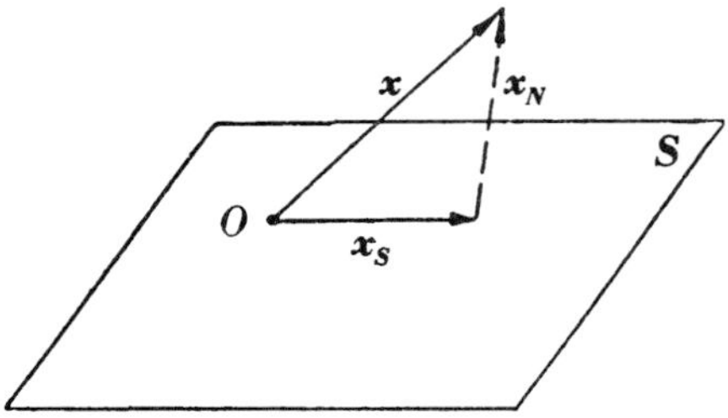

Fig. 5

Example. Let $\boldsymbol{R}$ be a three-dimensional euclidean vector space and $m = 2$. Let all vectors originate at a fixed point O. Then $\boldsymbol{S}$ is a plane passing through O; $\boldsymbol{x}_S$ is the orthogonal projection of $\boldsymbol{x}$ onto the plane $\boldsymbol{S}$; $\boldsymbol{x}_N$ is the perpendicular dropped from the endpoint of $\boldsymbol{x}$ onto the plane $\boldsymbol{S}$ (Fig. 5); and $h = |\boldsymbol{x}_N|$ is the distance of the endpoint of $\boldsymbol{x}$ from $\boldsymbol{S}$.

To establish the decomposition (15), we represent the required $\boldsymbol{x}_S$ in the form

$$\boldsymbol{x}_S = c_1\boldsymbol{x}_1 + c_2\boldsymbol{x}_2 + \cdots + c_m\boldsymbol{x}_m, \tag{16}$$

where $c_1, c_2, \ldots, c_m$ are complex numbers.[9]

[9] In the case of a euclidean space, $c_1, c_2, \ldots, c_m$ are real numbers.

To determine these numbers we shall start from the relations

$$(\boldsymbol{x}-\boldsymbol{x}_S\,,\ \boldsymbol{x}_k)=0 \qquad (k=1,\,2,\,\ldots,\,m). \tag{17}$$

When we substitute in (17) for $\boldsymbol{x}_S$ its expression (16), we obtain:

$$\left.\begin{array}{l}(\boldsymbol{x}_1\boldsymbol{x}_1)\,c_1+\cdots+(\boldsymbol{x}_m\boldsymbol{x}_1)\,c_m+(\boldsymbol{x}\boldsymbol{x}_1)\cdot(-1)=0\\ \cdots\cdots\cdots\cdots\cdots\cdots\cdots\\ (\boldsymbol{x}_1\boldsymbol{x}_m)\,c_1+\cdots+(\boldsymbol{x}_m\boldsymbol{x}_m)\,c_m+(\boldsymbol{x}\boldsymbol{x}_m)\cdot(-1)=0\\ \qquad\boldsymbol{x}_1c_1+\cdots+\qquad\boldsymbol{x}_mc_m+\quad\boldsymbol{x}_S\cdot(-1)=\boldsymbol{o}.\end{array}\right\} \tag{18}$$

Regarding this as a system of linear homogeneous equations with the non-zero solution $c_1, c_2, \ldots, c_m, -1$, we equate the determinant of the system to zero and obtain (after transposition with respect to the main diagonal):[10]

$$\begin{vmatrix}(\boldsymbol{x}_1\boldsymbol{x}_1)\ldots(\boldsymbol{x}_1\boldsymbol{x}_m) & \boldsymbol{x}_1\\ \cdots\cdots\cdots & \cdot\\ (\boldsymbol{x}_m\boldsymbol{x}_1)\ldots(\boldsymbol{x}_m\boldsymbol{x}_m) & \boldsymbol{x}_m\\ (\boldsymbol{x}\boldsymbol{x}_1)\ldots(\boldsymbol{x}\boldsymbol{x}_m) & \boldsymbol{x}_S\end{vmatrix}=\boldsymbol{o}. \tag{19}$$

When we separate from this determinant the term containing $\boldsymbol{x}_S$, we obtain (in a readily understandable notation):

$$\boldsymbol{x}_S=-\frac{\begin{vmatrix} & \boldsymbol{x}_1\\ G & \vdots\\ & \boldsymbol{x}_m\\ (\boldsymbol{x}\boldsymbol{x}_1)\ldots(\boldsymbol{x}\boldsymbol{x}_m) & 0\end{vmatrix}}{G}, \tag{20}$$

where $G=G(\boldsymbol{x}_1, \boldsymbol{x}_2, \ldots, \boldsymbol{x}_m)$ is the Gramian of the vectors $\boldsymbol{x}_1, \boldsymbol{x}_2, \ldots, \boldsymbol{x}_m$ (in virtue of the linear independence of these vectors, $G\neq 0$). From (15) and (20), we find:

$$\boldsymbol{x}_N=\boldsymbol{x}-\boldsymbol{x}_S=\frac{\begin{vmatrix} & \boldsymbol{x}_1\\ G & \vdots\\ & \boldsymbol{x}_m\\ (\boldsymbol{x}\boldsymbol{x}_1)\ldots(\boldsymbol{x}\boldsymbol{x}_m) & \boldsymbol{x}\end{vmatrix}}{G}. \tag{21}$$

---

[10] The determinant on the left-hand side of (19) is a vector whose i-th coordinate is obtained by replacing all the vectors $\boldsymbol{x}_1, \ldots, \boldsymbol{x}_m$, $\boldsymbol{x}_S$ in the last column by their i-th coordinates $(i=1, 2, \ldots, n)$; the coordinates are taken in an arbitrary basis. To justify the transition from (18) to (19), it is sufficient to replace the vectors $\boldsymbol{x}_1, \ldots, \boldsymbol{x}_m$, $\boldsymbol{x}_S$ by their i-th coordinates.

The formulas (20) and (21) express the projection $\boldsymbol{x}_S$ of $\boldsymbol{x}$ onto the subspace $\boldsymbol{S}$ and the projecting vector $\boldsymbol{x}_N$ in terms of the given vector $\boldsymbol{x}$ and the basis of $\boldsymbol{S}$.

2. We draw attention to another important formula. We denote by h the length of the vector $\boldsymbol{x}_N$. Then, by (15) and (21),

$$h^2 = (\boldsymbol{x}_N \boldsymbol{x}_N) = (\boldsymbol{x}_N \boldsymbol{x}) = \frac{\begin{vmatrix} & & (\boldsymbol{x}_1\boldsymbol{x}) \\ & G & \vdots \\ & & (\boldsymbol{x}_m\boldsymbol{x}) \\ (\boldsymbol{x}\boldsymbol{x}_1) & \dots\ (\boldsymbol{x}\boldsymbol{x}_m) & (\boldsymbol{x}\ \boldsymbol{x}) \end{vmatrix}}{G},$$

i.e.,

$$h^2 = \frac{G(\boldsymbol{x}_1, \boldsymbol{x}_2, \dots, \boldsymbol{x}_m, \boldsymbol{x})}{G(\boldsymbol{x}_1, \boldsymbol{x}_2, \dots, \boldsymbol{x}_m)}. \tag{22}$$

The quantity h can also be interpreted in the following way:

Let the vectors $\boldsymbol{x}_1, \boldsymbol{x}_2, \dots, \boldsymbol{x}_m, \boldsymbol{x}$ issue from a single point and construct on these vectors as edges an $(m+1)$-dimensional parallelepiped. Then h is the height of this parallelepiped measured from the end of the edge $\boldsymbol{x}$ to the base $\boldsymbol{S}$ that passes through the edges $\boldsymbol{x}_1, \boldsymbol{x}_2, \dots, \boldsymbol{x}_m$.

Let $\boldsymbol{y}$ be an arbitrary vector of $\boldsymbol{S}$ and $\boldsymbol{x}$ an arbitrary vector of $\boldsymbol{R}$. If all vectors start from the origin of coordinates of an n-dimensional point space, then $|\boldsymbol{x} - \boldsymbol{y}|$ and $|\boldsymbol{x} - \boldsymbol{x}_S|$ are equal to the value of the slant height and the height respectively from the endpoint of $\boldsymbol{x}$ to the hyperplane $\boldsymbol{S}$.[11] Therefore, when we set down that *the height is shorter than the slant height*, we have:[12]

$$h = |\boldsymbol{x} - \boldsymbol{x}_S| \leq |\boldsymbol{x} - \boldsymbol{y}|$$

(with equality only for $\boldsymbol{y} = \boldsymbol{x}_S$). Thus, among all vectors $\boldsymbol{y} \in \boldsymbol{S}$ the vector $\boldsymbol{x}_S$ deviates the least from the given vector $\boldsymbol{x} \in \boldsymbol{R}$. The quantity $h = \sqrt{\mathrm{N}(\boldsymbol{x} - \boldsymbol{x}_S)}$ is the *mean-square error* in the approximation $\boldsymbol{x} \approx \boldsymbol{x}_S$.[13]

§ 5. The Geometrical Meaning of the Gramian and Some Inequalities

1. We consider arbitrary vectors $\boldsymbol{x}_1, \boldsymbol{x}_2, \dots, \boldsymbol{x}_m$. Let us assume, to begin with, that they are linearly independent. In this case the Gramian formed from any of these vectors is different from zero. Then, when we set, in accordance with (22),

---

[11] See the example on p. 248.

[12] $\mathrm{N}(\boldsymbol{x} - \boldsymbol{y}) = \mathrm{N}(\boldsymbol{x}_N + \boldsymbol{x}_S - \boldsymbol{y}) = \mathrm{N}\boldsymbol{x}_N + \mathrm{N}(\boldsymbol{x}_S - \boldsymbol{y}) \geq \mathrm{N}(\boldsymbol{x}_N) = h^2$.

[13] As regards the application of metrized functional spaces to problems of approximation of functions, see [1].

$$\frac{G(\boldsymbol{x}_1, \boldsymbol{x}_2, \ldots, \boldsymbol{x}_{p+1})}{G(\boldsymbol{x}_1, \boldsymbol{x}_2, \ldots, \boldsymbol{x}_p)} = h_p^2 > 0 \qquad (p = 1, 2, \ldots, m-1), \tag{23}$$

and multiply these inequalities and the inequality

$$G(\boldsymbol{x}_1) = (\boldsymbol{x}_1\boldsymbol{x}_1) > 0, \tag{24}$$

we obtain

$$G(\boldsymbol{x}_1, \boldsymbol{x}_2, \ldots, \boldsymbol{x}_m) > 0.$$

Thus: *The Gramian of linearly independent vectors is positive; that of linearly dependent vectors is zero. Negative Gramians do not exist.*

Let us use the abbreviation $G_p = G(\boldsymbol{x}_1, \boldsymbol{x}_2, \ldots, \boldsymbol{x}_p)$ $(p = 1, 2, \ldots, m)$. Then, from (23) and (24), we have

$$\sqrt{G_1} = |\boldsymbol{x}_1| = V_1,$$
$$\sqrt{G_2} = V_1 h_1 = V_2,$$

where V_2 is the area of the parallelogram spanned by $\boldsymbol{x}_1$ and $\boldsymbol{x}_2$. Further,

$$\sqrt{G_3} = V_2 h_2 = V_3,$$

where V_3 is the volume of the parallelepiped spanned by $\boldsymbol{x}_1, \boldsymbol{x}_2, \boldsymbol{x}_3$. Continuing further, we find:

$$\sqrt{G_4} = V_3 h_3 = V_4$$

and, in general,

$$\sqrt{G_m} = V_{m-1} h_{m-1} = V_m. \tag{25}$$

It is natural to call V_m the volume of the m-dimensional parallelepiped spanned by the vectors $\boldsymbol{x}_1, \boldsymbol{x}_2, \ldots, \boldsymbol{x}_m$.[14]

We denote by $x_{1k}, x_{2k}, \ldots, x_{nk}$ the coordinates of $\boldsymbol{x}_k$ $(k = 1, 2, \ldots, m)$ in an orthonormal basis of $\boldsymbol{R}$ and set

$$X = \| x_{ik} \| \quad (i = 1, 2, \ldots, n;\ k = 1, 2, \ldots, m).$$

Then, in consequence of (10),

$$G_m = |X^{\mathsf{T}} \overline{X}|$$

and therefore (see formula (25)),

$$V_m^2 = G_m = \sum_{1 \le i_1 < i_2 < \cdots < i_m \le n} \operatorname{mod} \begin{vmatrix} x_{i_1 1} & x_{i_1 2} & \cdots & x_{i_1 m} \\ x_{i_2 1} & x_{i_2 2} & \cdots & x_{i_2 m} \\ \cdots & \cdots & \cdots & \cdots \\ x_{i_m 1} & x_{i_m 2} & \cdots & x_{i_m m} \end{vmatrix}^2. \tag{26}$$

[14] Formula (25) gives an inductive definition of the volume of an m-dimensional parallelepiped.

This equation has the following geometric meaning:

The square of the volume of a parallelepiped is equal to the sum of the squares of the volumes of its projections on all the m-dimensional coordinate subspaces. In particular, for $m = n$, it follows from (26) that

$$V_n = \operatorname{mod} \begin{vmatrix} x_{11} & x_{12} & \dots & x_{1n} \\ x_{21} & x_{22} & \dots & x_{2n} \\ \dots & \dots & \dots & \dots \\ x_{n1} & x_{n2} & \dots & x_{nn} \end{vmatrix}. \tag{27}$$

The formulas (20), (21), (22), (26), and (27) solve a number of fundamental metrical problems of n-dimensional unitary and n-dimensional euclidean analytical geometry.

2. Let us return to the decomposition (15). This has the immediate consequence:

$$(\boldsymbol{x}\boldsymbol{x}) = (\boldsymbol{x}_S + \boldsymbol{x}_N,\ \boldsymbol{x}_S + \boldsymbol{x}_N) = (\boldsymbol{x}_S,\ \boldsymbol{x}_S) + (\boldsymbol{x}_N,\ \boldsymbol{x}_N) \geqq (\boldsymbol{x}_N \boldsymbol{x}_N) = h^2 ,$$

which, in conjunction with (22), gives an inequality (for arbitrary vectors $\boldsymbol{x}_1, \boldsymbol{x}_2, \dots, \boldsymbol{x}_m, \boldsymbol{x}$)

$$G(\boldsymbol{x}_1, \boldsymbol{x}_2, \dots, \boldsymbol{x}_m, \boldsymbol{x}) \leqq G(\boldsymbol{x}_1, \boldsymbol{x}_2, \dots, \boldsymbol{x}_m)\, G(\boldsymbol{x}); \tag{28}$$

the equality sign holds if and only if $\boldsymbol{x}$ is orthogonal to $\boldsymbol{x}_1, \boldsymbol{x}_2, \dots, \boldsymbol{x}_m$.

From this we easily obtain the so-called *Hadamard inequality*

$$G(\boldsymbol{x}_1, \boldsymbol{x}_2, \dots, \boldsymbol{x}_m) \leqq G(\boldsymbol{x}_1)\, G(\boldsymbol{x}_2) \cdots G(\boldsymbol{x}_m), \tag{29}$$

where the equality sign holds if and only if the vectors $\boldsymbol{x}_1, \boldsymbol{x}_2, \dots, \boldsymbol{x}_m$ are pairwise orthogonal. The inequality (29) expresses the following fact, which is geometrically obvious:

The volume of a parallelepiped does not exceed the product of the lengths of its edges and is equal to it only when the parallelepiped is rectangular.

Hadamard's inequality can be put into its usual form by setting $m = n$ in (29) and introducing the determinant Δ formed from the coordinates $x_{1k}, x_{2k}, \dots, x_{nk}$ of the vectors $\boldsymbol{x}_k$ $(k = 1, 2, \dots, n)$ in some orthonormal basis:

$$\Delta = \begin{vmatrix} x_{11} & \dots & x_{1n} \\ \dots & \dots & \dots \\ x_{n1} & \dots & x_{nn} \end{vmatrix}.$$

Then it follows from (27) and (29) that

$$|\Delta|^2 \leqq \sum_{i=1}^{n} |x_{i1}|^2 \sum_{i=1}^{n} |x_{i2}|^2 \cdots \sum_{i=1}^{n} |x_{in}|^2. \tag{29'}$$

3.[15] We now turn to the inequality

$$G(\boldsymbol{x}_{1S}, \boldsymbol{x}_{2S}, \ldots, \boldsymbol{x}_{mS}) \leqq G(\boldsymbol{x}_1, \boldsymbol{x}_2, \ldots, \boldsymbol{x}_m) \tag{30}$$

If $G(\boldsymbol{x}_1, \boldsymbol{x}_2, \ldots, \boldsymbol{x}_m) \neq 0$, then the equality sign holds in (30) if and only if $\boldsymbol{x}_{iN} = 0$ $(i = 1, 2, \ldots, m)$. If $G(\boldsymbol{x}_1, \boldsymbol{x}_2, \ldots, \boldsymbol{x}_m) = 0$, then (30) implies, of course, that $G(\boldsymbol{x}_{1S}, \boldsymbol{x}_{2S}, \ldots, \boldsymbol{x}_{mS}) = 0$.

In virtue of (25), the inequality (30) expresses the following geometric fact.

The volume of the orthogonal projection of a parallelepiped onto a subspace **S** *does not exceed the volume of the given parallelepiped; these volumes are equal if and only if the projecting parallelepiped lies in* **S** *or has zero volume.*

We prove (30) by induction on m.

The first step $(m = 1)$ is trivial and yields the inequality

$$G(\boldsymbol{x}_{1S}) \leqq G(\boldsymbol{x}_1),$$

i.e., $|\boldsymbol{x}_{1S}| \leqq |\boldsymbol{x}_1|$ (see Fig. 5 on page 248).

We write the volume $\sqrt{G(\boldsymbol{x}_1, \boldsymbol{x}_2, \ldots, \boldsymbol{x}_m)}$ of our parallelepiped as the product of the 'base' $\sqrt{G(\boldsymbol{x}_1, \boldsymbol{x}_2, \ldots, \boldsymbol{x}_{m-1})}$ by the distance h of the vertex of $\boldsymbol{x}_m$ from the base:

$$\sqrt{G(\boldsymbol{x}_1, \boldsymbol{x}_2, \ldots, \boldsymbol{x}_{m-1})} \cdot h = \sqrt{G(\boldsymbol{x}_1, \boldsymbol{x}_2, \ldots, \boldsymbol{x}_m)}. \tag{31}$$

If we now go over on the left-hand side of (31) from the vectors $\boldsymbol{x}_i$ to their projections $\boldsymbol{x}_{iS}$ $(i = 1, 2, \ldots, m)$, then the first factor cannot increase, by the induction hypothesis, nor the second, by a simple geometric argument. But the product so obtained is the volume $\sqrt{G(\boldsymbol{x}_{1S}, \boldsymbol{x}_{2S}, \ldots, \boldsymbol{x}_{mS})}$ of the parallelepiped projected onto the subspace $\boldsymbol{S}$. Hence

$$\sqrt{G(\boldsymbol{x}_{1S}, \boldsymbol{x}_{2S}, \ldots, \boldsymbol{x}_{mS})} \leqq \sqrt{G(\boldsymbol{x}_1, \boldsymbol{x}_2, \ldots, \boldsymbol{x}_m)}.$$

and by squaring both sides, we obtain (30).

Our condition for the equality sign to hold follows immediately from the proof.

[15] Subsections 3 and 4 have been modified in accordance with a correction published by the author in 1954 (Uspehi Mat. Nauk, vol. 9, no. 3).

4. Now we shall establish a generalization of Hadamard's inequality which comprises both the inequalities (28) and (29):

$$G(\boldsymbol{x}_1, \boldsymbol{x}_2, \ldots, \boldsymbol{x}_m) \leqq G(\boldsymbol{x}_1, \ldots, \boldsymbol{x}_p)\, G(\boldsymbol{x}_{p+1}, \ldots, \boldsymbol{x}_m), \tag{32}$$

where the equality sign holds if and only if each vector $\boldsymbol{x}_1, \boldsymbol{x}_2, \ldots, \boldsymbol{x}_p$ is orthogonal to each of the vectors $\boldsymbol{x}_{p+1}, \ldots, \boldsymbol{x}_m$ or one of the determinants $G(\boldsymbol{x}_1, \boldsymbol{x}_2, \ldots, \boldsymbol{x}_p)$, $G(\boldsymbol{x}_{p+1}, \ldots, \boldsymbol{x}_m)$ vanishes.

The inequality (32) has the following geometric meaning:

The volume of a parallepiped does not exceed the product of the volumes of two complementary 'faces' and is equal to this product if and only if these faces are orthogonal or at least one of them has volume zero.

Let us prove the inequality (32). Let $p < m$. If $G(\boldsymbol{x}_1, \boldsymbol{x}_2, \ldots, \boldsymbol{x}_p) = 0$, then (32) holds with the equality sign. Let $G(\boldsymbol{x}_1, \boldsymbol{x}_2, \ldots, \boldsymbol{x}_p) \neq 0$. Then the p vectors $\boldsymbol{x}_1, \boldsymbol{x}_2, \ldots, \boldsymbol{x}_p$ are linearly independent and form a basis of a p-dimensional subspace $\boldsymbol{T}$ of $\boldsymbol{R}$. The set of all vectors $\boldsymbol{y}$ of $\boldsymbol{R}$ that are orthogonal to $\boldsymbol{T}$ are easily seen also to form a subspace of $\boldsymbol{R}$ (the so-called orthogonal complement of $\boldsymbol{T}$; for details, see § 8 of this Chapter). We denote it by $\boldsymbol{S}$, and then $\boldsymbol{R} = \boldsymbol{T} + \boldsymbol{S}$.

Since every vector of $\boldsymbol{S}$ is orthogonal to every vector of $\boldsymbol{T}$, we can go over, in the Gramian $G(\boldsymbol{x}_1, \boldsymbol{x}_2, \ldots, \boldsymbol{x}_m)$, whose square represents a certain volume, from the vectors $\boldsymbol{x}_{p+1}, \ldots, \boldsymbol{x}_m$ to their projections $\boldsymbol{x}_{p+1S}, \ldots, \boldsymbol{x}_{mS}$ onto the *subspace* $\boldsymbol{S}$:

$$G(\boldsymbol{x}_1, \ldots, \boldsymbol{x}_p, \boldsymbol{x}_{p+1}, \ldots, \boldsymbol{x}_m) = G(\boldsymbol{x}_1, \ldots, \boldsymbol{x}_p, \boldsymbol{x}_{p+1S}, \ldots, \boldsymbol{x}_{mS}).$$

The same arguments show that the Gramian on the right-hand side of this equation can be split:

$$G(\boldsymbol{x}_1, \ldots, \boldsymbol{x}_p, \boldsymbol{x}_{p+1S}, \ldots, \boldsymbol{x}_{mS}) = G(\boldsymbol{x}_1, \ldots, \boldsymbol{x}_p)\, G(\boldsymbol{x}_{p+1S}, \ldots, \boldsymbol{x}_{mS}).$$

If we now go back from the projections to the original vectors and use (30), then we obtain

$$G(\boldsymbol{x}_1, \ldots, \boldsymbol{x}_p)\, G(\boldsymbol{x}_{p+1S}, \ldots, \boldsymbol{x}_{mS}) \leqq G(\boldsymbol{x}_1, \ldots, \boldsymbol{x}_p)\, G(\boldsymbol{x}_{p+1}, \ldots, \boldsymbol{x}_m).$$

The equality sign holds in two cases: 1. When $G(\boldsymbol{x}_{p+1}, \ldots, \boldsymbol{x}_m) = 0$, for then it is obvious that $G(\boldsymbol{x}_{p+1S}, \ldots, \boldsymbol{x}_{mS}) = 0$; and 2. When $\boldsymbol{x}_{iS} = \boldsymbol{x}_i$ ($i = 1, 2, 3, \ldots, m$), i.e., when the vectors $\boldsymbol{x}_{p+1}, \ldots, \boldsymbol{x}_m$ belong to $\boldsymbol{S}$ or, what is the same, each vector $\boldsymbol{x}_{p+1}, \ldots, \boldsymbol{x}_m$ is orthogonal to every vector $\boldsymbol{x}_1, \boldsymbol{x}_2, \ldots, \boldsymbol{x}_p$ (the case $G(\boldsymbol{x}_1, \boldsymbol{x}_2, \ldots, \boldsymbol{x}_p) = 0$ has been considered at the beginning of the proof). By combining the last three relations we obtain the generalized Hadamard inequality (32) and the conditions for the equality sign to hold. This completes the proof.

5. The generalized Hadamard inequality (32) can also be put into analytic form.

Let $\sum_{i,k=1}^{n} h_{ik}x_i\bar{x}_k$ be an arbitrary positive-definite hermitian form. By regarding $x_1, x_2, \ldots, x_n$ as the coordinates, in a basis $\boldsymbol{e}_1, \boldsymbol{e}_2, \ldots, \boldsymbol{e}_n$, of a vector $\boldsymbol{x}$ in an n-dimensional space $\boldsymbol{R}$, we take $\sum_{i,k=1}^{n} h_{ik}x_i\bar{x}_k$ as the fundamental metric form of $\boldsymbol{R}$ (see p. 244). Then $\boldsymbol{R}$ becomes a unitary space. We apply the generalized Hadamard inequality to the basis vectors $\boldsymbol{e}_1, \boldsymbol{e}_2, \ldots, \boldsymbol{e}_n$:

$$G(\boldsymbol{e}_1, \boldsymbol{e}_2, \ldots, \boldsymbol{e}_n) \leqq G(\boldsymbol{e}_1, \ldots, \boldsymbol{e}_p)\, G(\boldsymbol{e}_{p+1}, \ldots, \boldsymbol{e}_n).$$

Setting $H = \| h_{ik} \|_1^n$ and noting that $(\boldsymbol{e}_i\boldsymbol{e}_k) = h_{ik}$ $(i, k = 1, 2, \ldots, n)$, we can rewrite the latter inequality as follows:

$$H\begin{pmatrix} 1 & 2 \ldots n \\ 1 & 2 \ldots n \end{pmatrix} \leqq H\begin{pmatrix} 1 & 2 \ldots p \\ 1 & 2 \ldots p \end{pmatrix} H\begin{pmatrix} p+1 \ldots n \\ p+1 \ldots n \end{pmatrix} \qquad (p < n). \tag{33}$$

Here the equality sign holds if and only if $h_{ik} = h_{ki} = 0$ $(i = 1, 2, \ldots, p;$ $k = p+1, \ldots, n)$.

The inequality (33) holds for the coefficient matrix $H = \| h_{ik} \|_1^n$ of an arbitrary positive-definite hermitian form. In particular, (33) holds if H is the real coefficient matrix of a positive-definite quadratic form $\sum_{i,k=1}^{n} h_{ik}\, x_i\, x_k$.[16]

6. We remind the reader of *Schwarz's inequality*:†

For arbitrary vectors $\boldsymbol{x}, \boldsymbol{y} \in \boldsymbol{R}$

$$|(\boldsymbol{x}\boldsymbol{y})|^2 \leqq \mathrm{N}\boldsymbol{x}\mathrm{N}\boldsymbol{y}, \tag{34}$$

and the equality sign holds only if the vectors $\boldsymbol{x}$ and $\boldsymbol{y}$ differ only by a scalar factor

The validity of Schwarz's inequality follows easily from the inequality established above

$$G(\boldsymbol{x}, \boldsymbol{y}) = \begin{vmatrix} (\boldsymbol{x}\boldsymbol{x}) & (\boldsymbol{x}\boldsymbol{y}) \\ (\boldsymbol{y}\boldsymbol{x}) & (\boldsymbol{y}\boldsymbol{y}) \end{vmatrix} \geqq 0.$$

By analogy with the scalar product of vectors in a three-dimensional euclidean space, we can introduce in an n-dimensional unitary space the

---

[16] An analytical approach to the generalized Hadamard inequality can be found in the book [17], § 8.

† In the Russian literature, this is known as Bunyakovskiĭ's inequality.

'angle' θ between the vectors x and y by defining[17]

$$\cos^2\theta = \frac{|(xy)|^2}{\mathrm{N}x\mathrm{N}y}.$$

From Schwarz's inequality it follows that θ is real.

§ 6. Orthogonalization of a Sequence of Vectors

1. The smallest subspace containing the vectors $x_1, x_2, \ldots, x_p$ will be denoted by $[x_1, x_2, \ldots, x_p]$. This subspace consists of all possible linear combinations $c_1 x_2 + c_2 x_2 + \cdots + c_p x_p$ of the vectors $x_1, x_2, \ldots, x_p$ ($c_1, c_2, c_3, \ldots, c_p$ are complex numbers.)[18] If $x_1, x_2, \ldots, x_p$ are linearly independent, then they form a basis of $[x_1, x_2, \ldots, x_p]$. In that case, the subspace is p-dimensional.

Two sequences of vectors

$$X: x_1, x_2, \ldots,$$
$$Y: y_1, y_2, \ldots$$

containing an equal number of vectors, finite or infinite, will be called *equivalent* if for all p

$$[x_1, x_2, \ldots, x_p] \equiv [y_1, y_2, \ldots, y_p] \quad (p = 1, 2, \ldots).$$

A sequence of vectors

$$X: x_1, x_2, \ldots$$

will be called *non-degenerate* if for every p the vectors $x_1, x_2, \ldots, x_p$ are linearly independent.

A sequence of vectors is called *orthogonal* if any two vectors of the sequence are orthogonal.

By *orthogonalization* of a sequence of vectors we mean a process of replacing the sequence by an equivalent orthogonal sequence.

THEOREM 2: *Every non-degenerate sequence of vectors can be orthogonalized. The orthogonalizing process leads to vectors that are uniquely determined to within scalar multiples.*

---

[17] In the case of a euclidean space, the angle θ between the vectors x and y is defined by the formula

$$\cos\theta = \frac{(xy)}{|x|\,|y|}.$$

[18] In the case of a euclidean space, these numbers are real.

Proof. 1) Let us prove the second part of the theorem first. Suppose that two orthogonal sequences $y_1, y_2, \ldots (Y)$ and $z_1, z_2, \ldots (Z)$ are equivalent to one and the same non-degenerate sequence $x_1, x_2, \ldots (X)$. Then Y and Z are equivalent to each other. Therefore for every p there exist numbers $c_{p1}, c_{p2}, \ldots, c_{pp}$ such that

$$z_p = c_{p1}y_1 + c_{p2}y_2 + \cdots + c_{p,p-1}y_{p-1} + c_{pp}y_p \qquad (p = 1, 2, \ldots).$$

When we form the scalar products of both sides of this equation by $y_1, y_2, \ldots, y_{p-1}$ and take account of the orthogonality of Y and of the relation

$$z_p \perp [z_1, z_2, \ldots, z_{p-1}] \equiv [y_1, y_2, \ldots, y_{p-1}],$$

we obtain $c_{p1} = c_{p2} = \cdots = c_{p\,p-1} = 0$, and therefore

$$z_p = c_{pp}y_p \qquad (p = 1, 2, \ldots).$$

2) A concrete form of the orthogonalizing process for an arbitrary non-degenerate sequence of vectors $x_1, x_2, \ldots$ (X) is given by the following construction.

Let

$$S_p \equiv [x_1, x_2, \ldots, x_p], \quad G_p = G(x_1, x_2, \ldots, x_p) \quad (p = 1, 2, \ldots).$$

We project the vector x_p orthogonally onto the subspace S_{p-1} $(p = 1, 2, \ldots)$:[19]

$$x_p = x_{pS_{p-1}} + x_{pN}, \; x_{pS_{p-1}} \in S_{p-1}, \; x_{pN} \perp S_{p-1} \qquad (p = 1, 2, \ldots).$$

We set

$$y_p = \lambda_p x_{pN} \qquad (p = 1, 2, \ldots; \; x_{1N} = x_1),$$

where λ_p $(p = 1, 2, \ldots)$ are arbitrary non-zero numbers.

Then it is easily seen that

$$Y\colon\; y_1, y_2, \ldots$$

is an orthogonal sequence equivalent to X. This proves Theorem 2.

By (21)

$$x_{pN} = \frac{\begin{vmatrix} & & x_1 \\ & G_{p-1} & \vdots \\ & & x_{p-1} \\ (x_p x_1) & \ldots\ (x_p x_{p-1}) & x_p \end{vmatrix}}{G_{p-1}} \qquad (p = 1, 2, \ldots; \; G_0 = 1).$$

---

[19] For $p = 1$ we set $x_{1S_0} = o$ and $x_{1N} = x_1$.

Setting $\lambda_p = G_{p-1}$ $(p = 1, 2, \ldots; G_0 = 1)$, we obtain the following formulas for the vectors of the orthogonalized sequence:

$$\boldsymbol{y}_1 = \boldsymbol{x}_1,\ \boldsymbol{y}_2 = \begin{vmatrix} (\boldsymbol{x}_1\boldsymbol{x}_1) & \boldsymbol{x}_1 \\ (\boldsymbol{x}_2\boldsymbol{x}_1) & \boldsymbol{x}_2 \end{vmatrix}, \ldots, \boldsymbol{y}_p = \begin{vmatrix} (\boldsymbol{x}_1\boldsymbol{x}_1) & \ldots & (\boldsymbol{x}_1\boldsymbol{x}_{p-1}) & \boldsymbol{x}_1 \\ \cdots & \cdots & \cdots & \cdots \\ (\boldsymbol{x}_{p-1}\boldsymbol{x}_1) & \ldots & (\boldsymbol{x}_{p-1}\boldsymbol{x}_{p-1}) & \boldsymbol{x}_{p-1} \\ (\boldsymbol{x}_p\boldsymbol{x}_1) & \ldots & (\boldsymbol{x}_p\boldsymbol{x}_{p-1}) & \boldsymbol{x}_p \end{vmatrix}, \ldots. \quad (35)$$

By (22),

$$\mathrm{N}\boldsymbol{y}_p = G_{p-1}^2 \mathrm{N}\boldsymbol{x}_{pN} = G_{p-1}^2 \cdot \frac{G_p}{G_{p-1}} = G_{p-1}G_p \qquad (p = 1, 2, \ldots; G_0 = 1). \quad (36)$$

Therefore, setting

$$\boldsymbol{z}_p = \frac{\boldsymbol{y}_p}{\sqrt{G_{p-1}G_p}} \qquad (p = 1, 2, \ldots), \quad (37)$$

we obtain an orthonormal sequence $\boldsymbol{Z}$ equivalent to the given sequence $\boldsymbol{X}$.

Example. In the space of real functions that are sectionally continuous in the interval $[-1, +1]$, we define the scalar product

$$(f, g) = \int_{-1}^{+1} f(x) g(x)\, dx.$$

We consider the non-degenerate sequence of 'vectors'

$$1, x, x^2, x^3, \ldots.$$

We orthogonalize this sequence by the formulas (35):

$$y_0 \equiv 1, \quad y_m = \begin{vmatrix} \frac{1}{1} & 0 & \frac{1}{3} & 0 & \frac{1}{5} & 0 \ldots & 1 \\ 0 & \frac{1}{3} & 0 & \frac{1}{5} & 0 & \frac{1}{7} \ldots & x \\ \frac{1}{3} & 0 & \frac{1}{5} & 0 & \frac{1}{7} & 0 \ldots & x^2 \\ \cdots & \cdots & \cdots & \cdots & \cdots & \cdots & \cdots \\ \cdots & \cdots & \cdots & \cdots & \cdots & \cdots & x^m \end{vmatrix} \qquad (m = 1, 2, \ldots).$$

These orthogonal polynomials coincide, apart from constant factors, with the well-known Legendre polynomials:[20]

$$P_0(x) = 1, P_m(x) = \frac{1}{2^m m!} \frac{d^m (x^2 - 1)^m}{dx^m} \qquad (m = 1, 2, \ldots).$$

The same sequence of powers $1, x, x^2, \ldots$ in a different metric

[20] See [12], p. 77ff.

$$(f, g) = \int_a^b f(x) g(x) \tau(x) dx$$

(where $\tau(x) \geqq 0$ for $a \leqq x \leqq b$) gives another sequence of orthogonal polynomials.

For example, if $a = -1$, $b = 1$ and $\tau(x) = \frac{1}{\sqrt{1-x^2}}$, then we obtain the Tchebyshev (Chebyshev) polynomials:

$$T_n(x) = \frac{1}{2^{n-1}} \cos (n \arccos x).$$

For $a = -\infty$, $b = +\infty$ and $\tau(x) = e^{-x^2}$ we obtain the hermitian polynomials, etc.[21]

2. We shall now take note of the so-called Bessel inequality for an orthonormal sequence of vectors $\boldsymbol{z}_1, \boldsymbol{z}_2, \ldots$ $(\boldsymbol{Z})$. Let $\boldsymbol{x}$ be an arbitrary vector. We denote by ξ_p the projection of $\boldsymbol{x}$ onto $\boldsymbol{z}_p$:

$$\xi_p = (\boldsymbol{x}\boldsymbol{z}_p) \qquad (p = 1, 2, \ldots).$$

Then the projection of $\boldsymbol{x}$ onto the subspace $\boldsymbol{S}_p = [\boldsymbol{z}_1, \boldsymbol{z}_2, \ldots, \boldsymbol{z}_p]$ can be represented in the form (see (20))

$$\boldsymbol{x}_{\boldsymbol{S}_p} = \xi_1 \boldsymbol{z}_1 + \xi_2 \boldsymbol{z}_2 + \cdots + \xi_p \boldsymbol{z}_p \qquad (p = 1, 2, \ldots).$$

But $\mathbf{N}\boldsymbol{x}_{\boldsymbol{S}_p} = |\xi_1|^2 + |\xi_2|^2 + \cdots + |\xi_p|^2 \leqq \mathbf{N}\boldsymbol{x}$. Therefore, for every p,

$$|\xi_1|^2 + |\xi_2|^2 + \cdots + |\xi_p|^2 \leqq \mathbf{N}\boldsymbol{x}. \tag{38}$$

This is *Bessel's inequality.*

In the case of a space of finite dimension n, this inequality has a completely obvious geometrical meaning. For $p = n$ it goes over into the theorem of Pythagoras

$$|\xi_1|^2 + |\xi_2|^2 + \cdots + |\xi_n|^2 = |\boldsymbol{x}|^2.$$

In the case of an infinite-dimensional space and an infinite sequence $\boldsymbol{Z}$, it follows from (38) that the series $\sum_{k=1}^{\infty} |\xi_k|^2$ converges and that

$$\sum_{k=1}^{\infty} |\xi_k|^2 \leqq \mathbf{N}\boldsymbol{x} = |\boldsymbol{x}|^2.$$

Let us form the series

---

[21] For further details see [12], Chapter II, § 9.

$$\sum_{k=1}^{\infty} \xi_k z_k .$$

For every p the p-th partial sum of this series,

$$\xi_1 z_1 + \xi_2 z_2 + \cdots + \xi_p z_p ,$$

is the projection x_{S_p} of x onto the subspace

$$S_p = [z_1, z_2, \ldots, z_p]$$

and is therefore the best approximation to the vector x in this subspace:

$$N(x - \sum_{k=1}^{p} \xi_k z_k) \leqq N(x - \sum_{k=1}^{p} c_k z_k),$$

where $c_1, c_2, \ldots, c_p$ are arbitrary complex numbers. Let us calculate the corresponding mean-square-deviation δ_p:

$$\delta_p^2 = N(x - \sum_{k=1}^{p} \xi_k z_k) = (x - \sum_{k=1}^{p} \xi_k z_k,\ \ x - \sum_{k=1}^{p} \xi_k z_k) = Nx - \sum_{k=1}^{p} |\xi_k|^2 .$$

Hence

$$\lim_{p \to \infty} \delta_p^2 = Nx - \sum_{k=1}^{\infty} |\xi_k|^2 .$$

If

$$\lim_{p \to \infty} \delta_p = 0 ,$$

then we say that the series $\sum_{k=1}^{\infty} \xi_k z_k$ *converges in the mean* (or converges with respect to the norm) to the vector x.

In this case we have an equality for the vector x in R (the theorem of Pythagoras in an infinite-dimensional space!):

$$Nx = |x|^2 = \sum_{k=1}^{\infty} |\xi_k|^2 . \tag{39}$$

If for every vector x of R the series $\sum_{k=1}^{\infty} \xi_k z_k$ converges in the mean to x, then the orthonormal sequence of vectors $z_1, z_2, \ldots$ is called *complete*. In this case, when we replace x in (39) by $x + y$ and use (39) three times, for $N(x + y)$, Nx, and Ny, then we easily obtain:

$$(xy) = \sum_{k=1}^{\infty} \xi_k \bar{\eta}_k \quad [\xi_k = (x z_k),\ \eta_k = (y z_k);\ k = 1, 2, \ldots]. \tag{40}$$

Example. We consider the space of all complex functions $f(t)$ (t is a real variable) that are sectionally continuous in the closed interval $[0, 2\pi]$. Let us define the norm of $f(t)$ by

$$\mathrm{N}f = \int_0^{2\pi} |f(t)|^2 \, dt .$$

Correspondingly, we have the formula

$$(f, g) = \int_0^{2\pi} f(t) \overline{g(t)} \, dt$$

for the scalar product of two functions $f(t)$ and $g(t)$.

We take the infinite sequence of functions

$$\frac{1}{\sqrt{2\pi}} e^{ikt} \qquad (k = 0, \pm 1, \pm 2, \ldots) .$$

These functions form an orthogonal sequence, because

$$\int_0^{2\pi} e^{i\mu t} e^{-i\nu t} \, dt = \int_0^{2\pi} e^{i(\mu-\nu)t} \, dt = \begin{cases} 0, & \text{for } \mu \neq \nu, \\ 2\pi, & \text{for } \mu = \nu . \end{cases}$$

The series

$$\sum_{k=-\infty}^{\infty} f_k e^{ikt} \qquad \left(f_k = \frac{1}{2\pi} \int_0^{2\pi} f(t) e^{-ikt} \, dt ; \quad (k = 0, \pm 1, \pm 2, \ldots \right)$$

converges in the mean to $f(t)$ in the interval $[0, 2\pi]$. This series is called the *Fourier series* of $f(t)$ and the coefficients f_k ($k = 0, \pm 1, \pm 2, \ldots$) are called the *Fourier coefficients* of $f(t)$.

In the theory of Fourier series it is proved that the system of functions e^{ikt} ($k = 0, \pm 1, \pm 2, \ldots$) is complete.[22]

The condition of completeness gives *Parseval's equality* (see (40))

$$\int_0^{2\pi} f(t) \overline{g(t)} \, dt = \sum_{k=-\infty}^{+\infty} \frac{1}{2\pi} \int_0^{2\pi} f(t) e^{-ikt} \, dt \int_0^{2\pi} \overline{g(t)} e^{ikt} \, dt .$$

If $f(t)$ is a real function, then f_0 is real, and f_k and f_{-k} are conjugate complex numbers. Setting

$$f_k = \frac{1}{2\pi} \int_0^{2\pi} f(t) e^{-ikt} \, dt = \frac{1}{2} (a_k + i b_k) ,$$

---

[22] See, for example, [12], Chapter II.

where

$$a_k = \frac{1}{\pi}\int_0^{2\pi} f(t)\cos kt\,dt, \quad b_k = \frac{1}{\pi}\int_0^{2\pi} f(t)\sin kt\,dt \quad (k = 0, 1, 2, \ldots),$$

we have

$$f_k e^{ikt} + f_{-k}e^{-ikt} = a_k \cos kt + b_k \sin kt \qquad (k = 1, 2, \ldots).$$

Therefore, for a real function $f(t)$ the Fourier series assumes the form

$$\frac{a_0}{2} + \sum_{k=1}^{\infty}(a_k \cos kt + b_k \sin kt) \quad \left(\begin{array}{l} a_k = \frac{1}{\pi}\int_0^{2\pi} f(t)\cos kt\,dt, \\ b_k = \frac{1}{\pi}\int_0^{2\pi} f(t)\sin kt\,dt, \end{array} \quad k = 0, 1, 2, \ldots \right).$$

§ 7. Orthonormal Bases

1. A basis of any finite-dimensional subspace $\boldsymbol{S}$ in a unitary or a euclidean space $\boldsymbol{R}$ is a non-degenerate sequence of vectors and therefore—by Theorem 2 of the preceding section—can be orthogonalized and normalized. Thus: *Every finite-dimensional subspace* $\boldsymbol{S}$ (*and, in particular, the whole space* $\boldsymbol{R}$ *if it is finite-dimensional*) *has an orthonormal basis.*

Let $\boldsymbol{e}_1, \boldsymbol{e}_2, \ldots, \boldsymbol{e}_n$ be an orthonormal basis of $\boldsymbol{R}$. We denote by $\boldsymbol{x}_1, \boldsymbol{x}_2, \boldsymbol{x}_3, \ldots, \boldsymbol{x}_n$ the coordinates of an arbitrary vector $\boldsymbol{x}$ in this basis:

$$\boldsymbol{x} = \sum_{k=1}^{n} x_k \boldsymbol{e}_k.$$

Multiplying both sides of this equation on the right by $\boldsymbol{e}_k$ and taking into account that the basis is orthonormal, we easily find:

$$x_k = (\boldsymbol{x}\boldsymbol{e}_k) \qquad (k = 1, 2, \ldots, n);$$

i.e., *in an orthonormal basis the coordinates of a vector are equal to its projections onto the corresponding basis vectors*:

$$\boldsymbol{x} = \sum_{k=1}^{n} (\boldsymbol{x}\boldsymbol{e}_k)\,\boldsymbol{e}_k. \tag{41}$$

Let $x_1, x_2, \ldots, x_n$ and $x_1', x_2', \ldots, x_n'$ be the coordinates of one and the same vector $\boldsymbol{x}$ in two different orthonormal bases $\boldsymbol{e}_1, \boldsymbol{e}_2, \ldots, \boldsymbol{e}_n$ and $\boldsymbol{e}_1', \boldsymbol{e}_2', \ldots, \boldsymbol{e}_n'$ of a unitary space $\boldsymbol{R}$. The formulas for the coordinate transformation have the form

$$x_i = \sum_{k=1}^{n} u_{ik} x'_k \qquad (i = 1, 2, \ldots, n). \tag{42}$$

Here the coefficients $u_{1k}, u_{2k}, \ldots, u_{nk}$ that form the k-th column of the matrix $U = \| u_{ik} \|_1^n$ are easily seen to be the coordinates of the vector e'_k in the basis $e_1, e_2, \ldots, e_n$. Therefore, when we write down the condition for the basis $e'_1, e'_2, \ldots, e'_n$ to be orthonormal in terms of coordinates (see (10)), we obtain the relations

$$\sum_{i=1}^{n} u_{ik} \bar{u}_{il} = \delta_{kl} = \begin{cases} 1, & \text{for } k = l, \\ 0, & \text{for } k \neq l. \end{cases} \tag{43}$$

A transformation (42) in which the coefficients satisfy the conditions (43) is called *unitary* and the corresponding matrix U is called a *unitary matrix*. Thus: *In an n-dimensional unitary space the transition from one orthonormal basis to another is effected by a unitary coordinate transformation.*

Let $\boldsymbol{R}$ be an n-dimensional euclidean space. The transition from one orthonormal basis of $\boldsymbol{R}$ to another is effected by a coordinate transformation

$$x_i = \sum_{k=1}^{n} v_{ik} x'_k \qquad (i = 1, 2, \ldots, n) \tag{44}$$

whose coefficients are connected by the relation

$$\sum_{i=1}^{n} v_{ik} v_{il} = \delta_{kl} \quad (k, l = 1, 2, \ldots, n). \tag{45}$$

Such a coordinate transformation is called *orthogonal* and the corresponding matrix V is called an *orthogonal matrix*.

2. We note an interesting matrix method of writing the orthogonalizing process. Let $A = \| a_{ik} \|_1^n$ be an arbitrary non-singular matrix ($|A| \neq 0$) with complex elements. We consider a unitary space $\boldsymbol{R}$ with an orthonormal basis $e_1, e_2, \ldots, e_n$ and define the linearly independent vectors $a_1, a_2, \ldots, a_n$ by the equations

$$a_k = \sum_{i=1}^{n} a_{ik} e_i \qquad (k = 1, 2, \ldots, n).$$

Let us perform the orthogonalizing process on the vectors $a_1, a_2, \ldots, a_n$. The orthonormal basis of $\boldsymbol{R}$ so obtained we shall denote by $u_1, u_2, \ldots, u_n$. Suppose we have

$$u_k = \sum_{i=1}^{n} u_{ik} e_i \qquad (k = 1, 2, \ldots, n).$$

Then

$$[\boldsymbol{a}_1, \boldsymbol{a}_2, \ldots, \boldsymbol{a}_p] = [\boldsymbol{u}_1, \boldsymbol{u}_2, \ldots, \boldsymbol{u}_p] \quad (p = 1, 2, \ldots, n),$$

i.e.,

$$\begin{aligned} \boldsymbol{a}_1 &= c_{11}\boldsymbol{u}_1, \\ \boldsymbol{a}_2 &= c_{12}\boldsymbol{u}_1 + c_{22}\boldsymbol{u}_2, \\ &\cdots\cdots\cdots\cdots \\ \boldsymbol{a}_n &= c_{1n}\boldsymbol{u}_1 + c_{2n}\boldsymbol{u}_2 + \cdots + c_{nn}\boldsymbol{u}_n, \end{aligned}$$

where the c_{ik} $(i, k = 1, 2, \ldots, n;\ i \leqq k)$ are certain complex numbers.

Setting $c_{ik} = 0$ for $i > k$, we have:

$$\boldsymbol{a}_k = \sum_{p=1}^{n} c_{pk}\boldsymbol{u}_p \qquad (k = 1, 2, \ldots, n).$$

When we go over to coordinates and introduce the upper triangular matrix $C = \| c_{ik} \|_1^n$ and the unitary matrix $U = \| u_{ik} \|_1^n$, we obtain

$$a_{ik} = \sum_{p=1}^{n} u_{ip}c_{pk} \qquad (i, k = 1, 2, \ldots, n),$$

or

$$A = UC. \tag{*}$$

According to this formula: ***Every non-singular matrix*** $A = \| a_{ik} \|_1^n$ *can be represented in the form of a product of a unitary matrix U and an upper triangular matrix C.*

Since the orthogonalizing process determines the vectors $\boldsymbol{u}_1, \boldsymbol{u}_2, \ldots, \boldsymbol{u}_n$ uniquely, apart from scalar multipliers $\varepsilon_1, \varepsilon_2, \ldots, \varepsilon_n$ $(| \varepsilon_i | = 1;\ i = 1, 2, \ldots, n)$, the factors U and C in (*) are uniquely determined apart from a diagonal factor $M = \{\varepsilon_1, \varepsilon_2, \ldots, \varepsilon_n\}$:

$$U = U_1 M, \qquad C = M^{-1}C_1.$$

This can also be shown directly.

Note 1. If A is a real matrix, the factors U and C in (*) can be chosen to be real. In this case, U is an orthogonal matrix.

Note 2. The formula (*) also remains valid for a singular matrix A $(| A | = 0)$. This can be seen by setting $A = \lim_{m \to \infty} A_m$, where $| A_m | \neq 0$ $(m = 1, 2, \ldots)$.

Then $A_m = U_m C_m$ $(m = 1, 2, \ldots)$. When we select from the sequence $\{U_m\}$ a convergent subsequence $\{U_{m_p}\}$ $(\lim_{p \to \infty} U_{m_p} = U)$ and proceed to the limit, then we obtain from the equation $A_{m_p} = U_{m_p} C_{m_p}$ for $p \to \infty$ the required decomposition $A = UC$. However, in the case $| A | = 0$ the factors U and C are no longer uniquely determined to within a diagonal factor M.

Note 3. Instead of (*) we can also obtain a formula

$$A = DW, \tag{**}$$

where D is a lower triangular matrix and W a unitary matrix. For when we apply the formula (*) that was established above to the transposed matrix A^{T}

$$A^{\mathsf{T}} = UC$$

and then set $W = U^{\mathsf{T}}$, $D = C^{\mathsf{T}}$, we obtain (**).[23]

§ 8. The Adjoint Operator

1. Let $\boldsymbol{A}$ be a linear operator in an n-dimensional unitary space.

DEFINITION 4: *A linear operator* $\boldsymbol{A}^*$ *is called adjoint to the operator* $\boldsymbol{A}$ *if and only if for any two vectors* $\boldsymbol{x}, \boldsymbol{y}$ *of* $\boldsymbol{R}$

$$(\boldsymbol{A}\boldsymbol{x}, \boldsymbol{y}) = (\boldsymbol{x}, \boldsymbol{A}^*\boldsymbol{y}). \tag{46}$$

We shall show that for every linear operator $\boldsymbol{A}$ there exists one and only one adjoint operator $\boldsymbol{A}^*$. To prove this, we take an orthonormal basis $\boldsymbol{e}_1, \boldsymbol{e}_2, \ldots, \boldsymbol{e}_n$ in $\boldsymbol{R}$. Then (see (41)) the required operator $\boldsymbol{A}^*$ and an arbitrary vector $\boldsymbol{y}$ of $\boldsymbol{R}$ must satisfy the equation

$$\boldsymbol{A}^*\boldsymbol{y} = \sum_{k=1}^{n} (\boldsymbol{A}^*\boldsymbol{y}, \boldsymbol{e}_k)\, \boldsymbol{e}_k .$$

By (46) this can be rewritten as follows:

$$\boldsymbol{A}^*\boldsymbol{y} = \sum_{k=1}^{n} (\boldsymbol{y}, \boldsymbol{A}\boldsymbol{e}_k)\, \boldsymbol{e}_k . \tag{47}$$

We now take (47) as the definition of an operator $\boldsymbol{A}^*$.

It is easy to verify that the operator $\boldsymbol{A}^*$ so defined is linear and satisfies (46) for arbitrary vectors $\boldsymbol{x}$ and $\boldsymbol{y}$ of $\boldsymbol{R}$. Moreover, (47) determines the operator $\boldsymbol{A}^*$ *uniquely*. Thus the existence and uniqueness of the adjoint operator $\boldsymbol{A}^*$ is established.

Let $\boldsymbol{A}$ be a linear operator in a unitary space and let $A = \| a_{ik} \|_1^n$ be the corresponding matrix in an orthonormal basis $\boldsymbol{e}_1, \boldsymbol{e}_2, \ldots, \boldsymbol{e}_n$. Then, by applying the formula (41) to the vector $\boldsymbol{A}\boldsymbol{e}_k = \sum_{i=1}^{n} a_{ik}\, \boldsymbol{e}_i$, we obtain

$$a_{ik} = (\boldsymbol{A}\boldsymbol{e}_k, \boldsymbol{e}_i) \quad (i, k = 1, 2, \ldots, n). \tag{48}$$

[23] From the fact that U is unitary it follows that U^{T} is unitary, since the condition (43), written in matrix form $U^{\mathsf{T}}\overline{U} = E$, implies that $U\overline{U}^{\mathsf{T}} = E$.

Now let $A^* = \| a^*_{ik} \|_1^n$ be the matrix corresponding to $\boldsymbol{A}^*$ in the same basis. Then, by (48),

$$a^*_{ik} = (\boldsymbol{A}^* \boldsymbol{e}_k, \boldsymbol{e}_i) \quad (i, k = 1, 2, \ldots, n). \tag{49}$$

From (48) and (49) it follows by (46) that

$$a^*_{ik} = \bar{a}_{ki} \quad (i, k = 1, 2, \ldots, n),$$

i.e.,

$$A^* = \bar{A}^{\mathsf{T}}.$$

The matrix A^* is the complex conjugate of the transpose of A. This matrix will be called the *adjoint* of A. (This is not to be confused with the adjoint of a matrix as defined on p. 82.)

Thus: *In an orthonormal basis adjoint matrices correspond to adjoint operators.*

The following properties of the adjoint operator follow from its definition:

1. $(\boldsymbol{A}^*)^* = \boldsymbol{A}$,
2. $(\boldsymbol{A} + \boldsymbol{B})^* = \boldsymbol{A}^* + \boldsymbol{B}^*$,
3. $(\alpha \boldsymbol{A})^* = \bar{\alpha} \boldsymbol{A}^*$ (α a scalar),
4. $(\boldsymbol{AB})^* = \boldsymbol{B}^* \boldsymbol{A}^*$.

2. We shall now introduce an important concept. Let $\boldsymbol{S}$ be an arbitrary subspace of $\boldsymbol{R}$. We denote by $\boldsymbol{T}$ the set of all vectors $\boldsymbol{y}$ of $\boldsymbol{R}$ that are orthogonal to $\boldsymbol{S}$. It is easy to see that $\boldsymbol{T}$ is a subspace of $\boldsymbol{R}$ and that every vector $\boldsymbol{x}$ of $\boldsymbol{R}$ can be represented uniquely in the form of a sum $\boldsymbol{x} = \boldsymbol{x}_S + \boldsymbol{x}_T$, where $\boldsymbol{x}_S \in \boldsymbol{S}$, $\boldsymbol{x}_T \in \boldsymbol{T}$, so that we have the resolution

$$\boldsymbol{R} = \boldsymbol{S} + \boldsymbol{T}, \quad \boldsymbol{S} \perp \boldsymbol{T}.$$

We obtain this resolution by applying the decomposition (15) to the arbitrary vector $\boldsymbol{x}$ of $\boldsymbol{R}$. $\boldsymbol{T}$ is called the *orthogonal complement* of $\boldsymbol{S}$. Obviously, $\boldsymbol{S}$ is the orthogonal complement of $\boldsymbol{T}$. We write $\boldsymbol{S} \perp \boldsymbol{T}$, meaning by this that each vector of $\boldsymbol{S}$ is orthogonal to every vector of $\boldsymbol{T}$.

Now we can formulate the fundamental property of the adjoint operator:

5. *If a subspace* $\boldsymbol{S}$ *is invariant with respect to* $\boldsymbol{A}$, *then the orthogonal complement* $\boldsymbol{T}$ *of the subspace is invariant with respect to* $\boldsymbol{A}^*$.

For let $\boldsymbol{x} \in \boldsymbol{S}$, $\boldsymbol{y} \in \boldsymbol{T}$. Then it follows from $\boldsymbol{A}\boldsymbol{x} \in \boldsymbol{S}$ that $(\boldsymbol{A}\boldsymbol{x}, \boldsymbol{y}) = 0$ and hence by (46) that $(\boldsymbol{x}, \boldsymbol{A}^*\boldsymbol{y}) = 0$. Since $\boldsymbol{x}$ is an arbitrary vector of $\boldsymbol{S}$, $\boldsymbol{A}^*\boldsymbol{y} \in \boldsymbol{T}$, and this is what we had to prove.

We introduce the following definition:

DEFINITION 5: *Two systems of vectors* $\boldsymbol{x}_1, \boldsymbol{x}_2, \ldots, \boldsymbol{x}_m$ *and* $\boldsymbol{y}_1, \boldsymbol{y}_2, \ldots, \boldsymbol{y}_m$ *are called bi-orthogonal if*

$$(\boldsymbol{x}_i \boldsymbol{y}_k) = \delta_{ik} \quad (i, k = 1, 2, \ldots, m), \tag{50}$$

where δ_{ik} *is the Kronecker symbol.*

Now we shall prove the following proposition:

6. *If* $\boldsymbol{A}$ *is a linear operator of simple structure, then the adjoint operator* $\boldsymbol{A}^*$ *is also of simple structure, and complete systems of characteristic vectors* $\boldsymbol{x}_1, \boldsymbol{x}_2, \ldots, \boldsymbol{x}_n$ *and* $\boldsymbol{y}_1, \boldsymbol{y}_2, \ldots, \boldsymbol{y}_n$ *of* $\boldsymbol{A}$ *and* $\boldsymbol{A}^*$ *can be chosen such that they are bi-orthogonal:*

$$\boldsymbol{A}\boldsymbol{x}_i = \lambda_i \boldsymbol{x}_i, \quad \boldsymbol{A}^*\boldsymbol{y}_i = \mu_i \boldsymbol{y}_i, \quad (\boldsymbol{x}_i \boldsymbol{y}_k) = \delta_{ik} \quad (i, k = 1, 2, \ldots, n).$$

For let $\boldsymbol{x}_1, \boldsymbol{x}_2, \ldots, \boldsymbol{x}_n$ be a complete system of characteristic vectors of $\boldsymbol{A}$. We use the notation

$$\boldsymbol{S}_k = [\boldsymbol{x}_1, \ldots, \boldsymbol{x}_{k-1}, \boldsymbol{x}_{k+1}, \ldots, \boldsymbol{x}_n] \quad (k = 1, 2, \ldots, n).$$

Consider the one-dimensional orthogonal complement $\boldsymbol{T}_k = [\boldsymbol{y}_k]$ to the $(n-1)$-dimensional subspace $\boldsymbol{S}_k$ $(k = 1, 2, \ldots, n)$. Then $\boldsymbol{T}_k$ is invariant with respect to $\boldsymbol{A}^*$:

$$\boldsymbol{A}^*\boldsymbol{y}_k = \mu_k \boldsymbol{y}_k, \quad \boldsymbol{y}_k \neq \boldsymbol{o} \quad (k = 1, 2, \ldots, n).$$

From $\boldsymbol{S}_k \perp \boldsymbol{y}_k$ it follows that $(\boldsymbol{x}_k \boldsymbol{y}_k) \neq 0$, because otherwise the vector $\boldsymbol{y}_k$ would have to be the null vector. Multiplying $\boldsymbol{x}_k, \boldsymbol{y}_k$ $(k = 1, 2, \ldots, n)$ by suitable numerical factors we obtain

$$(\boldsymbol{x}_i \boldsymbol{y}_k) = \delta_{ik} \quad (i, k = 1, 2, \ldots, n).$$

From the bi-orthogonality of the systems $\boldsymbol{x}_1, \boldsymbol{x}_2, \ldots, \boldsymbol{x}_n$ and $\boldsymbol{y}_1, \boldsymbol{y}_2, \ldots, \boldsymbol{y}_n$ it follows that the vectors of each system are linearly independent.

We mention one further proposition:

7. *If the operators* $\boldsymbol{A}$ *and* $\boldsymbol{A}^*$ *have a common characteristic vector, then the corresponding characteristic values are complex conjugates.*

For let $\boldsymbol{A}\boldsymbol{x} = \lambda \boldsymbol{x}$ and $\boldsymbol{A}^*\boldsymbol{x} = \mu \boldsymbol{x}$ $(\boldsymbol{x} \neq \boldsymbol{o})$. Then, setting $\boldsymbol{y} = \boldsymbol{x}$ in (46), we have $\lambda(\boldsymbol{x}, \boldsymbol{x}) = \bar{\mu}(\boldsymbol{x}, \boldsymbol{x})$ and hence $\lambda = \bar{\mu}$.

§ 9. Normal Operators in a Unitary Space

1. DEFINITION 6. *A linear operator* $\boldsymbol{A}$ *is called normal if it commutes with its adjoint:*

$$\boldsymbol{A}\boldsymbol{A}^* = \boldsymbol{A}^*\boldsymbol{A}. \tag{51}$$

DEFINITION 7. *A linear operator* $\boldsymbol{H}$ *is called hermitian if it is equal to its adjoint:*

$$\boldsymbol{H}^* = \boldsymbol{H}. \tag{52}$$

DEFINITION 8. *A linear operator* $\boldsymbol{U}$ *is called unitary if it is inverse to its adjoint:*

$$\boldsymbol{U}\boldsymbol{U}^* = \boldsymbol{E} \tag{53}$$

Note that a unitary operator can be regarded as an isometric operator in a hermitian space, i.e., as an operator preserving the metric.

For suppose that for arbitrary vectors $\boldsymbol{x}$ and $\boldsymbol{y}$ of $\boldsymbol{R}$

$$(\boldsymbol{U}\boldsymbol{x}, \boldsymbol{U}\boldsymbol{y}) = (\boldsymbol{x}, \boldsymbol{y}). \tag{54}$$

Then by (46)

$$(\boldsymbol{U}^*\boldsymbol{U}\boldsymbol{x}, \boldsymbol{y}) = (\boldsymbol{x}, \boldsymbol{y})$$

and therefore, since $\boldsymbol{y}$ is arbitrary,

$$\boldsymbol{U}^*\boldsymbol{U}\boldsymbol{x} = \boldsymbol{x},$$

i.e., $\boldsymbol{U}^*\boldsymbol{U} = \boldsymbol{E}$, or $\boldsymbol{U}^* = \boldsymbol{U}^{-1}$. Conversely, (53) implies (54).

From (53) and (54) it follows that 1. the product of two unitary operators is itself a unitary operator, 2. the unit operator $\boldsymbol{E}$ is unitary, and 3. the inverse of a unitary operator is also unitary. Therefore the set of all unitary operators is a group.[24] This is called the *unitary group.*

Hermitian operators and unitary operators are special cases of a normal operator.

2. We have

THEOREM 3: *Every linear operator* $\boldsymbol{A}$ *can be represented in the form*

$$\boldsymbol{A} = \boldsymbol{H}_1 + i\boldsymbol{H}_2, \tag{55}$$

where $\boldsymbol{H}_1$ *and* $\boldsymbol{H}_2$ *are hermitian operators* (the 'hermitian components' of $\boldsymbol{A}$). *The hermitian components are uniquely determined by* $\boldsymbol{A}$. *The operator* $\boldsymbol{A}$ *is normal if and only if its hermitian components* $\boldsymbol{H}_1$ *and* $\boldsymbol{H}_2$ *are permutable.*

---

[24] See footnote 13 on p. 18.

Proof. Suppose that (55) holds. Then

$$\boldsymbol{A}^* = \boldsymbol{H}_1 - i\boldsymbol{H}_2. \tag{56}$$

From (55) and (56) we have:

$$\boldsymbol{H}_1 = \frac{1}{2}(\boldsymbol{A} + \boldsymbol{A}^*), \quad \boldsymbol{H}_2 = \frac{1}{2i}(\boldsymbol{A} - \boldsymbol{A}^*). \tag{57}$$

Conversely, the formulas (57) define hermitian operators $\boldsymbol{H}_1$ and $\boldsymbol{H}_2$ connected with $\boldsymbol{A}$ by (55).

Now let $\boldsymbol{A}$ be a normal operator: $\boldsymbol{A}\boldsymbol{A}^* = \boldsymbol{A}^*\boldsymbol{A}$. Then it follows from (57) that $\boldsymbol{H}_1\boldsymbol{H}_2 = \boldsymbol{H}_2\boldsymbol{H}_1$. Conversely, from $\boldsymbol{H}_1\boldsymbol{H}_2 = \boldsymbol{H}_2\boldsymbol{H}_1$ it follows by (55) and (56) that $\boldsymbol{A}\boldsymbol{A}^* = \boldsymbol{A}^*\boldsymbol{A}$. This completes the proof.

The representation of an arbitrary linear operator $\boldsymbol{A}$ in the form (55) is an analogue to the representation of a complex number z in the form $x_1 + ix_2$, where x_1 and x_2 are real.

Suppose that in some orthonormal basis the operators $\boldsymbol{A}$, $\boldsymbol{H}$, and $\boldsymbol{U}$ correspond to the matrices A, H, and U. Then the operator equations

$$\boldsymbol{A}\boldsymbol{A}^* = \boldsymbol{A}^*\boldsymbol{A}, \quad \boldsymbol{H}^* = \boldsymbol{H}, \quad \boldsymbol{U}\boldsymbol{U}^* = \boldsymbol{E} \tag{58}$$

correspond to the matrix equations

$$AA^* = A^*A, \quad H^* = H, \quad UU^* = E. \tag{59}$$

Therefore we define a matrix as *normal* if it commutes with its adjoint, as *hermitian* if it is equal to its adjoint, and finally as *unitary* if it is inverse to its adjoint.

Then: *In an orthonormal basis a normal (hermitian, unitary) operator corresponds to a normal (hermitian, unitary) matrix.*

A hermitian matrix $H = \| h_{ik} \|_1^n$ is, by (59), characterized by the following relation among its elements:

$$h_{ki} = \bar{h}_{ik} \quad (i, k = 1, 2, \ldots, n);$$

i.e., a hermitian matrix is always the coefficient matrix of some hermitian form (see § 1).

A unitary matrix $U = \| u_{ik} \|_1^n$ is, by (59), characterized by the following relations among its elements:

$$\sum_{j=1}^{n} u_{ij}\bar{u}_{kj} = \delta_{ik} \quad (i, k = 1, 2, \ldots, n). \tag{60}$$

Since $UU^* = E$ implies that $U^*U = E$, from (60) there follow the equivalent relations:

$$\sum_{j=1}^{n} u_{ji}\bar{u}_{jk} = \delta_{ik} \quad (i, k = 1, 2, \ldots, n). \tag{61}$$

Equation (60) expresses the 'orthonormality' of the rows and equation (61) that of the columns of the matrix $U = \| u_{ik} \|_1^n$.[25]

A unitary matrix is the coefficient matrix of some unitary transformation (see § 7).

§ 10. The Spectra of Normal, Hermitian, and Unitary Operators

1. As a preliminary, we establish a property of permutable operators in the form of a lemma.

LEMMA 1: *Permutable operators* $\boldsymbol{A}$ *and* $\boldsymbol{B}$ ($\boldsymbol{AB} = \boldsymbol{BA}$) *always have a common characteristic vector.*

Proof. Let $\boldsymbol{x}$ be a characteristic vector of $\boldsymbol{A}$: $\boldsymbol{Ax} = \lambda \boldsymbol{x}$, $\boldsymbol{x} \neq \boldsymbol{o}$. Then, since $\boldsymbol{A}$ and $\boldsymbol{B}$ are permutable,

$$\boldsymbol{AB}^k\boldsymbol{x} = \lambda \boldsymbol{B}^k\boldsymbol{x} \quad (k = 0,\ 1,\ 2,\ \ldots). \tag{62}$$

Suppose that in the sequence of vectors

$$\boldsymbol{x},\ \boldsymbol{Bx},\ \boldsymbol{B}^2\boldsymbol{x},\ \ldots$$

the first p are linearly independent, while the $(p+1)$-th vector $\boldsymbol{B}^p\boldsymbol{x}$ is a linear combination of the preceding ones. Then $\boldsymbol{S} \equiv [\boldsymbol{x}, \boldsymbol{Bx}, \ldots, \boldsymbol{B}^{p-1}\boldsymbol{x}]$ is a subspace invariant with respect to $\boldsymbol{B}$, so that in this subspace $\boldsymbol{S}$ there exists a characteristic vector $\boldsymbol{y}$ of $\boldsymbol{B}$: $\boldsymbol{By} = \mu \boldsymbol{y}$, $\boldsymbol{y} \neq \boldsymbol{o}$. On the other hand, (62) shows that the vectors $\boldsymbol{x}, \boldsymbol{Bx}, \ldots, \boldsymbol{B}^{p-1}\boldsymbol{x}$ are characteristic vectors of $\boldsymbol{A}$ corresponding to one and the same characteristic value λ. Therefore every linear combination of these vectors, and in particular $\boldsymbol{y}$, is a characteristic vector of $\boldsymbol{A}$ corresponding to λ. Thus we have proved the existence of a common characteristic vector of the operators $\boldsymbol{A}$ and $\boldsymbol{B}$.

Let $\boldsymbol{A}$ be an arbitrary normal operator in an n-dimensional hermitian space $\boldsymbol{R}$. In that case $\boldsymbol{A}$ and $\boldsymbol{A}^*$ are permutable and therefore have a common characteristic vector $\boldsymbol{x}_1$. Then (see § 8, 7.)

[25] Thus, orthonormality of the columns of the matrix U is a consequence of the orthonormality of the rows, and vice versa.

$$Ax_1 = \lambda_1 x_1, \quad A^* x_1 = \bar{\lambda}_1 x_1 \qquad (x_1 \neq o).$$

We denote by S_1 the one-dimensional subspace containing the vector x_1 ($S_1 = [x_1]$) and by T_1 the orthogonal complement of S_1 in R:

$$R = S_1 + T_1, \quad S_1 \perp T_1.$$

Since S_1 is invariant with respect to A and A^*, T_1 is also invariant with respect to these operators (see § 8, 5.). Therefore, by Lemma 1, the permutable operators A and A^* have a common characteristic vector x_2 in T_1:

$$Ax_2 = \lambda_2 x_2, \quad A^* x_2 = \bar{\lambda}_2 x_2 \qquad (x_2 \neq o).$$

Obviously, $x_1 \perp x_2$. Setting $S_2 = [x_1, x_2]$ and

$$R = S_2 + T_2, \quad S_2 \perp T_2,$$

we establish in a similar way the existence of a common characteristic vector x_3 of A and A^* in T_2. Obviously $x_1 \perp x_3$ and $x_2 \perp x_3$. Continuing this process, we obtain n pairwise orthogonal common characteristic vectors $x_1, x_2, \ldots, x_n$ of A and A^*:

$$\left.\begin{aligned} Ax_k = \lambda_k x_k, \quad A^* x_k = \bar{\lambda}_k x_k \quad (x_k \neq o), \\ (x_i x_k) = 0, \quad \text{for} \quad i \neq k \end{aligned} \quad (i, k = 1, 2, \ldots, n). \right\} \tag{63}$$

The vectors $x_1, x_2, \ldots, x_n$ can be normalized without violating (63).

Thus we have proved that a normal operator always has a complete orthonormal system of characteristic vectors.[26]

Since $\lambda_k = \lambda_l$ always implies that $\bar{\lambda}_k = \bar{\lambda}_l$, it follows from (63) that:

1. *If A is a normal operator, every characteristic vector of A is a characteristic vector of the adjoint operator A^**, i.e., if A is a normal operator, then A and A^* have the same characteristic vectors.

Suppose now, conversely, that a linear operator A has a complete orthonormal system of characteristic vectors:

$$Ax_k = \lambda_k x_k, \quad (x_i x_k) = \delta_{ik} \qquad (i, k = 1, 2, \ldots, n).$$

We shall show that A is then a normal operator. For let us set:

$$y_l = A^* x_l - \bar{\lambda}_l x_l.$$

Then

$$\begin{aligned} (x_k y_l) &= (x_k, A^* x_l) - \lambda_l (x_k x_l) = (Ax_k, x_l) - \lambda_l (x_k x_l) \\ &= (\lambda_k - \lambda_l) \delta_{kl} = 0 \qquad (k, l = 1, 2, \ldots, n). \end{aligned}$$

Hence it follows that

[26] Here, and in what follows, we mean by a complete orthonormal system of vectors an orthonormal system of n vectors, where n is the dimension of the space.

$$y_l = A^* x_l - \bar{\lambda}_l x_l = o \qquad (l = 1, 2, \ldots, n),$$

i.e., that (63) holds.

But then

$$AA^* x_k = \lambda_k \bar{\lambda}_k x_k \text{ and } A^* A x_k = \lambda_k \bar{\lambda}_k x_k \qquad (k = 1, 2, \ldots, n),$$

or

$$AA^* = A^* A.$$

Thus we have obtained the following 'internal' (spectral) characterization of a normal operator A (apart from the 'external' one: $AA^* = A^*A$):

THEOREM 4: *A linear operator is normal if and only if it has a complete orthonormal system of characteristic vectors.*

In particular, we have shown that a normal operator is always of simple structure.

Let A be a normal operator with the characteristic values $\lambda_1, \lambda_2, \ldots, \lambda_n$. Using the Lagrange interpolation formula, we define two polynomials $p(\lambda)$ and $q(\lambda)$ by the conditions

$$p(\lambda_k) = \bar{\lambda}_k, \quad q(\bar{\lambda}_k) = \lambda_k \qquad (k = 1, 2, \ldots, n).$$

Then by (63)

$$A^* = p(A), \qquad A = q(A^*); \tag{64}$$

i.e.:

2. *If A is a normal operator, then each of the operators A and A^* can be represented as a polynomial in the other; these two polynomials are determined by the characteristic values of A.*

Let S be an invariant subspace of R for a normal operator A and let $R = S + T$, $S \perp T$. Then by § 8, 5., the subspace T is invariant with respect to A^*. But $A = q(A^*)$, where $q(\lambda)$ is a polynomial. Therefore T is also invariant with respect to A. Thus:

3. *If S is an invariant subspace with respect to a normal operator A and T is the orthogonal complement of S, then T is also an invariant subspace for A.*

2. Let us now discuss the spectrum of a hermitian operator. Since a hermitian operator H is a special form of a normal operator, by what we have proved it has a complete orthonormal system of characteristic vectors:

$$Hx_k = \lambda_k x_k, \quad (x_k x_l) = \delta_{kl} \qquad (k, l = 1, 2, \ldots, n). \tag{65}$$

From $H^* = H$ it follows that

$$\bar{\lambda}_k = \lambda_k \qquad (k = 1, 2, \ldots, n), \tag{66}$$

i.e., all the characteristic values of a hermitian operator $\boldsymbol{H}$ are real.

It is not difficult to see that, conversely, a normal operator with real characteristic values is always hermitian. For from (65), (66), and

$$\boldsymbol{H}^*\boldsymbol{x}_k = \lambda_k \boldsymbol{x}_k \qquad (k = 1, 2, \ldots, n)$$

it follows that

$$\boldsymbol{H}^*\boldsymbol{x}_k = \boldsymbol{H}\boldsymbol{x}_k \qquad (k = 1, 2, \ldots, n),$$

i.e.,

$$\boldsymbol{H}^* = \boldsymbol{H}.$$

We have obtained the following 'internal' characterization of a hermitian operator (apart from the 'external' one: $\boldsymbol{H}^* = \boldsymbol{H}$):

THEOREM 5: *A linear operator $\boldsymbol{H}$ is hermitian if and only if it has a complete orthonormal system of characteristic vectors with real characteristic values.*

Let us now discuss the spectrum of a unitary operator. Since a unitary operator $\boldsymbol{U}$ is normal, it has a complete orthonormal system of characteristic vectors

$$\boldsymbol{U}\boldsymbol{x}_k = \lambda_k \boldsymbol{x}_k, \quad (\boldsymbol{x}_k \boldsymbol{x}_l) = \delta_{kl} \qquad (k, l = 1, 2, \ldots, n), \tag{67}$$

where

$$\boldsymbol{U}^*\boldsymbol{x}_k = \bar{\lambda}_k \boldsymbol{x}_k \qquad (k = 1, 2, \ldots, n). \tag{68}$$

From $\boldsymbol{U}\boldsymbol{U}^* = \boldsymbol{E}$ we find:

$$\lambda_k \bar{\lambda}_k = 1. \tag{69}$$

Conversely, from (67), (68), and (69) it follows that $\boldsymbol{U}\boldsymbol{U}^* = \boldsymbol{E}$. Thus, among the normal operators a unitary operator is distinguished by the fact that all its characteristic values have modulus 1.

We have thus obtained the following 'internal' characterization of a unitary operator (apart from the 'external' one: $\boldsymbol{U}\boldsymbol{U}^* = \boldsymbol{E}$):

THEOREM 6: *A linear operator is unitary if and only if it has a complete orthonormal system of characteristic vectors with characteristic values of modulus* 1.

Since in an orthonormal basis a normal (hermitian, unitary) matrix corresponds to a normal (hermitian, unitary) operator, we obtain the following propositions:

THEOREM 4′: *A matrix A is normal if and only if it is unitarily similar to a diagonal matrix*:

$$A = U \| \lambda_i \delta_{ik} \|_1^n U^{-1} \quad (U^* = U^{-1}). \tag{70}$$

THEOREM 5′: *A matrix H is hermitian if and only if it is unitarily similar to a diagonal matrix with real diagonal elements:*

$$H = U \| \lambda_i \delta_{ik} \|_1^n U^{-1} \quad (U^* = U^{-1};\ \lambda_i = \bar{\lambda}_i;\ i = 1, 2, \ldots, n). \tag{71}$$

THEOREM 6′: *A matrix U is unitary if and only if it is unitarily similar to a diagonal matrix with diagonal elements of modulus* 1:

$$U = U_1 \| \lambda_i \delta_{ik} \|_1^n U_1^{-1} \quad (U_1^* = U_1^{-1};\ |\lambda_i| = 1;\ i = 1, 2, \ldots, n). \tag{72}$$

§ 11. Positive-Semidefinite and Positive-Definite Hermitian Operators

1. We introduce the following definition:

DEFINITION 9: *A hermitian operator* $\boldsymbol{H}$ *is called positive semidefinite if for every vector* $\boldsymbol{x}$ *of* $\boldsymbol{R}$

$$(\boldsymbol{Hx}, \boldsymbol{x}) \geqq 0,$$

and positive definite if for every vector $\boldsymbol{x} \neq \boldsymbol{o}$ *of* $\boldsymbol{R}$

$$(\boldsymbol{Hx}, \boldsymbol{x}) > 0.$$

If a vector $\boldsymbol{x}$ is given by its coordinates $x_1, x_2, \ldots, x_n$ in an arbitrary orthonormal basis, then $(\boldsymbol{Hx}, \boldsymbol{x})$, as is easy to see, is a hermitian form in the variables $x_1, x_2, \ldots, x_n$; and to a positive-semidefinite (positive-definite) operator there corresponds a positive-semidefinite (positive-definite) hermitian form (see § 1).

We choose an orthonormal basis $\boldsymbol{x}_1, \boldsymbol{x}_2, \ldots, \boldsymbol{x}_n$ of characteristic vectors of $\boldsymbol{H}$:

$$\boldsymbol{Hx}_k = \lambda_k \boldsymbol{x}_k, \quad (\boldsymbol{x}_k \boldsymbol{x}_l) = \delta_{kl} \qquad (k, l = 1, 2, \ldots, n). \tag{73}$$

Then, setting $\boldsymbol{x} = \sum_{k=1}^{n} \xi_k \boldsymbol{x}_k$, we have

$$(\boldsymbol{Hx}, \boldsymbol{x}) = \sum_{k=1}^{n} \lambda_k |\xi_k|^2 \qquad (k = 1, 2, \ldots, n).$$

Hence we easily deduce the 'internal' characterizations of positive-semidefinite and positive-definite operators:

THEOREM 7: *A hermitian operator is positive semidefinite (positive definite) if and only if all its characteristic values are non-negative (positive).*

From what we have shown, it follows that a positive-definite hermitian operator is non-singular and positive semidefinite.

Let $\boldsymbol{H}$ be a positive-semidefinite hermitian operator. The equation (73) holds for $\boldsymbol{H}$ with $\lambda_k \geqq 0$ $(k=1, 2, \ldots, n)$. We set $\varrho_k = \sqrt{\lambda_k} \geqq 0$ $(k=1, 2, 3, \ldots, n)$ and define a linear operator $\boldsymbol{F}$ by the equation

$$\boldsymbol{F}\boldsymbol{x}_k = \varrho_k \boldsymbol{x}_k \qquad (k=1, 2, \ldots, n). \tag{74}$$

Then $\boldsymbol{F}$ is also a positive-semidefinite operator and

$$\boldsymbol{F}^2 = \boldsymbol{H}. \tag{75}$$

We shall call the positive-semidefinite hermitian operator $\boldsymbol{F}$ connected with $\boldsymbol{H}$ by (75) the *arithmetical square root* of $\boldsymbol{H}$ and shall denote it by

$$\boldsymbol{F} = \sqrt{\boldsymbol{H}}.$$

If $\boldsymbol{H}$ is positive definite, then $\boldsymbol{F}$ is also positive definite.

We define the Lagrange interpolation polynomial $g(\lambda)$ by the equations

$$g(\lambda_k) = \varrho_k \left(= \sqrt{\lambda_k}\right) \quad (k=1, 2, \ldots, n). \tag{76}$$

Then from (73), (74), and (76) it follows that:

$$\boldsymbol{F} = g(\boldsymbol{H}). \tag{77}$$

The latter equation shows that $\sqrt{\boldsymbol{H}}$ *is a polynomial in* $\boldsymbol{H}$ *and is uniquely determined when the positive-semidefinite hermitian operator* $\boldsymbol{H}$ *is given* (the coefficients of $g(\lambda)$ depend on the characteristic values of $\boldsymbol{H}$).

2. Examples of positive-semidefinite hermitian operators are $\boldsymbol{A}\boldsymbol{A}^*$ and $\boldsymbol{A}^*\boldsymbol{A}$, where $\boldsymbol{A}$ is an arbitrary linear operator in the given space. Indeed, for an arbitrary vector $\boldsymbol{x}$,

$$(\boldsymbol{A}\boldsymbol{A}^*\boldsymbol{x}, \boldsymbol{x}) = (\boldsymbol{A}^*\boldsymbol{x}, \boldsymbol{A}^*\boldsymbol{x}) \geqq 0,$$
$$(\boldsymbol{A}^*\boldsymbol{A}\boldsymbol{x}, \boldsymbol{x}) = (\boldsymbol{A}\boldsymbol{x}, \boldsymbol{A}\boldsymbol{x}) \geqq 0.$$

If $\boldsymbol{A}$ is non-singular, then $\boldsymbol{A}\boldsymbol{A}^*$ and $\boldsymbol{A}^*\boldsymbol{A}$ are positive-definite hermitian operators.

The operators $\boldsymbol{A}\boldsymbol{A}^*$ and $\boldsymbol{A}^*\boldsymbol{A}$ are sometimes called the *left norm* and *right norm* of $\boldsymbol{A}$. $\sqrt{\boldsymbol{A}\boldsymbol{A}^*}$ and $\sqrt{\boldsymbol{A}^*\boldsymbol{A}}$ are called the *left modulus* and *right modulus* of $\boldsymbol{A}$.

For a normal operator the left and right norms, and hence the left and right moduli, are equal.[27]

---

[27] For a detailed study of normal operators, see [168]. In this paper necessary and sufficient conditions for the product of two normal operators to be normal are established.

§ 12. Polar Decomposition of a Linear Operator in a Unitary Space. Cayley's Formulas

1. We shall prove the following theorem:[28]

THEOREM 8: *Every linear operator A in a unitary space can be represented in the forms*

$$A = HU, \tag{78}$$

$$A = U_1 H_1, \tag{79}$$

where H, H_1 are positive-semidefinite hermitian operators and U, U_1 are unitary operators. A is normal if and only if in (78) (*or* (79)) *the factors H and U (H_1 and U_1) are permutable.*

Proof. From (78) and (79) it follows that H and H_1 are the left and right moduli, respectively, of A.

For

$$AA^* = HUU^*H = H^2,\ A^*A = H_1U_1^*U_1H_1 = H_1^2.$$

Note that it is sufficient to establish (78), since by applying this decomposition to A^* we obtain $A^* = HU$ and hence

$$A = U^{-1}H,$$

i.e., the decomposition (79) for A.

We begin by establishing (78) in the special case where A is non-singular ($|A| \neq 0$). We set:

$$H = \sqrt{AA^*} \text{ (here } |H|^2 = |A|^2 \neq 0),\ U = H^{-1}A$$

and verify that U is unitary:

$$UU^* = H^{-1}AA^*H^{-1} = H^{-1}H^2H^{-1} = E.$$

Note that in this case not only the first factor H in (78), but also the second factor U is uniquely determined by the non-singular operator A.

We now consider the general case where A may be singular.

First of all we observe that a complete orthonormal system of characteristic vectors of the right norm of A is always transformed by A into an orthogonal system of vectors. For let

$$A^*Ax_k = \varrho_k^2 x_k \quad [(x_k x_l) = \delta_{kl},\ \varrho_k \geqq 0;\ k, l = 1, 2, \ldots, n].$$

Then

$$(Ax_k, Ax_l) = (A^*Ax_k, x_l) = \varrho_k^2 \cdot (x_k x_l) = 0 \quad (k \neq l).$$

---

[28] See [168], p. 77.

Here

$$|\boldsymbol{Ax}_k|^2 = (\boldsymbol{Ax}_k, \boldsymbol{Ax}_k) = \varrho_k^2 \quad (k = 1, 2, \ldots, n).$$

Therefore there exists an orthonormal system of vectors $\boldsymbol{z}_1, \boldsymbol{z}_2, \ldots, \boldsymbol{z}_n$ such that

$$\boldsymbol{Ax}_k = \varrho_k \boldsymbol{z}_k \quad [(\boldsymbol{z}_k \boldsymbol{z}_l) = \delta_{kl};\ k, l = 1, 2, \ldots, n]. \tag{80}$$

We define linear operators $\boldsymbol{H}$ and $\boldsymbol{U}$ by the equations

$$\boldsymbol{Ux}_k = \boldsymbol{z}_k, \quad \boldsymbol{Hz}_k = \varrho_k \boldsymbol{z}_k. \tag{81}$$

From (80) and (81) we find:

$$\boldsymbol{A} = \boldsymbol{HU}.$$

Here $\boldsymbol{H}$ is, by (81), a positive-semidefinite hermitian operator, because it has a complete orthonormal system of characteristic vectors $\boldsymbol{z}_1, \boldsymbol{z}_2, \ldots, \boldsymbol{z}_n$ with non-negative characteristic values $\varrho_1, \varrho_2, \ldots, \varrho_n$; and $\boldsymbol{U}$ is a unitary operator, because it carries the orthonormal system of vectors $\boldsymbol{x}_1, \boldsymbol{x}_2, \ldots, \boldsymbol{x}_n$ into the orthonormal system $\boldsymbol{z}_1, \boldsymbol{z}_2, \ldots, \boldsymbol{z}_n$.

Thus we can take it as proved that an arbitrary linear operator $\boldsymbol{A}$ has decompositions (78) and (79), that the hermitian factors $\boldsymbol{H}$ and $\boldsymbol{H}_1$ are always uniquely determined by $\boldsymbol{x}$ (they are the left and right moduli of $\boldsymbol{A}$, respectively) and that the unitary factors $\boldsymbol{U}$ and $\boldsymbol{U}_1$ are uniquely determined only when $\boldsymbol{A}$ is non-singular.

From (78) we find easily:

$$\boldsymbol{AA}^* = \boldsymbol{H}^2, \quad \boldsymbol{A}^*\boldsymbol{A} = \boldsymbol{U}^{-1}\boldsymbol{H}^2\boldsymbol{U}. \tag{82}$$

If $\boldsymbol{A}$ is a normal operator ($\boldsymbol{AA}^* = \boldsymbol{A}^*\boldsymbol{A}$), then it follows from (82) that

$$\boldsymbol{H}^2\boldsymbol{U} = \boldsymbol{UH}^2. \tag{83}$$

Since $\boldsymbol{H} = \sqrt{\boldsymbol{H}^2} = g(\boldsymbol{H}^2)$ (see § 11), (83) shows that $\boldsymbol{U}$ and $\boldsymbol{H}$ commute. Conversely, if $\boldsymbol{H}$ and $\boldsymbol{U}$ commute, then it follows from (82) that $\boldsymbol{A}$ is normal. This completes the proof of the theorem.[29]

---

[29] If the characteristic values $\lambda_1, \lambda_2, \ldots, \lambda_n$ and $\varrho_1, \varrho_2, \ldots, \varrho_n$ of the linear operator $\boldsymbol{A}$ and its left modulus $\boldsymbol{H} = \sqrt{\boldsymbol{AA}^*}$ (by (82) $\varrho_1, \varrho_2, \ldots, \varrho_n$ are also the characteristic values of the right modulus $\boldsymbol{H}_1 = \sqrt{\boldsymbol{A}^*\boldsymbol{A}}$) are so numbered that

$$|\lambda_1| \geqq |\lambda_2| \geqq \cdots \geqq |\lambda_n|, \quad \varrho_1 \geqq \varrho_2 \geqq \cdots \geqq \varrho_n,$$

then (see [379], or [153] and [296]) the following *inequality of Weyl* holds:

$$|\lambda_1| \leqq \varrho_1, \quad |\lambda_1| + |\lambda_2| \leqq \varrho_1 + \varrho_2, \quad \ldots, \quad |\lambda_1| + \cdots + |\lambda_n| \leqq \varrho_1 + \cdots + \varrho_n.$$

It is hardly necessary to mention that together with the operator equations (78) and (79) the corresponding matrix equations hold.

The decompositions (78) and (79) are analogues to the representation of a complex number z in the form $z = ru$, where $r = |z|$ and $|u| = 1$.

2. Now let $\boldsymbol{x}_1, \boldsymbol{x}_2, \ldots, \boldsymbol{x}_n$ be a complete orthonormal system of characteristic vectors of the arbitrary unitary operator $\boldsymbol{U}$. Then

$$\boldsymbol{U}\boldsymbol{x}_k = e^{if_k}\boldsymbol{x}_k, \quad (\boldsymbol{x}_k\boldsymbol{x}_l) = \delta_{kl} \quad (k, l = 1, 2, \ldots, n). \tag{84}$$

where the f_k $(k = 1, 2, \ldots, n)$ are real numbers. We define a hermitian operator $\boldsymbol{F}$ by the equations

$$\boldsymbol{F}\boldsymbol{x}_k = f_k\boldsymbol{x}_k \quad (k = 1, 2, \ldots, n). \tag{85}$$

From (84) and (85) it follows that:[30]

$$\boldsymbol{U} = e^{i\boldsymbol{F}}. \tag{86}$$

Thus, a unitary operator $\boldsymbol{U}$ is always representable in the form (86), where $\boldsymbol{F}$ is a hermitian operator. Conversely, if $\boldsymbol{F}$ is a hermitian operator, then $\boldsymbol{U} = e^{i\boldsymbol{F}}$ is unitary.

The decompositions (78) and (79) together with (86) give the following equations:

$$\boldsymbol{A} = \boldsymbol{H}e^{i\boldsymbol{F}}, \tag{87}$$

$$\boldsymbol{A} = e^{i\boldsymbol{F}_1}\boldsymbol{H}_1 \tag{88}$$

where $\boldsymbol{H}, \boldsymbol{F}, \boldsymbol{H}_1$, and $\boldsymbol{F}_1$ are hermitian operators, with $\boldsymbol{H}$ and $\boldsymbol{H}_1$ positive semidefinite.

The decompositions (87) and (88) are analogues to the representation of a complex number z in the form $z = re^{i\varphi}$, where $r \geqq 0$ and φ are real numbers.

Note. In (86), the operator $\boldsymbol{F}$ is not uniquely determined by $\boldsymbol{U}$. For $\boldsymbol{F}$ is defined by means of the numbers f_k $(k = 1, 2, \ldots, n)$ and we can add to each of these numbers an arbitrary multiple of 2π without changing the original equations (84). By choosing these multiples of 2π suitably we can assume that $e^{if_k} = e^{if_l}$ always implies that $f_k = f_l$ $(1 \leqq k, l \leqq n)$. Then we can determine the interpolation polynomial $g(\lambda)$ by the equations

$$g(e^{if_k}) = f_k \quad (k = 1, 2, \ldots, n). \tag{89}$$

[30] $e^{i\boldsymbol{F}} = r(\boldsymbol{F})$, where $r(\lambda)$ is the Lagrange interpolation polynomial for the function $e^{i\lambda}$ at the places $f_1, f_2, \ldots, f_n$.

From (84), (85), and (89) it follows that

$$\boldsymbol{F} = g(\boldsymbol{U}) = g(e^{i\boldsymbol{F}}). \tag{90}$$

Similarly we can normalize the choice of $\boldsymbol{F}_1$ so that

$$\boldsymbol{F}_1 = h(\boldsymbol{U}_1) = h(e^{i\boldsymbol{F}_1}), \tag{91}$$

where $h(\lambda)$ is a polynomial.

By (90) and (91), the permutability of $\boldsymbol{H}$ and $\boldsymbol{U}$ ($\boldsymbol{H}_1$ and $\boldsymbol{U}_1$) implies that of $\boldsymbol{H}$ and $\boldsymbol{F}$ ($\boldsymbol{H}_1$ and $\boldsymbol{F}_1$), and vice versa. Therefore, by Theorem 8, $\boldsymbol{A}$ is normal if and only if in (87) $\boldsymbol{H}$ and $\boldsymbol{F}$ (or, in (88), $\boldsymbol{H}_1$ and $\boldsymbol{F}_1$) are permutable, provided the characteristic values of $\boldsymbol{F}$ (or $\boldsymbol{F}_1$) are suitably normalized.

The formula (86) is based on the fact that the functional dependence

$$\mu = e^{if} \tag{92}$$

carries n arbitrary numbers $f_1, f_2, \dots, f_n$ on the real axis into certain numbers $\mu_1, \mu_2, \dots, \mu_n$ on the unit circle $|\mu| = 1$, and vice versa.

The transcendental dependence (92) can be replaced by the rational dependence

$$\mu = \frac{1+if}{1-if}, \tag{93}$$

which carries the real axis $f = \bar{f}$ into the circle $|\mu| = 1$; here the point at infinity on the real axis goes over into the point $\mu = -1$. From (93), we find:

$$f = i\,\frac{1-\mu}{1+\mu}. \tag{94}$$

Repeating the arguments which have led us to the formula (86), we obtain from (93) and (94) the pair of inverse formulas:

$$\left.\begin{aligned} \boldsymbol{U} &= (\boldsymbol{E} + i\boldsymbol{F})(\boldsymbol{E} - i\boldsymbol{F})^{-1}, \\ \boldsymbol{F} &= i(\boldsymbol{E} - \boldsymbol{U})(\boldsymbol{E} + \boldsymbol{U})^{-1}, \end{aligned}\right\} \tag{95}$$

We have thus obtained *Cayley's formulas*. These formulas establish a one-to-one correspondence between arbitrary hermitian operators $\boldsymbol{F}$ and those unitary operators $\boldsymbol{U}$ that do not have the characteristic value -1.[31]

[31] The exceptional value -1 can be replaced by any number μ_0 ($|\mu_0| = 1$). For this purpose, we have to take instead of (93) a fractional-linear function mapping the real axis $f = \bar{f}$ onto the circle $|\mu| = 1$ and carrying the point $f = \infty$ into $\mu = \mu_0$. The formulas (94) and (95) can be modified correspondingly.

The formulas (86), (87), (88), and (95) are obviously valid when we replace all the operators by the corresponding matrices.

§ 13. Linear Operators in a Euclidean Space

1. We consider an n-dimensional euclidean space $\boldsymbol{R}$. Let $\boldsymbol{A}$ be a linear operator in $\boldsymbol{R}$.

DEFINITION 10: *The linear operator $\boldsymbol{A}^{\mathsf{T}}$ is called the transposed operator of $\boldsymbol{A}$ (or the transpose of $\boldsymbol{A}$) if for any two vectors $\boldsymbol{x}$ and $\boldsymbol{y}$ of $\boldsymbol{R}$:*

$$(\boldsymbol{A}\boldsymbol{x}, \boldsymbol{y}) = (\boldsymbol{x}, \boldsymbol{A}^{\mathsf{T}}\boldsymbol{y}). \tag{96}$$

The existence and uniqueness of the transposed operator is established in exactly the same way as was done in § 8 for the adjoint operator in a unitary space.

The transposed operator has the following properties:

1. $(\boldsymbol{A}^{\mathsf{T}})^{\mathsf{T}} = \boldsymbol{A}$,
2. $(\boldsymbol{A} + \boldsymbol{B})^{\mathsf{T}} = \boldsymbol{A}^{\mathsf{T}} + \boldsymbol{B}^{\mathsf{T}}$,
3. $(\alpha\boldsymbol{A})^{\mathsf{T}} = \alpha\boldsymbol{A}^{\mathsf{T}}$ (α a real number),
4. $(\boldsymbol{A}\boldsymbol{B})^{\mathsf{T}} = \boldsymbol{B}^{\mathsf{T}}\boldsymbol{A}^{\mathsf{T}}$.

We introduce a number of definitions.

DEFINITION 11: *A linear operator $\boldsymbol{A}$ is called normal if*

$$\boldsymbol{A}\boldsymbol{A}^{\mathsf{T}} = \boldsymbol{A}^{\mathsf{T}}\boldsymbol{A}.$$

DEFINITION 12: *A linear operator $\boldsymbol{S}$ is called symmetric if*

$$\boldsymbol{S}^{\mathsf{T}} = \boldsymbol{S}.$$

DEFINITION 13: *A symmetric operator $\boldsymbol{S}$ is called positive semidefinite if for every vector $\boldsymbol{x}$ of $\boldsymbol{R}$*

$$(\boldsymbol{S}\boldsymbol{x}, \boldsymbol{x}) \geqq 0.$$

DEFINITION 14: *A symmetric operator $\boldsymbol{S}$ is called positive definite if for every vector $\boldsymbol{x} \neq \boldsymbol{o}$ of $\boldsymbol{R}$*

$$(\boldsymbol{S}\boldsymbol{x}, \boldsymbol{x}) > 0.$$

DEFINITION 15: *A linear operator $\boldsymbol{K}$ is called skew-symmetric if*

$$\boldsymbol{K}^{\mathsf{T}} = -\boldsymbol{K}.$$

An arbitrary linear operator $\boldsymbol{A}$ can always be represented uniquely in the form

$$\boldsymbol{A} = \boldsymbol{S} + \boldsymbol{K}, \tag{97}$$

where $\boldsymbol{S}$ is symmetric and $\boldsymbol{K}$ is skew-symmetric.

For it follows from (97) that

$$\boldsymbol{A}^{\mathsf{T}} = \boldsymbol{S} - \boldsymbol{K}. \tag{98}$$

From (97) and (98) we have:

$$\boldsymbol{S} = \frac{1}{2}(\boldsymbol{A} + \boldsymbol{A}^{\mathsf{T}}), \quad \boldsymbol{K} = \frac{1}{2}(\boldsymbol{A} - \boldsymbol{A}^{\mathsf{T}}). \tag{99}$$

Conversely, (99) defines a symmetric operator $\boldsymbol{S}$ and a skew-symmetric operator $\boldsymbol{K}$ for which (97) holds.

$\boldsymbol{S}$ and $\boldsymbol{K}$ are called respectively the *symmetric component* and the *skew-symmetric component* of $\boldsymbol{A}$.

DEFINITION 16: *An operator $\boldsymbol{Q}$ is called orthogonal if it preserves the metric of the space, i.e., if for any two vectors $\boldsymbol{x}, \boldsymbol{y}$ of $\boldsymbol{R}$*

$$(\boldsymbol{Q}\boldsymbol{x}, \boldsymbol{Q}\boldsymbol{y}) = (\boldsymbol{x}, \boldsymbol{y}). \tag{100}$$

By (96), equation (100) can be written as: $(\boldsymbol{x}, \boldsymbol{Q}^{\mathsf{T}}\boldsymbol{Q}\boldsymbol{y}) = (\boldsymbol{x}, \boldsymbol{y})$. Hence

$$\boldsymbol{Q}^{\mathsf{T}}\boldsymbol{Q} = \boldsymbol{E}. \tag{101}$$

Conversely, (101) implies (100) (for arbitrary vectors $\boldsymbol{x}, \boldsymbol{y}$).[32] From (101) it follows that: $|\boldsymbol{Q}|^2 = 1$, i.e.,

$$|\boldsymbol{Q}| = \pm 1.$$

We shall call $\boldsymbol{Q}$ an orthogonal operator *of the first kind* (or *proper*) if $|\boldsymbol{Q}| = 1$ and *of the second kind* (or *improper*) if $|\boldsymbol{Q}| = -1$.

Symmetric, skew-symmetric, and orthogonal operators are special forms of a normal operator.

We consider an arbitrary orthonormal basis in the given euclidean space. Suppose that in this basis $\boldsymbol{A}$ corresponds to the matrix $A = \| a_{ik} \|_1^n$ (here all the a_{ik} are real numbers). The reader will have no difficulty in showing that the transposed operator $\boldsymbol{A}^{\mathsf{T}}$ corresponds in this basis to the transposed matrix $A^{\mathsf{T}} = \| a_{ik}^{\mathsf{T}} \|_1^n$, where $a_{ik}^{\mathsf{T}} = a_{ki}$ $(i, k = 1, 2, \ldots, n)$. Hence it follows that in an orthonormal basis a normal operator $\boldsymbol{A}$ corresponds to a normal

[32] The orthogonal operators in a euclidean space form a group, the so-called orthogonal group.

matrix A $(AA^{\mathsf{T}} = A^{\mathsf{T}}A)$, a symmetric operator $\boldsymbol{S}$ to a symmetric matrix $S = \| s_{ik} \|_1^n$ $(S^{\mathsf{T}} = S)$, a skew-symmetric operator K to a skew-symmetric matrix $K = \| k_{ij} \|_1^n$ $(K^{\mathsf{T}} = -K)$ and, finally, an orthogonal operator $\boldsymbol{Q}$ to an orthogonal matrix $Q = \| q_{ik} \|_1^n$ $(QQ^{\mathsf{T}} = E)$.[33]

Just as was done in § 8 for the adjoint operator, we can here make the following statement for the transposed operator:

If a subspace $\boldsymbol{S}$ *of* $\boldsymbol{R}$ *is invariant with respect to a linear operator* $\boldsymbol{A}$, *then the orthogonal complement* $\boldsymbol{T}$ *of* $\boldsymbol{S}$ *in* $\boldsymbol{R}$ *is invariant with respect to* $\boldsymbol{A}^{\mathsf{T}}$.

2. For the study of linear operators in a euclidean space $\boldsymbol{R}$, we extend $\boldsymbol{R}$ to a unitary space $\tilde{\boldsymbol{R}}$. This extension is made in the following way:

1. The vectors of $\boldsymbol{R}$ are called 'real' vectors.

2. We introduce 'complex' vectors $\boldsymbol{z} = \boldsymbol{x} + i\boldsymbol{y}$, where $\boldsymbol{x}$ and $\boldsymbol{y}$ are real, i.e., $\boldsymbol{x} \in \boldsymbol{R}, \boldsymbol{y} \in \boldsymbol{R}$.

3. The operations of addition of complex vectors and of multiplication by a complex number are defined in the natural way. Then the set of all complex vectors forms an n-dimensional vector space $\tilde{\boldsymbol{R}}$ over the field of complex numbers which contains $\boldsymbol{R}$ as a subspace.

4. In $\tilde{\boldsymbol{R}}$ we introduce a hermitian metric such that in $\boldsymbol{R}$ it coincides with the existing euclidean metric. The reader can easily verify that the required hermitian metric is given in the following way:

If $\boldsymbol{z} = \boldsymbol{x} + i\boldsymbol{y}$, $\boldsymbol{w} = \boldsymbol{u} + i\boldsymbol{v}$ $(\boldsymbol{x}, \boldsymbol{y}, \boldsymbol{u}, \boldsymbol{v} \in \boldsymbol{R})$, then

$$(\boldsymbol{z}\boldsymbol{w}) = (\boldsymbol{x}\boldsymbol{u}) + (\boldsymbol{y}\boldsymbol{v}) + i[(\boldsymbol{y}\boldsymbol{u}) - (\boldsymbol{x}\boldsymbol{v})].$$

Setting $\bar{\boldsymbol{z}} = \boldsymbol{x} - i\boldsymbol{y}$ and $\bar{\boldsymbol{w}} = \boldsymbol{u} - i\boldsymbol{v}$, we have

$$(\bar{\boldsymbol{z}}\,\bar{\boldsymbol{w}}) = \overline{(\boldsymbol{z}\boldsymbol{w})}.$$

If we choose a real basis, i.e., a basis of $\boldsymbol{R}$, then $\tilde{\boldsymbol{R}}$ will be the set of all vectors with complex coordinates and $\boldsymbol{R}$ the set of all vectors with real coordinates in this basis.

Every linear operator $\boldsymbol{A}$ in $\boldsymbol{R}$ extends uniquely to a linear operator in $\tilde{\boldsymbol{R}}$:

$$\boldsymbol{A}(\boldsymbol{x} + i\boldsymbol{y}) = \boldsymbol{A}\boldsymbol{x} + i\boldsymbol{A}\boldsymbol{y}.$$

3. Among all the linear operators of $\tilde{\boldsymbol{R}}$ those that are obtainable as the result of such an extension of operators of $\boldsymbol{R}$ can be characterized by the fact that they carry $\boldsymbol{R}$ into $\boldsymbol{R}$ $(\boldsymbol{A}\boldsymbol{R} \subset \boldsymbol{R})$. These operators are called *real*.

---

[33] The papers [138], [262a], [170b] are devoted to the study of the structure of orthogonal matrices. Orthogonal matrices, like orthogonal operators, are called proper and improper according as $|Q| = +1$ or $|Q| = -1$.

In a real basis real operators are determined by real matrices, i.e., matrices with real elements.

A real operator $\boldsymbol{A}$ carries conjugate complex vectors $\boldsymbol{z}=\boldsymbol{x}+i\boldsymbol{y}$, $\bar{\boldsymbol{z}}=\boldsymbol{x}-i\boldsymbol{y}$ $(\boldsymbol{x}, \boldsymbol{y} \in \boldsymbol{R})$ into conjugate complex vectors:

$$\boldsymbol{A}\boldsymbol{z}=\boldsymbol{A}\boldsymbol{x}+i\boldsymbol{A}\boldsymbol{y},\ \boldsymbol{A}\bar{\boldsymbol{z}}=\boldsymbol{A}\boldsymbol{x}-i\boldsymbol{A}\boldsymbol{y} \quad (\boldsymbol{A}\boldsymbol{x}, \boldsymbol{A}\boldsymbol{y} \in \boldsymbol{R}).$$

The secular equation of a real operator has real coefficients, so that when it has a root λ of multiplicity p it also has the root $\bar{\lambda}$ with the multiplicity p. From $\boldsymbol{A}\boldsymbol{z}=\lambda\boldsymbol{z}$ it follows that $\boldsymbol{A}\bar{\boldsymbol{z}}=\bar{\lambda}\bar{\boldsymbol{z}}$, i.e., to conjugate characteristic values there correspond conjugate characteristic vectors.

The two-dimensional space $[\boldsymbol{z}, \bar{\boldsymbol{z}}]$ has a real basis:

$$\boldsymbol{x}=\frac{1}{2}(\boldsymbol{z}+\bar{\boldsymbol{z}}),\quad \boldsymbol{y}=\frac{1}{2i}(\boldsymbol{z}-\bar{\boldsymbol{z}}).$$

We shall call the plane in $\boldsymbol{R}$ spanned by this basis an *invariant plane* of $\boldsymbol{A}$ corresponding to the pair of characteristic values λ, $\bar{\lambda}$.

Let $\lambda=\mu+i\nu$. Then it is easy to see that

$$\boldsymbol{A}\boldsymbol{x}=\mu\boldsymbol{x}-\nu\boldsymbol{y},$$

$$\boldsymbol{A}\boldsymbol{y}=\nu\boldsymbol{x}+\mu\boldsymbol{y}.$$

We consider a real operator $\boldsymbol{A}$ of simple structure with the characteristic values:

$$\lambda_{2k-1}=\mu_k+i\nu_k,\ \lambda_{2k}=\mu_k-i\nu_k,\ \lambda_l=\mu_l\ (k=1, 2, \dots, q;\ l=2q+1, \dots, n),$$

where μ_k, ν_k, μ_l are real and $\nu_k \neq 0$ $(k=1, 2, \dots, q)$.

Then the characteristic vectors $\boldsymbol{z}_1, \boldsymbol{z}_2, \dots, \boldsymbol{z}_n$ corresponding to these characteristic values can be chosen such that

$$\boldsymbol{z}_{2k-1}=\boldsymbol{x}_k+i\boldsymbol{y}_k,\quad \boldsymbol{z}_{2k}=\boldsymbol{x}_k-i\boldsymbol{y}_k,\quad \boldsymbol{z}_l=\boldsymbol{x}_l \tag{102}$$

$$(k=1, 2, \dots, q;\ l=2q+1, \dots, n).$$

The vectors

$$\boldsymbol{x}_1, \boldsymbol{y}_1, \boldsymbol{x}_2, \boldsymbol{y}_2, \dots, \boldsymbol{x}_q, \boldsymbol{y}_q, \boldsymbol{x}_{2q+1}, \dots, \boldsymbol{x}_n \tag{103}$$

form a basis of the euclidean space $\boldsymbol{R}$. Here

---

[34] If to the characteristic value λ of the real operator $\boldsymbol{A}$ there correspond the linearly independent characteristic vectors $\boldsymbol{z}_1, \boldsymbol{z}_2, \dots, \boldsymbol{z}_p$, then to the characteristic value $\bar{\lambda}$ there correspond the linearly independent characteristic vectors $\bar{\boldsymbol{z}}_1, \bar{\boldsymbol{z}}_2, \dots, \bar{\boldsymbol{z}}_p$.

$$\begin{aligned} \boldsymbol{A}\boldsymbol{x}_k &= \mu_k \boldsymbol{x}_k - \nu_k \boldsymbol{y}_k, \\ \boldsymbol{A}\boldsymbol{y}_k &= \nu_k \boldsymbol{x}_k + \mu_k \boldsymbol{y}_k, \\ \boldsymbol{A}\boldsymbol{x}_l &= \mu_l \boldsymbol{x}_l \end{aligned} \qquad \begin{pmatrix} k = 1, 2, \ldots, q \\ l = 2q+1, \ldots, n \end{pmatrix}. \tag{104}$$

In the basis (103) there corresponds to the operator $\boldsymbol{A}$ the real quasi-diagonal matrix

$$\left\{ \left\| \begin{matrix} \mu_1 & \nu_1 \\ -\nu_1 & \mu_1 \end{matrix} \right\|, \ldots, \left\| \begin{matrix} \mu_q & \nu_q \\ -\nu_q & \mu_q \end{matrix} \right\|, \mu_{2q+1}, \ldots, \mu_n \right\}. \tag{105}$$

Thus: *For every operator* $\boldsymbol{A}$ *of simple structure in a euclidean space there exists a basis in which* $\boldsymbol{A}$ *corresponds to a matrix of the form* (105). Hence it follows that: *A real matrix of simple structure is real-similar to a canonical matrix of the form* (105) :

$$A = T \left\{ \left\| \begin{matrix} \mu_1 & \nu_1 \\ -\nu_1 & \mu_1 \end{matrix} \right\|, \ldots, \left\| \begin{matrix} \mu_q & \nu_q \\ -\nu_q & \mu_q \end{matrix} \right\|, \mu_{2q+1}, \ldots, \mu_n \right\} T^{-1} \quad (T = \overline{T}). \tag{106}$$

The transposed operator $\boldsymbol{A}^{\mathsf{T}}$ of $\boldsymbol{A}$ in $\boldsymbol{R}$ upon extension becomes the adjoint operator $\boldsymbol{A}^*$ of $\boldsymbol{A}$ in $\tilde{\boldsymbol{R}}$. Therefore: *Normal, symmetric, skew-symmetric, and orthogonal operators in* $\boldsymbol{R}$ *after the extension become normal, hermitian, hermitian multiplied by i, and unitary real operators in* $\tilde{\boldsymbol{R}}$.

It is easy to show that *for a normal operator* $\boldsymbol{A}$ *in a euclidean space a canonical basis can be chosen as an orthonormal basis* (103) *for which* (104) *holds.*[35] *Therefore a real normal matrix is always real-similar and orthogonally-similar to a matrix of the form* (105) :

$$A = Q \left\{ \left\| \begin{matrix} \mu_1 & \nu_1 \\ -\nu_1 & \mu_1 \end{matrix} \right\|, \ldots, \left\| \begin{matrix} \mu_q & \nu_q \\ -\nu_q & \mu_q \end{matrix} \right\|, \mu_{2q+1}, \ldots, \mu_n \right\} Q^{-1} \tag{107}$$
$$(Q = Q^{\mathsf{T}\,-1} = \overline{Q}).$$

All the characteristic values of a symmetric operator $\boldsymbol{S}$ in a euclidean space are real, since after the extension the operator becomes hermitian. For a symmetric operator $\boldsymbol{S}$ we must set $q = 0$ in (104). Then we obtain:

$$\boldsymbol{S}\boldsymbol{x}_l = \mu_l \boldsymbol{x}_l \quad [(\boldsymbol{x}_k \boldsymbol{x}_l) = \delta_{kl};\ k, l = 1, 2, \ldots, n]. \tag{108}$$

A symmetric operator $\boldsymbol{S}$ *in a euclidean space always has an orthonormal system of characteristic vectors with real characteristic values.*[36] Therefore:

---

[35] The orthonormality of the basis (102) in the hermitian metric implies the orthonormality of the basis (103) in the corresponding euclidean metric.

[36] The symmetric operator $\boldsymbol{S}$ is positive semidefinite if in (108) all $\mu_l \geqq 0$ and positive definite if all $\mu_l > 0$.

A real symmetric matrix is always real-similar and orthogonally-similar to a diagonal matrix:

$$S = Q\{\mu_1, \mu_2, \ldots, \mu_n\}Q^{-1} \quad (Q = Q^{\mathsf{T}-1} = \overline{Q}). \tag{109}$$

All the characteristic values of a skew-symmetric operator $\boldsymbol{K}$ in a euclidean space are pure imaginary (after the extension the operator is i times a hermitian operator). For a skew-symmetric operator we must set in (104):

$$\mu_1 = \mu_2 = \cdots = \mu_q = \mu_{2q+1} = \cdots = \mu_n = 0$$

then the formulas assume the form

$$\begin{aligned} \boldsymbol{K}\boldsymbol{x}_k &= -\nu_k \boldsymbol{y}_k, \\ \boldsymbol{K}\boldsymbol{y}_k &= \nu_k \boldsymbol{x}_k, \qquad (k = 1, 2, \ldots, q;\; l - 2q + 1, \ldots, n). \\ \boldsymbol{K}\boldsymbol{x}_l &= \boldsymbol{o} \end{aligned} \tag{110}$$

Since $\boldsymbol{K}$ is a normal operator, the basis (103) can be assumed to be orthonormal. Thus: *Every real skew-symmetric matrix is real-similar and orthogonally-similar to a canonical skew-symmetric matrix:*

$$K = Q\left\{\left\|\begin{matrix} 0 & \nu_1 \\ -\nu_1 & 0 \end{matrix}\right\|, \ldots, \left\|\begin{matrix} 0 & \nu_q \\ -\nu_q & 0 \end{matrix}\right\|, 0, \ldots, 0\right\}Q^{-1} \quad (Q = Q^{\mathsf{T}-1} = \bar{Q}). \tag{111}$$

All the characteristic values of an orthogonal operator $\boldsymbol{Q}$ in a euclidean space are of modulus 1 (upon extension the operator becomes unitary). Therefore in the case of an orthogonal operator we must set in (104):

$$\mu_k^2 + \nu_k^2 = 1, \ \mu_l = \pm 1 \qquad (k = 1, 2, \ldots, q;\, l = 2q + 1, \ldots, n).$$

For this basis (103) can be assumed to be orthonormal. The formulas (104) can be represented in the form

$$\begin{aligned} \boldsymbol{Q}\boldsymbol{x}_k &= \boldsymbol{x}_k \cos\varphi_k - \boldsymbol{y}_k \sin\varphi_k, \\ \boldsymbol{Q}\boldsymbol{y}_k &= \boldsymbol{x}_k \sin\varphi_k + \boldsymbol{y}_k \cos\varphi_k, \\ \boldsymbol{Q}\boldsymbol{x}_l &= \pm\boldsymbol{x}_l \end{aligned} \qquad \begin{pmatrix} k = 1, 2, \ldots, q, \\ l = 2q + 1, \ldots, n \end{pmatrix}. \tag{112}$$

From what we have shown, it follows that: *Every real orthogonal matrix is real-similar and orthogonally-similar to a canonical orthogonal matrix:*

$$Q = Q_1\left\{\left\|\begin{matrix} \cos\varphi_1 & \sin\varphi_1 \\ -\sin\varphi_1 & \cos\varphi_1 \end{matrix}\right\|, \ldots, \left\|\begin{matrix} \cos\varphi_q & \sin\varphi_q \\ -\sin\varphi_q & \cos\varphi_q \end{matrix}\right\|, \pm 1, \ldots, \pm 1\right\}Q_1^{-1} \tag{113}$$

$$(Q_1 = Q_1^{\mathsf{T}-1} = \overline{Q}_1).$$

Example. We consider an arbitrary finite rotation around the point O in a three-dimensional space. It carries a directed segment $\overrightarrow{OA}$ into a directed segment $\overrightarrow{OB}$ and can therefore be regarded as an operator $\boldsymbol{Q}$ in a three-dimensional vector space (formed by all possible segments $\overrightarrow{OA}$). This operator is linear and orthogonal. Its determinant is $+1$, since $\boldsymbol{Q}$ does not change the orientation of the space.

Thus, $\boldsymbol{Q}$ is a proper orthogonal operator. For this operator the formulas (112) look as follows:

$$\begin{aligned} \boldsymbol{Q}\boldsymbol{x}_1 &= \boldsymbol{x}_1 \cos\varphi - \boldsymbol{y}_1 \sin\varphi, \\ \boldsymbol{Q}\boldsymbol{y}_1 &= \boldsymbol{x}_1 \sin\varphi + \boldsymbol{y}_1 \cos\varphi, \\ \boldsymbol{Q}\boldsymbol{x}_2 &= \pm\, \boldsymbol{x}_2 . \end{aligned}$$

From the equation $|\boldsymbol{Q}| = 1$ it follows that $\boldsymbol{Q}\boldsymbol{x}_2 = \boldsymbol{x}_2$. This means that all the points on the line through O in the direction of $\boldsymbol{x}_2$ remain fixed. Thus we have obtained the *Theorem of Euler-D'Alembert*:

Every finite rotation of a rigid body around a fixed point can be obtained as a finite rotation by an angle φ around some fixed axis passing through that point.

§ 14. Polar Decomposition of an Operator and the Cayley Formulas in a Euclidean Space

1. In § 12 we established the polar decomposition of a linear operator in a unitary space. In exactly the same way we obtain the polar decomposition of a linear operator in a euclidean space.

THEOREM 9. *Every linear operator $\boldsymbol{A}$ is representable in the form of a product*[37]

$$\boldsymbol{A} = \boldsymbol{S}\boldsymbol{Q} \tag{114}$$

$$\boldsymbol{A} = \boldsymbol{Q}_1\boldsymbol{S}_1 \tag{115}$$

where $\boldsymbol{S}$, $\boldsymbol{S}_1$ are positive-semidefinite symmetric and $\boldsymbol{Q}$, $\boldsymbol{Q}_1$ are orthogonal operators; here $\boldsymbol{S} = \sqrt{\boldsymbol{A}\boldsymbol{A}^{\mathsf{T}}} = g(\boldsymbol{A}\boldsymbol{A}^{\mathsf{T}})$, $\boldsymbol{S}_1 = \sqrt{\boldsymbol{A}^{\mathsf{T}}\boldsymbol{A}} = h(\boldsymbol{A}^{\mathsf{T}}\boldsymbol{A})$, where $g(\lambda)$ and $h(\lambda)$ are real polynomials.

$\boldsymbol{A}$ is a normal operator if and only if $\boldsymbol{S}$ and $\boldsymbol{Q}$ ($\boldsymbol{S}_1$ and $\boldsymbol{Q}_1$) are permutable.

Similar statements hold for matrices.

[37] As in Theorem 8, the operators $\boldsymbol{S}$ and $\boldsymbol{S}_1$ are uniquely determined by $\boldsymbol{A}$. If $\boldsymbol{A}$ is non-singular, then the orthogonal factors $\boldsymbol{Q}$ and $\boldsymbol{Q}_1$ are also uniquely determined.

Let us point out the geometrical content of the formulas (114) and (115). We let the vectors of an n-dimensional euclidean point space issue from the origin of the coordinate system. Then every vector is the radius vector of some point of the space. The orthogonal transformation realized by the operator $\boldsymbol{Q}$ (or $\boldsymbol{Q}_1$) is a 'rotation' in this space, because it preserves the euclidean metric and leaves the origin of the coordinate system fixed.[38] The symmetric operator $\boldsymbol{S}$ (or $\boldsymbol{S}_1$) represents a 'dilatation' of the n-dimensional space (i.e., a 'stretching' along n mutually perpendicular directions with stretching factors $\varrho_1, \varrho_2, \dots, \varrho_n$ that are, in general, distinct ($\varrho_1, \varrho_2, \dots, \varrho_n$ are arbitrary non-negative numbers)). According to the formulas (114) and (115), every linear homogeneous transformation of an n-dimensional euclidean space can be obtained by carrying out in succession some rotation and some dilatation (in any order).

2. Just as was done in the preceding section for a unitary operator, we now consider some representations of an orthogonal operator in a euclidean space $\boldsymbol{R}$.

Let $\boldsymbol{K}$ be an arbitrary skew-symmetric operator ($\boldsymbol{K}^{\mathsf{T}} = -\boldsymbol{K}$) and let

$$\boldsymbol{Q} = e^{\boldsymbol{K}}. \tag{116}$$

Then $\boldsymbol{Q}$ is a proper orthogonal operator. For

$$\boldsymbol{Q}^{\mathsf{T}} = e^{\boldsymbol{K}^{\mathsf{T}}} = e^{-\boldsymbol{K}} = \boldsymbol{Q}^{-1}$$

and

$$|\,\boldsymbol{Q}\,| = 1.$$[39]

Let us show that every proper orthogonal operator is representable in the form (116). For this purpose we take the corresponding orthogonal matrix Q. Since $|\,Q\,| = 1$, we have, by (113),[40]

---

[38] For $|\,\boldsymbol{Q}\,| = 1$ this is a proper rotation; but for $|\,\boldsymbol{Q}\,| = -1$ it is a combination of a rotation and a reflection in a coordinate plane.

[39] If $k_1, k_2, \dots, k_n$ are the characteristic values of $\boldsymbol{K}$, then $\mu_1 = e^{k_1}, \mu_2 = e^{k_2}, \dots, \mu_n = e^{k_n}$ are the characteristic values of $\boldsymbol{Q} = e^{\boldsymbol{K}}$; moreover

$$|\,\boldsymbol{Q}\,| = \mu_1\mu_2 \cdots \mu_n = e^{\sum_{i=1}^{n} k_i} = 1,$$

since

$$\sum_{i=1}^{n} k_i = 0.$$

[40] Among the characteristic values of a proper orthogonal matrix Q there is an even number equal to -1. The diagonal matrix $\begin{Vmatrix} -1 & 0 \\ 0 & -1 \end{Vmatrix}$ can be written in the form $\begin{Vmatrix} \cos\varphi & \sin\varphi \\ -\sin\varphi & \cos\varphi \end{Vmatrix}$ for $\varphi = \pi$.

$$Q = Q_1 \left\{ \begin{Vmatrix} \cos\varphi_1 & \sin\varphi_1 \\ -\sin\varphi_1 & \cos\varphi_1 \end{Vmatrix}, \ldots, \begin{Vmatrix} \cos\varphi_p & \sin\varphi_q \\ -\sin\varphi_p & \cos\varphi_q \end{Vmatrix}, +1, \ldots, +1 \right\} Q_1^{-1} \quad (117)$$

$$(Q_1 = (Q_1^{\mathsf{T}})^{-1} = \overline{Q}_1).$$

We define the skew-symmetric matrix K by the equation

$$K = Q_1 \left\{ \begin{Vmatrix} 0 & \varphi_1 \\ -\varphi_1 & 0 \end{Vmatrix}, \ldots, \begin{Vmatrix} 0 & \varphi_q \\ -\varphi_q & 0 \end{Vmatrix}, 0, \ldots, 0 \right\} Q_1^{-1}. \quad (118)$$

Since

$$\exp\left\{ \begin{Vmatrix} 0 & \varphi \\ -\varphi & 0 \end{Vmatrix} \right\} = \begin{Vmatrix} \cos\varphi & \sin\varphi \\ -\sin\varphi & \cos\varphi \end{Vmatrix},$$

it follows from (117) and (118) that

$$Q = e^K. \quad (119)$$

The matrix equation (119) implies the operator equation (116).

In order to represent an improper orthogonal operator we introduce a special operator $\boldsymbol{W}$ which is defined in an orthonormal basis $\boldsymbol{e}_1, \boldsymbol{e}_2, \ldots, \boldsymbol{e}_n$ by the equations

$$\boldsymbol{W}\boldsymbol{e}_1 = \boldsymbol{e}_1, \ldots, \boldsymbol{W}\boldsymbol{e}_{n-1} = \boldsymbol{e}_{n-1}, \boldsymbol{W}\boldsymbol{e}_n = -\boldsymbol{e}_n. \quad (120)$$

$\boldsymbol{W}$ is an improper orthogonal operator. If $\boldsymbol{Q}$ is an arbitrary improper orthogonal operator then $\boldsymbol{W}^{-1}\boldsymbol{Q}$ and $\boldsymbol{Q}\boldsymbol{W}^{-1}$ are proper and therefore representable in the form $e^{\boldsymbol{K}}$ and $e^{\boldsymbol{K}_1}$, where $\boldsymbol{K}$ and $\boldsymbol{K}_1$ are skew-symmetric operators. Hence we obtain the formulas for an improper orthogonal operator

$$\boldsymbol{Q} = \boldsymbol{W}e^{\boldsymbol{K}} = e^{\boldsymbol{K}_1}\boldsymbol{W}. \quad (121)$$

The basis $\boldsymbol{e}_1, \boldsymbol{e}_2, \ldots, \boldsymbol{e}_n$ in (120) can be chosen such that it coincides with the basis $\boldsymbol{x}_k, \boldsymbol{y}_k, \boldsymbol{x}_l$ ($k = 1, 2, \ldots, q$; $l = 2q + 1, \ldots, n$) in (110) and (112). The operator $\boldsymbol{W}$ so defined is permutable with $\boldsymbol{K}$; therefore the two formulas (121) merge into one

$$\boldsymbol{Q} = \boldsymbol{W}e^{\boldsymbol{K}} \quad (\boldsymbol{W} = \boldsymbol{W}^{\mathsf{T}} = \boldsymbol{W}^{-1}; \quad \boldsymbol{K}^{\mathsf{T}} = -\boldsymbol{K}, \boldsymbol{W}\boldsymbol{K} = \boldsymbol{K}\boldsymbol{W}). \quad (122)$$

Let us now turn to the Cayley formulas, which establish a connection between orthogonal and skew-symmetric operators in a euclidean space. The formula

$$\boldsymbol{Q} = (\boldsymbol{E} - \boldsymbol{K})(\boldsymbol{E} + \boldsymbol{K})^{-1}, \tag{123}$$

as is easily verified, carries the skew-symmetric operator $\boldsymbol{K}$ into the orthogonal operator $\boldsymbol{Q}$. (123) enables us to express $\boldsymbol{K}$ in terms of $\boldsymbol{Q}$:

$$\boldsymbol{K} = (\boldsymbol{E} - \boldsymbol{Q})(\boldsymbol{E} + \boldsymbol{Q})^{-1}. \tag{124}$$

The formulas (123) and (124) establish a one-to-one correspondence between the skew-symmetric operators and those orthogonal operators that do not have the characteristic value -1. Instead of (123) and (124) we can take the formulas

$$\boldsymbol{Q} = -(\boldsymbol{E} - \boldsymbol{K})(\boldsymbol{E} + \boldsymbol{K})^{-1}, \tag{125}$$

$$\boldsymbol{K} = (\boldsymbol{E} + \boldsymbol{Q})(\boldsymbol{E} - \boldsymbol{Q})^{-1}. \tag{126}$$

In this case the number $+1$ plays the role of the exceptional value.

3. The polar decomposition of a real matrix in accordance with Theorem 9 enables us to obtain the fundamental formulas (107), (109), (111), and (113) without embedding the euclidean space in a unitary space, as was done above. This second approach to the fundamental formulas is based on the following theorem:

THEOREM 10: *If two real normal matrices are similar,*

$$B = T^{-1}AT \quad (AA^{\mathsf{T}} = A^{\mathsf{T}}A,\ BB^{\mathsf{T}} = B^{\mathsf{T}}B,\ A = \overline{A},\ B = \overline{B}), \tag{127}$$

then they are real-similar and orthogonally-similar:

$$B = Q^{-1}AQ \quad (Q = \overline{Q} = Q^{\mathsf{T}-1}). \tag{128}$$

Proof: Since the normal matrices A and B have the same characteristic values, there exists a polynomial $g(\lambda)$ (see 2. on p. 272) such that

$$A^{\mathsf{T}} = g(A),\ B^{\mathsf{T}} = g(B).$$

Therefore the equation

$$g(B) = T^{-1}g(A)T,$$

which is a consequence of (127), can be written as follows:

$$B^{\mathsf{T}} = T^{-1}A^{\mathsf{T}}T. \tag{129}$$

When we go over to the transposed matrices in this equation, we obtain:

$$B = T^{\mathsf{T}}AT^{\mathsf{T}-1}. \tag{130}$$

A comparison of (127) with (130) shows that

$$TT^{\mathsf{T}}A = ATT^{\mathsf{T}}. \tag{131}$$

Now we make use of the polar decomposition of T:

$$T = SQ, \tag{132}$$

where $S = \sqrt{TT^{\mathsf{T}}} = h(TT^{\mathsf{T}})$ ($h(\lambda)$ a polynomial) is symmetric and Q is real and orthogonal. Since A, by (131), is permutable with TT^{T}, it is also permutable with $S = h(TT^{\mathsf{T}})$. Therefore, when we substitute the expression for T from (132) in (127), we have:

$$B = Q^{-1}S^{-1}ASQ = Q^{-1}AQ.$$

This completes the proof.

Let us consider the real canonical matrix

$$\left\{ \begin{Vmatrix} \mu_1 & \nu_1 \\ -\nu_1 & \mu_1 \end{Vmatrix}, \ldots, \begin{Vmatrix} \mu_q & \nu_q \\ -\nu_q & \mu_q \end{Vmatrix}, \quad \mu_{2q+1}, \ldots, \mu_n \right\}. \tag{133}$$

The matrix (133) is normal and has the characteristic values $\mu_1 \pm i\nu_1, \ldots, \mu_q \pm i\nu_q, \mu_{2q+1}, \ldots, \mu_n$. Since normal matrices are of simple structure, every normal matrix having the same characteristic values is similar (and by Theorem 10 real-similar and orthogonally-similar) to the matrix (133). Thus we arrive at the formula (107).

The formulas (109), (111), and (113) are obtained in exactly the same way.

§ 15. Commuting Normal Operators

In § 10 we have shown that two commuting operators $\boldsymbol{A}$ and $\boldsymbol{B}$ in an n-dimensional unitary space $\boldsymbol{R}$ always have a common characteristic vector. By mathematical induction we can show that this statement is true not only for two, but for any finite number, of commuting operators. For given m pairwise commuting operators $\boldsymbol{A}_1, \boldsymbol{A}_2, \ldots, \boldsymbol{A}_m$ the first $m-1$ of which have a common characteristic vector $\boldsymbol{x}$, by repeating verbatim the argument of Lemma 1 (p. 270) (for $\boldsymbol{A}$ we take any $\boldsymbol{A}_i$ ($i = 1, 2, \ldots, m-1$) and for $\boldsymbol{B}$ we take $\boldsymbol{A}_m$), we obtain a vector $\boldsymbol{y}$ which is a common characteristic vector of $\boldsymbol{A}_1, \boldsymbol{A}_2, \ldots, \boldsymbol{A}_m$.

This statement is even true for an infinite set of commuting operators, because such a set can only contain a finite number ($\leq n^2$) of linearly independent operators, and a common characteristic value of the latter is a common characteristic value of all the operators of the given set.

2. Now suppose that an arbitrary finite or infinite set of pairwise commuting *normal* operators $\boldsymbol{A}, \boldsymbol{B}, \boldsymbol{C}, \ldots$ is given. They all have a common characteristic vector $\boldsymbol{x}_1$. We denote by $\boldsymbol{T}_1$ the $(n-1)$-dimensional subspace consisting of all vectors of $\boldsymbol{R}$ that are orthogonal to $\boldsymbol{x}_1$. By § 10, 3. (p. 272), the subspace $\boldsymbol{T}_1$ is invariant with respect to $\boldsymbol{A}, \boldsymbol{B}, \boldsymbol{C}, \ldots$. Therefore all these operators have a common characteristic vector $\boldsymbol{x}_2$ in $\boldsymbol{T}_1$. We consider the orthogonal complement $\boldsymbol{T}_2$ of the plane $[\boldsymbol{x}_1, \boldsymbol{x}_2]$ and select in it a vector $\boldsymbol{x}_3$, etc. Thus we obtain an orthogonal system $\boldsymbol{x}_1, \boldsymbol{x}_2, \ldots, \boldsymbol{x}_n$ of common characteristic vectors of $\boldsymbol{A}, \boldsymbol{B}, \boldsymbol{C}, \ldots$. These vectors can be normalized. Hence we have proved:

Theorem 11: *If a finite or infinite set of pairwise commuting normal operators* $\boldsymbol{A}, \boldsymbol{B}, \boldsymbol{C}, \ldots$ *in a unitary space* $\boldsymbol{R}$ *is given, then all these operators have a complete orthonormal system of common characteristic vectors* $\boldsymbol{z}_1, \boldsymbol{z}_2, \ldots, \boldsymbol{z}_n$:

$$\boldsymbol{A}\boldsymbol{z}_i = \lambda_i \boldsymbol{z}_i, \quad \boldsymbol{B}\boldsymbol{z}_i = \lambda'_i \boldsymbol{z}_i, \quad \boldsymbol{C}\boldsymbol{z}_i = \lambda''_i \boldsymbol{z}_i, \ \ldots \ [(\boldsymbol{z}_i \boldsymbol{z}_k) = \delta_{ik};\ i, k = 1, 2, \ldots, n]. \tag{134}$$

In matrix form, this theorem reads as follows:

Theorem 11′: *If a finite or infinite set of pairwise commuting normal matrices* $A, B, C, \ldots$ *is given, then all these matrices can be carried by one and the same unitary transformation into diagonal form, i.e., there exists a unitary matrix* U *such that*

$$\left.\begin{aligned} A = U\{\lambda_1, \ldots, \lambda_n\}U^{-1}, \quad B = U\{\lambda'_1, \ldots, \lambda'_n\}U^{-1}, \\ C = U\{\lambda''_1, \ldots, \lambda''_n\}U^{-1}, \ldots \quad (U = U^{*-1}). \end{aligned}\right\} \tag{135}$$

Now suppose that commuting normal operators in a euclidean space $\boldsymbol{R}$ are given. We denote by $\boldsymbol{A}, \boldsymbol{B}, \boldsymbol{C}, \ldots$ the linearly independent ones among them (their number is finite). We embed $\boldsymbol{R}$ (under preservation of the metric) in a unitary space $\tilde{\boldsymbol{R}}$, as was done in § 13. Then by Theorem 11, the operators $\boldsymbol{A}, \boldsymbol{B}, \boldsymbol{C}, \ldots$ have a complete orthonormal system of common characteristic vectors $\boldsymbol{z}_1, \boldsymbol{z}_2, \ldots, \boldsymbol{z}_n$ in $\tilde{\boldsymbol{R}}$, i.e., (134) is satisfied.

We consider an arbitrary linear combination of $\boldsymbol{A}, \boldsymbol{B}, \boldsymbol{C}, \ldots$:

$$\boldsymbol{P} = \alpha \boldsymbol{A} + \beta \boldsymbol{B} + \gamma \boldsymbol{C} + \cdots.$$

For arbitrary real values $\alpha, \beta, \gamma, \ldots$ $\boldsymbol{P}$ is a real ($\boldsymbol{P}\boldsymbol{R} \subset \boldsymbol{R}$) normal operator in $\tilde{\boldsymbol{R}}$ and

$$\left.\begin{aligned} \boldsymbol{P}\boldsymbol{z}_j = \Lambda_j \boldsymbol{z}_j, \quad \Lambda_j = \alpha\lambda_j + \beta\lambda'_j + \gamma\lambda''_j + \cdots \\ [(\boldsymbol{z}_j \boldsymbol{z}_k) = \delta_{jk}; \quad j, k = 1, 2, \ldots, n]. \end{aligned}\right\} \tag{136}$$

The characteristic values Λ_j ($j = 1, 2, \ldots, n$) of $\boldsymbol{P}$ are linear forms in $\alpha, \beta, \gamma, \ldots$. Since $\boldsymbol{P}$ is real, these forms can be split into pairs of complex conjugates and real ones; with a suitable numbering of the characteristic vectors, we have

$$\Lambda_{2k-1}=M_k+iN_k,\ \Lambda_{2k}=M_k-iN_k,\ \Lambda_l=M_l \qquad (137)$$
$$(k=1,\ 2,\ \ldots,\ q;\quad l=2q+1,\ \ldots,\ n),$$

where M_k, N_k, and M_l are real linear forms in α, β, γ, ...

We may assume that in (136) the corresponding vectors $\boldsymbol{z}_{2k-1}$ and $\boldsymbol{z}_{2k}$ are complex conjugates, and the $\boldsymbol{z}_l$ real:

$$\boldsymbol{z}_{2k-1}=\boldsymbol{x}_k+i\boldsymbol{y}_k,\ \boldsymbol{z}_{2k}=\boldsymbol{x}_k-i\boldsymbol{y}_k,\ \boldsymbol{z}_l=\boldsymbol{x}_l \qquad (138)$$
$$(k=1,\ 2,\ \ldots,\ q;\quad l=2q+1,\ \ldots,\ n).$$

But then, as is easy to see, the real vectors

$$\boldsymbol{x}_k,\ \boldsymbol{y}_k,\ \boldsymbol{x}_l\quad (k=1,\ 2,\ \ldots,\ q;\ l=2q+1,\ \ldots,\ n) \qquad (139)$$

form an orthonormal basis of $\boldsymbol{R}$. In this canonical basis we have:[41]

$$\left.\begin{aligned}\boldsymbol{P}\boldsymbol{x}_k&=M_k\boldsymbol{x}_k-N_k\boldsymbol{y}_k,\\ \boldsymbol{P}\boldsymbol{y}_k&=N_k\boldsymbol{x}_k+M_k\boldsymbol{y}_k,\\ \boldsymbol{P}\boldsymbol{x}_l&=M_l\boldsymbol{x}_l\end{aligned}\right.\quad \begin{pmatrix}k=1,\ 2,\ \ldots,\ q,\\ l=2q+1,\ \ldots,\ n\end{pmatrix}. \qquad (140)$$

Since all the operators of the given set are obtained from $\boldsymbol{P}$ for special values of α, β, γ, ... the basis (139), which does not depend on these parameters, is a common canonical basis for all the operators. Thus we have proved:

THEOREM 12: *If an arbitrary set of commuting normal linear operators in a euclidean space $\boldsymbol{R}$ is given, then all these operators have a common orthonormal canonical basis $\boldsymbol{x}_k$, $\boldsymbol{y}_k$, $\boldsymbol{x}_l$:*

$$\left.\begin{aligned}\boldsymbol{A}\boldsymbol{x}_k&=\mu_k\boldsymbol{x}_k-\nu_k\boldsymbol{y}_k, & \boldsymbol{B}\boldsymbol{x}_k&=\mu_k'\boldsymbol{x}_k-\nu_k'\boldsymbol{y}_k,\ \ldots,\\ \boldsymbol{A}\boldsymbol{y}_k&=\nu_k\boldsymbol{x}_k+\mu_k\boldsymbol{y}_k, & \boldsymbol{B}\boldsymbol{y}_k&=\nu_k'\boldsymbol{x}_k+\mu_k'\boldsymbol{y}_k,\ \ldots,\\ \boldsymbol{A}\boldsymbol{x}_l&=\mu_l\boldsymbol{x}_l; & \boldsymbol{B}\boldsymbol{x}_l&=\mu_l'\boldsymbol{x}_l,\ \ldots.\end{aligned}\right\} \qquad (141)$$

We give the matrix form of Theorem 12:

THEOREM 12′: *Every set of commuting normal real matrices $A, B, C, \ldots$ can be carried by one and the same real orthogonal transformation Q into canonical form*

$$\left.\begin{aligned}A&=Q\left\{\begin{Vmatrix}\mu_1 & \nu_1\\ -\nu_1 & \mu_1\end{Vmatrix},\ \ldots,\ \begin{Vmatrix}\mu_q & \nu_q\\ -\nu_q & \mu_q\end{Vmatrix},\ \mu_{2q+1},\ \ldots,\ \mu_n\right\}Q^{-1},\\ B&=Q\left\{\begin{Vmatrix}\mu_1' & \nu_1'\\ -\nu_1' & \mu_1'\end{Vmatrix},\ \ldots,\ \begin{Vmatrix}\mu_q' & \nu_q'\\ -\nu_q' & \mu_q'\end{Vmatrix},\ \mu_{2q+1}',\ \ldots,\ \mu_n'\right\}Q^{-1},\\ &\ldots\ldots\ldots\ldots\ldots\ldots\ldots\ldots\ldots\end{aligned}\right\} \qquad (142)$$

[41] The equation (140) follows from (136), (137), and (138).

Note. If one of the operators $\boldsymbol{A}, \boldsymbol{B}, \boldsymbol{C}, \ldots$ (matrices $A, B, C, \ldots$)—say $\boldsymbol{A}$ (A)—is symmetric, then in the corresponding formulas (141) ((142)) all the ν are zero. In the case of skew-symmetry, all the μ are zero. In the case where $\boldsymbol{A}$ is an orthogonal operator (A an orthogonal matrix), we have $\mu_k = \cos \varphi_k, \nu_k = \sin \varphi_k, \mu_l = \pm 1$ $(k = 1, 2, \ldots, q;\ l = 2q + 1, \ldots, n)$.

CHAPTER X

QUADRATIC AND HERMITIAN FORMS

§ 1. Transformation of the Variables in a Quadratic Form

1. A *quadratic form* is a homogeneous polynomial of the second degree in n variables $x_1, x_2, \ldots, x_n$. A quadratic form always has a representation

$$\sum_{i,k=1}^{n} a_{ik}x_i x_k \qquad (a_{ik} = a_{ki};\ i, k = 1, 2, \ldots, n),$$

where $A = \| a_{ik} \|_1^n$ is a symmetric matrix.

If we denote the column matrix $(x_1, x_2, \ldots, x_n)$ by x and denote the quadratic form by

$$A(x, x) = \sum_{i,k=1}^{n} a_{ik}x_i x_k, \tag{1}$$

then we can write:[1]

$$A(x, x) = x^{\mathsf{T}} A x. \tag{2}$$

If $A = \| a_{ik} \|_1^n$ is a real symmetric matrix, then the form (1) is called *real*. In this chapter we shall mainly be concerned with real quadratic forms.

The determinant $| A | = | a_{ik} |_1^n$ is called the *discriminant* of the quadratic form $A(x, x)$. The form is called *singular* if its discriminant is zero.

To every quadratic form there corresponds a *bilinear form*

$$A(x, y) = \sum_{i,k=1}^{n} a_{ik}x_i y_k, \tag{3}$$

or

$$A(x, y) = x^{\mathsf{T}} A y \qquad (x = (x_1, \ldots, x_n),\ y = (y_1, \ldots, y_n)). \tag{4}$$

If $x^1, x^2, \ldots, x^l, y^1, y^2, \ldots, y^m$ are column matrices and $c_1, c_2, \ldots, c_l, d_1, d_2, \ldots, d_m$ are scalars, then by the bilinearity of $A(x, y)$ (see (4)),

[1] The sign ^{T} denotes transposition. In (2) the quadratic form is represented as a product of three matrices: the row x^{T}, the square matrix A, and the column x.

$$A\left(\sum_{i=1}^{l} c_i x^i,\ \sum_{j=1}^{m} d_j y^j\right) = \sum_{i=1}^{l}\sum_{j=1}^{m} c_i d_j A(x^i,\ y^j)\,. \tag{5}$$

If $\boldsymbol{A}$ is an operator in an n-dimensional euclidean space and if in some orthonormal basis $\boldsymbol{e}_1, \boldsymbol{e}_2, \ldots, \boldsymbol{e}_n$ this symmetric operator corresponds to the matrix $A = \| a_{ik} \|_1^n$, then for arbitrary vectors

$$\boldsymbol{x} = \sum_{i=1}^{n} x_i \boldsymbol{e}_i, \qquad \boldsymbol{y} = \sum_{i=1}^{n} y_i \boldsymbol{e}_i$$

we have the identity[2]

$$A(x, y) = (\boldsymbol{Ax}, \boldsymbol{y}) = (\boldsymbol{x}, \boldsymbol{Ay}).$$

In particular,

$$A(x, x) = (\boldsymbol{Ax}, \boldsymbol{x}) = (\boldsymbol{x}, \boldsymbol{Ax}),$$

where

$$a_{ik} = (\boldsymbol{Ae}_i, \boldsymbol{e}_k) \qquad (i, k = 1, 2, \ldots, n).$$

2. Let us see how the coefficient matrix of the form changes under a transformation of the variables:

$$x_i = \sum_{k=1}^{n} t_{ik} \xi_k \ (i = 1, 2, \ldots, n). \tag{6}$$

In matrix notation, this transformation looks as follows:

$$x = T\xi. \tag{6'}$$

Here x, ξ are column matrices: $x = (x_1, x_2, \ldots, x_n)$ and $\xi = (\xi_1, \xi_2, \ldots, \xi_n)$; and T is the transforming matrix: $T = \| t_{ik} \|_1^n$.

Substituting the expression for x in (2), we obtain from (6′):

$$A(x,\ x) = \xi^{\mathsf{T}} T^{\mathsf{T}} A T \xi = \xi^{\mathsf{T}} \tilde{A} \xi = \tilde{A}(\xi,\ \xi),$$

where

$$\tilde{A} = T^{\mathsf{T}} A T. \tag{7}$$

The formula (7) expresses the coefficient matrix $\tilde{A} = \| \tilde{a}_{ik} \|_1^n$ of the transformed form $\tilde{A}(\xi, \xi) = \sum_{i,k=1}^{n} \tilde{a}_{ik} \xi_i \xi_k$ in terms of the coefficient matrix of the original form $A = \| a_{ik} \|_1^n$ and the transformation matrix $T = \| t_{ik} \|_1^n$.

It follows from (7) that under a transformation the discriminant of the form is multiplied by the square of the determinant of the transformation:

---

[2] In $A(x, y)$, the parentheses form part of the notation; in $(\boldsymbol{Ax}, \boldsymbol{y})$ and $(\boldsymbol{x}, \boldsymbol{Ay})$, they denote the scalar product.

$$|\tilde{A}| = |A|\,|T|^2. \tag{8}$$

In what follows we shall make use exclusively of non-singular transformations of the variables ($|T| \neq 0$). Under such transformations, as is clear from (7), the rank of the coefficient matrix remains unchanged (the rank of A is the same as that of $\tilde{A}$).[3] The rank of the coefficient matrix is usually called the *rank of the quadratic form.*

DEFINITION 1: *Two symmetric matrices A and $\tilde{A}$ connected as in formula* (7), *with* $|T| \neq 0$, *are called congruent.*

Thus, a whole class of congruent symmetric matrices is associated with every quadratic form. As mentioned above, all these matrices have one and the same rank, the rank of the form. The rank is an invariant for the given class of matrices. In the real case, a second invariant is the so-called 'signature' of the quadratic form. We shall now proceed to introduce this concept.

§ 2. Reduction of a Quadratic Form to a Sum of Squares. The Law of Inertia

1. A real quadratic form $A(x, x)$ can be represented in an infinite number of ways in the form

$$A(x, x) = \sum_{i=1}^{r} a_i X_i^2, \tag{9}$$

where $a_i \neq 0$ $(i = 1, 2, \ldots, r)$ and

$$X_i = \sum_{k=1}^{n} \alpha_{ki} x_k \qquad (i = 1, 2, \ldots, r)$$

are *linearly independent* real linear forms in the variables $x_1, x_2, \ldots, x_n$ (so that $r \leq n$).

Let us consider a non-singular transformation of the variables under which the first r of the new variables $\xi_1, \xi_2, \ldots, \xi_n$ are connected with $x_1, x_2, \ldots, x_n$ by the formulas[4]

$$\xi_i = X_i \qquad (i = 1, 2, \ldots, r)$$

Then, in the new variables,

[3] See p. 17.

[4] We obtain the necessary transformation by adjoining to the system of linear forms $X_1, \ldots, X_r$ such linear forms $X_{r+1}, \ldots, X_n$ that the forms X_j $(j = 1, 2, \ldots, n)$ are linearly independent and then setting $\xi_j = X_j$ $(j = 1, 2, \ldots, n)$.

$$A(x, x) = \tilde{A}(\xi, \xi) = \sum_{i=1}^{r} a_i \xi_i^2$$

and therefore $\tilde{A} = \{a_1, a_2, \ldots, a_r, 0, \ldots, 0\}$. But the rank of $\tilde{A}$ is r. Hence: *The number of squares in the representation* (9) *is always equal to the rank of the form.*

2. We shall show that not only is the total number of squares invariant in the various representations of $A(x, x)$ in the form (9), but also so is the number of positive[5] (and, hence, the number of negative) squares.

THEOREM 1 (The Law of Inertia for Quadratic Forms): *In a representation of a real quadratic form $A(x, x)$ as a sum of independent squares*[6]

$$A(x, x) = \sum_{i=1}^{r} a_i X_i^2, \tag{9}$$

the number of positive and the number of negative squares are independent of the choice of the representation.

Proof. Let us assume that we have, in addition to (9), another representation of $A(x, x)$ in the form of a sum of independent squares

$$A(x, x) = \sum_{i=1}^{r} b_i Y_i^2$$

and that

$$a_1 > 0, a_2 > 0, \ldots, a_g > 0, a_{g+1} < 0, \ldots, a_r < 0,$$

$$b_1 > 0, b_2 > 0, \ldots, b_h > 0, b_{h+1} < 0, \ldots, b_r < 0,$$

Suppose that $g \neq h$, say $g < h$. Then in the identity

$$\sum_{i=1}^{r} a_i X_i^2 = \sum_{i=1}^{r} b_i Y_i^2 \tag{10}$$

we give to the variables $x_1, x_2, \ldots, x_n$ values that satisfy the system of $r-(h-g)$ equations

$$X_1 = 0, \quad X_2 = 0, \quad \ldots, X_g = 0, \quad Y_{h+1} = 0, \quad \ldots \quad Y_r = 0, \tag{11}$$

[5] By the number of positive (negative) squares in (9) we mean the number of positive (or negative) a_i.

[6] By a sum of independent squares we mean a sum of the form (9) in which all $a_i \neq 0$ and the forms $X_1, X_2, \ldots, X_r$ are linearly independent.

and for which at least one of the forms $X_{g+1}, \ldots, X_r$ does not vanish.[7] For these values of the variables the left-hand side of the identity is

$$\sum_{j=g+1}^{r} a_j X_j^2 < 0,$$

and the right-hand side is

$$\sum_{k=1}^{h} b_k Y_k^2 \geqq 0.$$

Thus, the assumption $g \neq h$ has led to a contradiction, and the theorem is proved.

DEFINITION 2: *The difference σ between the number π of positive squares and the number ν of negative squares in the representation of $A(x, x)$ is called the signature of the form $A(x, x)$.* (Notation: $\sigma = \sigma[A(x, x)]$).

The rank r and the signature σ determine the numbers π and ν uniquely, since

$$r = \pi + \nu, \quad \sigma = \pi - \nu.$$

Note that in (9) the positive factor $\sqrt{|a_i|}$ can be absorbed into the form X_i $(i = 1, 2, \ldots, r)$. Then (9) assumes the form

$$A(x, x) = X_1^2 + X_2^2 + \cdots + X_\pi^2 - X_{\pi+1}^2 - \cdots - X_r^2. \tag{12}$$

Setting[8] $\xi_i = X_i$ $(i = 1, 2, \ldots, r)$, we reduce $A(x, x)$ to the canonical form

$$\tilde{A}(\xi, \xi) = \xi_1^2 + \xi_2^2 + \cdots + \xi_\pi^2 - \xi_{\pi+1}^2 - \cdots - \xi_r^2. \tag{13}$$

Hence we deduce from Theorem 1 that: *Every real symmetric matrix A is congruent to a diagonal matrix in which the diagonal elements are* $+1, -1,$ *or* 0:

$$A = T^{\mathsf{T}} \{ \underbrace{+1, \ldots, +1}_{\pi}, \underbrace{-1, \ldots, -1}_{\nu}, 0, \ldots, 0 \} T. \tag{14}$$

In the next section we shall give a rule for determining the signature from the coefficients of the quadratic form.

---

[7] Such values exist, since otherwise the equations $X_{g+1} = 0, \ldots, X_r = 0$ and hence all the equations $X_1 = 0, X_2 = 0, \ldots, X_r = 0$ would be consequences of the $r - (h - g)$ equations (11). This is impossible, because the linear forms $X_1, X_2, \ldots, X_r$ are independent.

[8] See footnote 4.

§ 3. The Methods of Lagrange and Jacobi of Reducing a Quadratic Form to a Sum of Squares

It follows from the preceding section that in order to determine the rank and the signature of a form it is sufficient to reduce it in any way to a sum of independent squares.

We shall describe here two reduction methods: that of Lagrange and that of Jacobi.

1. *Lagrange's Method.* Let a quadratic form

$$A(x, x) = \sum_{i,k=1}^{n} a_{ik} x_i x_k$$

be given.

We consider two cases:

1) For some g $(1 \leqq g \leqq n)$ the diagonal coefficient a_{gg} is not equal to zero. Then we set

$$A(x, x) = \frac{1}{a_{gg}} \left(\sum_{k=1}^{n} a_{gk} x_k \right)^2 + A_1(x, x) \tag{15}$$

and convince ourselves by direct verification that the quadratic form $A_1(x, x)$ does not contain the variable x_g. This method of separating out a square form in a quadratic form is always applicable when there is a non-zero diagonal element in the matrix $A = \| a_{ik} \|_1^n$.

2) $a_{gg} = 0$ and $a_{hh} = 0$, but $a_{gh} \neq 0$. Then we set:

$$A(x, x) = \frac{1}{2a_{hg}} \left[\sum_{k=1}^{n} (a_{gk} + a_{hk}) x_k \right]^2 - \frac{1}{2a_{hg}} \left[\sum_{k=1}^{n} (a_{gk} - a_{hk}) x_k \right]^2 + A_2(x, x). \tag{16}$$

The forms

$$\sum_{k=1}^{n} a_{gk} x_k, \qquad \sum_{k=1}^{n} a_{hk} x_k \tag{17}$$

are linearly independent, since the first contains x_h but not x_g, and the second contains x_g but not x_h. Therefore, in (16), the forms within the brackets are linearly independent (as sum and difference, respectively, of the independent linear forms (17)).

Therefore we have separated out two independent squares in $A(x, x)$. Each of these squares contains x_g and x_h, whereas $A_2(x, x)$ does not contain these variables, as is easy to verify.

By successive application of a combination of the methods 1) and 2), we can always reduce the form $A(x, x)$ by means of rational operations to a sum of squares. Moreover, the squares so obtained are linearly independent, since at each stage the square that is separated out contains an unknown that does not occur in the subsequent squares.

Note that the basic formulas (15) and (16) can be written as follows

$$A(x, x) = \frac{1}{4a_{gg}}\left(\frac{\partial A}{\partial x_g}\right)^2 + A_1(x, x), \tag{15'}$$

$$A(x, x) = \frac{1}{8a_{gh}}\left[\left(\frac{\partial A}{\partial x_g} + \frac{\partial A}{\partial x_h}\right)^2 - \left(\frac{\partial A}{\partial x_g} - \frac{\partial A}{\partial x_h}\right)^2\right] + A_2(x, x). \tag{16'}$$

Example.

$$A(x, x) = 4x_1^2 + x_2^2 + x_3^2 + x_4^2 - 4x_1x_2 - 4x_1x_3 + 4x_1x_4 + 4x_2x_3 - 4x_2x_4 .$$

We apply formula (15′) with $g = 1$:

$$\begin{aligned} A(x, x) &= \frac{1}{16}(8x_1 - 4x_2 - 4x_3 + 4x_4)^2 + A_1(x, x) \\ &= (2x_1 - x_2 - x_3 + x_4)^2 + A_1(x, x), \end{aligned}$$

where

$$A_1(x, x) = 2x_2x_3 - 2x_2x_4 + 2x_3x_4 .$$

We apply formula (16′) with $g = 2$ and $h = 3$:

$$\begin{aligned} A_1(x, x) &= \frac{1}{8}(2x_2 + 2x_3)^2 - \frac{1}{8}(2x_3 - 2x_2 - 4x_4)^2 + A_2(x, x) \\ &= \frac{1}{2}(x_2 + x_3)^2 - \frac{1}{2}(x_3 - x_2 - 2x_4)^2 + A_2(x, x), \end{aligned}$$

where

$$A_2(x, x) = 2x_4^2.$$

Finally,

$$A(x, x) = (2x_1 - x_2 - x_3 + x_4)^2 + \frac{1}{2}(x_2 + x_3)^2 - \frac{1}{2}(x_3 - x_2 - 2x_4)^2 + 2x_4^2,$$

$$r = 4, \quad \sigma = 2.$$

2. *Jacobi's Method.* We denote the rank of $A(x, x) = \sum_{i,k=1}^{n} a_{ik}x_ix_k$ by r and assume that

$$D_k = A\begin{pmatrix} 1 & 2 & \dots & k \\ 1 & 2 & \dots & k \end{pmatrix} \neq 0 \qquad (k = 1, 2, \dots, r).$$

Then the symmetric matrix $A = \| a_{ik} \|_1^n$ can be reduced to the form

$$G=\left\|\begin{matrix} g_{11} & g_{12} & \cdot & \cdots & g_{1n} \\ 0 & g_{22} & \cdots & \cdots & g_{2n} \\ \cdot & \cdots & \cdots & \cdots & \cdot \\ 0 & 0 & \cdots & g_{rr} \cdots & g_{rn} \\ 0 & 0 & \cdots & 0 \cdots & 0 \\ \cdot & \cdots & \cdots & \cdots & \cdot \\ 0 & 0 & \cdots & 0 \cdots & 0 \end{matrix}\right\| \tag{18}$$

by Gauss's elimination algorithm (see Chapter II, § 1).

The elements of G are expressed in terms of the elements of A by the well-known formulas[9]

$$g_{pq}=\frac{A\begin{pmatrix}1 & 2 & \dots & p-1 & p\\ 1 & 2 & \dots & p-1 & q\end{pmatrix}}{A\begin{pmatrix}1 & 2 & \dots & p-1\\ 1 & 2 & \dots & p-1\end{pmatrix}} \quad (q=p,\, p+1,\, \dots,\, n;\ p=1, 2,\, \dots,\, r). \tag{19}$$

In particular,

$$g_{pp}=\frac{D_p}{D_{p-1}} \qquad (p=1,\, 2,\, \dots,\, r;\ D_0=1)\,. \tag{20}$$

In Chapter II, § 4 (formula (55) on page 41) we have shown that

$$A=G^{\mathsf{T}}\widehat{D}G\,, \tag{21}$$

where $\widehat{D}$ is the diagonal matrix:

$$\widehat{D}=\left\{\frac{1}{D_1},\, \frac{D_1}{D_2},\, \dots,\, \frac{D_{r-1}}{D_r},\, 0,\, \dots,\, 0\right\}=\left\{\frac{1}{g_{11}},\, \frac{1}{g_{22}},\, \dots,\, \frac{1}{g_{rr}},\, 0,\, \dots,\, 0\right\}. \tag{22}$$

Without infringing (21) we may replace some of the zeros in the last $n-r$ rows of G by arbitrary elements. By such a replacement we can make G into a non-singular upper triangular matrix

$$T=\left\|\begin{matrix} g_{11} & g_{12} & \cdots & \cdots & g_{1n} \\ 0 & g_{22} & \cdots & \cdots & g_{2n} \\ \cdot & \cdots & \cdots & \cdots & \cdot \\ 0 & 0 \dots g_{rr} & & \cdots & g_{rn} \\ 0 & 0 \dots 0 & * & \cdots & * \\ \cdot & \cdots & \cdots & \cdots & \cdot \\ 0 & 0 & \cdots & \cdots & * \end{matrix}\right\| \qquad (|T|\neq 0)\,. \tag{23}$$

[9] See Chapter II, § 2.

The equation (21) can then be rewritten:

$$A = T^{\mathsf{T}} \widehat{D} T. \tag{24}$$

From this equation it follows that the quadratic form[10]

$$\widehat{D}(\xi, \xi) = \sum_{k=1}^{r} \frac{D_{k-1}}{D_k} \xi_k^2 = \sum_{k=1}^{r} \frac{\xi_k^2}{g_{kk}}$$

$$(\xi = (\xi_1, \xi_2, \ldots, \xi_n); \quad D_0 = 1)$$

goes over into the form $A(x, x)$ under the transformation

$$\xi = Tx$$

Since

$$\xi_k = X_k, \quad X_k \equiv g_{kk}x_k + g_{k,k+1}x_{k+1} + \cdots + g_{kn}x_n \qquad (k = 1, \ldots, r), \tag{25}$$

we have *Jacobi's Formula*[11]

$$A(x, x) = \sum_{k=1}^{r} \frac{D_{k-1}}{D_k} X_k^2 = \sum_{k=1}^{r} \frac{X_k^2}{g_{kk}} \qquad (D_0 = 1). \tag{26}$$

This formula gives a representation of $A(x, x)$ in the form of a sum of r independent squares.[12]

Jacobi's formula is often given in another form.

Instead of X_k $(k = 1, 2, \ldots, r)$, the linearly independent forms

$$Y_k = D_{k-1} X_k \qquad (k = 1, 2, \ldots, r; \; D_0 = 1) \tag{27}$$

are introduced. Then Jacobi's formula (26) can be written as:

$$A(x, x) = \sum_{k=1}^{r} \frac{Y_k^2}{D_{k-1} D_k}. \tag{28}$$

Here

$$Y_k = c_{kk}x_k + c_{k,k+1}x_{k+1} + \cdots + c_{kn}x_n \qquad (k = 1, 2, \ldots, r) \tag{29}$$

where

---

[10] We regard $\widehat{D}(\xi, \xi)$ as a quadratic form in the n variables $\xi_1, \xi_2, \ldots, \xi_n$.

[11] Another approach to Jacobi's formula, which does not depend on (21), can be found, for example, in [17], pp. 43-44.

[12] The independence of the squares in Jacobi's formula follows from the fact that the form $A(x, x)$ is of rank r. But we can also convince ourselves directly of the independence of the forms $X_1, X_2, \ldots, X_r$. For, according to (20), $g_{kk} = \dfrac{D_k}{D_{k-1}} \neq 0$ and therefore X_k contains the variable x_k, which does not occur in the forms $X_{k+1}, \ldots, X_r$ $(k = 1, 2, 3, \ldots, r)$. Hence $X_1, X_2, \ldots, X_r$ are linearly independent forms.

$$c_{kq}=A\begin{pmatrix}1 & 2 & \dots & k-1 & k\\ 1 & 2 & \dots & k-1 & q\end{pmatrix}\qquad (q=k,\ k+1,\ \dots,\ n;\ k=1,\ 2,\ \dots,\ r). \tag{30}$$

Example.

$$A(x,\ x)=x_1^2+3x_2^2-3x_4^2-4x_1x_2+2x_1x_3-2x_1x_4-6x_2x_3+8x_2x_4+2x_3x_4.$$

We reduce the matrix

$$A=\begin{Vmatrix}1 & -2 & 1 & -1\\ -2 & 3 & -3 & 4\\ 1 & -3 & 0 & 1\\ -1 & 4 & 1 & -3\end{Vmatrix}$$

to the Gaussian form

$$G=\begin{Vmatrix}1 & -2 & 1 & -1\\ 0 & -1 & -1 & 2\\ 0 & 0 & 0 & 0\\ 0 & 0 & 0 & 0\end{Vmatrix}.$$

Hence $r=2$, $g_{11}=1$, $g_{22}=-1$.

Jacobi's formula (26) yields:

$$A(x,\ x)=(x_1-2x_2+x_3-x_4)^2-(-x_2-x_3+2x_4)^2.$$

Jacobi's formula (28) yields the following theorem:

THEOREM 2 (Jacobi). *If for the quadratic form*

$$A(x,\ x)=\sum_{i,\,k=1}^{n}{}' a_{ik}x_ix_k$$

of rank r the inequality

$$D_k=A\begin{pmatrix}1 & 2 & \dots & k\\ 1 & 2 & \dots & k\end{pmatrix}\neq 0\qquad (k=1,\ 2,\ \dots,\ r), \tag{31}$$

holds, then the number π of positive squares and the number ν of negative squares of $A(x, x)$ coincide, respectively, with the number P of permanences of sign and the number V of variations of sign in the sequence

$$1,\ D_1,\ D_2,\ \dots,\ D_r, \tag{32}$$

i.e., $\pi=P(1, D_1, D_2, \dots, D_r)$, $\nu=V(1, D_1, D_2, \dots, D_r)$, *and the signature*

$$\sigma=r-2V(1,\ D_1,\ D_2,\ \dots,\ D_r). \tag{33}$$

Note 1. *If in the sequence* $1, D_1, \ldots, D_r \neq 0$ *there are zeros, but not three in succession, then the signature can be determined by the use of the formula*

$$\sigma = r - 2V(1, D_1, D_2, \ldots, D_r)$$

omitting the zero D_k *provided* $D_{k-1}D_{k+1} \neq 0$, *and setting*

$$V(D_{k-1}, D_k, D_{k+1}, D_{k+2}) = \begin{cases} 1, \text{ when } \dfrac{D_{k+2}}{D_{k-1}} < 0, \\ 2, \text{ when } \dfrac{D_{k+2}}{D_{k-1}} > 0 \end{cases} \tag{34}$$

if $D_k = D_{k+1} = 0$.

We state this rule without proof.[13]

Note 2. *When three consecutive zeros occur in* $D_1, D_2, \ldots, D_{r-1}$, *then the signature of the quadratic form cannot be immediately determined by Jacobi's Theorem.* In this case, the signs of the non-zero D_k do not determine the signature of the form. This is shown by the following example:

$$A(x, x) = 2a_1x_1x_4 + a_2x_2^2 + a_3x_3^2 \qquad (a_1a_2a_3 \neq 0).$$

Here

$$D_1 = D_2 = D_3 = 0, \quad D_4 = -a_1^2a_2a_3 \neq 0.$$

But

$$\nu = \begin{cases} 1, \text{ when } a_2 > 0,\ a_3 > 0, \\ 3, \text{ when } a_2 < 0,\ a_3 < 0. \end{cases}$$

In both cases, $D_4 < 0$.

Note 3. *If* $D_1 \neq 0, \ldots, D_{r-1} \neq 0$, *but* $D_r = 0$, *then the signs of* $D_1, D_2, \ldots, D_{r-1}$ *do not determine the signature of the form.* As a corroborating example, we can take the form

$$ax_1^2 + ax_2^2 + bx_3^2 + 2ax_1x_2 + 2ax_2x_3 + 2ax_1x_3 = a(x_1 + x_2 + x_3)^2 + (b - a)x_3^2.$$

§ 4. Positive Quadratic Forms

1. In this section we deal with the special, but important, class of positive quadratic forms.

DEFINITION 3: *A real quadratic form* $A(x, x) = \sum_{i,k=1}^{n} a_{ik}x_ix_k$ *is called positive (negative) semidefinite if for arbitrary real values of the variables*:

$$A(x, x) \geqq 0 \qquad (\leqq 0). \tag{35}$$

---

[13] This rule was found in the case of a single zero D_k by Gundenfinger and for two successive zeros D_k by Frobenius [162].

Definition 4: *A real quadratic form* $A(x, x) = \sum_{i,k=1}^{n} a_{ik}x_i x_k$ *is called positive (negative) definite if for arbitrary values of the variables, not all zero,* $(x \neq o)$

$$A(x, x) > 0 \qquad (< 0). \tag{36}$$

The class of positive (negative) definite forms is part of the class of positive (negative) semidefinite forms.

Let $A(x, x)$ be a positive-semidefinite form. We represent it in the form of a sum of linearly independent squares:

$$A(x, x) = \sum_{i=1}^{r} a_i X_i^2, \tag{37}$$

In this representation, all the squares must be positive:

$$a_i > 0 \qquad (i = 1, 2, \ldots, r). \tag{38}$$

For if any a_i were negative, then we could select values of $x_1, x_2, \ldots, x_n$ for which

$$X_1 = \cdots = X_{i-1} = X_{i+1} = \cdots = X_r = 0, \quad X_i \neq 0.$$

But then $A(x, x)$ would have a negative value for these values of the variables, and by assumption this is impossible. It is clear that, conversely, it follows from (37) and (38) that the form $A(x, x)$ is positive semidefinite.

Thus, a positive semidefinite quadratic form is characterized by the equations $\sigma = r$ $(\pi = r, \nu = 0)$.

Now let $A(x, x)$ be a positive-definite form. Then $A(x, x)$ is also positive semidefinite. Therefore it is representable in the form (37), where all the a_i $(i = 1, 2, \ldots, r)$ are positive. From the positive definiteness it follows that $r = n$. For if $r < n$, we could find values of $x_1, x_2, \ldots, x_n$, not all zero, such that all the X_i would be zero. But then by (37) $A(x, x) = 0$ for $x \neq o$, and this contradicts (36).

It is easy to see that, conversely, if in (37) $r = n$ and all the $a_1, a_2, \ldots, a_n$ are positive, then $A(x, x)$ is a positive-definite form.

In other words: *A positive-semidefinite form is positive definite if and only if it is not singular.*

2. The following theorem gives a criterion for positive definiteness in the form of inequalities which the coefficients of the form must satisfy. We shall use the notation of the preceding section for the sequence of the principal minors of A:

$$D_1 = a_{11}, \quad D_2 = \begin{vmatrix} a_{11} & a_{12} \\ a_{21} & a_{22} \end{vmatrix}, \quad \ldots, \quad D_n = \begin{vmatrix} a_{11} & a_{12} & \ldots & a_{1n} \\ a_{21} & a_{22} & \ldots & a_{2n} \\ \cdot & \cdot & \cdot & \cdot \\ a_{n1} & a_{n2} & \ldots & a_{nn} \end{vmatrix}.$$

THEOREM 3: *A quadratic form is positive definite if and only if*

$$D_1 > 0, D_2 > 0, \ldots, D_n > 0. \tag{39}$$

Proof. The sufficiency of the conditions (39) follows immediately from Jacobi's formula (28). The necessity of (39) is established as follows. From the fact that $A(x, x) = \sum_{i,k=1}^{n} a_{ik}x_i x_k$ is positive definite, it follows that the 'restricted' forms[14]

$$A_p(x, x) = \sum_{i,k=1}^{p} a_{ik}x_i x_k \qquad (p = 1, 2, \ldots, n)$$

are also positive definite. But then all these forms must be non singular, i.e.,

$$D_p = |A_p| \neq 0 \qquad (p = 1, 2, \ldots, n).$$

We are now in a position to apply Jacobi's formula (28) (for $r = n$). Since all the squares on the right-hand side of the formula must be positive, we have

$$D_1 > 0, \; D_1 D_2 > 0, \; D_2 D_3 > 0, \; \ldots, \; D_{n-1} D_n > 0.$$

Hence the inequality (39) follows, and the theorem is proved.

Since every principal minor of A can be brought into the top left corner by a suitable numbering of the variables, we have the

COROLLARY: *In a positive-definite quadratic form $A(x, x) = \sum_{i,k=1}^{n} a_{ik}x_i x_k$, all the principal minors of the coefficient matrix are positive:*[15]

$$A\begin{pmatrix} i_1 \, i_2 \ldots i_p \\ i_1 \, i_2 \ldots i_p \end{pmatrix} > 0 \;\; (1 \leq i_1 < i_2 < \cdots < i_p \leq n; \; p = 1, 2, \ldots, n).$$

Note. If the successive principal minors are non-negative,

$$D_1 \geqq 0, D_2 \geqq 0, \ldots, D_n \geqq 0, \tag{40}$$

---

[14] The form $A_p(x, x)$ is obtained from $A(x, x)$ if we set in the latter

$$x_{p+1} = \cdots = x_n = 0 \;\; (p = 1, 2, \ldots, n).$$

[15] Thus, when the successive principal minors of a real symmetric matrix are positive, all the remaining principal minors are then also positive.

it does *not* follow that $A(x, x)$ is positive semidefinite. For, the form

$$a_{11}x_1^2 + 2a_{12}x_1x_2 + a_{22}x_2^2$$

in which $a_{11} = a_{12} = 0$, $a_{22} < 0$ satisfies (40), but is not positive semidefinite.

However, we have the following theorem.

THEOREM 4: *A quadratic form* $A(x, x) = \sum_{i,k=1}^{n} a_{ik}x_ix_k$ *is positive semidefinite if and only if all the principal minors of its coefficient matrix are non-negative:*

$$A\begin{pmatrix} i_1 & i_2 & \dots & i_p \\ i_1 & i_2 & \dots & i_p \end{pmatrix} \geqq 0 \quad (1 \leqq i_1 < i_2 < \cdots < i_p \leqq n;\ p = 1, 2, \dots, n). \quad \textbf{(41)}$$

Proof. We introduce the auxiliary form

$$A_\varepsilon(x, x) = A(x, x) + \varepsilon \sum_{i=1}^{n} x_i^2 \qquad (\varepsilon > 0).$$

Obviously $\lim_{\varepsilon \to 0} A_\varepsilon(x, x) = A(x, x)$.

The fact that $A(x, x)$ is positive semidefinite implies that $A_\varepsilon(x, x)$ is positive definite, so that we have the inequality (cf. Corollary to Theorem 3):

$$A_\varepsilon\begin{pmatrix} i_1 & i_2 & \dots & i_p \\ i_1 & i_2 & \dots & i_p \end{pmatrix} > 0 \quad (1 \leqq i_1 < i_2 < \cdots < i_p \leqq n;\ p = 1, 2, \dots, n).$$

Proceeding to the limit for $\varepsilon \to 0$, we obtain (41).

Suppose, conversely, that (41) holds. Then we have

$$A_\varepsilon\begin{pmatrix} i_1 & i_2 & \dots & i_p \\ i_1 & i_2 & \dots & i_p \end{pmatrix} = \varepsilon^p + \cdots \geqq \varepsilon^p > 0 \quad (1 \leqq i_1 < i_2 < \cdots < i_p \leqq n;\ p = 1, 2, \dots, n).$$

But then (by Theorem 3), $A_\varepsilon(x, x)$ is positive definite

$$A_\varepsilon(x, x) > 0 \quad (x \neq o).$$

Proceeding to the limit for $\varepsilon \to 0$ we obtain:

$$A(x, x) \geqq 0.$$

This completes the proof.

The conditions for a form to be negative semidefinite and negative definite are obtained from (39) and (41), respectively, when these inequalities are applied to $-A(x, x)$.

THEOREM 5: *A quadratic form $A(x, x)$ is negative definite if and only if the following inequalities hold:*

$$D_1 < 0, D_2 > 0, D_3 < 0, \ldots, (-1)^n D_n > 0. \tag{42}$$

THEOREM 6: *A quadratic form $A(x, x)$ is negative semidefinite if and only if the following inequalities hold:*

$$(-1)^p A\begin{pmatrix} i_1 & i_2 & \ldots & i_p \\ i_1 & i_2 & \ldots & i_p \end{pmatrix} \geqq 0 \quad (1 \leqq i_1 < i_2 < \cdots < i_p \leqq n;\ p = 1, 2, \ldots, n). \tag{43}$$

§ 5. Reduction of a Quadratic Form to Principal Axes

1. We consider an arbitrary real quadratic form

$$A(x, x) = \sum_{i, k=1}^{n} a_{ik} x_i x_k .$$

Its coefficient matrix $A = \| a_{ik} \|_1^n$ is real and symmetric. Therefore (see Chapter IX, § 13) it is orthogonally similar to a real diagonal matrix Λ, i.e., there exists a real orthogonal matrix Q such that

$$\Lambda = Q^{-1} A Q \quad (\Lambda = \| \lambda_i \delta_{ik} \|_1^n, \quad QQ^{\mathsf{T}} = E). \tag{44}$$

Here $\lambda_1, \lambda_2, \ldots, \lambda_n$ are the characteristic values of A.

Since for an orthogonal matrix $Q^{-1} = Q^{\mathsf{T}}$, it follows from (43) that under the orthogonal transformation of the variables

$$x = Q\xi \qquad (QQ^{\mathsf{T}} = E) \tag{45}$$

or, in greater detail,

$$x_i = \sum_{k=1}^{n} q_{ik} \xi_k \quad \left(\sum_{j=1}^{n} q_{ij} q_{kj} = \delta_{ik};\ i, k = 1, 2, \ldots, n \right), \tag{45'}$$

the form $A(x, x)$ goes over into

$$\Lambda(\xi, \xi) = \sum_{i=1}^{n} \lambda_i \xi_i^2 . \tag{46}$$

THEOREM 7: *Every real quadratic form* $A(x, x) = \sum_{i,k=1}^{n} a_{ik}x_i x_k$ *can be reduced to the canonical form* (46) *by an orthogonal transformation, where* $\lambda_1, \lambda_2, \ldots, \lambda_n$ *are the characteristic values of* $A = \| a_{ik} \|_1^n$.

The reduction of the quadratic form $A(x, x)$ to the canonical form (46) is called *reduction to principal axes.* The reason for this name is that the equation of a central hypersurface of the second order

$$\sum_{i,k=1}^{n} a_{ik}x_i x_k = c \quad (c = \text{const.} \neq 0) \tag{47}$$

under the orthogonal transformation (45′) of the variables assumes the canonical form

$$\sum_{i=1}^{n} \varepsilon_i \frac{\xi_i^2}{a_i^2} = 1 \quad \left(\frac{\varepsilon_i}{a_i^2} = \frac{\lambda_i}{c};\quad \varepsilon_i = \pm 1;\quad i = 1, 2, \ldots, n \right). \tag{48}$$

If we regard $x_1, x_2, \ldots, x_n$ as coordinates in an orthonormal basis in an n-dimensional euclidean space, then $\xi_1, \xi_2, \ldots, \xi_n$ are the coordinates in a new orthonormal basis of the same space, and the 'rotation'[16] of the axes is brought about by the orthogonal transformation (45). The new coordinate axes are axes of symmetry of the central surface (47) and are usually called its *principal axes.*

2. It follows from (46) that *the rank* r *of* $A(x, x)$ *is equal to the number of non-zero characteristic values of* A *and the signature* σ *is equal to the difference between the number of positive and the number of negative characteristic values of* A.

Hence, in particular, we have the following proposition:

If under a continuous change of the coefficients of a quadratic form the rank remains unchanged, then the signature also remains unchanged.

Here we have started from the fact that a continuous change of the coefficients produces a continuous change of the characteristic values. The signature can only change when some characteristic value changes sign. But then at some intermediate stage this characteristic value must pass through zero, and this results in a change of the rank of the form.

---

[16] If $|Q| = -1$, then (45) is a combination of a rotation with a reflection (see p. 287). However, the reduction to principal axes can always be effected by a proper orthogonal matrix ($|Q| = 1$). This follows from the fact that, without changing the canonical form, we can perform the additional transformation

$$\xi_i = \xi_i^{\mathsf{T}} \quad (i = 1, 2, \ldots, n-1), \quad \xi_n = -\xi_n^{\mathsf{T}}.$$

§ 6. Pencils of Quadratic Forms

1. In the theory of small oscillations it is necessary to consider simultaneously two quadratic forms one of which gives the potential, and the other the kinetic energy, of the system. The second form is always positive definite.

The study of a system of two such forms is the object of this section.

Two real quadratic forms

$$A(x, x) = \sum_{i,k=1}^{n} a_{ik}x_i x_k \quad \text{and} \quad B(x, x) = \sum_{i,k=1}^{n} b_{ik}x_i x_k$$

determine the *pencil* of forms $A(x, x) - \lambda B(x, x)$ (λ is a parameter).

If the form $B(x, x)$ is positive definite, the pencil $A(x, x) - \lambda B(x, x)$ is then called *regular.*

The equation

$$|A - \lambda B| = 0$$

is called the *characteristic equation of the pencil of forms* $A(x,x) - \lambda B(x,x)$.

We denote by λ_0 some root of this equation. Since the matrix $A - \lambda_0 B$ is singular, there exists a column $z = (z_1, z_2, \ldots, z_n) \neq o$ such that $(A - \lambda_0 B)z = o$, or

$$Az = \lambda_0 Bz \quad (z \neq o).$$

The number λ_0 will be called a *characteristic value of the pencil* $A(x, x) - \lambda B(x, x)$ and z a corresponding *principal column* or '*principal vector*' of the pencil. The following theorem holds:

THEOREM 8: *The characteristic equation*

$$|A - \lambda B| = 0$$

of a regular pencil of forms $A(x, x) - \lambda B(x, x)$ *always has* n *real roots* λ_k *with the corresponding principal vectors* $z^k = (z_{1k}, z_{2k}, \ldots, z_{nk})$ $(k = 1, 2, \ldots, n)$:

$$Az^k = \lambda_k Bz^k \qquad (k = 1, 2, \ldots, n). \tag{49}$$

These principal vectors z^k *can be chosen such that the relations*

$$B(z^i, z^k) = \delta_{ik} \qquad (i, k = 1, 2, \ldots, n) \tag{50}$$

are satisfied.

Proof. We observe that (49) can be written as:

$$B^{-1}Az^k = \lambda_k z^k \ (k = 1, 2, \dots, n). \tag{51}$$

Thus, our theorem states that the matrix

$$D = B^{-1}A \tag{52}$$

1. has simple structure, 2. has real characteristic values, and 3. has characteristic columns (vectors) $z^1, z^2, \dots, z^n$ corresponding to these characteristic values and satisfying the relations (50).[17]

In order to prove these three statements, we introduce an n-dimensional vector space $\boldsymbol{R}$ over the field of real numbers. In this space we fix a basis $\boldsymbol{e}_1, \boldsymbol{e}_2, \dots, \boldsymbol{e}_n$ and introduce a scalar product of two arbitrary vectors

$$\boldsymbol{x} = \sum_{i=1}^{n} x_i \boldsymbol{e}_i, \qquad \boldsymbol{y} = \sum_{i=1}^{n} y_i \boldsymbol{e}_i$$

by means of the positive-definite bilinear form $B(x, y)$:

$$(\boldsymbol{xy}) = B(x, y) = \sum_{i,k=1}^{n} b_{ik} x_i y_k = x^{\mathsf{T}} B y \tag{53}$$

and hence the square of the length of a vector $\boldsymbol{x}$ by means of the form $B(x, x)$:

$$(\boldsymbol{xx}) = B(x, x) = x^{\mathsf{T}} B x\,, \tag{53'}$$

where x and y are columns $x = (x_1, x_2, \dots, x_n)$, $y = (y_1, y_2, \dots, y_n)$.

It is easy to verify that the metric so introduced satisfies the postulates 1.-5. (p. 243) and is, therefore, euclidean.

We have obtained an n-dimensional euclidean space $\boldsymbol{R}$, but the original basis $\boldsymbol{e}_1, \boldsymbol{e}_2, \dots, \boldsymbol{e}_n$ is, in general, not orthonormal. To the matrices A, B, and $D = B^{-1}A$ there correspond in this basis linear operators in $\boldsymbol{R}$: $\boldsymbol{A}$, $\boldsymbol{B}$, and $\boldsymbol{D} = \boldsymbol{B}^{-1}\boldsymbol{A}$.[18]

---

[17] If D were a symmetric matrix, then the properties 1. and 2. would follow immediately from properties of a symmetric operator (Chapter IX, p. 284). However, D, as a product of two symmetric matrices, is not necessarily itself symmetric, since $D = B^{-1}A$ and $D^{\mathsf{T}} = AB^{-1}$.

[18] Since the basis $\boldsymbol{e}_1, \boldsymbol{e}_2, \dots, \boldsymbol{e}_n$ is not orthonormal, the operators $\boldsymbol{A}$ and $\boldsymbol{B}$ to which, in this basis, the symmetric matrices A and B correspond, are not necessarily symmetric themselves.

We shall show that $\boldsymbol{D}$ is a symmetric operator in $\boldsymbol{R}$ (see Chapter IX, § 13).[19] Indeed, for arbitrary vectors $\boldsymbol{x}$ and $\boldsymbol{y}$ with the coordinate columns $x=(x_1, x_2, \ldots, x_n)$ and $y=(y_1, y_2, \ldots, y_n)$ we have, by (52) and (53),

$$(\boldsymbol{D}\boldsymbol{x}, \boldsymbol{y})=(Dx)^{\mathsf{T}}By=x^{\mathsf{T}}D^{\mathsf{T}}By=x^{\mathsf{T}}AB^{-1}By=x^{\mathsf{T}}Ay$$

and

$$(\boldsymbol{x}, \boldsymbol{D}\boldsymbol{y})=x^{\mathsf{T}}BDy=x^{\mathsf{T}}BB^{-1}Ay=x^{\mathsf{T}}Ay,$$

i.e.,

$$(\boldsymbol{D}\boldsymbol{x}, \boldsymbol{y})=(\boldsymbol{x}, \boldsymbol{D}\boldsymbol{y}).$$

The symmetric operator $\boldsymbol{D}=\boldsymbol{B}^{-1}\boldsymbol{A}$ has real characteristic values $\lambda_1, \lambda_2, \lambda_3, \ldots, \lambda_n$ and a complete orthonormal system of characteristic vectors $\boldsymbol{z}^1, \boldsymbol{z}^2, \boldsymbol{z}^3, \ldots, \boldsymbol{z}^n$ (see p. 284, Chapter IX):

$$\boldsymbol{B}^{-1}\boldsymbol{A}\boldsymbol{z}^k=\lambda_k\boldsymbol{z}^k \qquad (k=1, 2, \ldots, n), \tag{54}$$

$$(\boldsymbol{z}^i\boldsymbol{z}^k)=\delta_{ik} \qquad (i, k=1, 2, \ldots, n). \tag{54'}$$

Let $z^k=(z_{1k}, z_{2k}, \ldots, z_{nk})$ be the coordinate column of $\boldsymbol{z}^k$ $(k=1, 2, \ldots, n)$ in the basis $\boldsymbol{e}_1, \boldsymbol{e}_2, \ldots, \boldsymbol{e}_n$. Then the equations (54) can be written in the form (51) or (49) and the relations (54′), by (53), yield the equation (50).

This completes the proof.

Note that it follows from (50) that the columns $z^1, z^2, \ldots, z^n$ are linearly independent. For suppose that

$$\sum_{k=1}^{n} c_k z^k=0. \tag{55}$$

Then for every i $(1 \leqq i \leqq n)$, by (50),

$$0=B\left(z^i, \sum_{k=1}^{n} c_k z^k\right)=\sum_{k=1}^{n} c_k B(z^i, z^k)=c_i.$$

Then all the c_i $(i=1, 2, \ldots, n)$ in (55) are zero and there is no linear dependence among the columns $z^1, z^2, \ldots, z^n$.

A square matrix formed from principal columns $z^1, z^2, \ldots, z^n$ satisfying the relations (50)

$$Z=(z^1, z^2, \ldots, z^n)=\| z_{ik} \|_1^n$$

will be called a *principal matrix* for the pencil of forms $A(x, x)-\lambda B(x, x)$.

---

[19] Hence D is similar to some symmetric matrix.

The principal matrix Z is non-singular ($|Z| \neq 0$), because its columns are linearly independent.

The equation (50) can be written as follows:

$$z^{i^{\mathsf{T}}} B z^k = \delta_{ik} \qquad (i, k = 1, 2, \ldots, n). \tag{56}$$

Moreover, when we multiply both sides of (49) on the left by the row matrix $z^{i^{\mathsf{T}}}$, we obtain:

$$z^{i^{\mathsf{T}}} A z^k = \lambda_k z^{i^{\mathsf{T}}} B z^k = \lambda_k \delta_{ik} \qquad (i, k = 1, 2, \ldots, n). \tag{57}$$

By introducing the principal matrix $Z = (z^1, z^2, \ldots, z^n)$, we can represent (56) and (57) in the form

$$Z^{\mathsf{T}} A Z = \| \lambda_k \delta_{ik} \|_1^n, \qquad Z^{\mathsf{T}} B Z = E. \tag{58}$$

The formulas (58) show that the non-singular transformation

$$x = Z\xi \tag{59}$$

reduces the quadratic forms $A(x, x)$ and $B(x, x)$ simultaneously to sums of squares:

$$\sum_{k=1}^{n} \lambda_k \xi_k^2 \text{ and } \sum_{k=1}^{n} \xi_k^2. \tag{60}$$

This property of (59) characterizes a principal matrix Z. For suppose that the transformation (59) reduces the forms $A(x, x)$ and $B(x, x)$ simultaneously to the canonical forms (60). Then (58) holds, and hence (56) and (57) holds for Z. (58) implies that Z is non-singular ($|Z| \neq 0$). We rewrite (57) as follows:

$$z^{i^{\mathsf{T}}} (A z^k - \lambda_k B z^k) = o \qquad (i = 1, 2, \ldots, n), \tag{61}$$

where k has an arbitrary fixed value ($1 \leqq k \leqq n$). The system of equations (61) can be contracted into the single equation

$$Z^{\mathsf{T}} (A z_k - \lambda_k B z^k) = O;$$

hence, since Z^{T} is non-singular,

$$A z^k - \lambda_k B z^k = O;$$

i.e., for every k (49) holds. Therefore Z is a principal matrix. Thus we have proved the following theorem:

THEOREM 9: *If* $Z = \| z_{ik} \|_1^n$ *is a principal matrix of a regular pencil of forms* $A(x, x) - \lambda B(x, x)$, *then the transformation*

$$x = Z\xi \tag{62}$$

reduces the forms $A(x, x)$ *and* $B(x, x)$ *simultaneously to sums of squares*

$$\sum_{k=1}^{n} \lambda_k \xi_k^2, \quad \sum_{k=1}^{n} \xi_k^2, \tag{63}$$

where $\lambda_1, \lambda_2, \ldots, \lambda_n$ *are the characteristic values of the pencil* $A(x, x) - \lambda B(x, x)$ *corresponding to the columns* $z^1, z^2, \ldots, z^n$ *of* Z.

Conversely, if some transformation (62) *simultaneously reduces* $A(x, x)$ *and* $B(x, x)$ *to the form* (63), *then* $Z = \| z_{ik} \|_1^n$ *is a principal matrix of the regular pencil of forms* $A(x, x) - \lambda B(x, x)$.

Sometimes the characteristic property of the transformation (62) formulated in Theorem 9 is used for the construction of a principal matrix and the proof of Theorem 8.[20] For this purpose, we first of all carry out a transformation of variables $x = Ty$ that reduces the form $B(x, x)$ to the 'unit' sum of squares $\sum_{k=1}^{n} y_k^2$ (which is always possible, since $B(x, x)$ is positive definite). Then $A(x, x)$ is carried into a certain form $A_1(y, y)$. Now the form $A_1(y, y)$ is reduced to the form $\sum_{k=1}^{n} \lambda_k \xi_k^2$ by an orthogonal transformation $y = Q\xi$ (reduction to principal axes!). Then, obviously,[21] $\sum_{k=1}^{n} y_k^2 = \sum_{k=1}^{n} \xi_k^2$. Thus the transformation $x = Z\xi$, where $Z = TQ$, reduces the two given forms to (63). Afterwards it turns out (as we have shown on p. 313) that the columns $z^1, z^2, \ldots, z^n$ of Z satisfy the relations (49) and (50).

In the special case where $B(x, x)$ is the unit form, i.e., $B(x, x) = \sum_{k=1}^{n} x_k^2$, so that $B = E$, the characteristic equation of the pencil $A(x, x) - \lambda B(x, x)$ coincides with the characteristic equation of A, and the principal vectors of the pencil are characteristic vectors of A. In this case the relations (50) can be written as follows:

$$z^{i\mathsf{T}} z^k = \delta_{ik} \ (i, k = 1, 2, \ldots, n)$$

and they express the orthonormality of the columns $z^1, z^2, \ldots, z^n$.

---

[20] See [17], pp. 56-57.

[21] An orthogonal transformation does not alter a sum of squares of the variables, because $(Qx)^{\mathsf{T}} Qx = x^{\mathsf{T}} x$.

2. Theorems 8 and 9 admit of an intuitive geometric interpretation. We introduce a euclidean space $\boldsymbol{R}$ with the basis $\boldsymbol{e}_1, \boldsymbol{e}_2, \ldots, \boldsymbol{e}_n$ and the fundamental metric form $B(x, x)$ just as was done for the proof of Theorem 8. In $\boldsymbol{R}$ we consider a central hypersurface of the second order whose equation is

$$A(x, x) \equiv \sum_{i,k=1}^{n} a_{ik} x_i x_k = c. \tag{64}$$

After the coordinate transformation $x = Z\xi$, where $Z = \| z_{ik} \|_1^n$ is a principal matrix of the pencil $A(x, x) - \lambda B(x, x)$, the new basis vectors are the vectors $z^1, z^2, \ldots, z^n$ whose coordinates in the old basis form the columns of Z, i.e., the principal vectors of the pencil. These vectors form an orthonormal basis in which the equation of the hypersurface (64) has the form

$$\sum_{k=1}^{n} \lambda_k \xi_k^2 = c. \tag{65}$$

Therefore the principal vectors $z^1, z^2, \ldots, z^n$ of the pencil coincide in direction with the principal axes of the hypersurface (64), and the characteristic values $\lambda_1, \lambda_2, \ldots, \lambda_n$ of the pencil determine the lengths of the semi-axes:

$$\lambda_k = \pm \frac{c}{a_k^2} \quad (k = 1, 2, \ldots, n).$$

Thus, the task of determining the characteristic values and the principal vectors of a regular pencil of forms $A(x, x) - \lambda B(x, x)$ is equivalent to the task of reducing the equation (64) of a central hypersurface of the second order to principal axes, provided the equation of the hypersurface is given in a general skew coordinate system[22] in which the 'unit sphere' has the equation $B(x, x) = 1$.

Example. Given the equation of a surface of the second order

$$2x^2 - 2y^2 - 3z^2 - 10yz + 2xz - 4 = 0 \tag{66}$$

in a general skew coordinate system in which the equation of the unit sphere is

$$2x^2 + 3y^2 + 2z^2 + 2xz = 1, \tag{67}$$

it is required to reduce equation (66) to principal axes.

In this case

$$A = \begin{Vmatrix} 2 & 0 & 1 \\ 0 & -2 & -5 \\ 1 & -5 & -3 \end{Vmatrix}, \qquad B = \begin{Vmatrix} 2 & 0 & 1 \\ 0 & 3 & 0 \\ 1 & 0 & 2 \end{Vmatrix}.$$

[22] I.e., a skew coordinate system with distinct units of lengths along the axes.

The characteristic equation of the pencil $| A - \lambda B | = 0$ has the form

$$\begin{vmatrix} 2-2\lambda & 0 & 1-\lambda \\ 0 & -2-3\lambda & -5 \\ 1-\lambda & -5 & -3-2\lambda \end{vmatrix} = 0. \tag{68}$$

This equation has three roots: $\lambda_1 = 1$, $\lambda_2 = 1$, $\lambda_3 = -4$.

We denote the coordinates of a principal vector corresponding to the characteristic value 1 by u, v, w. The values of u, v, w are determined from the system of homogeneous equations whose coefficients are the elements of the determinant (68) for $\lambda = 1$:

$$0 \cdot u + 0 \cdot v + 0 \cdot w = 0,$$
$$0 \cdot u - 5v \quad - 5w \ = 0,$$
$$0 \cdot u - 5v \quad - 5w \ = 0.$$

In fact we have only one relation

$$v + w = 0.$$

To the characteristic value $\lambda = 1$ there must correspond two orthonormal principal vectors. The coordinates of the first can be chosen arbitrarily, provided they satisfy the relation $v + w = 0$.

We set

$$u = 0,\ v,\ w = -v.$$

We take the coordinates of the second principal vector in the form

$$u',\ v',\ w' = -v'$$

and write down the condition for orthogonality $(B(z^1, z^2) = 0)$:

$$2uu' + 3vv' + 2ww' + uw' + u'w = 0.$$

Hence we find: $u' = 5v'$. Thus, the coordinates of the second principal vector are

$$u' = 5v',\ v',\ w' = -v'.$$

Similarly, by setting $\lambda = -4$ in the characteristic determinant, we find for the corresponding principal vector:

$$u'',\ v'' = -u'',\ w'' = -2u''.$$

The values of v, v', and u'' are determined from the condition that the coordinates of a principal vector must satisfy the equation of the unit sphere ($B(x, x) = 1$), i.e., (67). Hence we find:

$$v = \frac{1}{\sqrt{5}}, \qquad v' = \frac{1}{3\sqrt{5}}, \qquad u'' = -\frac{1}{3}.$$

Therefore the principal matrix has the form

$$Z = \begin{Vmatrix} 0 & \frac{\sqrt{5}}{3} & -\frac{1}{3} \\ \frac{1}{\sqrt{5}} & \frac{1}{3\sqrt{5}} & \frac{1}{3} \\ -\frac{1}{\sqrt{5}} & -\frac{1}{3\sqrt{5}} & \frac{2}{3} \end{Vmatrix},$$

and the corresponding coordinate transformation ($x = Z\xi$) reduces the equations (66) and (67) to the canonical form

$$\xi_1^2 + \xi_2^2 - 4\xi_3^2 - 4 = 0, \quad \xi_1^2 + \xi_2^2 + \xi_3^2 = 1$$

The first equation can also be written as follows:

$$\frac{\xi_1^2}{4} + \frac{\xi_2^2}{4} - \frac{\xi_3^2}{1} = 1.$$

This is the equation of a one-sheet hyperboloid of rotation with real semi-axes equal to 2, and an imaginary one equal to 1. The coordinates of the endpoint of the axis of rotation is determined by the third column of Z, i.e., $-1/3, 1/3, 2/3$. The coordinates of the endpoints of the other two orthogonal axes are given by the first and second columns.

§ 7. Extremal Properties of the Characteristic Values of a Regular Pencil of Forms[23]

1. Suppose that two quadratic forms are given

$$A(x, x) = \sum_{i,k=1}^{n} a_{ik} x_i x_k \quad \text{and} \quad B(x, x) = \sum_{i,k=1}^{n} b_{ik} x_i x_k,$$

of which $B(x, x)$ is positive definite. We number the characteristic values of the regular pencil of forms $A(x, x) - \lambda B(x, x)$ in non-descending order:

$$\lambda_1 \leqq \lambda_2 \leqq \cdots \leqq \lambda_n. \tag{69}$$

[23] In the exposition of this section, we follow the book [17], § 10.

The principal vectors[24] corresponding to these characteristic values are denoted, as before, by $z^1, z^2, \ldots, z^n$:

$$z^k = (z_{1k}, z_{2k}, \ldots, z_{nk}) \quad (k = 1, 2, \ldots, n).$$

Let us determine the least value (minimum) of the ratio of the forms $\frac{A(x, x)}{B(x, x)}$ considering all possible values of the variables, not all equal to zero $(x \neq o)$. For this purpose it is convenient to go over to new variabes $\xi_1, \xi_2, \ldots, \xi_n$ by means of the transformation

$$x = Z\xi \quad (x_i = \sum_{k=1}^{n} z_{ik}\xi_k;\ i = 1, 2, \ldots, n),$$

where $Z = \| z_{ik} \|_1^n$ is a principal matrix of the pencil $A(x, x) - \lambda B(x, x)$. In the new variables the ratio of the forms is represented (see (63)) by

$$\frac{A(x, x)}{B(x, x)} = \frac{\lambda_1 \xi_1^2 + \lambda_2 \xi_2^2 + \cdots + \lambda_n \xi_n^2}{\xi_1^2 + \xi_2^2 + \cdots + \xi_n^2}. \tag{70}$$

On the real axis we take the n points $\lambda_1, \lambda_2, \ldots, \lambda_n$. We ascribe to these points non-negative masses $m_1 = \xi_1^2$, $m_2 = \xi_2^2$, $\ldots$, $m_n = \xi_n^2$, respectively. Then, by (70), the quotient $\frac{A(x, x)}{B(x, x)}$ is the coordinate of the center of these masses. Therefore

$$\lambda_1 \leqq \frac{A(x, x)}{B(x, x)} \leqq \lambda_n.$$

Let us, for the time being, ignore the second part of the inequality and investigate when the equality sign holds in the first part. For this purpose, we group together the equal characteristic values in (69):

$$\lambda_1 = \cdots = \lambda_{p_1} < \lambda_{p_1+1} = \cdots = \lambda_{p_1+p_2} < \cdots. \tag{71}$$

The center of mass can coincide with the least value λ_1 only if all the masses are zero except at this point, i.e., when

$$\xi_{p_1+1} = \cdots = \xi_n = 0.$$

In this case the corresponding x is a linear combination of the principal columns $z^1, z^2, \ldots, z^{p_1}$.[25] Therefore all these columns correspond to the characteristic value λ_1, so that x is also a principal column (vector) for $\lambda = \lambda_1$.

---

[24] Here we use the term 'principal vector' in the sense of a principal column of the pencil (see p. 310). Throughout this section, having the geometric interpretation in mind, we often call a column, a vector.

[25] From $x = Z\xi$ it follows that $x = \sum_{k=1}^{n} \xi_k z^k$.

We have proved:

THEOREM 10: *The smallest characteristic value of the regular pencil $A(x, x) - \lambda B(x, x)$ is the minimum of the ratio of the forms $A(x, x)$ and $B(x, x)$*

$$\lambda_1 = \min \frac{A(x, x)}{B(x, x)}. \tag{72}$$

and this minimum is only assumed for principal vectors of the characteristic value λ_1.

2. In order to give an analogous 'minimal' characteristic for the next characteristic value λ_2, we restrict ourselves to all the vectors orthogonal to z^1, i.e., to those that satisfy the equation[26]

$$B(z^1, x) = 0.$$

For these vectors,

$$\frac{A(x, x)}{B(x, x)} = \frac{\lambda_2 \xi_2^2 + \cdots + \lambda_n \xi_n^2}{\xi_2^2 + \cdots + \xi_n^2}$$

and therefore

$$\min \frac{A(x, x)}{B(x, x)} = \lambda_2 \ (B(z^1, x) = 0).$$

Here the equality sign holds only for those vectors orthogonal to z^1 that are principal vectors for the characteristic value λ_2.

Proceeding to the subsequent characteristic values, we eventually obtain the following theorem:

THEOREM 11: *For every p $(1 \leqq p \leqq n)$ the p-th characteristic value λ_p in (69) is the minimum of the ratio of the forms*

$$\lambda_p = \min \frac{A(x, x)}{B(x, x)}, \tag{73}$$

provided that the variable vector x is orthogonal to the first $p - 1$ orthonormal principal vectors $z^1, z^2, \ldots, z^{p-1}$:

---

[26] Here, and in what follows, we shall mean by the orthogonality of two vectors (columns) x, y that the equation $B(x, y) = 0$ holds. This is in complete agreement with the geometric interpretation given in the preceding section. We shall regard the quantities $x_1, x_2, \ldots, x_n$ as the coordinates of a vector x in some basis of a euclidean space in which the square of the length (the norm) is given by the positive-definite form $B(x, x) = \sum_{i,k=1}^{n} b_{ik} x_i x_k$. In this metric the vectors $z^1, z^2, \ldots, z^n$ form an orthonormal basis. Therefore, if the vector $x = \sum_{k=1}^{n} \xi_k z^k$ is orthogonal to one of the z^k, then the corresponding $\xi_k = 0$.

$$B(z^1, x) = 0, \ \ldots, \ B(z^{p-1}, x) = 0. \tag{74}$$

Moreover, the minimum is assumed only for those vectors that satisfy the condition (74) *and are at the same time principal vectors for the characteristic value* λ_p.

3. The characterization of λ_p given in Theorem 11 has the disadvantage that it is connected with the preceding principal vectors $z^1, z^2, \ldots, z^{p-1}$ and can therefore be used only when these vectors are known. Moreover, there is a certain arbitrariness in the choice of these vectors.

In order to give a characterization of λ_p $(p = 1, 2, \ldots, n)$ free from these defects, we introduce the concept of *constraint* imposed on the variables $x_1, x_2, \ldots, x_n$.

Suppose that linear forms in the variables $x_1, x_2, \ldots, x_n$ are given:

$$L_k(x) = l_{1k}x_1 + l_{2k}x_2 + \cdots + l_{nk}x_n \quad (k = 1, 2, \ldots, h). \tag{74'}$$

We shall say that the variables $x_1, x_2, \ldots, x_n$ or (what is the same) the vector x is subject to h constraints $L_1, L_2, \ldots, L_h$ if only such values of the variables are considered that satisfy the system of equations

$$L_k(x) = 0 \quad (k = 1, 2, \ldots, h). \tag{74''}$$

Preserving the notation (74′) for arbitrary linear forms we introduce a specialized notation for the 'scalar product' of x with the principal vectors $z^1, z^2, \ldots, z^n$:

$$\widetilde{L}_k(x) = B(z^k, x) \quad (k = 1, 2, \ldots, n).^{27} \tag{75}$$

Furthermore, when the variable vector is subject to the constraints (74′′) we shall denote $\min \dfrac{A(x, x)}{B(x, x)}$ as follows:

$$\mu\left(\frac{A}{B}; L_1, L_2, \ldots, L_h\right).$$

In this notation, (73) is written as follows:

$$\lambda_p = \mu\left(\frac{A}{B}; \widetilde{L}_1, \widetilde{L}_2, \ldots, \widetilde{L}_{p-1}\right) \quad (p = 1, 2, \ldots, n). \tag{76}$$

We consider the constraints

$$L_1(x) = 0, \ \ldots, \ L_{p-1}(x) = 0 \tag{77}$$

and

$$\widetilde{L}_{p+1}(x) = 0, \ \ldots, \ \widetilde{L}_n(x) = 0. \tag{78}$$

---

[27] $\widetilde{L}_k(x) = z^{k\mathsf{T}}Bx = \widetilde{l}_{1k}x_1 + \widetilde{l}_{2k}x_2 + \cdots + \widetilde{l}_{nk}x_n$, where $\widetilde{l}_{1k}, \widetilde{l}_{2k}, \ldots, \widetilde{l}_{nk}$ are the elements of the row matrix $z^{k\mathsf{T}}B$ $(k = 1, 2, \ldots, n)$.

Since the number of constraints (77) and (78) is less than n, there exists a vector $x^{(1)} \neq o$ satisfying all these constraints. Since the constraints (78) express the orthogonality of x to the principal vectors $z^{p+1}, \ldots, z^n$, the corresponding coordinates of $x^{(1)}$ are $\xi_{p+1} = \cdots = \xi_n = 0$. Therefore, by (70),

$$\frac{A(x^{(1)}, x^{(1)})}{B(x^{(1)}, x^{(1)})} = \frac{\lambda_1 \xi_1^2 + \cdots + \lambda_p \xi_p^2}{\xi_1^2 + \cdots + \xi_p^2} \leq \lambda_p .$$

But then

$$\mu\left(\frac{A}{B};\ L_1, L_2, \ldots, L_{p-1}\right) \leq \frac{A(x^{(1)}, x^{(1)})}{B(x^{(1)}, x^{(1)})} \leq \lambda_p .$$

This inequality in conjunction with (76) shows that for variable constraints $L_1, L_2, \ldots, L_{p-1}$ the value of μ remains less than or equal to λ_p and becomes λ_p if the specialized constraints $\tilde{L}_1, \tilde{L}_2, \ldots, \tilde{L}_{p-1}$ are taken.

Thus we have proved:

THEOREM 12: *If we consider the minimum of the ratio of the two forms* $\frac{A(x, x)}{B(x, x)}$ *for $p-1$ arbitrary, but variable, constraints $L_1, L_2, \ldots, L_{p-1}$, then the maximum of these minima is equal to λ_p:*

$$\lambda_p = \max \mu\left(\frac{A}{B};\ L_1, L_2, \ldots, L_{p-1}\right) \quad (p = 1, \ldots, n). \tag{79}$$

Theorem 12 gives a 'maximal-minimal' characterization of $\lambda_1, \lambda_2, \ldots, \lambda_n$ in contrast to the 'minimal' characterization which we discussed in Theorem 11.

4. Note that when in the pencil $A(x, x) - \lambda B(x, x)$ the form $A(x, x)$ is replaced by $-A(x, x)$, all the characteristic values of the pencil change sign, but the corresponding principal vectors remain unchanged. Thus, the characteristic values of the pencil $-A(x, x) - \lambda B(x, x)$ are

$$-\lambda_n \leq -\lambda_{n-1} \leq \cdots \leq -\lambda_1 .$$

Moreover, by using the notation

$$\nu\left(\frac{A}{B};\ L_1, L_2, \ldots, L_h\right) = \max \frac{A(x, x)}{B(x, x)} \tag{80}$$

when the variable vector is subject to the constraints $L_1, L_2, \ldots, L_h$, we can write:

$$\mu\left(-\frac{A}{B};\ L_1, L_2, \ldots, L_h\right) = -\nu\left(\frac{A}{B};\ L_1, L_2, \ldots, L_h\right)$$

and

$$\max \mu\left(-\frac{A}{B};\ L_1, L_2, \ldots, L_h\right) = -\min \nu\left(\frac{A}{B};\ L_1, L_2, \ldots, L_h\right).$$

Therefore, by applying Theorems 10, 11, and 12 to the ratio $-\frac{A(x, x)}{B(x, x)}$, we obtain instead of (72), (76), and (79) the formulas

$$\lambda_n = \max \frac{A(x,x)}{B(x,x)},$$

$$\left.\begin{array}{l}\lambda_{n-p+1} = \nu\left(\frac{A}{B};\, \tilde{L}_n,\, \tilde{L}_{n-1},\, \ldots,\, \tilde{L}_{n-p+2}\right)\\[2ex] \lambda_{n-p+1} = \min \nu\left(\frac{A}{B}; L_1, L_2, \ldots, L_{p-1}\right),\end{array}\right\} \quad (p = 2, \ldots, n).$$

These formulas establish the 'maximal' and the 'minimal-maximal' properties, respectively, of $\lambda_1, \lambda_2, \ldots, \lambda_n$, which we formulate in the following theorem:

THEOREM 13: *Suppose that to the characteristic values*

$$\lambda_1 \leqq \lambda_2 \leqq \cdots \leqq \lambda_n$$

of the regular pencil of forms $A(x,x) - \lambda B(x,x)$ there correspond the linearly independent principal vectors of the pencil $z^1, z^2, \ldots, z^n$. Then:

1) *The largest characteristic value λ_n is the maximum of the ratio of the forms $\frac{A(x,x)}{B(x,x)}$:*

$$\lambda_n = \max \frac{A(x,x)}{B(x,x)}, \tag{81}$$

and this maximum is assumed only for principal vectors of the pencil corresponding to the characteristic value λ_n.

2) *The characteristic value p-th from the end λ_{n-p+1} $(2 \leqq p \leqq n)$ is the maximum of the same ratio of the forms*

$$\lambda_{n-p+1} = \max \frac{A(x,x)}{B(x,x)} \tag{82}$$

provided that the variable vector x is subject to the constraints:[28]

$$B(z^n, x) = 0,\ B(z^{n-1}, x) = 0,\ \ldots,\ B(z^{n-p+2}, x) = 0, \tag{83}$$

i.e.,

$$\lambda_{n-p+1} = \nu\left(\frac{A}{B};\, \tilde{L}_n,\, \tilde{L}_{n-1},\, \ldots,\, \tilde{L}_{n-p+2}\right); \tag{84}$$

this maximum is assumed only for principal vectors of the pencil corresponding to the characteristic value λ_{n-p+1} and satisfying the constraints (83).

---

[28] In a euclidean space with a metric form $B(x,x)$, the condition (83) expresses the fact that the vector x is orthogonal to the principal vectors $z^{n-p+2}, \ldots, z^n$. See footnote 26.

3) *If in the maximum of the ratio of the forms* $\frac{A(x,x)}{B(x,x)}$ *with the constraints*

$$L_1(x)=0, \ \ldots, \ L_{p-1}(x)=0 \qquad (2 \leqq p \leqq n)$$

$(2 \leqq p \leqq n)$ *the constraints are varied, then the least value (minimum) of this maximum is equal to* λ_{n-p+1}:

$$\lambda_{n-p+1} = \min \nu\left(\frac{A}{B}; L_1, L_2, \ldots, L_{p-1}\right). \tag{85}$$

5. Let

$$L_1^0(x)=0, \ L_2^0(x)=0, \ \ldots, \ L_h^0(x)=0. \tag{86}$$

be h *independent* constraints.[29] Then we can express h of the variables $x_1, x_2, \ldots, x_n$ by the remaining variables, which we denote by $v_1, v_2, \ldots, v_{n-h}$. Therefore, when the constraints (86) are imposed, the regular pencil of forms $A(x,x) - \lambda B(x,x)$ goes over into the pencil $A^0(v,v) - \lambda B^0(v,v)$, where $B^0(v,v)$ is again a positive-definite form (only in $n-h$ variables). The regular pencil so obtained has $n-h$ real characteristic values

$$\lambda_1^0 \leqq \lambda_2^0 \leqq \cdots \leqq \lambda_{n-h}^0. \tag{87}$$

Subject to the constraints (86) we can express all the variables in terms of $n-h$ independent ones $v_1, v_2, \ldots, v_{n-h}$ in various ways. However, the characteristic values (87) are independent of this arbitrariness and have completely definite values. This follows, for example, from the maximal-minimal property of the characteristic values

$$\lambda_1^0 = \min \frac{A^0(v,v)}{B^0(v,v)} = \mu\left(\frac{A}{B}; L_1^0, L_2^0, \ldots, L_h^0\right) \tag{88}$$

and, in general,

$$\begin{aligned} \lambda_p^0 &= \max \mu\left(\frac{A^0}{B^0}; L_1, L_2, \ldots, L_{p-1}\right) \\ &= \max \mu\left(\frac{A}{B}; L_1^0, \ldots, L_h^0, L_1, \ldots, L_{p-1}\right), \end{aligned} \tag{89}$$

where in (89) only the constraints $L_1, L_2, \ldots, L_{p-1}$ are allowed to vary.

[29] The constraints (86) are independent when the linear forms $L_1^0(x), L_2^0(x), \ldots, L_h^0(x)$ on the left-hand sides of (86) are independent.

The following theorem holds:

THEOREM 14: *If $\lambda_1 \leqq \lambda_2 \leqq \dots \leqq \lambda_n$ are the characteristic values of the regular pencil of forms $A(x,x) - \lambda B(x,x)$ and $\lambda_1^0 \leqq \lambda_2^0 \leqq \cdots \leqq \lambda_{n-h}^0$ are the characteristic values of the same pencil subject to h independent constraints, then*

$$\lambda_p \leqq \lambda_p^0 \leqq \lambda_{p+h} \quad (p = 1, 2, \dots, n-h). \tag{90}$$

Proof. The inequality $\lambda_p \leqq \lambda_p^0$ $(p = 1, 2, \dots, n-h)$ follows easily from (79) and (89). For when new constraints are added, the value of the minimum $\mu\left(\frac{A}{B}; L_1, \dots, L_{p-1}\right)$ increases or remains the same. Therefore

$$\mu\left(\frac{A}{B}; L_1, \dots, L_{p-1}\right) \leqq \mu\left(\frac{A}{B}; L_1^0, \dots, L_h^0; L_1, \dots, L_{p-1}\right).$$

Hence

$$\lambda_p = \max \mu\left(\frac{A}{B}; L_1, \dots, L_{p-1}\right) \leqq \lambda_p^0 = \max \mu\left(\frac{A}{B}; L_1^0, \dots, L_h^0, L_1, \dots, L_{p-1}\right).$$

The second part of the inequality (90) holds in view of the relations

$$\begin{aligned}\lambda_p^0 &= \max \mu\left(\frac{A}{B}; L_1^0, \dots, L_h^0; L_1, \dots, L_{p-1}\right)\\ &\leqq \max \mu\left(\frac{A}{B}; L_1, \dots, L_{p-1}, L_p, \dots, L_{p+h-1}\right) = \lambda_{p+h}.\end{aligned}$$

Here not only are $L_1, \dots, L_{p-1}$ varied, on the right-hand side, but $L_p, \dots, L_{p+h-1}$ also; on the left-hand side the latter are replaced by the fixed constraints $L_1^0, L_2^0, \dots, L_h^0$.

This completes the proof.

6. Suppose that two regular pencils of forms

$$A(x,x) - \lambda B(x,x), \quad \tilde{A}(x,x) - \lambda \tilde{B}(x,x) \tag{91}$$

are given and that for every $x \neq o$,

$$\frac{A(x,x)}{B(x,x)} \leqq \frac{\tilde{A}(x,x)}{\tilde{B}(x,x)}.$$

Then obviously,

$$\max \mu\left(\frac{A}{B};\, L_1, L_2, \ldots, L_{p-1}\right) \leqq \max \mu\left(\frac{\tilde{A}}{\tilde{B}};\, L_1, L_2, \ldots, L_{p-1}\right)$$
$$(p = 1, 2, \ldots, n).$$

Therefore, if we denote by $\lambda_1 \leqq \lambda_2 \leqq \ldots \leqq \lambda_n$ and $\tilde{\lambda}_1 \leqq \tilde{\lambda}_2 \leqq \ldots \leqq \tilde{\lambda}_n$, respectively, the characteristic values of the pencils (91), then we have:

$$\lambda_p \leqq \tilde{\lambda}_p \quad (p = 1, 2, \ldots, n).$$

Thus, we have proved the following theorem:

THEOREM 15: *If two regular pencils of forms $A(x, x) - \lambda B(x, x)$ and $\tilde{A}(x, x) - \lambda \tilde{B}(x, x)$ with the characteristic values $\lambda_1 \leqq \lambda_2 \leqq \ldots \leqq \lambda_n$ and $\tilde{\lambda}_1 \leqq \tilde{\lambda}_2 \leqq \ldots \leqq \tilde{\lambda}_n$ are given, then the identical relation*

$$\frac{A(x, x)}{B(x, x)} \leqq \frac{\tilde{A}(x, x)}{\tilde{B}(x, x)} \tag{92}$$

implies that

$$\lambda_p \leqq \tilde{\lambda}_p \quad (p = 1, 2, \ldots, n). \tag{93}$$

Let us consider the special case where, in (92), $B(x, x) \equiv \tilde{B}(x, x)$. In this case, the difference $\tilde{A}(x, x) - A(x, x)$ is a positive-semidefinite quadratic form and can therefore be expressed as a sum of independent positive squares:

$$\tilde{A}(x, x) = A(x, x) + \sum_{i=1}^{r} [X_i(x)]^2.$$

Then, when the r independent constraints

$$X_1(x) = 0,\ X_2(x) = 0,\ \ldots,\ X_r(x) = 0$$

are imposed, the forms $A(x, x)$ and $\tilde{A}(x, x)$ coincide, and the pencils $A(x, x) - \lambda B(x, x)$ and $\tilde{A}(x, x) - \lambda B(x, x)$ have the same characteristic values

$$\lambda_1^0 \leqq \lambda_2^0 \leqq \cdots \leqq \lambda_{n-r}^0.$$

Applying Theorem 14 to both pencils $A(x, x) - \lambda B(x, x)$ and $\tilde{A}(x, x) - \lambda B(x, x)$, we have:

$$\tilde{\lambda}_p \leqq \lambda_p^0 \leqq \lambda_{p+r} \quad (p = 1, 2, \ldots, n - r).$$

In conjunction with the inequality (93), this leads to the following theorem:

THEOREM 16: *If* $\lambda_1 \leqq \lambda_2 \leqq \dots \leqq \lambda_n$ *and* $\tilde{\lambda}_1 \leqq \tilde{\lambda}_2 \leqq \dots \leqq \tilde{\lambda}_n$ *are the characteristic values of two regular pencils of forms* $A(x,x) - \lambda B(x,x)$ *and* $\tilde{A}(x,x) - \lambda B(x,x)$, *where*

$$\tilde{A}(x,x) = A(x,x) + \sum_{i=1}^{r} [X_i(x)]^2,$$

and $X_i(x)$ $(i = 1, 2, \dots, r)$ *are independent linear forms, then the following inequalities hold:*[30]

$$\lambda_p \leqq \tilde{\lambda}_p \leqq \lambda_{p+r} \quad (p = 1, 2, \dots, n). \tag{94}$$

In exactly the same way the following theorem is proved:

THEOREM 17: *If* $\lambda_1 \leqq \lambda_2 \leqq \dots \leqq \lambda_n$ *and* $\tilde{\lambda}_1 \leqq \tilde{\lambda}_2 \leqq \dots \leqq \tilde{\lambda}_n$ *are the characteristic values of the regular pencil of forms* $A(x,x) - \lambda B(x,x)$ *and* $A(x,x) - \lambda \tilde{B}(x,x)$, *where the form* $\tilde{B}(x,x)$ *is obtained from* $B(x,x)$ *by adding r positive squares, then the following inequalities hold:*[31]

$$\lambda_{p-r} \leqq \tilde{\lambda}_p \leqq \lambda_p \quad (p = 1, 2, \dots, n). \tag{95}$$

Note. In Theorems 16 and 17 we can claim that for some p we have, respectively $\lambda_p < \tilde{\lambda}_p$ and $\tilde{\lambda}_p < \lambda_p$, provided of course that $r \neq 0$.[32]

§ 8. Small Oscillations of a System with n Degrees of Freedom

The results of the two preceding sections have important applications in the theory of small oscillations of a mechanical system with n degrees of freedom.

1. We consider the free oscillations of a conservative mechanical system with n degrees of freedom near a stable position of equilibrium. We shall give the deviation of the system from the position of equilibrium by means of independent generalized coordinates $q_1, q_2, \dots, q_n$. The position of equilibrium itself corresponds to zero values of these coordinates: $q_1 = 0$, $q_2 = 0, \dots, q_n = 0$. Then the kinetic energy of the system is represented as a quadratic form in the generalized velocities $\dot{q}_1, \dot{q}_2, \dots, \dot{q}_n$:[33]

$$T = \sum_{i,k=1}^{n} b_{ik}(q_1, q_2, \dots, q_n)\, \dot{q}_i \dot{q}_k.$$

---

[30] The second parts of these inequalities hold for $p \leqq n - r$ only.

[31] The first parts of the inequalities hold for $p > r$.

[32] See [17], pp. 71-73.

[33] A dot denotes the derivative with respect to time.

Expanding the coefficients $b_{ik}(q_1, q_2, \ldots, q_n)$ as power series in $q_1, q_2, \ldots, q_n$

$$b_{ik}(q_1, q_2, \ldots, q_n) = b_{ik} + \cdots \qquad (i, k = 1, 2, \ldots, n)$$

and keeping only the constant terms b_{ik}, since the deviations $q_1, q_2, \ldots, q_n$ are small, we then have:

$$T = \sum_{i,k=1}^{n} b_{ik}\dot{q}_i\dot{q}_k \qquad (b_{ik} = b_{ki}\,;\ i, k = 1, 2, \ldots, n).$$

The kinetic energy is always positive, and is zero only for zero velocities $\dot{q}_1 = \dot{q}_2 = \ldots = \dot{q}_n = 0$. Therefore $\sum_{i,k=1}^{n} b_{ik}\dot{q}_i\dot{q}_k$ is a positive-definite form.

The potential energy of the system is a function of the coordinates: $P(q_1, q_2, \ldots, q_n)$. Without loss of generality, we can take

$$P_0 = P(0, 0, \ldots, 0) = 0.$$

Then, expanding the potential energy as a power series in $q_1, q_2, \ldots, q_n$, we obtain:

$$P = \sum_{i=1}^{n} a_i q_i + \sum_{i,k=1}^{n} a_{ik} q_i q_k + \cdots.$$

Since in a position of equilibrium the potential energy always has a stationary value, we have

$$a_i = \left.\frac{\partial P}{\partial q_i}\right|_0 = 0 \qquad (i = 1, 2, \ldots, n).$$

Keeping only the terms of the second order in $q_1, q_2, \ldots, q_n$, we have

$$P = \sum_{i,k=1}^{n} a_{ik} q_i q_k \qquad (a_{ik} = a_{ki};\ i, k = 1, 2, \ldots, n).$$

Thus, the potential energy P and the kinetic energy T are determined by two quadratic forms:

$$P = \sum_{i,k=1}^{n} a_{ik} q_i q_k, \quad T = \sum_{i,k=1}^{n} b_{ik}\dot{q}_i\dot{q}_k, \tag{96}$$

the second of which is positive definite.

We now write down the differential equations of motion in the form of Lagrange's equations of the second kind:[34]

$$\frac{d}{dt}\frac{\partial T}{\partial \dot{q}_i} - \frac{\partial T}{\partial q_i} = -\frac{\partial P}{\partial q_i} \qquad (i = 1, 2, \ldots, n). \tag{97}$$

[34] See, for example, G. K. Suslow (Suslov), *Theoretische Mechanik*, § 191.

When we substitute for T and P their expressions from (96), we obtain:

$$\sum_{k=1}^{n} b_{ik}\ddot{q}_k + \sum_{k=1}^{n} a_{ik}q_k = 0 \qquad (i = 1, 2, \ldots, n). \tag{98}$$

We introduce the real symmetric matrices

$$A = \| a_{ik} \|_1^n \text{ and } B = \| b_{ik} \|_1^n$$

and the column matrix $q = (q_1, q_2, \ldots, q_n)$ and write the system of equations (98) in the following matrix form:

$$B\ddot{q} + Aq = o. \tag{98'}$$

We shall seek solutions of (98) in the form of harmonic oscillations

$$q_1 = v_1 \sin(\omega t + \alpha),\ q_2 = v_2 \sin(\omega t + \alpha),\ \ldots,\ q_n = v_n \sin(\omega t + \alpha),$$

in matrix notation:

$$q = v \sin(\omega t + \alpha). \tag{99}$$

Here $v = (v_1, v_2, \ldots, v_n)$ is the constant-amplitude column (constant-amplitude 'vector'), ω is the frequency, and α is the initial phase of the oscillation.

Substituting the expression (99) for q in (98′) and cancelling $\sin(\omega t + \alpha)$, we obtain:

$$Av = \lambda Bv \qquad (\lambda = \omega^2).$$

But this equation is the same as (49). Therefore the required amplitude vector is a principal vector, and the square of the frequency $\lambda = \omega^2$ is the corresponding characteristic value of the regular pencil of forms $A(x, x) - \lambda B(x, x)$.

We subject the potential energy to an additional restriction by postulating that the function $P(q_1, q_2, \ldots, q_n)$ in a position of equilibrium shall have a strict minimum.[35]

Then, by a theorem of Dirichlet,[36] the position of equilibrium is stable. On the other hand, our assumption means that the quadratic form $P = A(q, q)$ is also positive definite.

By Theorem 8, the regular pencil of forms $A(x, x) - \lambda B(x, x)$ has real characteristic values $\lambda_1, \lambda_2, \ldots, \lambda_n$ and n corresponding principal characteristic vectors $v^1, v^2, \ldots, v_n$ ($v^k = (v_{1k}, v_{2k}, \ldots, v_{nk})$; $k = 1, 2, \ldots, n$) satisfying the condition

---

[35] I.e., that the value of P_0 in the position of equilibrium is less than all other values of the function in some neighborhood of the position of equilibrium.

[36] See G. K. Suslow (Suslov), *Theoretische Mechanik*, § 210.

$$B(v^i, v^k) = \sum_{\mu,\nu=1}^{n} b_{\mu\nu} v_{\mu i} v_{\nu k} = \delta_{ik} \qquad (i, k = 1, 2, \ldots, n). \tag{100}$$

From the fact that $A(x, x)$ is positive definite it follows that all the characteristic values of the pencil $A(x, x) - \lambda B(x, x)$ are positive:[37]

$$\lambda_k > 0 \qquad (k = 1, 2, \ldots, n).$$

But then there exist n harmonic oscillations[38]

$$v^k \sin(\omega_k t + \alpha_k) \qquad (\omega_k^2 = \lambda_k,\ k = 1, 2, \ldots, n), \tag{101}$$

whose amplitude vectors $v^k = (v_{1k}, v_{2k}, \ldots, v_{nk})$ $(k = 1, 2, \ldots, n)$ satisfy the conditions of 'orthonormality' (100).

Since the equation (98′) is linear, every oscillation can be obtained by a superposition of the harmonic oscillations (101):

$$q = \sum_{k=1}^{n} A_k \sin(\omega_k t + \alpha_k)\, v^k, \tag{102}$$

where A_k and α_k are arbitrary constants. For, whatever the values of these constants, the expression (102) is a solution of (98′). On the other hand, the arbitrary constants can be made to satisfy the following initial conditions:

$$q\,|_{t=0} = q_0, \quad \dot{q}\,|_{t=0} = \dot{q}_0.$$

For from (102) we find:

$$q_0 = \sum_{k=1}^{n} A_k \sin\alpha_k v^k, \quad \dot{q}_0 = \sum_{k=1}^{n} \omega_k A_k \cos\alpha_k v^k. \tag{103}$$

Since the principal columns $v^1, v^2, \ldots, v^n$ are always linearly independent, the values $A_k \sin\alpha_k$ and $\omega_k \cos\alpha_k$ $(k = 1, 2, \ldots, n)$, and hence the constants A_k and α_k $(k = 1, 2, \ldots, n)$, are uniquely determined from (103).

The solution (102) of our system of differential equations can be written more conveniently:

$$q_i = \sum_{k=1}^{n} A_k \sin(\omega_k t + \alpha_k)\, v_{ik}. \tag{104}$$

Note that we could also derive the formulas (102) and (104) starting from Theorem 9. For let us consider a non-singular transformation of the

[37] This follows, for example, from the representation (63).

[38] Here the initial phases α_k $(k = 1, 2, \ldots, n)$ are arbitrary constants.

variables with the matrix $V = \| v_{ik} \|_1^n$ that reduces the two forms $A(x, x)$ and $B(x, x)$ simultaneously to the canonical form (63). Setting

$$q_i = \sum_{k=1}^{n} v_{ik} \theta_k \qquad (i = 1, 2, \ldots, n) \tag{105}$$

or, more briefly,

$$q = V\theta \quad (\theta = (\theta_1, \theta_2, \ldots, \theta_n)) \tag{106}$$

and observing that $\dot{q} = V\dot{\theta}$, we have:

$$P = A(q, q) = \sum_{i=1}^{n} \lambda_k \theta_k^2, \quad T = B(\dot{q}, \dot{q}) = \sum_{k=1}^{n} \dot{\theta}_k^2 . \tag{107}$$

The coordinates $\theta_1, \theta_2, \ldots, \theta_n$ in which the potential and kinetic energies have a representation as in (107) are called *principal coordinates.*

We now make use of Lagrange's equations of the second kind (98) and substitute the expressions (107) for P and T. We obtain:

$$\ddot{\theta}_k + \lambda_k \theta_k = 0 \qquad (k = 1, 2, \ldots, n). \tag{108}$$

Since $A(q, q)$ is positive definite, all the numbers $\lambda_1, \lambda_2, \ldots, \lambda_n$ are positive and can be represented in the form

$$\lambda_k = \omega_k^2 \qquad (\omega_k > 0;\ k = 1, 2, \ldots, n). \tag{109}$$

From (108) and (109), we find:

$$\theta_k = A_k \sin(\omega_k t + \alpha_k) \qquad (k = 1, 2, \ldots, n). \tag{110}$$

When we substitute these expressions for θ_k in (105), we again obtain the formulas (104) and therefore (102). The values v_{ik} $(i, k = 1, 2, \ldots, n)$ in both methods are the same, because the matrix $V = \| v_{ik} \|_1^n$ in (106) is, by Theorem 9, a principal matrix of the regular pencil of forms $A(x, x) - \lambda B(x, x)$.

2. We also mention a mechanical interpretation of Theorems 14 and 15.

We number the frequencies $\omega_1, \omega_2, \ldots, \omega_n$ of the given mechanical system in non-descending order:

$$0 < \omega_1 \leqq \omega_2 \leqq \cdots \leqq \omega_n .$$

The disposition of the corresponding characteristic values $\lambda_k = \omega_k^2$ $(k = 1, 2, 3, \ldots, n)$ of the pencil $A(x, x) - \lambda B(x, x)$ is then also determined:

$$\lambda_1 \leqq \lambda_2 \leqq \cdots \leqq \lambda_n .$$

We impose h independent finite stationary constraints[39] on the given system. Since the deviations $q_1, q_2, \ldots, q_n$ are supposed to be small, these connections can be assumed to be linear in $q_1, q_2, \ldots, q_n$:

$$L_1(q) = 0, \; L_2(q) = 0, \; \ldots, \; L_h(q) = 0.$$

After the constraints are imposed, our system has $n - h$ degrees of freedom. The frequencies of the system,

$$\omega_1^0 \leqq \omega_2^0 \leqq \cdots \leqq \omega_{n-h}^0,$$

are connected with the characteristic values $\lambda_1^0 \leqq \lambda_2^0 \leqq \cdots \leqq \lambda_{n-h}^0$ of the pencil $A(x, x) - \lambda B(x, x)$, subject to the constraints $L_1, L_2, \ldots, L_h$, by the relations $\lambda_j^0 = \omega_j^{02}$ $(j = 1, 2, \ldots, n - h)$. Therefore Theorem 14 immediately implies that

$$\omega_j \leqq \omega_j^0 \leqq \omega_{j+h} \qquad (j = 1, 2, \ldots, n - h).$$

Thus: *When h constraints are imposed, the frequencies of a system can only increase, but the value of the new j-th frequency ω_j^0 cannot exceed the value of the previous $(j + h)$-th frequency ω_{j+h}.*

In exactly the same way, we can assert on the basis of Theorem 15 that: *With increasing rigidity of the system, i.e., with an increase of the form $A(q, q)$ for the potential energy (without a change in $B(\dot{q}, \dot{q})$), the frequencies can only increase; and with increasing inertia of the system, i.e., with an increase of the form $B(\dot{q}, \dot{q})$ for the kinetic energy (without a change in $A(q, q)$), the frequencies can only decrease.*

Theorems 16 and 17 lead to an additional sharpening of this proposition.[40]

§ 9. Hermitian Forms[41]

1. All the results of §§ 1-7 of this chapter that were established for quadratic forms can be extended to hermitian forms.

We recall[42] that a *hermitian form* is an expression

---

[39] A finite stationary constraint is expressed by an equation $f(q_1, q_2, \ldots, q_n) = 0$, where $f(q_1, q_2, \ldots, q_n)$ is some function of the generalized coordinates.

[40] The reader can find an account of the oscillatory properties of elastic oscillations of a system with n degrees of freedom in [17], Chapter III.

[41] In the preceding sections, all the numbers and variables were real. In this section, the numbers are complex and the variables assume complex values.

[42] See Chapter IX, § 2.

$$H(x, x) = \sum_{i,k=1}^{n} h_{ik} x_i \bar{x}_k \qquad (h_{ik} = \bar{h}_{ki};\ i, k = 1, 2, \ldots, n). \tag{111}$$

To the hermitian form (111) there corresponds the following *bilinear hermitian form*:

$$H(x, y) = \sum_{i,k=1}^{n} h_{ik} x_i \bar{y}_k ; \tag{112}$$

moreover,

$$H(y, x) = \overline{H(x, y)} \tag{113}$$

and, in particular,

$$H(x, x) = \overline{H(x, x)} \tag{113'}$$

i.e., the hermitian form $H(x, x)$ assumes real values only.

The coefficient matrix $H = \| h_{ik} \|_1^n$ of the hermitian form is hermitian, i.e., $H^* = H$.[43]

By means of the matrix $H = \| h_{ik} \|_1^n$ we can represent $H(x, y)$ and, in particular, $H(x, x)$ in the form of a product of three matrices, a row, a square, and a column matrix:[44]

$$H(x, y) = x^{\mathsf{T}} H \bar{y}, \quad H(x, x) = x^{\mathsf{T}} H \bar{x}. \tag{114}$$

If

$$x = \sum_{i=1}^{m} c_i u^i, \quad y = \sum_{k=1}^{p} d_k v^k, \tag{115}$$

where u^i, v^k are column matrices and c_i, d_k are complex numbers ($i = 1, 2, 3, \ldots, m$; $k = 1, 2, \ldots, p$), then

$$H(x, y) = \sum_{i=1}^{m} \sum_{k=1}^{p} c_i \bar{d}_k H(u^i, v^k). \tag{116}$$

We subject the variables $x_1, x_2, \ldots, x_n$ to the linear transformation

$$x_i = \sum_{k=1}^{n} t_{ik} \xi_k \qquad (i = 1, 2, \ldots, n) \tag{117}$$

---

[43] A matrix symbol followed by an asterisk * denotes the matrix that is obtained from the given one by transposition and replacement of all the elements by their complex conjugates ($H^* = \bar{H}^{\mathsf{T}}$).

[44] Here

$$x = (x_1, x_2, \ldots, x_n), \quad \bar{x} = (\bar{x}_1, \bar{x}_2, \ldots, \bar{x}_n), \quad y = (y_1, y_2, \ldots, y_n), \quad \bar{y} = (\bar{y}_1, \bar{y}_2, \ldots, \bar{y}_n);$$

the sign ^{T} denotes transposition.

or, in matrix notation,

$$x = T\xi \qquad (T = \| t_{ik} \|_1^n). \tag{117'}$$

After the transformation, $H(x, x)$ assumes the form

$$\widetilde{H}(\xi, \xi) = \sum_{i,k=1}^{n} \widetilde{h}_{ik}\xi_i\bar{\xi}_k ,$$

where the new coefficient matrix $\widetilde{H} = \| \widetilde{h}_{ik} \|_1^n$ is connected with the old coefficient matrix $H = \| h_{ik} \|_1^n$ by the formula

$$\widetilde{H} = T^{\mathsf{T}} H \overline{T}. \tag{118}$$

This is immediately clear when, in the second of the formulas (114), x is replaced by $T\xi$.

If we set $T = \overline{W}$, then we can rewrite (118) as follows:

$$\widetilde{H} = W^* H W. \tag{119}$$

From the formula (118) it follows that H and $\widetilde{H}$ have the same rank provided the transformation (117) is non-singular ($|T| \neq 0$). The rank of H is called the *rank of the form* $H(x, x)$.

The determinant $|H|$ is called the *discriminant* of $H(x, x)$. From (118) we obtain the formula for the transformation of the discriminant on transition to new variables:

$$|\widetilde{H}| = |H|\,|T|\,|\overline{T}|.$$

A hermitian form is called *singular* if its discriminant is zero. Obviously, a singular form remains singular under any transformation of the variables (117).

A hermitian form $H(x, x)$ can be represented in infinitely many ways in the form

$$H(x, x) = \sum_{i=1}^{r} a_i X_i \overline{X}_i, \tag{120}$$

where $a_i \neq 0$ $(i = 1, 2, \ldots, r)$ are real numbers and

$$X_i = \sum_{k=1}^{n} a_{ik} x_k \quad (i = 1, 2, \ldots, r)$$

are independent complex linear forms in the variables $x_1, x_2, \ldots, x_n$.[45]

---

[45] Therefore $r \leq n$.

We shall call the right-hand side of (120) a *sum of linearly independent squares*[46] and every term in the sum a *positive* or a *negative* square according as $a_i > 0$ or < 0. Just as for quadratic forms, the number r in (120) is equal to the rank of the form $H(x, x)$.

THEOREM 18 (The Law of Inertia for Hermitian Forms): *In the representation of a hermitian form $H(x, x)$ as a sum of linearly independent squares,*

$$H(x, x) = \sum_{i=1}^{r} a_i X_i \bar{X}_i ,$$

the number of positive squares and the number of negative squares do not depend on the choice of the representation.

The proof is completely analogous to the proof of Theorem 1 (p. 297).

The difference σ between the number π of positive squares and the number ν of negative squares in (120) is called the *signature* of the hermitian form $H(x, x)$: $\sigma = \pi - \nu$.

Lagrange's method of reduction of quadratic forms to sums of squares can also be used for hermitian forms, only the fundamental formulas (15) and (16) on p. 299 must then be replaced by the formulas[47]

$$H(x, x) = \frac{1}{h_{gg}} \left| \sum_{k=1}^{n} h_{kg} x_k \right|^2 + H_1(x, x), \tag{121}$$

$$H(x, x) = \frac{1}{2} \left\{ \left| \sum_{k=1}^{n} \left(h_{kf} + \frac{h_{kg}}{h_{fg}} \right) x_k \right|^2 - \left| \sum_{k=1}^{n} \left(h_{kf} - \frac{h_{kg}}{h_{fg}} \right) x_k \right|^2 \right\} + H_2(x, x). \tag{122}$$

Let us proceed to establish Jacobi's formula for a hermitian form $H(x, x) = \sum_{i,k=1}^{n} h_{ik} x_i \bar{x}_k$ of rank r. Here, as in the case of a quadratic form, we assume that

$$D_k = H \begin{pmatrix} 1 & 2 & \dots & k \\ 1 & 2 & \dots & k \end{pmatrix} \neq 0 \qquad (k = 1, 2, \dots, r). \tag{123}$$

This inequality enables us to use Theorem 2 of Chapter II (p. 38) on the representation of an arbitrary square matrix in the form of a product of three matrices: a lower triangular matrix F, a diagonal matrix D, and an upper triangular matrix L. We apply this theorem to the matrix $H = \| h_{ik} \|_1^n$

[46] This terminology is connected with the fact that $X_i \bar{X}_i$ is the square of the modulus of X_i ($X_i \bar{X}_i = |X_i|^2$).

[47] The formula (121) is applicable when $h_{gg} \neq 0$; and (122), when $h_{ff} = h_{gg} = 0$, $h_{fg} \neq 0$.

and obtain

$$H = F\left\{D_1, \frac{D_2}{D_1}, \ldots, \frac{D_r}{D_{r-1}}, 0, \ldots, 0\right\} L, \tag{124}$$

where $F = \| f_{ik} \|_1^n$, $L = \| l_{ik} \|_1^n$, and

$$f_{jk} = \frac{1}{D_k} H\begin{pmatrix} 1 & \ldots & k-1 & j \\ 1 & \ldots & k-1 & k \end{pmatrix}, \quad l_{kj} = \frac{1}{D_k} H\begin{pmatrix} 1 & \ldots & k-1 & k \\ 1 & \ldots & k-1 & j \end{pmatrix} \tag{125}$$

$$(j = k,\ k+1, \ldots, n;\ k = 1, 2, \ldots, r),$$

$$f_{ik} = l_{ki} = 0 \qquad (i < k;\ i, k = 1, 2, \ldots, n). \tag{126}$$

Since $H = \| h_{ik} \|_1^n$ is a hermitian matrix, it follows from these equations that

$$f_{ik} = \bar{l}_{ki} \quad \begin{pmatrix} i \geqq k;\ k = 1, 2, \ldots, r;\ i = 1, 2, \ldots, n, \\ i < k;\ i, k = 1, 2, \ldots, n \end{pmatrix}. \tag{127}$$

Since all the elements in the last $n-r$ columns of F and the last $n-r$ rows of L can be chosen arbitrarily,[48] we choose these elements such that 1) the relations (127) hold for all i, k

$$f_{ik} = \bar{l}_{ik} \quad (i, k = 1, 2, \ldots, n)$$

and 2) $|F| = |L| \neq 0$. Then

$$F = L^*, \tag{128}$$

and (124) assumes the form

$$H = L^*\left\{D_1, \frac{D_2}{D_1}, \ldots, \frac{D_r}{D_{r-1}}, 0, \ldots, 0\right\} L. \tag{129}$$

Setting

$$T = \| t_{ik} \|_1^n = \bar{L}, \tag{130}$$

we write (129) as follows:

$$H = T^{\mathsf{T}}\left\{D_1, \frac{D_2}{D_1}, \ldots, \frac{D_r}{D_{r-1}}, 0, \ldots, 0\right\} \bar{T} \qquad (|T| \neq 0). \tag{131}$$

A comparison of this formula with (118) shows that the hermitian form

$$\sum_{k=1}^{n} \frac{D_k}{D_{k-1}} \xi_k \bar{\xi}_k \qquad (D_0 = 1) \tag{132}$$

under the transformation of the variables

---

[48] These elements, in fact, drop out of the right-hand side of (124), because the last $n-r$ diagonal elements of D are zero.

$$\xi_i = \sum_{k=1}^{n} t_{ik} x_k \qquad (i = 1, 2, \ldots, n)$$

goes over into $H(x, x)$, i.e., that Jacobi's formula holds:

$$H(x, x) = \sum_{k=1}^{n} \frac{D_k}{D_{k-1}} X_k \overline{X}_k \qquad (D_0 = 1), \tag{133}$$

where

$$X_k = x_k + t_{k,k+1} x_{k+1} + \cdots + t_{kn} x_n \qquad (k = 1, 2, \ldots, r) \tag{134}$$

and

$$t_{kj} = \frac{1}{D_k} H\begin{pmatrix} 1 & 2 & \ldots & k-1 & j \\ 1 & 2 & \ldots & k-1 & k \end{pmatrix} \quad (j = k+1, \ldots, n;\ k = 1, 2, \ldots, r). \tag{135}$$

The linear forms $X_1, X_2, \ldots, X_r$ are independent, since X_k contains the variable x_k which does not occur in the subsequent forms $X_{k+1}, \ldots, X_r$.

When we introduce, in place of $X_1, X_2, \ldots, X_r$, the linearly independent forms

$$Y_k = D_k X_k \qquad (k = 1, 2, \ldots, r), \tag{136}$$

we can write Jacobi's formula (133) in the form

$$H(x, x) = \sum_{k=1}^{r} \frac{Y_k \overline{Y}_k}{D_{k-1} D_k} \qquad (D_0 = 1). \tag{137}$$

According to Jacobi's formula (137), the number of negative squares in the representation of $H(x, x)$ is equal to the number of variations of sign in the sequence $1, D_1, D_2, \ldots, D_r$

$$\nu = V(1, D_1, D_2, \ldots, D_r),$$

so that the signature σ of $H(x, x)$ is determined by the formula

$$\sigma = r - 2V(1, D_1, D_2, \ldots, D_r). \tag{138}$$

All the remarks about the special cases that may occur, made for quadratic forms (§ 3), automatically carry over to hermitian forms.

DEFINITION 5: *A hermitian form* $H(x, x) = \sum_{i,k=1}^{n} h_{ik} x_i \bar{x}_k$ *is called positive (negative) semidefinite if for arbitrary values of the variables* $x_1, x_2, x_3, \ldots, x_n$, *not all equal to zero,*

$$H(x, x) \geqq 0 \qquad (\leqq 0).$$

DEFINITION 6: *A hermitian form* $H(x, x) = \sum_{i,k=1}^{n} h_{ik}x_i\bar{x}_k$ *is called positive (negative) definite if for arbitrary values of the variables* $x_1, x_2, \ldots, x_n$, *not all equal to zero,*

$$H(x, x) > 0 \qquad (< 0).$$

THEOREM 19: *A hermitian form* $H(x, x) = \sum_{i,k=1}^{n} h_{ik}x_i\bar{x}_k$ *is positive definite if and only if the following inequalities hold:*

$$D_k = H\begin{pmatrix}1 & 2 & \ldots & k\\ 1 & 2 & \ldots & k\end{pmatrix} > 0 \qquad (k = 1, 2, \ldots, n). \tag{139}$$

THEOREM 20: *A hermitian form* $H(x, x) = \sum_{i,k=1}^{n} h_{ik}x_i\bar{x}_k$ *is positive semidefinite if and only if all the principal minors of* $H = \| h_{ik} \|_1^n$ *are non-negative:*

$$H\begin{pmatrix}i_1 & i_2 & \ldots & i_p\\ i_1 & i_2 & \ldots & i_p\end{pmatrix} \geqq 0$$
$$(i_1, i_2, \ldots, i_p = 1, 2, \ldots, n;\, p = 1, 2, \ldots, n). \tag{140}$$

The proofs of Theorems 19 and 20 are completely analogous to the proofs of Theorems 3 and 4 for quadratic forms.

The conditions for a hermitian form $H(x, x)$ to be negative definite or semidefinite are obtained by applying (139) and (140) to the form $-H(x, x)$.

From Theorem 5′ of Chapter IX (p. 274), we obtain the *Theorem on the reduction of a hermitian form to principal axes:*

THEOREM 21: *Every hermitian form* $H(x, x) = \sum_{i,k=1}^{n} h_{ik}x_i\bar{x}_k$ *can be reduced by a unitary transformation of the variables*

$$x = U\xi \qquad (UU^* = E) \tag{141}$$

to the canonical form

$$\Lambda(\xi, \xi) = \sum_{i=1}^{n} \lambda_i\xi_i\bar{\xi}_i\,, \tag{142}$$

where $\lambda_1, \lambda_2, \ldots, \lambda_n$ *are the characteristic values of the matrix* $H = \| h_{ik} \|_1^n$.

Theorem 21 follows from the formula

$$H = U \,\|\, \lambda_i\delta_{ik} \,\|\, U^{-1} = T^{\mathsf{T}} \,\|\, \lambda_i\delta_{ik} \,\|\, \bar{T} \quad (U^{\mathsf{T}} = \bar{U}^{-1} = T)\,. \tag{143}$$

Let $H(x, x) = \sum_{i,k=1}^{n} h_{ik}x_i\bar{x}_k$ and $G(x, x) = \sum_{i,k=1}^{n} g_{ik}x_i\bar{x}_k$ be two hermitian forms. We shall study the pencil of hermitian forms $H(x, x) - \lambda G(x, x)$

(λ is a real parameter). This pencil is called *regular* if $G(x, x)$ is positive definite. By means of the hermitian matrices $H = \| h_{ik} \|_1^n$ and $G = \| g_{ik} \|_1^n$ we form the equation

$$| H - \lambda G | = 0.$$

This equation is called the *characteristic equation of the pencil of hermitian forms.* Its roots are called the *characteristic values of the pencil.*

If λ_0 is a characteristic value of the pencil, then there exists a column $z = (z_1, z_2, \ldots, z_n) \neq o$ such that

$$Hz = \lambda_0 z.$$

We shall call the column z a *principal column* or *principal vector of the pencil* $H(x, x) - \lambda G(x, x)$ corresponding to the characteristic value λ_0.

Then the following theorem holds:

THEOREM 22: *The characteristic equation of a regular pencil of hermitian forms $H(x, x) - \lambda G(x, x)$ has n real roots $\lambda_1, \lambda_2, \ldots, \lambda_n$. To these roots there correspond n principal vectors $z^1, z^2, \ldots, z^n$ satisfying the conditions of 'orthonormality':*

$$G(z^i, z^k) = \delta_{ik} \qquad (i, k = 1, 2, \ldots, n).$$

The proof is completely analogous to the proof of Theorem 8.

All extremal properties of the characteristic values of a regular pencil of quadratic forms remain valid for hermitian forms.

Theorems 10-17 remain valid if the term 'quadratic form' is replaced throughout by the term 'hermitian form.' The proofs of the theorems are then unchanged.

§ 10. Hankel Forms

1. Let $s_0, s_1, \ldots, s_{2n-2}$ be a sequence of numbers. We form, by means of these numbers, a quadratic form in n variables

$$S(x, y) = \sum_{i,k=0}^{n-1} s_{i+k} x_i x_k. \tag{144}$$

This is called a *Hankel form.* The matrix $S = \| s_{i+k} \|_0^{n-1}$ corresponding to this form is called a *Hankel matrix.* It has the form

$$S=\begin{Vmatrix} s_0 & s_1 & s_2 & \dots & s_{n-1} \\ s_1 & s_2 & s_3 & \dots & s_n \\ s_2 & s_3 & s_4 & \dots & s_{n+1} \\ \cdot & \cdot & \cdot & \cdot & \cdot \\ s_{n-1} & s_n & s_{n+1} & \dots & s_{2n-2} \end{Vmatrix}.$$

We denote the sequence of principal minors of S by $D_1, D_2, \dots, D_n$:

$$D_p = | s_{i+k} |_0^{p-1} \qquad (p=1,2,\dots,n).$$

In this section we shall derive the fundamental results of Frobenius about the rank and signature of real Hankel forms.[49]

We begin by proving two lemmas.

LEMMA 1: *If the first h rows of the Hankel matrix $S=\| s_{i+k} \|_0^{n-1}$ are linearly independent, but the first $h+1$ rows linearly dependent, then*

$$D_h \neq 0.$$

Proof. We denote the first $h+1$ rows of S by $R_1, R_2, \dots, R_h, R_{h+1}$. By assumption, $R_1, R_2, \dots, R_h$ are linearly independent and R_{h+1} is expressed linearly in terms of them:

$$R_{h+1} = \sum_{j=1}^{h} \alpha_j R_{h-j+1}$$

or

$$s_q = \sum_{j=1}^{h} \alpha_j s_{q-j} \qquad (q=h, h+1, \dots, h+n-1). \tag{145}$$

We write down the matrix formed from the first h rows $R_1, R_2, \dots, R_h$ of S:

$$\begin{Vmatrix} s_0 & s_1 & s_2 & \dots & s_{n-1} \\ s_1 & s_2 & s_3 & \dots & s_n \\ \cdot & \cdot & \cdot & \cdot & \cdot \\ s_{h-1} & s_h & s_{h+1} & \dots & s_{h+n-2} \end{Vmatrix}. \tag{146}$$

This matrix is of rank h. On the other hand, by (145) every column of the matrix can be expressed linearly in terms of the preceding h columns and hence in the terms of the first h columns. But since the rank of (146) is h, these first h columns of (146) must then be linearly independent, i.e.,

$$D_h \neq 0.$$

This proves the lemma.

[49] See [162].

LEMMA 2: *If in the matrix* $S = \| s_{i+k} \|_0^{n-1}$, *for a certain* h $(< n)$,

$$D_h \neq 0, \quad D_{h+1} = \cdots = D_n = 0 \tag{147}$$

and

$$t_{ik} = \frac{S\begin{pmatrix} 1 \ldots h & h+i+1 \\ 1 \ldots h & h+k+1 \end{pmatrix}}{S\begin{pmatrix} 1 \ldots h \\ 1 \ldots h \end{pmatrix}} = \frac{1}{D_h} \begin{vmatrix} & & s_{h+k} \\ D_h & & \vdots \\ & & s_{2h+k-1} \\ s_{h+i} & \cdots \; s_{2h+i-1} & s_{2h+i+k} \end{vmatrix}, \tag{148}$$

$$(i, k, = 0, 1, \ldots, n-h-1)$$

then the matrix $T = \| t_{ik} \|_0^{n-h-1}$ *is also a Hankel matrix and all its elements above the secondary diagonal are zero, i.e., there exist numbers* $t_{n-h-1}, \ldots, t_{2n-2h-2}$ *such that*

$$t_{ik} = t_{i+k} \qquad (i, k = 0, 1, \ldots, n-h-1;\ t_0 = t_1 = \cdots = t_{n-h-2} = 0).$$

Proof. We introduce the matrices

$$T_p = \| t_{ik} \|_0^{p-1} \qquad (p = 1, 2, \ldots\ n-h).$$

In this notation $T = T_{n-h}$.

We shall show that every T_p $(p = 1, 2, \ldots, n-h)$ is a Hankel matrix and that $t_{ik} = 0$ for $i + k \leqq p - 2$. The proof is by induction with respect to p.

For the matrix T_1, our assertion is trivial; for T_2, it is obvious, since

$$T_2 = \left\| \begin{matrix} t_{00} & t_{01} \\ t_{10} & t_{11} \end{matrix} \right\|, \quad t_{01} = t_{10} \quad \text{(because } S \text{ is symmetric) and} \quad t_{00} = \frac{D_{h+1}}{D_h} = 0.$$

Let us assume that our assertion is true for the matrices T_p $(p < n-h)$; we shall show that it is also true for $T_{p+1} = \| t_{ik} \|_0^p$. From the assumption it follows that there exist numbers $t_{p-1}, t_p, \ldots, t_{2p-2}$ such that with $t_0 = \ldots = t_{p-2} = 0$

$$T_p = \| t_{i+k} \|_0^{p-1}.$$

Here

$$| T_p | = t_{p-1}^p. \tag{149}$$

On the other hand, using Sylvester's determinant identity (see (28) on page 32), we find:

$$| T_p | = \frac{D_{h+p}}{D_h} = 0. \tag{150}$$

Comparing (149) with (150), we obtain

$$t_{p-1} = 0. \tag{151}$$

Furthermore from (148)

$$t_{ik} = s_{2h+i+k} + \frac{1}{D_h} \begin{vmatrix} & & s_{h+k} \\ & D_h & \vdots \\ & & s_{2h+k-1} \\ s_{h+i} & \cdots \; s_{2h+i-1} & 0 \end{vmatrix}. \tag{152}$$

By the preceding lemma, it follows from (147) that the $(h+1)$-th row of the matrix $S = \| s_{i+k} \|_0^{n-1}$ is linearly dependent on the first h rows:

$$s_q = \sum_{g=1}^{h} \alpha_g s_{q-g} \qquad (q = h,\, h+1,\, \ldots,\, h+n-1). \tag{153}$$

Let $i, k \leqq p \leqq i+k \leqq 2p-1$. Then one of the numbers i or k is less than p. Without loss of generality, we assume that $i < p$. Then, when we expand, by (153), the last column of the determinant of the right-hand side of (152) and use the relations (152) again, we shall have

$$\begin{aligned} t_{ik} &= s_{2h+i+k} + \sum_{g=1}^{h} \frac{\alpha_g}{D_h} \begin{vmatrix} & & s_{h+k-g} \\ & D_h & \vdots \\ & & s_{2h+k-g-1} \\ s_{h+i} & \cdots \; s_{2h+i-1} & 0 \end{vmatrix} \\ &= s_{2h+i+k} + \sum_{g=1}^{h} \alpha_g \left(t_{i,k-g} - s_{2h+i+k-g}\right). \end{aligned} \tag{154}$$

By the induction hypothesis (151) holds, and since in (154) $i < p$, $k - g < p$ and $i + k - g \leqq 2p - 2$, we have $t_{i,k-g} = t_{i+k-g}$. Therefore, for $i + k < p$ all the $t_{ik} = 0$, and for $p \leqq i + k \leqq 2p - 1$ the value of t_{ik}, by (154), depends on $i + k$ only.

Thus, T_{p+1} is a Hankel matrix, and all its elements $t_0, t_1, \ldots, t_{p-1}$ above the second diagonal are zero.

This proves the lemma.

Using Lemma 2, we shall prove the following theorem:

THEOREM 23: *If the Hankel matrix* $S = \| s_{i+k} \|_0^{n-1}$ *has rank* r *and if for some* h $(< r)$

$$D_h \neq 0, \quad D_{h+1} = \cdots = D_r = 0,$$

then the principal minor of order r *formed from the first* h *and the last* $r - h$ *rows and columns of* S *is not zero*:

$$D^{(r)} = S\begin{pmatrix} 1 & \dots & h & n-r+h+1 & n-r+h+2 & \dots & n \\ 1 & \dots & h & n-r+h+1 & n-r+h+2 & \dots & n \end{pmatrix} \neq 0.$$

Proof. By the preceding lemma, the matrix

$$T = \| t_{ik} \|_0^{n-h-1} \qquad \left(t_{ik} = \frac{S\begin{pmatrix} 1 & \dots & h & h+i+1 \\ 1 & \dots & h & h+k+1 \end{pmatrix}}{S\begin{pmatrix} 1 & \dots & h \\ 1 & \dots & h \end{pmatrix}} \quad (i, k = 0, 1, \dots, n-h+1) \right)$$

is a Hankel matrix in which all the elements above the second diagonal are zero. Therefore

$$|T| = t_{0,\, n-h-1}^{n-h}.$$

On the other hand,[50] $|T| = \dfrac{D_n}{D_h} = 0$. Therefore $t_{0,\, n-h-1} = 0$, and the matrix T has the form

$$T = \left\| \begin{array}{ccccc} 0 & \cdot & \cdots & \cdot & 0 \\ \cdot & & & \cdot & u_{n-h-1} \\ \vdots & & \cdot^{\cdot^{\cdot}} & & \vdots \\ \cdot & \cdot & & & u_2 \\ 0 & u_{n-h-1} & \cdots & u_2 & u_1 \end{array} \right\|.$$

The rank of T must be $r - h$.[51] Therefore for $r < n - 1$ in the matrix T the elements $u_{r-h+1} = \dots = u_{n-h+1} = 0$, and

[50] By Sylvester's determinant identity (see (28) on p. 32).

[51] From Sylvester's identity it follows that all the minors of T in which the order exceeds $r - h$ are zero. On the other hand, S contains some non-vanishing minors of order r bordering D_h. Hence it follows that the corresponding minor of order $r - h$ of T is different from zero.

$$T=\left\|\begin{matrix} 0 & \cdots & \cdots & \cdots & 0 \\ \vdots & & & & \vdots \\ \vdots & & & & 0 \\ \vdots & & & & u_{r-h} \\ \vdots & & & & \vdots \\ 0 \ldots 0 & u_{r-h} & \ldots & & u_1 \end{matrix}\right\| \quad (u_{r-h}\neq 0).$$

But then, by Sylvester's identity (see page 32),

$$D^{(r)}=D_hT\begin{pmatrix} n-r+1 \ldots n-h \\ n-r+1 \ldots n-h \end{pmatrix}=D_hu_{r-h}^{r-h}\neq 0,$$

and this is what we had to prove.

Let us consider a real[52] Hankel form $S(x,x)=\sum_{i,k=0}^{n-1}{}' s_{i+k}x_ix_k$ of rank r. We denote by π, ν, and σ, respectively, the number of positive and of negative squares and the signature of the form:

$$\pi+\nu=r, \quad \sigma=\pi-\nu=r-2\nu.$$

By the theorem of Jacobi (p. 303) these values can be determined from the signs of the successive minors

$$D_0=1,\ D_1,\ D_2,\ \ldots,\ D_{r-1},\ D_r \tag{155}$$

by the formulas

$$\left.\begin{aligned} &\pi=P(1,D_1,\ldots D_r), \quad \nu=V(1,D_1,\ldots,D_r), \\ &\sigma=P(1,D_1,\ldots,D_r)-V(1,D_1,\ldots D_r)=r-2V(1,D_1,\ldots,D_r). \end{aligned}\right\} \tag{156}$$

These formulas become inapplicable when the last term in (155) or any three consecutive terms are zero (see § 3). However, as Frobenius has shown, for Hankel forms there is a rule that enables us to use the formulas (156) in the general case:

THEOREM 24 (Frobenius): *For a real Hankel form* $S(x,x)=\sum_{i,k=0}^{n-1} s_{i+k}x_ix_k$ *of rank* r *the values of* π, ν, *and* σ *can be determined by the formulas* (156) *provided that*

---

[52] In the preceding Lemmas 1 and 2 and in Theorem 23, the ground field can be taken as an arbitrary number field—in particular, the field of complex or of real numbers.

1) *for*

$$D_h \neq 0,\ D_{h+1} = \cdots = D_r = 0 \quad (h < r) \tag{157}$$

D_r is replaced by $D^{(r)}$, where

$$D^{(r)} = S\begin{pmatrix} 1 & \ldots & h & n-r+h+1 & \ldots & n \\ 1 & \ldots & h & n-r+h+1 & \ldots & n \end{pmatrix} \neq 0;$$

2) *in any group of p consecutive zero determinants*

$$(D_h \neq 0) \quad D_{h+1} = D_{h+2} = \cdots = D_{h+p} = 0 \quad (D_{h+p+1} \neq 0) \tag{158}$$

a sign is attributed to the zero determinants according to the formula

$$\operatorname{sign} D_{h+j} = (-1)^{\frac{j(j-1)}{2}} \operatorname{sign} D_h. \tag{159}$$

The values of P, V, and $P - V$ corresponding to the group (158) *are then:*[53]

	p odd	p even
$P_{h,p} = P(D_h, D_{h+1}, \ldots, D_{h+p+1})$	$\frac{p+1}{2}$	$\frac{p+1+\varepsilon}{2}$
$V_{h,p} = V(D_h, D_{h+1}, \ldots, D_{h+p+1})$	$\frac{p+1}{2}$	$\frac{p+1-\varepsilon}{2}$
$P_{h,p} - V_{h,p}$	0	ε

; (160)

$$\varepsilon = (-1)^{\frac{p}{2}} \operatorname{sign} \frac{D_{h+p+1}}{D_h}.$$

Proof. To begin with we consider the case where $D_r \neq 0$. Then the forms $S(x, x) = \sum_{i,k=0}^{n-1} s_{i+k} x_i x_k$ and $S_r(x, x) = \sum_{i,k=0}^{r-1} s_{i+k} x_i x_k$ have not only the same rank r, but also the same signature σ. For let $S(x, x) = \sum_{i=1}^{r} \varepsilon_i Z_i^2$, where the Z_i are real linear forms and $\varepsilon_i = \pm 1$ $(i = 1, 2, \ldots, r)$. We set $x_{r+1} = \ldots = x_{n-1} = 0$. Then the forms $S(x, x)$ and Z_i go over, respectively, into $S_r(x, x)$ and $\widehat{Z}_i$ $(i = 1, 2, \ldots, r)$; and $S_r(x, x) = \sum_{i=1}^{r} \varepsilon_i \widehat{Z}_i^2$, i.e., $S_r(x, x)$ has

[53] The formulas (159) and (160) are also applicable to (157), but we have to set $p = r - h - 1$ and interpret D_{h+p+1} not as $D_r = 0$, but as $D^{(r)} \neq 0$.

the same number of positive and negative squares as $S(x, x)$.[54] Thus the signature of $S_r(x, x)$ is σ.

We now vary the parameters $s_0, s_1, \ldots, s_{2r-2}$ continuously in such a way that for the new parameter values $s_0^*, s_1^*, \ldots, s_{2r-2}^*$ all the terms of the sequence[55]

$$1, D_1^*, D_2^*, \ldots, D_r^* \quad (D_q^* = | s_{i+k}^* |_0^{q-1};\ q = 1, 2, \ldots, r)$$

are different from zero and that in the process of variation none of the non-zero determinants (155) vanishes.[56]

Since the rank of $S_r(x, x)$ does not change during the variation, its signature also remains unchanged (see p. 309). Therefore

$$\sigma = P(1, D_1^*, \ldots, D_r^*) - V(1, D_1^*, \ldots, D_r^*). \tag{161}$$

If $D_i \neq 0$ for some i, then sign $D_i^* =$ sign D_i. Therefore the whole problem reduces to determining the variations in sign among those D_i^* that correspond to $D_i = 0$. More accurately, for every group of the form (158) we have to determine

$$P(D_h^*, D_{h+1}^*, \ldots, D_{h+p+1}^*) - V(D_h^*, D_{h+1}^*, \ldots, D_{h+p}^*, D_{h+p+1}^*).$$

For this purpose we set:

$$t_{ik} = \frac{1}{D_h} \begin{vmatrix} & & s_{h+k} \\ D_h & & \vdots \\ & & s_{2h+k-1} \\ s_{h+i} \cdots s_{2h+i-1} & & s_{2h+i+k} \end{vmatrix} \qquad (i, k = 0, 1, \ldots, p).$$

By Lemma 2, the matrix $T = \| t_{ik} \|_0^p$ is a Hankel matrix and all its elements above the second diagonal are zero, so that T has the form

---

[54] The linear forms $\hat{Z}_1, \hat{Z}_2, \ldots, \hat{Z}_r$ are linearly independent, because the quadratic form $S(x, x) = \sum_{i=1}^{r} \varepsilon_i \hat{Z}_i^2$ is of rank r $(D_r \neq 0)$.

[55] In this section, the asterisk * does not indicate the adjoint matrix.

[56] Such a variation of the parameter is always possible, because in the space of the parameters $s_0, s_1, \ldots, s_{2r-2}$ an equation of the form $D_i = 0$ determines a certain algebraic hypersurface. If a point lies in some such hypersurfaces, then it can always be approximated by arbitrarily close points that do not lie in these hypersurfaces.

$$T = \left\| \begin{matrix} 0 & \cdot & \cdot & \cdot & 0 & t_p \\ \cdot & & & & \cdot & * \\ \cdot & & & \cdot & & \cdot \\ \cdot & & \cdot & & & \cdot \\ \cdot & \cdot & & & & \cdot \\ 0 & & & & & \cdot \\ t_p & * & \cdots & & & * \end{matrix} \right\| . \tag{162}$$

We denote the successive minors of T by $\widehat{D}_1, \widehat{D}_2, \ldots, \widehat{D}_{p+1}$:

$$\widehat{D}_q = | t_{ik} |_0^{q-1} \quad (q = 1, 2, \ldots, p+1).$$

Together with T, we consider the matrix

$$T^* = \| t_{ik}^* \|_0^p ,$$

where

$$t_{ik}^* = \frac{1}{D_h^*} \left| \begin{matrix} & & & s_{h+k}^* \\ & D_h^* & & \vdots \\ & & & s_{2h+k-1}^* \\ s_{h+i}^* & \cdots & s_{2h+i-1}^* & s_{2h+i+k}^* \end{matrix} \right| \qquad (i, k = 0, 1, \ldots, p)$$

and the corresponding determinants

$$\widehat{D}_q^* = | t_{ik}^* |_0^{q-1} \quad (q = 1, 2, \ldots, p+1).$$

By Sylvester's determinant identity,

$$D_{h+q}^* = D_h^* \widehat{D}_q^* \qquad (q = 1, 2, \ldots, p+1).$$

Therefore

$$\begin{aligned} P(D_h^*, D_{h+1}^*, \ldots, D_{h+p+1}^*) - V(D_h^*, D_{h+1}^*, \ldots, D_{h+p+1}^*) \\ = P(1, \widehat{D}_1^*, \ldots, \widehat{D}_{p+1}^*) - V(1, \widehat{D}_1^*, \ldots, \widehat{D}_{p+1}^*) = \widehat{\sigma}^*, \end{aligned} \tag{163}$$

where $\widehat{\sigma}^*$ is the signature of the form

$$T^*(x, x) = \sum_{i,k=0}^{p} t_{ik}^* x_i x_k .$$

Together with $T^*(x, x)$, we consider the forms

$$T(x, x) = \sum_{i,k=0}^{p} t_{i+k} x_i x_k \quad \text{and} \quad T^{**}(x, x) = t_p (x_0 x_p + x_1 x_{p-1} + \cdots + x_p x_0).$$

The matrix T^{**} is obtained from T (see (162)) when we replace in the latter all the elements above the second diagonal by zeros. We denote the signatures of $T(x, x)$ and $T^{**}(x, x)$ by $\hat{\sigma}$ and $\hat{\sigma}^{**}$. Since $T^{*}(x, x)$ and $T^{**}(x, x)$ are obtained from $T(x, x)$ by variations of the coefficients during which the rank of the form does not change ($| T^{**} | = | T | = \frac{D_{h+p+1}}{D_h} \neq 0$, $| T^{*} | = \frac{D^{*}_{h+p+1}}{D^{*}_h} \neq 0$), the signatures of $T(x, x)$, $T^{*}(x, x)$, and $T^{**}(x, x)$ must also be equal:

$$\hat{\sigma} = \hat{\sigma}^{*} = \hat{\sigma}^{**}. \tag{164}$$

But

$$T^{**}(x, x) = \begin{cases} 2t_p (x_0 x_{2k-1} + \cdots + x_{k-1} x_k) & \text{for odd } p, \\ t_p [2 (x_0 x_{2k} + \cdots + x_{k-1} x_{k+1}) + x_k^2] & \text{for even } p \end{cases}$$

Since every product of the form $x_\alpha x_\beta$ with $\alpha \neq \beta$ can be replaced by a difference of squares $\left(\frac{x_\alpha + x_\beta}{2}\right)^2 - \left(\frac{x_\alpha - x_\beta}{2}\right)^2$, we can obtain a decomposition of $T^{**}(x, x)$ into independent real squares and we have

$$\hat{\sigma}^{**} = \begin{cases} 0 & \text{for odd } p, \\ \operatorname{sign} t_p & \text{for even } p. \end{cases} \tag{165}$$

On the other hand, from (162),

$$\frac{D_{h+p+1}}{D_h} = | T | = (-1)^{\frac{p(p+1)}{2}} t_p^{p+1}. \tag{166}$$

From (163), (164), (165), and (166), it follows that:

$$\begin{aligned} P(D^{*}_h, D^{*}_{h+1}, \ldots, D^{*}_{h+p+1}) - V(D^{*}_h, D^{*}_{h+1}, \ldots, D^{*}_{h+p+1}) \\ = \begin{cases} 0 & \text{for odd } p, \\ \varepsilon & \text{for even } p \end{cases} \end{aligned} \tag{167}$$

where

$$\varepsilon = (-1)^{\frac{p}{2}} \operatorname{sign} \frac{D_{h+p+1}}{D_h}.$$

Since

$$P(D^{*}_{h+1}, D^{*}_{h+2}, \ldots, D^{*}_{h+p+1}) + V(D^{*}_{h+1}, D^{*}_{h+2}, \ldots, D^{*}_{h+p+1}) = p + 1, \tag{168}$$

the table (160) can be deduced from (167) and (168).

Now let $D_r = 0$. Then for some $h < r$

$$D_h \neq 0,\ D_{h+1} = \cdots = D_r = 0.$$

In this case, by Theorem 23,

$$D^{(r)} = S\begin{pmatrix} 1 \ldots h & n-r+h+1 \ldots n \\ 1 \ldots h & n-r+h+1 \ldots n \end{pmatrix} \neq 0.$$

The case to be considered reduces to the preceding case by renumbering the variables in the quadratic form $S(x, x) = \sum_{i,k=0}^{n-1} s_{i+k} x_i x_k$. We set:

$$\tilde{x}_0 = x_0, \quad \ldots, \quad \tilde{x}_{h-1} = x_{h-1}, \quad \tilde{x}_h = x_{n-r+h}, \quad \ldots, \quad \tilde{x}_{r-1} = x_{n-1},$$
$$\tilde{x}_r = x_h, \quad \ldots, \quad \tilde{x}_{n-1} = x_{n-r+h-1}.$$

Then $S(x, x) = \sum_{i,k=0}^{n-1} \tilde{s}_{i+k}\, x_i x_k$.

Starting from the structure of the matrix T on page 346 and using the relations

$$\hat{D}_j = \frac{D_{h+j}}{D_h}, \quad \hat{\tilde{D}}_j = \frac{\tilde{D}_{h+j}}{D_h} \qquad (j = 1, 2, \ldots, n-h)$$

obtained from Sylvester's determinant identity, we find that the sequence $1, \tilde{D}_1, \tilde{D}_2, \ldots, \tilde{D}_n$ is obtained from $1, D_1, D_2, \ldots, D_n$ by replacing the single element D_r by $D^{(r)}$.

We leave it to the reader to verify that the table (160) corresponds to the attribution of signs to the zero determinants given by (159).

This completes the proof of the theorem.

Note. It follows from (166) that for odd p (p is the number of zero determinants in the group (158))

$$\operatorname{sign} \frac{D_{h+p+1}}{D_h} = (-1)^{\frac{p+1}{2}}.$$

In particular, for $p = 1$ we have $D_h D_{h+2} < 0$. In this case, we can omit D_{h+1} in computing $V(1, D_1, \ldots, D_r)$, thus obtaining Gundenfinger's rule. In exactly the same way, we obtain Frobenius' rule (see page 304) from (160) for $p = 2$.

BIBLIOGRAPHY

BIBLIOGRAPHY

Items in the Russian language are indicated by *

PART A. Textbooks, Monographs, and Surveys

[1] ACHIESER (Akhieser), N. J., *Theory of Approximation.* New York: Ungar, 1956. [Translated from the Russian.]

[2] AITKEN, A. C., *Determinants and matrices.* 9th ed., Edinburgh: Oliver and Boyd, 1956.

[3] BELLMAN, R., *Stability Theory of Differential Equations.* New York: McGraw-Hill, 1953.

*[4] BERNSTEIN, S. N., *Theory of Probability.* 4th ed., Moscow: Gostekhizdat, 1946.

[5] BODEWIG, E., *Matrix Calculus.* 2nd ed., Amsterdam: North Holland, 1959.

[6] CAHEN, G., *Éléments du calcul matriciel.* Paris: Dunod, 1955.

*[7] CHEBOTARËV, N. G., and MEĬMAN, N. N., *The problem of Routh-Hurwitz for polynomials and integral functions.* Trudy Mat. Inst. Steklov., vol. 26 (1949).

*[8] CHEBYSHEV, P. L., *Complete collected works.* vol. III. Moscow: Izd. AN SSSR, 1948.

*[9] CHETAEV, N. G., *Stability of motion.* Moscow: Gostekhizdat, 1946.

[10] COLLATZ, L., *Eigenwertaufgaben mit technischen Anwendungen.* Leipzig: Akad. Velags., 1949.

[11] ——— *Eigenwertprobleme und ihre numerische Behandlung.* New York: Chelsea, 1948.

[12] COURANT, R. and HILBERT, D., *Methods of Mathematical Physics,* vol. I. Trans. and revised from the German original. New York: Interscience, 1953.

*[13] ERUGIN, N. R., *The method of Lappo-Danilevskiĭ in the theory of linear differential equations.* Leningrad: Leningrad University, 1956.

*[14] FADDEEV, D. K. and SOMINSKIĬ, I. S., *Problems in higher algebra.* 2nd ed., Moscow, 1949; 5th ed. Moscow: Gostekhizdat, 1954.

[15] FADDEEVA, V. N., *Computational methods of linear algebra.* New York: Dover Publications, 1959. [Translated from the Russian.]

[16] FRAZER, R. A., DUNCAN, W. J., and COLLAR, A., *Elementary Matrices and Some Applications to Dynamics and Differential Equations.* Cambridge: Cambridge University Press, 1938.

*[17] GANTMACHER (Gantmakher), F. R. and KREĬN, M. G., *Oscillation matrices and kernels and small vibrations of dynamical systems.* 2nd ed., Moscow: Gostekhizdat, 1950. [A German translation is in preparation.]

[18] GRÖBNER, W., *Matrizenrechnung.* Munich: Oldenburg, 1956.

[19] HAHN, W., *Theorie und Anwendung der direkten Methode von Lyapunov* (Ergebnisse der Mathematik, Neue Folge, Heft 22). Berlin: Springer, 1959. [Contains an extensive bibliography.]

[20] INCE, E. L., *Ordinary Differential Equations.* New York: Dover, 1948.

[21] JUNG, H., *Matrizen und Determinanten. Eine Einführung.* Leipzig, 1953.

[22] KLEIN, F., *Vorlesungen über höhere Geometrie.* 3rd ed., New York: Chelsea, 1949.

[23] KOWALEWSKI, G., *Einführung in die Determinantentheorie.* 3rd ed., New York: Chelsea, 1949.

*[24] KREĬN, M. G., *Fundamental propositions in the theory of λ-zone stability of a canonical system of linear differential equations with periodic coefficients.* Moscow: Moscow Academy, 1955.

*[25] KREĬN, M. G. and NAĬMARK, M. A., *The method of symmetric and hermitian forms in the theory of separation of roots of algebraic equations.* Kharkov: GNTI, 1936.

*[26] KREĬN, M. G. and RUTMAN, M. A., *Linear operators leaving a cone in a Banach space invariant.* Uspehi Mat. Nauk, vol. 3 no. 1, (1948).

*[27] KUDRYAVCHEV, L. D., *On some mathematical problems in the theory of electrical networks.* Uspehi Mat. Nauk, vol. 3 no. 4 (1948).

*[28] LAPPO-DANILEVSKIĬ, I. A., *Theory of functions of matrices and systems of linear differential equations.* Moscow, 1934.

[29] ——— *Mémoires sur la théorie des systemes des équations différentielles linéaires.* 3 vols., Trudy Mat. Inst. Steklov. vols. 6-8 (1934-1936). New York: Chelsea, 1953.

[30] LEFSCHETZ, S., *Differential Equations: Geometric Theory.* New York: Interscience, 1957.

[31] LICHNEROWICZ, A., *Algèbre et analyse linéaires.* 2nd ed., Paris: Masson, 1956.

[32] LYAPUNOV (Liapounoff), A. M., *Le Problème général de la stabilité du mouvement* (Annals of Mathematics Studies, No. 17). Princeton: Princeton Univ. Press, 1949.

[33] MACDUFFEE, C. C., *The Theory of Matrices.* New York: Chelsea, 1946.

[34] ——— *Vectors and matrices.* La Salle: Open Court, 1943.

*[35] MALKIN, I. G., *The method of Lyapunov and Poincaré in the theory of non-linear oscillations.* Moscow: Gostekhizdat, 1949.

[36] ——— *Theory of stability of motion.* Moscow: Gostekhizdat, 1952. [A German translation is in preparation.]

[37] MARDEN, M., *The geometry of the zeros of a polynomial in a complex variable* (Mathematical Surveys, No. 3). New York: Amer. Math. Society, 1949.

*[38] MARKOV, A. A., *Collected works.* Moscow, 1948.

*[39] MEĬMAN, N. N., *Some problems in the disposition of roots of polynomials.* Uspehi Mat. Nauk, vol. 4 (1949).

[40] MIRSKY, L., *An Introduction to Linear Algebra.* Oxford: Oxford University Press, 1955.

*[41] NAĬMARK, Y. I., *Stability of linearized systems.* Leningrad: Leningrad Aeronautical Engineering Academy, 1949.

[42] PARODI, M., *Sur quelques propriétés des valeurs caractéristiques des matrices carrées* (Mémorial des Sciences Mathématiques, vol. 118), Paris: Gauthiers-Villars, 1952.

[43] PERLIS, S., *Theory of Matrices.* Cambridge. (Mass.): Addison-Wesley, 1952.

[44] PICKERT, G., *Normalformen von Matrizen* (Enz. Math. Wiss., Band I, Teil B. Heft 3, Teil I). Leipzig: Teubner, 1953.

*[45] POTAPOV, V. P., *The multiplicative structure of J-inextensible matrix functions.* Trudy Moscow Mat. Soc., vol. 4 (1955).

*[46] ROMANOVSKIĬ, V. I., *Discrete Markov chains.* Moscow: Gostekhizdat, 1948.

[47] ROUTH, E. J., *A treatise on the stability of a given state of motion.* London: Macmillan, 1877.

[48] ——— *The advanced part of a Treatise on the Dynamics of a Rigid Body.* 6th ed., London: Macmillan, 1905; repr., New York: Dover, 1959.

[49] SCHLESINGER, L., *Vorlesungen über lineare Differentialgleichungen.* Berlin, 1908.

[50] ——— *Einführung in die Theorie der gewöhnlichen Differentialgleichungen auf funktionentheoretischer Grundlage.* Berlin, 1922.

[51] SCHMEIDLER, W., *Vortrage über Determinanten und Matrizen mit Anwendungen in Physik und Technik.* Berlin: Akademie-Verlag, 1949.

[52] SCHREIER, O. and SPERNER, E., *Vorlesungen über Matrizen.* Leipzig: Teubner, 1932. [A slightly revised version of this book appears as Chapter V of [53].]

[53] ——— *Introduction to Modern Algebra and Matrix Theory.* New York: Chelsea, 1958.

[54] SCHWERDTFEGER, H., *Introduction to Linear Algebra and the Theory of Matrices.* Groningen: Noordhoff, 1950.

[55] SHOHAT, J. A. and TAMARKIN, J. D., *The problem of moments* (Mathematical Surveys, No. 1). New York: Amer. Math. Society, 1943.

[56] SMIRNOW, W. I. (Smirnov, V. I.), *Lehrgang der höheren Mathematik,* Vol. III. Berlin, 1956. [This is a translation of the 13th Russian edition.]

[57] SPECHT, W., *Algebraische Gleichungen mit reellen oder komplexen Koeffizienten* (Enz. Math. Wiss., Band I, Teil B, Heft 3, Teil II). Stuttgart: Teubner, 1958.

[58] STIELTJES, T. J., *Oeuvres Complètes.* 2 vols., Groningen: Noordhoff.

[59] STOLL, R. R., *Linear Algebra and Matrix Theory.* New York: McGraw-Hill, 1952.

[60] THRALL, R. M. and TORNHEIM, L., *Vector spaces and matrices.* New York: Wiley, 1957.

[61] TURNBULL, H. W., *The Theory of Determinants, Matrices and Invariants.* London: Blackie, 1950.

[62] TURNBULL, H. W. and AITKEN, A. C., *An Introduction to the Theory of Canonical Matrices.* London: Blackie, 1932.

[63] VOLTERRA, V. et HOSTINSKY, B., *Opérations infinitésimales linéaires.* Paris: Gauthiers-Villars, 1938.

[64] WEDDERBURN, J. H. M., *Lectures on matrices.* New York: Amer. Math. Society, 1934.

[65] WEYL, H., *Mathematische Analyse des Raumproblems.* Berlin, 1923. [A reprint is in preparation: Chelsea, 1960.]

[66] WINTNER, A., *Spektraltheorie der unendlichen Matrizen.* Leipzig, 1929.

[67] ZURMÜHL, R., *Matrizen.* Berlin, 1950.

PART B. Papers

[101] AFRIAT, S., *Composite matrices,* Quart. J. Math. vol. 5, pp. 81-89 (1954).

*[102] AIZERMAN (Aisermann), M. A., *On the computation of non-linear functions of several variables in the investigation of the stability of an automatic regulating system,* Avtomat. i Telemeh. vol. 8 (1947).

[103] AISERMANN, M. A. and F. R. GANTMACHER, *Determination of stability by linear approximation of a periodic solution of a system of differential equations with discontinuous right-hand sides,* Quart. J. Mech. Appl. Math. vol. 11, pp. 385-98 (1958).

[104] Aitken, A. C., *Studies in practical mathematics. The evaluation, with applications, of a certain triple product matrix.* Proc. Roy. Soc. Edinburgh vol. 57, (1936-37).

[105] Amir Moéz Ali, R., *Extreme properties of eigenvalues of a hermitian transformation and singular values of the sum and product of linear transformations,* Duke Math. J. vol. 23, pp. 463-76 (1956).

*[106] Artashenkov, P. V., *Determination of the arbitrariness in the choice of a matrix reducing a system of linear differential equations to a system with constant coefficients.* Vestnik Leningrad. Univ., Ser. Mat., Phys. i Chim., vol. 2, pp. 17-29 (1953).

*[107] Arzhanych, I. S., *Extension of Krylov's method to polynomial matrices,* Dokl. Akad. Nauk SSSR, Vol. 81, pp. 749-52 (1951).

*[108] Azbelev, N. and R. Vinograd, *The process of successive approximations for the computation of eigenvalues and eigenvectors,* Dokl. Akad. Nauk., vol. 83, pp. 173-74 (1952).

[109] Baker, H. F., *On the integration of linear differential equations,* Proc. London Math. Soc., vol. 35, pp. 333-78 (1903).

[110] Barankin, E. W., *Bounds for characteristic roots of a matrix,* Bull. Amer. Math. Soc., vol. 51, pp. 767-70 (1945).

[111] Bartsch, H., *Abschätzungen für die Kleinste charakteristische Zahl einer positiv-definiten hermitschen Matrix,* Z. Angew. Math. Mech., vol. 34, pp. 72-74 (1954).

[112] Bellman, R., *Notes on matrix theory,* Amer. Math. Monthly, vol. 60, pp. 173-75, (1953); vol. 62, pp. 172-73, 571-72, 647-48 (1955); vol. 64, pp. 189-91 (1957).

[113] Bellman, R. and A. Hoffman, *On a theorem of Ostrowski,* Arch. Math., vol. 5, pp. 123-27 (1954).

[114] Bendat, J. and S. Silverman, *Monotone and convex operator functions,* Trans. Amer. Math. Soc., vol. 79, pp. 58-71 (1955).

[115] Berge, C., *Sur une propriété des matrices doublement stochastiques,* C. R. Acad. Sci. Paris, vol. 241, pp. 269-71 (1955).

[116] Birkhoff, G., *On product integration,* J. Math. Phys., vol. 16, pp. 104-32 (1937).

[117] Birkhoff, G. D., *Equivalent singular points of ordinary linear differential equations,* Math. Ann., vol. 74, pp. 134-39 (1913).

[118] Bott, R. and R. Duffin, *On the algebra of networks,* Trans. Amer. Math. Soc., vol. 74, pp. 99-109 (1953).

[119] Brauer, A., *Limits for the characteristic roots of a matrix,* Duke Math. J., vol. 13, pp. 387-95 (1946); vol. 14, pp. 21-26 (1947); vol. 15, pp. 871-77 (1948); vol. 19, pp. 73-91, 553-62 (1952); vol. 22, pp. 387-95 (1955).

[120] ——— *Über die Lage der charakteristischen Wurzeln einer Matrix,* J. Reine Angew. Math., vol. 192, pp. 113-16 (1953).

[121] ——— *Bounds for the ratios of the coordinates of the characteristic vectors of a matrix,* Proc. Nat. Acad. Sci. U.S.A., vol. 41, pp. 162-64 (1955).

[122] ——— *The theorems of Ledermann and Ostrowski on positive matrices,* Duke Math. J., vol. 24, pp. 265-74 (1957).

[123] Brenner, J., *Bounds for determinants,* Proc. Nat. Acad. Sci. U.S.A., vol. 40, pp. 452-54 (1954); Proc. Amer. Math. Soc., vol. 5, pp. 631-34 (1954); vol. 8, pp. 532-34 (1957); C. R. Acad. Sci. Paris, vol. 238, pp. 555-56 (1954).

[124] Bruijn, N., *Inequalities concerning minors and eigenvalues,* Nieuw Arch. Wisk., vol. 4, pp. 18-35 (1956).

[125] Bruijn, N. and G. Szekeres, *On some exponential and polar representatives of matrices,* Nieuw Arch. Wisk., vol. 3, pp. 20-32 (1955).

*[126] BULGAKOV, B. V., *The splitting of rectangular matrices,* Dokl. Akad. Nauk SSSR, vol. 85, pp. 21-24 (1952).

[127] CAYLEY, A., *A memoir on the theory of matrices,* Phil. Trans. London, vol. 148, pp. 17-37 (1857); Coll. Works, vol. 2, pp. 475-96.

[128] COLLATZ, L., *Einschliessungssatz für die charakteristischen Zahlen von Matrizen,* Math. Z., vol. 48, pp. 221-26 (1942).

[129] ——— *Über monotone systeme linearen Ungleichungen,* J. Reine Angew. Math., vol. 194, pp. 193-94 (1955).

[130] CREMER, L., *Die Verringerung der Zahl der Stabilitätskriterien bei Voraussetzung positiven koeffizienten der charakteristischen Gleichung,* Z. Angew. Math. Mech., vol. 33, pp. 222-27 (1953).

*[131] DANILEVSKIĬ, A. M., *On the numerical solution of the secular equation,* Mat. Sb., vol. 2, pp. 169-72 (1937).

[132] DILIBERTO, S., *On systems of ordinary differential operations.* In: *Contributions to the Theory of Non-linear Oscillations,* vol. I, edited by S. Lefschetz (Annals of Mathematics Studies, No. 20). Princeton: Princeton Univ. Press (1950), pp. 1-38.

*[133] DMITRIEV, N. A. and E. B. DYNKIN, *On the characteristic roots of stochastic matrices,* Dokl. Akad. Nauk SSSR, vol. 49, pp. 159-62 (1945).

*[133a] ——— *Characteristic roots of Stochastic Matrices,* Izv. Akad. Nauk, Ser. Fiz-Mat., vol. 10, pp. 167-94 (1946).

[134] DOBSCH, O., *Matrixfunktionen beschränkter Schwankung,* Math. Z., vol. 43, pp. 353-88 (1937).

*[135] DONSKAYA, L. I., *Construction of the solution of a linear system in the neighborhood of a regular singularity in special cases,* Vestnik Leningrad. Univ., vol. 6 (1952).

*[136] ——— *On the structure of the solution of a system of linear differential equations in the neighbourhood of a regular singularity,* Vestnik Leningrad. Univ., vol. 8, pp. 55-64 (1954).

*[137] DUBNOV, Y. S., *On simultaneous invariants of a system of affinors,* Trans. Math. Congress in Moscow 1927, pp. 236-37.

*[138] ——— *On doubly symmetric orthogonal matrices,* Bull. Ass. Inst. Univ. Moscow, pp. 33-35 (1927).

*[139] ——— *On Dirac's matrices,* Uč. zap. Univ. Moscow, vol. 2, pp. 2, 43-48 (1934).

*[140] DUBNOV, Y. S. and V. K. IVANOV, On the reduction of the degree of affinor *polynomials,* Dokl. Akad. Nauk SSSR, vol. 41, pp. 99-102 (1943).

[141] DUNCAN, W., *Reciprocation of triply-partitioned matrices,* J. Roy. Aero. Soc., vol. 60, pp. 131-32 (1956).

[142] EGERVÁRY, E., *On a lemma of Stieltjes on matrices,* Acta. Sci. Math., vol. 15, pp. 99-103 (1954).

[143] ——— *On hypermatrices whose blocks are commutable in pairs and their application in lattice-dynamics,* Acta Sci. Math., vol. 15, pp. 211-22 (1954).

[144] EPSTEIN, M. and H. FLANDERS, *On the reduction of a matrix to diagonal form,* Amer. Math. Monthly, vol. 62, pp. 168-71 (1955).

*[145] ERSHOV, A. P., *On a method of inverting matrices,* Dokl. Akad. Nauk SSSR, vol. 100, pp. 209-11 (1955).

[146] ERUGIN, N. P., *Sur la substitution exposante pour quelques systèmes irreguliers,* Mat. Sb., vol. 42, pp. 745-53 (1935).

*[147] ——— *Exponential substitutions of an irregular system of linear differential equations,* Dokl. Akad. Nauk SSSR, vol. 17, pp. 235-36 (1935).

*[148] ——— *On Riemann's problem for a Gaussian system,* Uč. Zap. Ped. Inst., vol. 28, pp. 293-304 (1939).

*[149] FADDEEV, D. K., *On the transformation of the secular equation of a matrix,* Trans. Inst. Eng. Constr., vol. 4, pp. 78-86 (1937).

[150] FAEDO, S., *Un nuove problema di stabilità per le equationi algebriche a coefficienti reali,* Ann. Scuola Norm. Sup. Pisa, vol. 7, pp. 53-63 (1953).

*[151] FAGE, M. K., *Generalization of Hadamard's determinant inequality,* Dokl. Akad. Nauk SSSR, vol. 54, pp. 765-68 (1946).

*[152] ——— *On symmetrizable matrices,* Uspehi Mat. Nauk, vol. 6, no. 3, pp. 153-56 (1951).

[153] FAN, K., *On a theorem of Weyl concerning eigenvalues of linear transformations,* Proc. Nat. Acad. Sci. U.S.A., vol. 35, pp. 652-55 (1949); vol. 36, pp. 31-35 (1950).

[154] ——— *Maximum properties and inequalities for the eigenvalues of completely continuous operators,* Proc. Nat. Acad. Sci. U.S.A., vol. 37, pp. 760-66 (1951).

[155] ——— *A comparison theorem for eigenvalues of normal matrices,* Pacific J. Math., vol. 5, pp. 911-13 (1955).

[156] ——— *Some inequalities concerning positive-definite Hermitian matrices,* Proc. Cambridge Philos. Soc., vol. 51, pp. 414-21 (1955).

[157] ——— *Topological proofs for certain theorems on matrices with non-negative elements,* Monatsh. Math., vol. 62, pp. 219-37 (1958).

[158] FAN, K. and A. HOFFMAN, *Some metric inequalities in the space of matrices,* Proc. Amer. Math. Soc., vol. 6, pp. 111-16 (1958).

[159] FAN, K. and G. PALE, *Imbedding conditions for Hermitian and normal matrices,* Canad. J. Math., vol. 9, pp. 298-304 (1957).

[160] FAN, K. and J. TODD, *A determinantal inequality,* J. London Math. Soc., vol. 30, pp. 58-64 (1955).

[161] FROBENIUS, G., *Über lineare substitutionen und bilineare Formen,* J. Reine Angew. Math., vol. 84, pp. 1-63 (1877).

[162] ——— *Über das Trägheitsgesetz der quadratischen Formen,* S.-B. Deutsch. Akad. Wiss. Berlin. Math.-Nat. Kl., 1894, pp. 241-56, 407-31.

[163] ——— *Über die cogredienten transformationen der bilinearer Formen,* S.-B. Deutsch. Akad. Wiss. Berlin. Math.-Nat. Kl., 1896, pp. 7-16.

[164] ——— *Über die vertauschbaren Matrizen,* S.-B. Deutsch. Akad. Wiss. Berlin. Math.-Nat. Kl., 1896, pp. 601-614.

[165] ——— *Über Matrizen aus positiven Elementen,* S.-B. Deutsch. Akad. Wiss. Berlin. Math-Nat. Kl. 1908, pp. 471-76; 1909, pp. 514-18.

[166] ——— *Über Matrizen aus nicht negativen Elementen,* S.-B. Deutsch. Akad. Wiss. Berlin Math.-Nat. Kl., 1912, pp. 456-77.

*[167] GANTMACHER, F. R., *Geometric theory of elementary divisors after Krull,* Trudy Odessa Gos. Univ. Mat., vol. 1, pp. 89-108 (1935).

*[168] ——— *On the algebraic analysis of Krylov's method of transforming the secular equation,* Trans. Second Math. Congress, vol. II, pp. 45-48 (1934).

[169] ——— *On the classification of real simple Lie groups,* Mat. Sb., vol. 5, pp. 217-50 (1939).

*[170] GANTMACHER, F. R. and M. G. KREĬN, *On the structure of an orthogonal matrix,* Trans. Ukrain, Acad. Sci. Phys.-Mat. Kiev (Trudy fiz.-mat. otdela VUAN, Kiev), 1929, pp. 1-8.

*[171] ——— *Normal operators in a hermitian space,* Bull. Phys-Mat. Soc. Univ. Kasan (Izvestiya fiz.-mat. ob-va pri Kazanskom universitete), IV, vol. 1, ser. 3, pp. 71-84 (1929-30).

*[172] ——— *On a special class of determinants connected with Kellogg's integral kernels*, Mat. Sb., vol. 42, pp. 501-8 (1935).

[173] ——— *Sur les matrices oscillatoires et complètement non-négatives*, Compositio Math., vol. 4, pp. 445-76 (1937).

[174] GANTSCHI, W., *Bounds of matrices with regard to an hermitian metric*, Compositio Math., vol. 12, pp. 1-16 (1954).

*[175] GELFAND, I. M. and V. B. LIDSKIĬ, *On the structure of the domains of stability of linear canonical systems of differential equations with periodic coefficients*, Uspehi. Mat. Nauk, vol. 10, no. 1, pp. 3-40 (1955).

[176] GERSHGORIN, S. A., *Über die Abgrenzung der Eigenwerte einer Matrix*, Izv. Akad. Nauk SSSR, Ser. Fiz.-Mat., vol. 6, pp. 749-54 (1931).

[177] GODDARD, L., *An extension of a matrix theorem of A. Brauer*, Proc. Int. Cong. Math. Amsterdam, 1954, vol. 2, pp. 22-23.

[178] GOHEEN, H. E., *On a lemma of Stieltjes on matrices*, Amer. Math. Monthly, vol. 56, pp. 328-29 (1949).

*[179] GOLUBCHIKOV, A. F., *On the structure of the automorphisms of the complex simple Lie groups*, Dokl. Akad. Nauk SSSR, vol. 27, pp. 7-9 (1951).

*[180] GRAVE, D. A., *Small oscillations and some propositions in algebra*, Izv. Akad. Nauk SSSR, Ser. Fiz.-Mat., vol. 2, pp. 563-70 (1929).

*[181] GROSSMAN, D. P., *On the problem of a numerical solution of systems of simultaneous linear algebraic equations*, Uspehi Mat. Nauk, vol. 5, no. 3, pp. 87-103 (1950).

[182] HAHN, W., *Eine Bemerkung zur zweiten Methode von Lyapunov*, Math. Nachr., vol. 14, pp. 349-54 (1956).

[183] ——— *Über die Anwendung der Methode von Lyapunov auf Differenzengleichungen*, Math. Ann., vol. 136, pp. 430-41 (1958).

[184] HAYNSWORTH, E., *Bounds for determinants with dominant main diagonal*, Duke Math. J., vol. 20, pp. 199-209 (1953).

[185] ——— *Note on bounds for certain determinants*, Duke Math. J., vol. 24, pp. 313-19 (1957).

[186] HELLMANN, O., *Die Anwendung der Matrizanten bei Eigenwertaufgaben*, Z. Angew. Math. Mech., vol. 35, pp. 300-15 (1955).

[187] HERMITE, C., *Sur le nombre des racines d'une équation algébrique comprise entre des limites données*, J. Reine Angew. Math., vol. 52, pp. 39-51 (1856).

[188] HJELMSLER, J., *Introduction à la théorie des suites monotones*, Kgl. Danske Vid. Selbsk. Forh. 1914, pp. 1-74.

[189] HOFFMAN, A. and O. TAUSSKY, *A characterization of normal matrices*, J. Res. Nat. Bur. Standards, vol. 52, pp. 17-19 (1954).

[190] HOFFMAN, A. and H. WIELANDT, *The variation of the spectrum of a normal matrix*, Duke Math. J., vol. 20, pp. 37-39 (1953).

[191] HORN, A., *On the eigenvalues of a matrix with prescribed singular values*, Proc. Amer. Math. Soc., vol. 5, pp. 4-7 (1954).

[192] HOTELLING, H., *Some new methods in matrix calculation*, Ann. Math. Statist., vol. 14, pp. 1-34 (1943).

[193] HOUSEHOLDER, A. S., *On matrices with non-negative elements*, Monatsh. Math., vol. 62, pp. 238-49 (1958).

[194] HOUSEHOLDER, A. S. and F. L. BAUER, *On certain methods for expanding the characteristic polynomial*, Numer. Math., vol. 1, pp. 29-35 (1959).

[195] HSU, P. L., *On symmetric, orthogonal, and skew-symmetric matrices*, Proc. Edinburgh Math. Soc., vol. 10, pp. 37-44 (1953).

[196] ——— *On a kind of transformation of matrices,* Acta Math. Sinica, vol. 5, pp. 333-47 (1955).

[197] HUA, L.-K., *On the theory of automorphic functions of a matrix variable,* Amer. J. Math., vol. 66, pp. 470-88; 531-63 (1944).

[198] ——— *Geometries of matrices,* Trans. Amer. Math. Soc., vol. 57, pp. 441-90 (1945).

[199] ——— *Orthogonal classification of Hermitian matrices,* Trans. Amer. Math. Soc., vol. 59, pp. 508-23 (1946).

*[200] ——— *Geometries of symmetric matrices over the real field,* Dokl. Akad. Nauk SSSR, vol. 53, pp. 95-98; 195-96 (1946).

*[201] ——— *Automorphisms of the real symplectic group,* Dokl. Akad. Nauk SSSR, vol. 53, pp. 303-306 (1946).

[202] ——— *Inequalities involving determinants,* Acta Math. Sinica, vol. 5, pp. 463-70 (1955).

*[203] HUA, L.-K. and B. A. ROSENFELD, *The geometry of rectangular matrices and their application to the real projective and non-euclidean geometries,* Izv. Higher Ed. SSSR, Matematika, vol. 1, pp. 233-46 (1957).

[204] HURWITZ, A., *Über die Bedingungen, unter welchen eine Gleichung nur Wurzeln mit negativen reellen Teilen besitzt,* Math. Ann., vol. 46, pp. 273-84 (1895).

[205] INGRAHAM, M. H., *On the reduction of a matrix to its rational canonical form,* Bull. Amer. Math. Soc., vol. 39, pp. 379-82 (1933).

[206] IONESCU, D., *O identitate importantá si descompunere a unei forme bilineare into sumá de produse,* Gaz. Mat. Ser. Fiz. A. 7, vol. 7, pp. 303-312 (1955).

[207] ISHAK, M., *Sur les spectres des matrices,* Sém. P. Dubreil et Ch. Pisot, Fac. Sci. Paris, vol. 9, pp. 1-14 (1955/56).

*[208] KAGAN, V. F., *On some number systems arising from Lorentz transformations,* Izv. Ass. Inst. Moscow Univ. 1927, pp. 3-31.

*[209] KARPELEVICH, F. I., *On the eigenvalues of a matrix with non-negative elements,* Izv. Akad. Nauk SSSR Ser. Mat., vol. 15, pp. 361-83 (1951).

[210] KHAN, N. A., *The characteristic roots of a product of matrices,* Quart. J. Math., vol. 7, pp. 138-43 (1956).

*[211] KHLODOVSKIĬ, I. N., *On the theory of the general case of Krylov's transformation of the secular equation,* Izv. Akad. Nauk, Ser. Fiz.-Mat., vol. 7, pp. 1076-1102 (1933).

*[212] KOLMOGOROV, A. N., *Markov chains with countably many possible states,* Bull. Univ. Moscow (A), vol. 1:3 (1937).

*[213] KOTELYANSKIĬ, D. M., *On monotonic matrix functions of order n,* Trans. Univ. Odessa, vol. 3, pp. 103-114 (1941).

*[214] ——— *On the theory of non-negative and oscillatory matrices,* Ukrain. Mat. Z., vol. 2, pp. 94-101 (1950).

*[215] ——— *On some properties of matrices with positive elements,* Mat. Sb., vol. 31, pp. 497-506 (1952).

*[216] ——— *On a property of matrices of symmetric signs,* Uspehi Mat. Nauk, vol. 8, no. 4, pp. 163-67 (1953).

*[217] ——— *On some sufficient conditions for the spectrum of a matrix to be real and simple,* Mat. Sb., vol. 36, pp. 163-68 (1955).

*[218] ——— *On the influence of Gauss' transformation on the spectra of matrices,* Uspehi Mat. Nauk, vol. 9, no. 3, pp. 117-21 (1954).

*[219] ——— *On the distribution of points on a matrix spectrum,* Ukrain. Mat. Z., vol. 7, pp. 131-33 (1955).

*[220] ——— *Estimates for determinants of matrices with dominant main diagonal*, Izv. Akad. Nauk SSSR, Ser. Mat., vol. 20, pp. 137-44 (1956).

*[221] KOVALENKO, K. R. and M. G. KREĬN, *On some investigations of Lyapunov concerning differential equations with periodic coefficients*, Dokl. Akad. Nauk SSSR, vol. 75, pp. 495-99 (1950).

[222] KOWALEWSKI, G., *Natürliche Normalformen linearer Transformationen*, Leipz. Ber., vol. 69, pp. 325-35 (1917).

*[223] KRASOVSKIĬ, N. N., *On the stability after the first approximation*, Prikl. Mat. Meh., vol. 19, pp. 516-30 (1955).

*[224] KRASNOSEL'SKIĬ, M. A. and M. G. KREĬN, *An iteration process with minimal deviations*, Mat. Sb., vol. 31, pp. 315-34 (1952).

[225] KRAUS, F., *Über konvexe Matrixfunktionen*, Math. Z., vol. 41, pp. 18-42 (1936).

*[226] KRAVCHUK, M. F., *On the general theory of bilinear forms*, Izv. Polyt. Inst. Kiev, vol. 19, pp. 17-18 (1924).

*[227] ——— *On the theory of permutable matrices*, Zap. Akad. Nauk Kiev, Ser. Fiz.-Mat., vol. 1:2, pp. 28-33 (1924).

*[228] ——— *On a transformation of quadratic forms*, Zap. Akad. Nauk Kiev, Ser. Fiz.-Mat., vol. 1:2, pp. 87-90 (1924).

*[229] ——— *On quadratic forms and linear transformations*, Zap. Akad. Nauk Kiev, Ser. Fiz.-Mat., vol. 1:3, pp. 1-89 (1924).

*[230] ——— *Permutable sets of linear transformations*, Zap. Agr. Inst. Kiev, vol. 1, pp. 25-58 (1926).

[231] ——— *Über vertauschbare Matrizen*, Rend. Circ. Mat. Palermo, vol. 51, pp. 126-30 (1927).

*[232] ——— *On the structure of permutable groups of matrices*, Trans. Second. Mat. Congress 1934, vol. 2, pp. 11-12.

*[233] KRAVCHUK, M. F. and Y. S. GOL'DBAUM, *On groups of commuting matrices*, Trans. Av. Inst. Kiev, 1929, pp. 73-98; 1936, pp. 12-23.

*[234] ——— *On the equivalence of singular pencils of matrices*, Trans. Av. Inst. Kiev, 1936, pp. 5-27.

*[235] KREĬN, M. G., *Addendum to the paper 'On the structure of an orthogonal matrix,'* Trans. Fiz.-Mat. Class. Akad. Nauk Kiev, 1931, pp. 103-7.

*[236] ——— *On the spectrum of a Jacobian form in connection with the theory of torsion oscillations of drums*, Mat. Sb., vol. 40, pp. 455-66 (1933).

*[237] ——— *On a new class of hermitian forms*, Izv. Akad. Nauk SSSR, Ser. Fiz.-Mat., vol. 9, pp. 1259-75 (1933).

*[238] ——— *On the nodes of harmonic oscillations of mechanical systems of a special type*, Mat. Sb., vol. 41, pp. 339-48 (1934).

[239] ——— *Sur quelques applications des noyaux de Kellog aux problèmes d'oscillations*, Proc. Charkov Mat. Soc. (4), vol. 11, pp. 3-19 (1935).

[240] ——— *Sur les vibrations propres des tiges dont l'une des extrémités est encastrée et l'autre libre*, Proc. Charkov. Mat. Soc. (4), vol. 12, pp. 3-11 (1935).

*[241] ——— *Generalization of some results of Lyapunov on linear differential equations with periodic coefficients*, Dokl. Akad. Nauk SSSR, vol. 73, pp. 445-48 (1950).

*[242] ——— *On an application of the fixed-point principle in the theory of linear transformations of spaces with indefinite metric*, Uspehi Mat. Nauk, vol. 5, no. 2, pp. 180-90 (1950).

*[243] ——— *On an application of an algebraic proposition in the theory of monodromy matrices*, Uspehi Mat. Nauk, vol. 6, no. 1, pp. 171-77 (1951).

*[244] ——— *On some problems concerning Lyapunov's ideas in the theory of stability*, Uspehi Mat. Nauk, vol. 3, no. 3, pp. 166-69 (1948).

*[245] ——— *On the theory of integral matrix functions of exponential type*, Ukrain. Mat. Z., vol. 3, pp. 164-73 (1951).

*[246] ——— *On some problems in the theory of oscillations of Sturm systems*, Prikl. Mat. Meh., vol. 16, pp. 555-68 (1952).

*[247] KREĬN, M. G. and M. A. NAĬMARK (Neumark), *On a transformation of the Bézoutian leading to Sturm's theorem*, Proc. Charkov Mat. Soc., (4), vol. 10, pp. 33-40 (1933).

*[248] ——— *On the application of the Bézoutian to problems of the separation of roots of algebraic equations*, Trudy Odessa Gos. Univ. Mat., vol. 1, pp. 51-69 (1935).

[249] KRONECKER, L., *Algebraische Reduction der Schaaren bilinearer Formen*, S.-B. Akad. Berlin 1890, pp. 763-76.

[250] KRULL, W., *Theorie und Anwendung der verallgemeinerten Abelschen Gruppen*, S.-B. Akad. Heidelberg 1926, p. 1.

*[251] KRYLOV, A. N., *On the numerical solution of the equation by which the frequency of small oscillations is determined in technical problems*, Izv. Akad. Nauk SSSR Ser. Fiz.-Mat., vol. 4, pp. 491-539 (1931).

[252] LAPPO-DANILEVSKIĬ, I. A., *Résolution algorithmique des problèmes réguliers de Poincaré et de Riemann*, J. Phys. Mat. Soc. Leningrad, vols. 2:1, pp. 94-120; 121-54 (1928).

[253] ——— *Théorie des matrices satisfaisantes à des systèmes des équations differentielles linéaires a coefficients rationnels arbitraires*, J. Phys. Mat. Soc. Leningrad, vols. 2:2, pp. 41-80 (1928).

*[254] ——— *Fundamental problems in the theory of systems of linear differential equations with arbitrary rational coefficients*, Trans. First Math. Congr., ONTI, 1936, pp. 254-62.

[255] LEDERMANN, W., *Reduction of singular pencils of matrices*, Proc. Edinburgh Math. Soc., vol. 4, pp. 92-105 (1935).

[256] ——— *Bounds for the greatest latent root of a positive matrix*, J. London Math. Soc., vol. 25, pp. 265-68 (1950).

*[257] LIDSKIĬ, V. B., *On the characteristic roots of a sum and a product of symmetric matrices*, Dokl. Akad. Nauk SSSR, vol. 75, pp. 769-72 (1950).

*[258] ——— *Oscillation theorems for canonical systems of differential equations*, Dokl. Akad. Nauk SSSR, vol. 102, pp. 111-17 (1955).

[259] LIÉNARD, and CHIPART, *Sur la signe de la partie réelle des racines d'une équation algébrique*, J. Math. Pures Appl. (6), vol. 10, pp. 291-346 (1914).

*[260] LIPIN, N. V., *On regular matrices*, Trans. Inst. Eng 8. Transport, vol. 9, p. 105 (1934).

*[261] LIVSHITZ, M. S. and V. P. POTAPOV, *The multiplication theorem for characteristic matrix functions*, Dokl. Akad. Nauk SSSR, vol. 72, pp. 164-73 (1950).

*[262] LOPSHITZ, A. M., *Vector solution of a problem on doubly symmetric matrices*, Trans. Math. Congress Moscow, 1927, pp. 186-87.

*[263] ——— *The characteristic equation of lowest degree for affinors and its application to the integration of differential equations*, Trans. Sem. Vectors and Tensors, vols. 2/3 (1935).

*[264] ——— *A numerical method of determining the characteristic roots and characteristic planes of a linear operator*, Trans. Sem. Vectors and Tensors, vol. 7, pp. 233-59 (1947).

*[265] ——— *An extremal theorem for a hyper-ellipsoid and its application to the solution of a system of linear algebraic equations,* Trans. Sem. Vectors and Tensors, vol. 9, pp. 183-97 (1952).

[266] LÖWNER, K., *Über monotone Matrixfunktionen,* Math. Z., vol. 38, pp. 177-216 (1933); vol. 41, pp. 18-42 (1936).

[267] ——— *Some classes of functions defined by difference on differential inequalities,* Bull. Amer. Math. Soc., vol. 56, pp. 308-19 (1950).

*[268] LUSIN, N. N., *On Krylov's method of forming the secular equation,* Izv. Akad. Nauk SSSR, Ser. Fiz.-Mat., vol. 7, pp. 903-958 (1931).

*[269] ——— *On some properties of the displacement factor in Krylov's method,* Izv. Akad. Nauk SSSR, Ser. Fiz.-Mat., vol. 8, pp. 596-638; 735-62; 1065-1102 (1932).

*[270] ——— *On the matrix theory of differential equations,* Avtomat. i Telemeh, vol. 5, pp. 3-66 (1940).

*[271] LYUSTERNIK, L. A., *The determination of eigenvalues of functions by an electric scheme,* Electričestvo, vol. 11, pp. 67-8 (1946).

*[272] ——— *On electric models of symmetric matrices,* Uspehi Mat. Nauk, vol. 4, no. 2, pp. 198-200 (1949).

*[273] LYUSTERNIK, L. A. and A. M. PROKHOROV, *Determination of eigenvalues and functions of certain operators by means of an electrical network,* Dokl. Akad Nauk SSSR, vol. 55, pp. 579-82; Izv. Akad. Nauk SSSR, Ser. Mat., vol. 11, pp. 141-45 (1947).

[274] MARCUS, M., *A remark on a norm inequality for square matrices,* Proc. Amer. Math. Soc., vol. 6, pp. 117-19 (1955).

[275] ——— *An eigenvalue inequality for the product of normal matrices,* Amer. Math. Monthly, vol. 63, pp. 173-74 (1956).

[276] ——— *A determinantal inequality of H. P. Robertson,* II, J. Washington Acad. Sci., vol. 47, pp. 264-66 (1957).

[277] ——— *Convex functions of quadratic forms,* Duke Math. J., vol. 24, pp. 321-26 (1957).

[278] MARCUS, M. and J. L. MCGREGOR, *Extremal properties of Hermitian matrices,* Canad. J. Math., vol. 8, pp. 524-31 (1956).

[279] MARCUS, M. and B. N. MOYLS, *On the maximum principle of Ky Fan,* Canad. J. Math., vol. 9, pp. 313-20 (1957).

[280] ——— *Maximum and minimum values for the elementary symmetric functions of Hermitian forms,* J. London Math. Soc., vol. 32, pp. 374-77 (1957).

*[281] MAYANTS, L. S., *A method for the exact determination of the roots of secular equations of high degree and a numerical analysis of their dependence on the parameters of the corresponding matrices,* Dokl. Akad. Nauk SSSR, vol. 50, pp. 121-24 (1945).

[282] MIRSKY, L., *An inequality for positive-definite matrices,* Amer. Math. Monthly, vol. 62, pp. 428-30 (1955).

[283] ——— *The norm of adjugate and inverse matrices,* Arch. Math., vol. 7, pp. 276-77 (1956).

[284] ——— *The spread of a matrix,* Mathematika, vol. 3, pp. 127-30 (1956).

[285] ——— *Inequalities for normal and Hermitian matrices,* Duke Math. J., vol. 24, pp. 591-99 (1957).

[286] MITROVIC, D., *Conditions graphiques pour que toutes les racines d'une équation algébrique soient à parties réelles négatives,* C. R. Acad. Sci. Paris, vol. 240, pp. 1177-79 (1955).

[287] MORGENSTERN, D., *Eine Verschärfung der Ostrowskischen Determinantenabschätzung,* Math. Z., vol. 66, pp. 143-46 (1956).

[288] MOTZKIN, T. and O. TAUSSKY, *Pairs of matrices with property L.*, Trans. Amer. Math. Soc., vol. 73, pp. 108-14 (1952); vol. 80, pp. 387-401 (1954).

*[289] NEĬGAUS (Neuhaus), M. G. and V. B. LIDSKIĬ, *On the boundedness of the solutions of linear systems of differential equations with periodic coefficients*, Dokl. Akad. Nauk SSSR, vol. 77, pp. 183-93 (1951).

[290] NEUMANN, J., *Approximative of matrices of high order*, Portugal. Math., vol. 3, pp. 1-62 (1942).

*[291] NUDEL'MAN, A. A. and P. A. SHVARTSMAN, *On the spectrum of the product of unitary matrices*, Uspehi Mat. Nauk, vol. 13, no. 6, pp. 111-17 (1958).

[292] OKAMOTO, M., *On a certain type of matrices with an application to experimental design*, Osaka Math. J., vol. 6, pp. 73-82 (1954).

[293] OPPENHEIM, A., *Inequalities connected with definite Hermitian forms*, Amer. Math. Monthly, vol. 61, pp. 463-66 (1954).

[294] ORLANDO, L., *Sul problema di Hurwitz relativo alle parti reali delle radici di un' equatione algebrica*, Math. Ann., vol. 71, pp. 233-45 (1911).

[295] OSTROWSKI, A., *Bounds for the greatest latent root of a positive matrix*, J. London Math. Soc., vol. 27, pp. 253-56 (1952).

[296] ——— *Sur quelques applications des fonctions convexes et concaves au sens de I. Schur*, J. Math. Pures Appl., vol. 31, pp. 253-92 (1952).

[297] ——— *On nearly triangular matrices*, J. Res. Nat. Bur. Standards, vol. 52, pp. 344-45 (1954).

[298] ——— *On the spectrum of a one-parametric family of matrices*, J. Reine Angew. Math., vol. 193, pp. 143-60 (1954).

[299] ——— *Sur les déterminants à diagonale dominante*, Bul. Soc. Math. Belg., vol. 7, pp. 46-51 (1955).

[300] ——— *Note on bounds for some determinants*, Duke Math. J., vol. 22, pp. 95-102 (1955).

[301] ——— *Über Normen von Matrizen*, Math. Z., vol. 63, pp. 2-18 (1955).

[302] ——— *Über die Stetigkeit von charakteristischen Wurzeln in Abhängigkeit von den Matrizenelementen*, Jber. Deutsch. Math. Verein., vol. 60, pp. 40-42 (1957).

*[303] PAPKOVICH, P. F., *On a method of computing the roots of a characteristic determinant*, Prikl. Mat. Meh., vol. 1, pp. 314-18 (1933).

[304] PAPULIS, A., *Limits on the zeros of a network determinant*, Quart. Appl. Math., vol. 15, pp. 193-94 (1957).

[305] PARODI, M., *Remarques sur la stabilité*, C. R. Acad. Sci. Paris, vol. 228, pp. 51-2; 807-8; 1198-1200 (1949).

[306] ——— *Sur une propriété des racines d'une équation qui intervient en mécanique*, C. R. Acad. Sci. Paris, vol. 241, pp. 1019-21 (1955).

[307] ——— *Sur la localisation des valeurs caractéristiques des matrices dans le plan complexe*, C. R. Acad. Sci. Paris, vol. 242, pp. 2617-18 (1956).

[308] PEANO, G., *Intégration par series des équations différentielles linéaires*, Math. Ann., vol. 32, pp. 450-56 (1888).

[309] PENROSE, R., *A generalized inverse for matrices*, Proc. Cambridge Philos. Soc., vol. 51, pp. 406-13 (1955).

[310] ——— *On best approximate solutions of linear matrix equations*, Proc. Cambridge Philos. Soc., vol. 52, pp. 17-19 (1956).

[311] PERFECT, H., *On matrices with positive elements*, Quart. J. Math., vol. 2, pp. 286-90 (1951).

[312] ——— *On positive stochastic matrices with real characteristic roots*, Proc. Cambridge Philos. Soc., vol. 48, pp. 271-76 (1952).

[313] ——— *Methods of constructing certain stochastic matrices*, Duke Math. J., vol. 20, pp. 395-404 (1953); vol. 22, pp. 305-11 (1955).

[314] ——— *A lower bound for the diagonal elements of a non-negative matrix*, J. London Math. Soc., vol. 31, pp. 491-93 (1956).

[315] PERRON, O., *Jacobischer Kettenbruchalgorithmus*, Math. Ann., vol. 64, pp. 1-76 (1907).

[316] ——— *Über Matrizen*, Math. Ann., vol. 64, pp. 248-63 (1907).

[317] ——— *Über Stabilität und asymptotisches Verhalten der Lösungen eines Systems endlicher Differenzengleichungen*, J. Reine Angew. Math., vol. 161, pp. 41-64 (1929).

[318] PHILLIPS, H. B., *Functions of matrices*, Amer. J. Math., vol. 41, pp. 266-78 (1919).

*[319] PONTRYAGIN, L. S., *Hermitian operators in a space with indefinite metric*, Izv. Akad. Nauk SSSR, Ser. Mat., vol. 8, pp. 243-80 (1944).

*[320] POTAPOV, V. P., *On holomorphic matrix functions bounded in the unit circle*, Dokl. Akad. Nauk SSSR, vol. 72, pp. 849-53 (1950).

[321] RASCH, G., *Zur Theorie und Anwendung der Produktintegrals*, J. Reine Angew. Math., vol. 171, pp. 65-119 (19534).

[322] REICHARDT, H., *Einfarbe Herleitung der Jordanschen Normalform*, Wiss. Z. Humboldt-Univ. Berlin. Math.-Nat. Reihe, vol. 6, pp. 445-47 (1953/54).

*[323] RECHTMAN-OL'SHANSKAYA, P. G., *On a theorem of Markov*, Uspehi Mat. Nauk, vol. 12, no. 3, pp. 181-87 (1957).

[324] RHAM, G. DE, *Sur un théorème de Stieltjes relatif à certains matrices*, Acad. Serbe Sci. Publ. Inst. Math., vol. 4, pp. 133-54 (1952).

[325] RICHTER, H., *Über Matrixfunktionen*, Math. Ann., vol. 122, pp. 16-34 (1950).

[326] ——— *Bemerkung zur Norm der Inversen einer Matrix*, Arch. Math., vol. 5, pp. 447-48 (1954).

[327] ——— *Zur Abschätzung von Matrizennormen*, Math. Nachr., vol. 18, pp. 178-87 (1958).

[328] ROMANOVSKIĬ, V. I., *Un théorème sur les zéros des matrices non-négatives*, Bull. Soc. Math. France, vol. 61, pp. 213-19 (1933).

[329] ——— *Recherches sur les chaînes de Markoff*, Acta Math., vol. 66, pp. 147-251 (1935).

[330] ROTH, W., *On the characteristic polynomial of the product of two matrices*, Proc. Amer. Math. Soc., vol. 5, pp. 1-3 (1954).

[331] ——— *On the characteristic polynomial of the product of several matrices*, Proc. Amer. Math. Soc., vol. 7, pp. 578-82 (1956).

[332] ROY, S., *A useful theorem in matrix theory*, Proc. Amer. Math. Soc., vol. 5, pp. 635-38 (1954).

*[333] SAKHNOVICH, L. A., *On the limits of multiplicative integrals*, Uspehi Mat. Nauk, vol. 12 no. 3, pp. 205-11 (1957).

*[334] SARYMSAKOV, T. A., *On sequences of stochastic matrices*, Dokl. Akad. Nauk, vol. 47, pp. 331-33 (1945).

[335] SCHNEIDER, H., *An inequality for latent roots applied to determinants with dominant principal diagonal*, J. London Math. Soc., vol. 28, pp. 8-20 (1953).

[336] ——— *A pair of matrices with property P*, J. Amer. Math. Monthly, vol. 62, pp. 247-49 (1955).

[337] ——— *A matrix problem concerning projections*, Proc. Edinburgh Math. Soc., vol. 10, pp. 129-30 (1956).

[338] ——— *The elementary divisors, associated with* 0, *of a singular M-matrix*, Proc. Edinburgh Math. Soc., vol. 10, pp. 108-22 (1956).

[339] SCHOENBERG, J., *Über variationsvermindernde lineare transformationen*, Math. Z., vol. 32, pp. 321-28 (1930).

[340] ——— *Zur abzählung der reellen wurzeln algebraischer gleichungen*, Math. Z., vol. 38, p. 546 (1933).

[341] SCHOENBERG, I. J., and A. WHITNEY, *A theorem on polygons in n dimensions with application to variation diminishing linear transformations*, Compositio Math., vol. 9, pp. 141-60 (1951).

[342] SCHUR, I., *Über die charakteristischen wurzeln einer linearen substitution mit einer anwendung auf die theorie der integralgleichungen*, Math. Ann., vol. 66, pp. 488-510 (1909).

*[343] SEMENDYAEV, K. A., *On the determination of the eigenvalues and invariant manifolds of matrices by means of iteration*, Prikl. Matem. Meh., vol. 3, pp. 193-221 (1943).

*[344] SEVAST'YANOV, B. A., *The theory of branching random processes*, Uspehi Mat. Nauk, vol. 6, no. 6, pp. 46-99 (1951).

*[345] SHIFFNER, L. M., *The development of the integral of a system of differential equations with regular singularities in series of powers of the elements of the differential substitution*, Trudy Mat. Inst. Steklov. vol. 9, pp. 235-66 (1935).

*[346] ——— *On the powers of matrices*, Mat. Sb., vol. 42, pp. 385-94 (1935).

[347] SHODA, K., *Über mit einer matrix vertauschbare matrizen*, Math. Z., vol. 29, pp. 696-712 (1929).

*[348] SHOSTAK, P. Y., *On a criterion for the conditional definiteness of quadratic forms in n linearly independent variables and on a sufficient condition for a conditional extremum of a function of n variables*, Uspehi Mat. Nauk, vol. 8, no. 4, pp. 199-206 (1954).

*[349] SHREĬDER, Y. A., *A solution of systems of linear algebraic equations*, Dokl. Akad. Nauk, vol. 76, pp. 651-55 (1950).

*[350] SHTAERMAN (Steiermann), I. Y., *A new method for the solution of certain algebraic equations which have application to mathematical physics*, Z. Mat., Kiev, vol. 1, pp. 83-89 (1934); vol. 4, pp. 9-20 (1934).

*[351] SHTAERMAN (Steiermann), I. Y. and N. I. AKHIESER (Achieser), *On the theory of quadratic forms*, Izv. Polyteh., Kiev, vol. 19, pp. 116-23 (1934).

*[352] SHURA-BURA, M. R., *An estimate of error in the numerical computation of matrices of high order*, Uspehi Mat. Nauk, vol. 6, no. 4, pp. 121-50 (1951).

*[353] SHVARTSMAN (Schwarzmann), A. P., *On Green's matrices of self-adjoint differential operators*, Proc. Odessa Univ. Matematika, vol. 3, pp. 35-77 (1941).

[354] SIEGEL, C. L., *Symplectic Geometry*, Amer. J. Math., vol. 65, pp. 1-86 (1943).

*[355] SKAL'KINA, M. A., *On the preservation of asymptotic stability on transition from differential equations to the corresponding difference equations*, Dokl. Akad. Nauk SSSR, vol. 104, pp. 505-8, (1955).

*[356] SMOGORZHEVSKIĬ, A. S., *Sur les matrices unitaires du type de circulants*, J. Mat. Circle Akad. Nauk Kiev, vol. 1, pp. 89-91 (1932).

*[356a] SMOGORZHEVSKIĬ, A. S. and M. F. KRAVCHUK, *On orthogonal transformations*, Zap. Inst. Ed. Kiev, vol. 2, pp. 151-56 (1927).

[357] STENZEL, H., *Über die Darstellbarkeit einer Matrix als Produkt von zwei symmetrischen Matrizen*, Math. Z., vol. 15, pp. 1-25 (1922).

[358] STÖHR, A., *Oszillationstheoreme für die Eigenvektoren speziellen Matrizen*, J. Reine Angew. Math., vol. 185, pp. 129-43 (1943).

*[359] SULEĬMANOVA, K. R., *Stochastic matrices with real characteristic values*, Dokl. Akad. Nauk SSSR, vol. 66, pp. 343-45 (1949).

*[360] ——— *On the characteristic values of stochastic matrices,* Uč. Zap. Moscow Ped. Inst., Ser. 71, Math., vol. 1, pp. 167-97 (1953).

*[361] SULTANOV, R. M., *Some properties of matrices with elements in a non-commutative ring,* Trudy Mat. Sectora Akad. Nauk Baku, vol. 2, pp. 11-17 (1946).

*[362] SUSHKEVICH, A. K., *On matrices of a special type,* Uč. Zap. Univ. Charkov, vol. 10, pp. 1-16 (1937).

[363] SZ-NAGY, B., *Remark on S. N. Roy's paper 'A useful theorem in matrix theory,'* Proc. Amer. Math. Soc., vol. 7, p. 1 (1956).

[364] TA LI, *Die Stabilitätsfrage bei Differenzengleichungen,* Acta Math., vol. 63, pp. 99-141 (1934).

[365] TAUSSKY, O., *Bounds for characteristic roots of matrices,* Duke Math. J., vol. 15, pp. 1043-44 (1948).

[366] ——— *A determinantal inequality of H. P. Robertson,* I, J. Washington Acad. Sci., vol. 47, pp. 263-64 (1957).

[367] ——— *Commutativity in finite matrices,* Amer. Math. Monthly, vol. 64, pp. 229-35 (1957).

[368] TOEPLITZ, O., *Das algebraische Analogon zu einem Satz von Fejér,* Math. Z., vol. 2, pp. 187-97 (1918).

[369] TURNBULL, H. W., *On the reduction of singular matrix pencils,* Proc. Edinburgh Math. Soc., vol. 4, pp. 67-76 (1935).

*[370] TURCHANINOV, A. S., *On some applications of matrix calculus to linear differential equations,* Uč. Zap. Univ. Odessa, vol. 1, pp. 41-48 (1921).

*[371] VERZHBITSKIĬ, B. D., *Some problems in the theory of series compounded from several matrices,* Mat. Sb., vol. 5, pp. 505-12 (1939).

*[372] VILENKIN, N. Y., *On an estimate for the maximal eigenvalue of a matrix,* Uč. Zap. Moscow Ped. Inst., vol. 108, pp. 55-57 (1957).

[373] VIVIER, M., *Note sur les structures unitaires et paraunitaires,* C. R. Acad. Sci. Paris, vol. 240, pp. 1039-41 (1955).

[374] VOLTERRA, V., *Sui fondamenti della teoria delle equazioni differenziali lineari,* Mem. Soc. Ital. Sci. (3), vol. 6, pp. 1-104 (1887); vol. 12, pp. 3-68 (1902).

[375] WALKER, A. and J. WESTON, *Inclusion theorems for the eigenvalues of a normal matrix,* J. London Math. Soc., vol. 24, pp. 28-31 (1949).

[376] WAYLAND, H., *Expansions of determinantal equations into polynomial form,* Quart. Appl. Math., vol. 2, pp. 277-306 (1945).

[377] WEIERSTRASS, K., *Zur theorie der bilinearen und quadratischen Formen,* Monatsh. Akad. Wiss. Berlin, 1867, pp. 310-38.

[378] WELLSTEIN, J., *Über symmetrische, alternierende und orthogonale Normalformen von Matrizen,* J. Reine Angew. Math., vol. 163, pp. 166-82 (1930).

[379] WEYL, H., *Inequalities between the two kinds of eigenvalues of a linear transformation,* Proc. Nat. Acad. Sci., vol. 35, pp. 408-11 (1949).

[380] WEYR, E., *Zur Theorie der bilinearen Formen,* Monatsh. f. Math. und Physik, vol. 1, pp. 163-236 (1890).

[381] WHITNEY, A., *A reduction theorem for totally positive matrices,* J. Analyse Math., vol. 2, pp. 88-92 (1952).

[382] WIELANDT, H., *Ein Einschliessungssatz für charakteristische Wurzeln normaler Matrizen,* Arch. Math., vol. 1, pp. 348-52 (1948/49).

[383] ——— *Die Einschliessung von Eigenwerten normaler Matrizen,* Math. Ann. vol. 121, pp. 234-41 (1949).

[384] ——— *Unzerlegbare nicht-negative Matrizen,* Math. Z., vol. 52, pp. 642-48 (1950).

[385] ——— *Lineare Scharen von Matrizen mit reellen Eigenwerten*, Math. Z., vol. 53, pp. 219-25 (1950).

[386] ——— *Pairs of normal matrices with property L*, J. Res. Nat. Bur. Standards, vol. 51, pp. 89-90 (1953).

[387] ——— *Inclusion theorems for eigenvalues*, Nat. Bur. Standards, Appl. Math. Sci., vol. 29, pp. 75-78 (1953).

[388] ——— *An extremum property of sums of eigenvalues*, Proc. Amer. Math. Soc., vol. 6, pp. 106-110 (1955).

[389] ——— *On eigenvalues of sums of normal matrices*, Pacific J. Math., vol. 5, pp. 633-38 (1955).

[390] WINTNER, A., *On criteria for linear stability*, J. Math. Mech., vol. 6, pp. 301-9 (1957).

[391] WONG, Y., *An inequality for Minkowski matrices*, Proc. Amer. Math. Soc., vol. 4, pp. 137-41 (1953).

[392] ——— *On non-negative valued matrices*, Proc. Nat. Acad. Sci. U.S.A., vol. 40, pp. 121-24 (1954).

*[393] YAGLOM, I. M., *Quadratic and skew-symmetric bilinear forms in a real symplectic space*, Trudy Sem. Vect. Tens. Anal. Moscow, vol. 8, pp. 364-81 (1950).

*[394] YAKUBOVICH, V. A., *Some criteria for the reducibility of a system of differential equations*, Dokl. Akad. Nauk SSSR, vol. 66, pp. 577-80 (1949).

*[395] ZEITLIN (Tseĭtlin), M. L., *Application of the matrix calculus to the synthesis of relay-contact schemes*, Dokl. Akad. Nauk SSSR, vol. 86, pp. 525-28 (1952).

*[396] ZIMMERMANN (Tsimmerman), G. K., *Decomposition of the norm of a matrix into products of norms of its rows*, Nauč. Zap. Ped. Inst. Nikolaevsk, vol. 4, pp. 130-35 (1953).

INDEX

INDEX

[Numbers in italics refer to Volume Two]

ISBN: 0-8218-1393-5 (Set)
ISBN: 0-8218-1376-5 (Vol. I)

9 780821 813768

CHEL/131.H

This is an excellent and unusual textbook on the application of the theory of matrices. ... The text includes many chapters of interest to applied mathematicians.

—***Zentralblatt MATH***

This book is part of a two-volume set (the second volume is published by the AMS as volume 133 in the same series). Written by one of Russia's leading mathematicians, this treatise provides us, in easily accessible form, a coherent account of matrix theory with a view toward applications in mathematics, theoretical physics, statistics, electrical engineering, etc. The individual chapters have been kept as far as possible independent of each other, so that the reader acquainted with the contents of Chapter 1 can proceed immediately to chapters of special interest. In this volume the reader will find the analytic theory of elementary divisors, the study of various matrix equations, and the theory of pencils of quadratic and Hermitian forms.